MW00843841

Species Conservation and Management

EDITED BY

H. REŞIT AKÇAKAYA

MARK A. BURGMAN

OSKAR KINDVALL

CHRIS C. WOOD

PER SJÖGREN-GULVE

JEFF S. HATFIELD

MICHAEL A. McCARTHY

Species Conservation and Management

Case Studies

2004

OXFORD
UNIVERSITY PRESS

Oxford New York
Auckland Bangkok Buenos Aires Cape Town Chennai
Dar es Salaam Delhi Hong Kong Istanbul Karachi Kolkata
Kuala Lumpur Madrid Melbourne Mexico City Mumbai Nairobi
São Paulo Shanghai Taipei Tokyo Toronto

Published by Oxford University Press, Inc.,
198 Madison Avenue, New York, New York 10016

www.oup.com

Library of Congress Cataloging-in-Publication Data
Species conservation and management : case studies /
H. Reşit Akçakaya . . . [et al.].
p. cm.
ISBN 0-19-516646-9
1. Population viability analysis—Case studies. 2. Nature
conservation—Computer simulation—Case studies. I. Akçakaya, H. R.
QH352.5.S64 2004
577.8'8—dc22 2003014902

9 8 7 6 5 4 3 2 1

Printed in the United States of America
on acid-free paper

Contents

V Birds

VI Mammals

Contributing Authors

H. Reşit Akçakaya
Applied Biomathematics
Setauket, New York, USA

María J. Albert
Universidad Rey Juan Carlos
Mo'stoles (Madrid), Spain

Mathew W. Alldredge
North Carolina State University
Raleigh, North Carolina, USA

Mahboob Ansari
Conservator, Sindh Wildlife Department
Karachi, Pakistan

Tony D. Auld
NSW National Parks and
Wildlife Service
Hurstville, New South Wales, Australia

Sarah A. Bekessy
School of Botany, University of Melbourne
Parkville, Victoria, Australia

Sven-Åke Berglind
Department of Conservation
Biology and Genetics
Uppsala University
Uppsala, Sweden

Karl-Olof Bergman
Department of Biology
Linköping University
Linköping, Sweden

C. Can Bilgin
Biology Department
Middle East Technical University
Ankara, Turkey

Kevin J. Bonham
Sandy Bay, Tasmania, Australia

David R. Breininger
Dynamac
Kennedy Space Center
Florida, USA

Barry W. Brook
Key Center for Tropical Wildlife
Management
Northern Territory University
Darwin, Australia

Mark A. Burgman
School of Botany
University of Melbourne
Parkville 3010, Victoria, Australia

Bruce Burns
School of Botany
University of Melbourne
Parkville 3010, Victoria, Australia

Steven X. Cadrin
NOAA Northeast Fisheries
Science Center
Woods Hole, Massachusetts, USA

Milani Chaloupka
School of Economics
University of Queensland,
Brisbane, Queensland, Australia

Marco Cortes
School of Botany
University of Melbourne
Parkville 3010, Victoria, Australia

Nicholas Cottone
Department of Biology
St. Joseph's Univcrsity
Philadelphia, Pennsylvania, USA

David D. Diamond
Missouri Resource Assessment
Partnership
Columbia, Missouri, USA

David Draper
Museu, Laboratório e Jardim Botânico
da Universidade de Lisboa
Lisbon, Portugal

Martin Drechsler
UFZ-Center for Environmental Research
Leipzig-Halle, Germany

Gregory J. Ettl
Department of Biology
St. Joseph's University
Philadelphia, Pennsylvania
and Warren Wilson College
Asheville, North Carolina, USA

Laurie Fajardo
Centro de Ecología
Instituto Venezolano de Investigaciones
Científicas
Caracas, Venezuela

Julian C. Fox
School of Botany
University of Melbourne
Parkville, Victoria, Australia

Caihong Fu
Science Branch, Fisheries and
Oceans Canada
Pacific Biological Station
Nanaimo, British Columbia, Canada

Leonardo Gallo
School of Botany
University of Melbourne
Parkville 3010, Victoria, Australia

Leah R. Gerber
Deptartment of Biology
Arizona State University
Tempe, Arizona, USA

James P. Gibbs
College of Environmental Science
and Forestry
State University of New York
Syracuse, New York, USA

Mauro Gonzalez
School of Botany
University of Melbourne
Parkville 3010, Victoria, Australia

Paul C. Griffin
Wildlife Biology Program
University of Montana
Missoula, Montana, USA

Anthony D. Griffiths
Key Center for Tropical Wildlife
Management
Northern Territory University
Darwin, Australia

Richard A. Griffiths
Durrell Institute of Conservation and Ecology (DICE)
University of Kent
Canterbury, Kent, UK

Rhidian Harrington
Zoology Department
La Trobe University
Bundoora 3086, Victoria, Australia

Deborah R. Hart
NOAA Northeast Fisheries Science Center
Woods Hole, Massachusetts, USA

Jeff S. Hatfield
USGS Patuxent Wildlife Research Center
Laurel, Maryland, USA

Jonas Hedin
Department of Animal Ecology
Lund University
Lund, Sweden

Selina Heppell
Department of Fisheries and Wildlife
Oregon State University
Corvallis, Oregon, USA

Ileana Herrera
Centro de Ecología, Instituto Venezolano de Investigaciones Científicas
Caracas, Venezuela

Pablo Inchausti
Laboratoire d'Ecologie
Ecole Normale Supérieure
Paris, France

José M. Iriondo
Universidad Politécnica
Madrid, Spain

Fernanda Izquierdo
School of Botany
University of Melbourne
Parkville 3010, Victoria, Australia

David Keith
NSW National Parks and Wildlife Service
Hurstville 2220, New South Wales, Australia

Oskar Kindvall
Swedish Species Information Center
Uppsala, Sweden

Antonio Lara
School of Botany
University of Melbourne
Parkville 3010, Victoria, Australia

Michael L. Legare
Mattamuskeet National Wildlife Refuge
Swan Quarter, North Carolina, USA

Roel R. Lopez
Department of Wildlife and Fisheries Sciences
Texas A&M University
College Station, Texas, USA

Michael A. McCarthy
School of Botany
University of Melbourne
and Royal Botanic Gardens
Victoria, Australia

Michelle McClure
National Marine Fisheries Service
Seattle, Washington, USA

Paul McElhany
National Marine Fisheries Service
Seattle, Washington, USA

Peter W. Menkhorst
Department of Natural Resources and Environment
East Melbourne, Victoria, Australia

Robert Mesibov
Queen Victoria Museum and Art Gallery
Launceston, Tasmania, Australia

L. Scott Mills
Wildlife Biology Program
University of Montana
Missoula, Montana USA

E. J. Milner-Gulland
Department of Environmental Science and Technology
Imperial College, London, UK

David Morgan
Department of Zoology
University of Melbourne
Parkville, Victoria, Australia

Eric Morgan
Department of Biological Sciences
University of Warwick
Warwick, UK

Adrian C. Newton
School of Botany
University of Melbourne
Parkville, Victoria, Australia

Simon Nicol
Department of Natural Resources and Environment
Heidelberg, Victoria, Australia

Kristian Shawn Omland
School of Natural Resources
University of Vermont
Burlington, Vermont, USA

Nicolas Perrin
Laboratory for Conservation Biology
Institute of Ecology
University of Lausanne
Lausanne, Switzerland

Andrea Premoli
School of Botany
University of Melbourne
Parkville, Victoria, Australia

Andrew H. Price
Texas Parks and Wildlife Department
Austin, Texas, USA

Bruce R. Quin
Department of Natural Resources and Environment
Woori Yallock, Victoria, Australia

Thomas Ranius
Department of Entomology
Swedish University of Agricultural Sciences
Uppsala, Sweden

Helen M. Regan
NCEAS
University of California Santa Barbara
Santa Barbara, California, USA

Tracey J. Regan
School of Botany
University of Melbourne
Parkville, Victoria, Australia

Annedis Reyes
Provita
Apdo. 47552
Caracas, Venezuela

Jon Paul Rodríguez
Centro de Ecología
Instituto Venezolano de Investigaciones Científicas
Caracas, Venezuela

Karen V. Root
Department of Biological Sciences
Bowling Green State University
Bowling Green, Ohio, USA

Mary Ruckelshaus
National Marine Fisheries Service
Seattle, Washington, USA

Sébastien Sachot
Centre de conservation de la faune et de la nature
St-Sulpice, Switzerland

Ada Sánchez
Centro de Ecología
Instituto Venezolano de Investigaciones Científicas
Caracas, Venezuela

Jake Schweigert
Science Branch, Fisheries and Oceans Canada
Pacific Biological Station
Nanaimo, British Columbia, Canada

Zeynep Sezen
Department of Biology
The Pennsylvania University
University Park, Pennsylvania, USA

Blok Shaikenov
Institute of Zoology
Ministry of Education
Almaty, Kazakhstan

W. Gregory Shriver
Marsh-Billings-Rockefeller National Historical Park
Woodstock, Vermont, USA

Per Sjögren-Gulve
The Swedish Environmental Protection Agency
Stockholm, Sweden

Ian J. Smales
Conservation and Research Department
Zoological Parks and Gardens
Healesville, Victoria, Australia

R. Kent Smedbol
Fisheries and Oceans Canada Biological Station
St. Andrews, New Brunswick, Canada

Rebecca B. Smith
Dynamac
Kennedy Space Center, Florida, USA

Robert L. Stephenson
Fisheries and Oceans Canada Biological Station
St. Andrews, New Brunswick, Canada

Charles Todd
Department of Natural Resources and Environment
Heidelberg, Victoria, Australia

Paul Torgerson
Department of Veterinary Microbiology and Parasitology
University College Dublin
Dublin, Ireland

C. Diane True
Missouri Resource Assessment Partnership
Columbia, Missouri, USA

Josef Wanzenböck
Institute for Limnology
Austrian Academy of Sciences
Mondsee, Austria

Henri Weimerskirch
Centre National de la Recherche Scientifique
Villiers en Bois, France

Chris C. Wood
Science Branch, Fisheries and Oceans Canada
Pacific Biological Station
Nanaimo, British Columbia, Canada

Kuniko Yamada
School of Botany
University of Melbourne
Parkville, Victoria 3010, Australia

Species Conservation and Management

1

Using Models for Species Conservation and Management

An Introduction

H. REŞİT AKÇAKAYA

Quantitative methods, especially modeling and population viability analysis (PVA) are increasingly important tools in the conservation and management of species. The increase in the use of these quantitative methods has increased the importance of developing and applying models correctly. There are several books on using models in species conservation and management (e.g., Burgman et al. 1993, Akçakaya et al. 1999, Sjögren-Gulve and Ebenhard 2000, Beissinger and McCullough 2002, Morris and Doak 2002). However, there is a shortage of examples and case studies, especially for species other than birds and mammals. This book is a collection of case studies of models applied to a variety of species. Each chapter uses one or several models to address conservation and management issues related to a particular species. These models are available on the CD-ROM included with the book.

The models are implemented in the population modeling and viability analysis software RAMAS GIS 4.0 (Akçakaya 2002). The CD-ROM contains a demonstration version of the software. An appendix describes how to install the program and open, inspect, and run the models. The chapters are organized into sections based on major taxonomic groups and represent a wide range of life histories. Each section starts with an introductory chapter that discusses the life history characteristics of that taxonomic group from the modeling point of view, and then presents an overview of the chapters in the section. This chapter includes two short introductions—to PVA and modeling and to RAMAS GIS—and discusses limitations of, and alternatives to, these approaches.

Short Introduction to Modeling and PVA

Modeling is a process of building simple, abstract representations (e.g., as mathematical equations) of complex systems (e.g., a biological population) to gain insights into how the system works, to predict how it will behave in the future, to guide further investigations, and to make decisions about how it can be managed. Population viability analysis (PVA) is a process of using species-specific data and models to evaluate the threats faced by species in terms of their risks of extinction or decline, as well as chances for their recovery (Boyce 1992, Akçakaya and Sjögren-Gulve 2000). This book consists of examples of models used in conservation and management of species, but it is not an introductory textbook. This section provides only a brief introduction to modeling and PVA; for a more detailed introduction, see the books mentioned at the beginning of this chapter.

Uses of PVA

Results of viability analyses are used for four main objectives. First, they are often used to identify parameters that have the most effect on viability, and thus to guide fieldwork, especially in cases where data are uncertain or lacking. Many of the chapters in this book evaluate model sensitivity to various parameters.

Second, they are used to assess vulnerability of species to extinction. Viability results may be used to categorize or rank threatened species with respect to their predicted risk of extinction. Such categories may be used in combination with other criteria such as cultural priorities, economic imperatives, and taxonomic uniqueness, to set conservation policies and priorities. Several chapters in this book (e.g., 32 and 34) estimate risks of decline and extinction as a way of determining the conservation status of a species.

Third, model results are used to assess the impact of human activities by comparing results of models with and without the population-level consequences of the human activity. Many of the chapters in this book are concerned with evaluating the effect of various factors on the viability of populations and metapopulations. The factors considered are harvest, including poaching, hunting, and fishing (Chapters 5, 21, 22, 23, 30, 32, 41, 42), habitat loss (10, 13, 19, 26, 40, 44), fragmentation (14, 27), disease (4, 8, 16), road mortality (27, 40), seed predation (3), introduced species (4), Allee effects (6), droughts (7, 25), volcanic eruptions and fire (4), and global change (16, 35). Fourth, model results are used to evaluate management options and consider other conservation measures, such as habitat restoration and management (Chapters 10, 12, 13, 15, 18, 20, 33, 34, 35, 39); population management, including introduction, translocation, and sowing (7, 12, 25, 29, 36, 41); fire management (3, 6, 8, 10, 28); habitat protection, including restricting access and impact (7, 10, 33); connectivity (11, 33); pollution control (12); on-site disease treatment (8); predator control (3); and reducing effect on humans (16).

Advantages and Limitations of PVA

Compared to other alternatives for making conservation decisions, quantitative methods such as PVA have several advantages (Akçakaya and Sjögren-Gulve 2000), including internal consistency, transparency, relative freedom from linguistic ambiguities, and ability to integrate diverse types information and uncertainties (see next section in this chapter). Another important advantage of PVA is its rigor. A PVA can be replicated by different researchers, its assumptions can be explicitly stated and enumerated, and its results can be validated (McCarthy et al. 2001). For example, in a collective comparison of the historic trajectories of 21 populations with the results of the PVAs for these populations, Brook et al. (2000) validated PVAs in terms of their predictions of abundance and risks of decline. Brook et al. (2000) estimated the parameters from the first half of each data set and used the second half to evaluate model performance. They found that the predicted risk of decline closely matched observed outcomes; that despite some substantial variation in individual assessments, there was no significant or important bias; and that population size projections did not differ significantly or substantially from reality. Further, the predictions of five PVA software packages they tested were highly concordant. PVA results can also be validated for single models by comparing predicted values with those observed or measured in the field (e.g., McCarthy and Broome 2000, McCarthy et al. 2001).

PVAs can be time consuming, they appear to demand data that often are unavailable, and they require understanding of population dynamics, age- and stage-structured models, density dependence, dispersal, and the ways in which uncertainty in these processes may be carried through chains of calculations. People who build the models need to understand such things as the nuances of choosing different kinds of statistical distributions and the need to partition measurement error from natural variation in data and model assumptions. Some skill is required to marry the amount of available data and the questions that need to be answered with model structures and assumptions. Nevertheless, these investments are necessary if conservation biologists are to translate data and ecological understanding into repeatable and coherent management decisions. The predictions for individual cases may generate depressingly large confidence intervals. But the only real difference between PVA and other tools is that the alternatives are confronted with the same challenges but submerge these demands under assumptions that may not be apparent, even to the person making the assessment. This book explores how to use PVAs even when data are scarce, so that what is known is transparent. It demonstrates how robust decisions may emerge despite uncertainty.

Dealing with Uncertainty

Uncertainty is a prevalent feature of ecological data. A simple way to incorporate data uncertainties into a PVA is to estimate ranges (lower and upper bounds, instead of point estimates) of model parameters, and to build multiple models for alternative structures (e.g., types of density dependence). Combining the results of these models gives a range of estimates of extinction or decline risk and other results.

If the uncertainty in the model is such that the answer the model gives to a specific question is too uncertain to be useful, then there are several alternatives: (1) change the question and ask a different one (e.g., a more specific question, one that pertains to a shorter time horizon, or one that focuses on relative rather than absolute results); (2) reduce uncertainty by collecting more data, based on a sensitivity analysis with the preliminary model; (3) try a simpler model with fewer parameters; (4) use the model only as an exploratory (not decision-making) tool; or (5) don't use models. The risk in (5) is that you turn to an alternative that is subject to the same demands for data and understanding but that gives the appearance of providing a solution by ignoring uncertainty or by making hidden assumptions.

Short Introduction to Modeling with RAMAS GIS 4.0

The models presented in the book are implemented in RAMAS GIS software for developing either single-population or metapopulation models. In addition to the large variety of environmental impacts and conservation measures considered, a wide range of demographic and life history characteristics are modeled in various chapters. As a result, different chapters use different sets of model features, options, and parameters. To provide a common introduction to these features, and to avoid repetition of model features in the following chapters, the basic components of the models are summarized in this section. Please consult the help files on the CD-ROM for detailed information on any option or feature.

Demographic Structure

The main characteristic of most of the models in this book is demographic structure, which refers to the way individuals in a population are grouped into a number of classes. Such models are also called frequency-based, and they differ from individual-based models (which follow each organism) and occupancy models (which follow only which patches are occupied, but not the abundances of populations). For a discussion of other types of models and alternatives to this program, see the section on "Alternative Approaches."

Unstructured (scalar) models represent each population's abundance at a given time step with a single number (the total number of individuals in the population). An example of this is the European Mudminnow model (Chapter 18). Other models represent the population's abundance with a set of numbers—one for each age class or stage. In addition, there may be separate stages for males and females. For example, in the Sindh Ibex model (Chapter 42), there is one age class (named "juvenile") for 0-year-old males and females and there are separate age classes for 1-year-old and older males and females. For such a model, the stages are defined in the Stages dialog (under the Model menu of the Metapopulation Model subprogram), which also includes stage-specific information such as the proportion of breeders in each age class or stage.

The Sex Structure dialog is used to specify whether males, females, or both sexes are modeled; it also includes information about the number of female stages, the mating system (monogamous, polygynous, or polyandrous), and the degree of polygamy. For a monogamous mating system, fecundity is based on minimum of the number of males and females in breeding classes. For polygynous and polyandrous mating systems, fecundity also depends on the degree of polygamy. This is represented with the average number of mates for males and females in polygynous and polyandrous mating systems, respectively. For example, if the number of females per male is specified as 2.0, then fecundity is based on the minimum of the number of females, or 2.0 times the number of males, at each time step.

Survival rates and fecundities, or the rate of transition among stages, are specified in the Stage Matrix dialog (under the Model menu). Each row (and column) of the stage matrix corresponds to one age class or stage. The element at the ith row and jth column of the matrix represents the rate of transition from stage j to stage i, including survival, growth, and reproduction. For more information about stage-based or matrix models, see Caswell (1989), Burgman et al. (1993), Akçakaya et al. (1999), and Akçakaya (2000a).

To use the stage matrix to make projections of population size, it is also necessary to specify the initial number of individuals in each age class. In RAMAS GIS, this is done in two steps for flexibility. First, the total initial abundance (of all stages or age classes in a patch) is entered in the Populations dialog. Second, the relative abundances of stages are entered in the Initial Abundances dialog. The stage matrix and initial abundances are sufficient to make a projection of the population's structure and abundance, but for many other calculations, it is necessary to specify which entries in the stage matrix are survivals and which are fecundities. The Constraints Matrix (accessed with a button from the Stage Matrix dialog) is used to specify the proportion of each stage matrix element that is a survival rate (as opposed to a fecundity).

Variability

Fluctuations in environmental factors such as weather cause unpredictable changes in a population's parameters (e.g., survival rates and fecundities). This type of variation is called *environmental stochasticity*. In RAMAS GIS, environmental stochasticity is modeled by sampling the vital rates (fecundity and survival rates) and other model parameters (carrying capacity or dispersal rate) in each time step of the simulation from random distributions with given means and standard deviations. For example, the amount of random variability in stage matrix elements is specified in the Standard Deviation Matrix dialog. Variability in the carrying capacity is specified as the standard deviation of K parameters (under the "Density dep." tab in the Populations dialog), and variability in dispersal rates is specified in the Stochasticity dialog.

Extreme environmental events that adversely affect large proportions of a population are called *catastrophes*. Catastrophes may in some cases be considered to be a source of environmental variation that is independent of the normal year-to-year fluctuations in population parameters. In RAMAS GIS, there may be up to two catastrophes (which may be correlated; see the Stochasticity dialog).

Each catastrophe has the following parameters, which are specified either in the Catastrophes dialog or (if they are population-specific) under the "Catastrophes" tab in the Populations dialog. Extent is either local (all populations affected independently) or regional (all populations affected at the same time). Probability can be a constant number or a time series representing a function of the number of years since the last catastrophe. Each catastrophe affects abundances, vital rates (i.e., stage matrix elements), carrying capacities, and/or dispersal rates. When a catastrophe occurs, its effect depends on stage-specific multipliers and local multipliers. Local multipliers can also be a function of the number of years since the last catastrophe (e.g., if the severity of fire depends on fuel accumulation). Catastrophes can spread among populations, either by dispersers (i.e., disease) or probabilistically (e.g., fire); the parameters for these modes of spread are specified in the Advanced Catastrophe Settings dialog (accessed with a button from the Catastrophes dialog).

When the number of individuals gets to be very small, there is another source of variation that becomes important even if the population growth rate were to remain constant. This variation is called *demographic stochasticity* and is modeled by sampling both the number of survivors and dispersers from binomial distributions and the number of offspring (recruits) from Poisson distributions (Akçakaya 1991), if the Use Demographic Stochasticity box (in the Stochasticity dialog) is checked.

Density Dependence

As a population grows, increasingly scarce resources limit the growth potential of the population. Such dependence of population growth on "density" (or abundance) of the organisms can substantially affect the dynamics of the populations and its risks of decline and extinction (Ginzburg et al. 1990).

In RAMAS GIS, density dependence is incorporated by allowing the matrix elements (vital rates) to decrease as population size increases. Thus, density dependence is a relationship that describes the growth of a population as a function of its density or abundance (see Figure 1.1 for an example).

Density dependence in RAMAS GIS is specified with the following parameters, specified either in the Density Dependence dialog, or (if they are population-specific) under the "Density dep." tab in the Populations dialog):

1. The abundance on which the density dependence is based (i.e., the horizontal axis in Figure 1.1). The simplest choice is the total number of individuals in all stages. Density dependence can also be based on the abundance in a subset of stages. The stages that are the basis for density dependence are selected using the Basis for DD parameter in the Stages dialog.
2. The vital rates affected (i.e., the vertical axis in Figure 1.1). The choices are either survivals or fecundities, or both.
3. The form of the density dependence function (i.e., the shape of the curve in Figure 1.1). The simplest type of density dependence involves truncating total abundance at a ceiling. Other types include Scramble (Ricker or logistic) and Contest (Beverton-Holt).
4. The parameters of the density dependence function. The basic parameters are the maximum rate of growth when density is low (R_{max}), and the carrying capacity (K). Other related parameters include the standard deviation of K, the temporal trend in K, and the effect of catastrophes on K.

In addition, RAMAS GIS accepts user-defined functions for density dependence. These are specified by the user as a DLL in the Density Dependence dialog. Several examples are provided with the program, together with their source codes.

Population growth may also be affected negatively as population size reaches very low levels. The factors that cause such a decline (e.g., difficulty in finding a mate or disruption of social functions) are collectively called *Allee effects*. Modeling Allee effects involves selecting one of the built-in functions that incorporate Allee effects and specifying an "Allee" parameter. A simpler way is to use a local extinction threshold greater than zero; see the parameters "local threshold" (in the Populations dialog), and "when abundance is below local threshold" (in the Stochasticity dialog).

Genetics

Effects of genetics can be modeled by making the survivals or fecundities (or both) a function of the inbreeding coefficient, which, in turn, is a function of the effective population size (Burgman and Lamont 1992, Mills and Smouse 1994). In RAMAS GIS, effects of inbreeding can be incorporated by using the user-defined density dependence function with Inbreeding.DLL. This feature is described in the source code of this program, Inbreeding.DPR (this is a text file that can be opened in Notepad or WordPad).

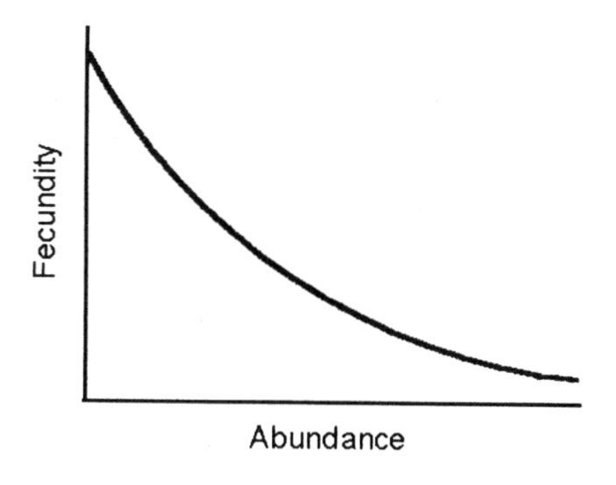

Figure 1.1 Example of density dependence: fecundity as a function of abundance.

Spatial Structure

A set of populations of the same species in the same general geographic area is called a *metapopulation.* Many species exist as metapopulations because of the patchy structure of their habitats, which may be natural or caused by human-induced habitat fragmentation. The dynamics of a metapopulation or a species depends not only on the factors discussed above (e.g., variability and density dependence) but also on the spatial variation in these factors and on other factors that characterize interactions among these populations. The additional factors that operate at the metapopulation or species level include the number and geographic configuration of habitat patches and the dispersal and spatial correlation among these patches.

In RAMAS GIS, spatial structure of a metapopulation is specified in three different ways.

Population-Specific Parameters

In the Populations dialog, the location of each population is entered with its x- and y-coordinates. Also in this dialog, many of the model parameters just described can be entered for each population, summarizing the spatial variability in these parameters. For example, each population may be assigned to a different stage matrix or to a different set of standard deviations. The probability and effect of catastrophes and the parameters related to density dependence may also be specific to the population. Many of these parameters, including the location, carrying capacity, and size of the populations, may be based on the spatial distribution of the suitable habitat (see the discussion on "Habitat Relationships").

Correlation

Spatial correlation refers to the similarity (synchrony) of environmentally induced fluctuations in different populations. Independent (uncorrelated) fluctuations in survival and fecundity decrease the likelihood that all populations go extinct at the same time, compared to a case where the fluctuations were dependent—that is, synchronous (Akçakaya and Ginzburg 1991, Burgman et al. 1993, LaHaye et al. 1994). In most metapopulations, fluctuations in demographic rates are caused by factors such as rainfall and temperature that are often correlated even at relatively large distances.

In RAMAS GIS, spatial correlations among populations are specified as a matrix in the Correlations dialog. Spatial correlation is often a function of the distance among the populations. Thus, the program offers an easy way to fill this matrix as a function of distances. The function is

$$C_{ij} = \mathbf{a} \exp(-D_{ij}^{c}/\mathbf{b})$$

where C_{ij} is the coefficient of correlation between the ith and jth populations; $\boldsymbol{a}$, $\boldsymbol{b}$, and $\boldsymbol{c}$ are the function parameters; and D_{ij} is the distance between the two populations.

Dispersal

Dispersal among populations may lead to recolonization of empty patches (i.e., extinct populations) by immigration from extant populations. In RAMAS GIS, dispersal refers to movement of organisms from one population to another. It is equivalent to "migration" in other metapopulation models. It does not mean back-and-forth seasonal movement between two locations, and it does not refer to dispersal within a population (within a habitat patch). Dispersal rates are specified as the proportion (*not as the total number*) of dispersing individuals per time step from one population to another.

Dispersal rates depend on many factors—for example, species-specific characteristics such as the mode of seed dispersal, motility of individuals, ability and propensity to disperse, and so on. Dispersal rate between any two populations may depend on a number of population-specific characteristics, including the distance between the populations, the type of habitat used during dispersal, and the density of the source population.

In RAMAS GIS, various input parameters are used to model dispersal:

1. In the Dispersal dialog, a dispersal matrix gives the proportion of individuals in each population that move (disperse) to other populations. Dispersal is often a function of the distance among the populations. Thus, the program offers an easy way to fill this matrix as a function of distances. The function is

$$\begin{aligned} M_{ij} &= a \exp(-D_{ij}^{c}/b), && \text{if } D_{ij} \leq D_{max} \\ &= 0, && \text{if } D_{ij} > D_{max} \end{aligned}$$

   where M_{ij} is the proportion of individuals in the source population j that move to the target population i; a, b, c, and D_{max} are the function parameters; and D_{ij} is the distance between the two populations.
2. Rate of dispersal *out of* a population can be defined as a function of the (source) population's abundance, and rate of dispersal *into* a population can be defined as a function of the (target) population's carrying capacity (see density dependent dispersal parameters under the "Density dep." tab in the Populations dialog).
3. Dispersal rates can be made stage- or age-specific using the relative dispersal parameter in the Stages dialog.
4. Dispersal rates can be specified to randomly fluctuate, using the CV for dispersal parameter, or the number of dispersers can fluctuate due to demographic stochasticity (see the Stochasticity dialog).
5. Human-mediated dispersal (translocation of individuals from high-density populations to empty or low-density patches, and introduction of individuals to patches or metapopulations that were previously inhabited by the same species) can be modeled in the Population Management dialog.

Habitat Relationships

Modeling spatial structure with a metapopulation model requires identifying discrete populations and specifying their location, size, and other characteristics. RAMAS GIS provides functions for determining these aspects of the spatial structure of a metapopulation based on habitat maps (Akçakaya 2000b).

The habitat maps are entered in the Input Maps dialog under the Model menu of the Spatial Data subprogram. These maps are combined into a habitat suitability map using the habitat suitability function in the Habitat Relationships dialog.

The habitat suitability (HS) map (which can be viewed under the Results menu) is then used to calculate the spatial structure of the metapopulation, based on two parameters in the Habitat Relationships dialog. The "habitat suitability threshold" is the minimum HS value below which the habitat is not suitable for reproduction and survival (although individuals may disperse or migrate through a habitat that has a lower HS than this threshold). "Neighborhood distance" is used to identify nearby cells that belong to the same patch. For an animal species, the neighborhood distance parameter may represent the foraging distance.

The program identifies clusters of suitable cells in the HS map. Suitable cells (as defined by the threshold parameter) that are separated by a distance less than or equal to the neighborhood distance are regarded to be in the same patch. Thus, neighborhood distance determines the spatial separation above which suitable cells (or clusters of suitable cells) are considered to belong to separate patches. A small neighborhood distance means that the species perceives the landscape as more patchy. Given the same HS map, either a higher threshold HS or a smaller neighborhood distance (or both) will result in a greater number of smaller patches and thus a more patchy landscape. The result of this process is that groups of cells are combined into patches that are based on the species-specific parameters.

In addition to the location and number of habitat patches, the HS map (and other maps) can be used to calculate parameters of the metapopulation model. For example, the carrying capacity of each patch can be based on the total HS in that patch (sum of the HS values of all cells in the patch), incorporating both the size of the patch and the quality of the habitat within the patch. Other parameters, such as fecundities and survival rates, may be based on the average HS (total HS divided by the area of the patch). These functions are specified in the Link to Metapopulation dialog. Parameters that are not based on habitat maps are specified in the Default Population dialog, and those that describe relationships among the populations are specified in the Dispersal and Correlation dialogs.

After the program identifies habitat patches and calculates the patch-specific parameters, the results are saved as a metapopulation model file (see under the File menu), which can be opened in the Metapopulation Model subprogram.

Examples of habitat-based metapopulation models include Chapters 4, 7, 11, 15, 16, 34, and 39, as well as Akçakaya and Atwood (1997), Akçakaya and Raphael (1998), and Broadfoot et al. (2001).

Habitat Dynamics

In many metapopulations, the landscape in which the species lives is not static but is changing due to natural processes (such as succession) and human activities (such as timber harvest, development, and road building). This results in a dynamic habitat in which patches may increase or decrease in size, split into or merge with other patches, and even disappear (due or habitat loss) or appear (due to habitat growth). In some cases change in the habitat can be predicted, based on long-term development or resource-use plans or on landscape models. These predictions can be summarized as a time series of maps—for example, one vegetation cover map for each year or decade.

If this kind of map series is available, RAMAS GIS allows the calculation of time-series of carrying capacities and/or vital rates for each population. This can be used to model the effects of changes in habitat in time (e.g., habitat loss due to planned logging, or habitat increase due to forest growth, etc.). To do this, for each time step, the maps for that time step are used with the Spatial Data subprogram (as described in the discussion on "Habitat Relationships"). The results are combined using the Habitat Dynamics subprogram.

Alternative Approaches and Methods

For most conservation questions, there are several alternatives that may provide answers or support a decision. First, there are alternatives to PVA ranging from expert opinion and rules of thumb to complex ecosystem models. When the question involves the viability of species and populations, these alternatives are often vague, less able to deal with uncertainty, and less transparent about their reliability, and they do not use all the available information (Akçakaya and Sjögren-Gulve 2000, Brook et al. 2002). They may be useful when time and resources are very limited or when the skills to build PVAs are unavailable. Some of them can contribute to the development of a PVA; for example, a structured method of eliciting expert opinion may help reduce parameter uncertainties.

Second, there are alternatives to the frequency-based models that can be developed in RAMAS GIS. These include occupancy models and individual-based models. For an overview and comparison of these types of models, as well as for guidance about selecting the appropriate model, see Akçakaya and Sjögren-Gulve (2000), Sjögren-Gulve and Hanski (2000), Akçakaya (2000a), and Lacy (2000).

Third, there are alternative methods of developing demographically structured models. These include using other generic programs such as VORTEX (Lacy 1993), RAMAS Stage (Ferson 1990), ALEX (Possingham and Davies 1995), INMAT (Mills and Smouse 1994), and others. Although VORTEX uses individual-based modeling techniques, its user interface has a fixed, age-based demographic structure. For comparisons of older versions of these programs and their results, see Lindenmeyer et al. (1995) and Brook et al. (2000). Another alternative is writing a specific, custom-made program. This is discussed in the next section.

Generic versus Custom-Made Programs

One of the decisions in developing a PVA model is whether to write a program specifically designed for the species and question at hand or to use a generic program such as RAMAS GIS. The main advantage of "custom-designed" programs is that they can be very specific for a particular situation. The main advantages of generic programs are reliability, generality, ease of use, cost, documentation, and transparency.

"Reliability" is one of major advantages of programs used and tested by a large number of modelers (for example, all the authors in this book also acted as beta-testers for version 4 of the program). "Generality" means supporting a large variety of model structures and features. A generic program like RAMAS is considerably more general and adaptable than a program written for a specific species. It is important to note that RAMAS GIS is not a model but a modeling platform. It allows many different types of models—from an unstructured, single-population exponential growth model to a sex- and stage-structured metapopulation model with density dependence and explicit spatial structure based on GIS data. "Ease of use" refers to the various features that make a program easy and convenient to use, including detailed manuals and help files, an interactive user interface, and large variety of results. "Cost" refers to the time and resources needed to develop a model; generic programs are often more cost-effective, because there is no need to spend time and resources for programming and testing. "Transparency" means that the users know exactly what the program does during a simulation. For the demonstration version of RAMAS included with this book, the help files provide a detailed description of the methods used. In addition, a PDF file summarizes the source code of the program in the form of pseudo-code, and several publications describe the methods used in the program (see references in this chapter). In contrast, programs written for a specific case can sometimes be "black boxes" or transparent only to their authors, because of their complexity and because resources are often not available for developing detailed documentation.

Conclusions

This book intends to provide examples that will guide both novice and experienced modelers to build models in the RAMAS GIS platform that represent their species as they would wish. The breadth of examples is intended to display the full range of the platform's capabilities. It also serves the more general purpose of demonstrating the utility of PVAs in a very wide range of circumstances, from data-rich to data-poor circumstances, and in circumstances in which the results support decisions, assist the development of new data collection effects, and resolve differences among stakeholders by making the consequences of what is believed to be true transparent.

References

Akçakaya, H. R. 1991. A method for simulating demographic stochasticity. *Ecological Modelling* 54: 133–136.

Akçakaya, H. R. 2000a. Population viability analyses with demographically and spatially structured models. *Ecological Bulletins* 48: 23–38.

Akçakaya, H. R. 2000b. Viability analyses with habitat-based metapopulation models. *Population Ecology* 42: 45–53.

Akçakaya, H. R. 2002. *RAMAS GIS: linking spatial data with population viability analysis* (version 4.0). Applied Biomathematics, Setauket, N.Y.

Akçakaya, H. R., and Atwood, J. L. 1997. A habitat-based metapopulation model of the California Gnatcatcher. *Conservation Biology* 11: 422–434.

Akçakaya, H. R., and Ginzburg, L. R. 1991. Ecological risk analysis for single and multiple populations. Pages 73–87 in A. Seitz and V. Loeschcke (eds.), *Species conservation: a population-biological approach*. Birkhauser Verlag, Basel.

Akçakaya, H. R., and Raphael, M. G. 1998. Assessing human impact despite uncertainty: viability of the Northern Spotted Owl metapopulation in the northwestern U.S. *Biodiversity and Conservation* 7: 875–894.

Akçakaya H. R., and Sjögren-Gulve, P. 2000. Population viability analysis in conservation planning: an overview. *Ecological Bulletins* 48: 9–21.

Akçakaya, H. R., Burgman, M. A., and Ginzburg L. R. 1999. *Applied population ecology: principles and computer exercises using RAMAS EcoLab 2.0.* 2nd ed. Sinauer Associates, Sunderland, Mass.

Beissinger, S. R., and McCullough D. R. (eds). 2002. *Population viability analysis.* University of Chicago Press, Chicago.

Boyce, M. S. 1992. Population viability analysis. *Annual Review of Ecology and Systematics* 23: 481–506.

Broadfoot, J. D., Rosatte, R. C., and O'Leary, D. T. 2001. Raccoon and skunk population models for urban disease control planning in Ontario, Canada. *Ecological Applications* 11: 295–303.

Brook, B. W., O'Grady, J. J., Chapman, A. P., Burgman, M. A., Akçakaya, H. R., and Frankham, R. 2000. Predictive accuracy of population viability analysis in conservation biology. *Nature* 404: 385–387.

Brook, B. W., Burgman, M. A., Akçakaya, H. R., O'Grady, J. J., and Frankham, R. 2002. Critiques of PVA ask the wrong questions: throwing the heuristic baby out with the numerical bathwater. *Conservation Biology* 16: 262–263.

Burgman, M. A., and Lamont, B. B. 1992. A stochastic model for the viability of *Banksia cuneata* populations: environmental, demographic and genetic effects. *Journal of Applied Ecology* 29: 719–727.

Burgman, M. A., Ferson, S., and Akçakaya, H. R. 1993. *Risk assessment in conservation biology*. Chapman and Hall, London.

Caswell, H. 1989. *Matrix population models: construction, analysis, and interpretation*. Sinauer Associates, Sunderland, Mass.

Ferson, S. 1990. *RAMAS Stage: generalized stage-based modeling for population dynamics.* Applied Biomathematics, Setauket, N.Y.

Ginzburg, L. R., Ferson, S., and Akçakaya, H. R. 1950. Reconstructibility of density dependence and the conservative assessment of extinction risks. *Conservation Biology* 4: 63–70. 1990.

Lacy, R. C. 1993. VORTEX: a computer simulation model for population viability analysis. *Wildlife Research* 20: 45–65

Lacy, R. C. 2000. Considering threats to the viability of small populations with individual-based models. *Ecological Bulletins* 48: 39–51.

LaHaye, W. S., Gutierrez, R. J., and Akçakaya, H. R. 1994. Spotted owl meta-population dynamics in southern California. *Journal of Animal Ecology* 63: 775–785.

Lindenmeyer, D., Burgman, M., Akçakaya, H. R., Lacy, R., and Possingham, H. 1995. A review of generic computer programs ALEX, RAMAS/space and VORTEX for modelling the viability of wildlife metapopulations. *Ecological Modelling* 82: 161–174.

McCarthy, M. A., and Broome, L. S. 2000. A method for validating stochastic models of population viability: a case study of the mountain pygmy-possum (*Burramys parvus*). *Journal of Animal Ecology* 69: 599–607.

McCarthy, M. A., Possingham, H. P., Day, J. R., and Tyre,A. J. 2001. Testing the accuracy of population viability analysis. *Conservation Biology* 15: 1030–1038.

Mills, L. S., and P. E. Smouse. 1994. Demographic consequences of inbreeding in remnant populations. *American Naturalist* 144: 412–431.

Morris, W. F., and Doak, D. F. 2002. *Quantitative conservation biology: theory and practice of population viability analysis*. Sinauer Associates, Sunderland, Mass.

Possingham, H. P., and Davies, I. 1995. ALEX: a model for the viability analysis of spatially structured populations. *Biological Conservation* 73: 143–150

Sjögren-Gulve, P., and Ebenhard, T. (eds). 2000. *The use of population viability analyses in conservation planning*. *Ecological Bulletins* 48 (special volume).

Sjögren-Gulve, P., and Hanski, I. 2000. Metapopulation viability analysis using occupancy models. *Ecological Bulletins* 48: 53–71.

PART I

PLANTS

2

Strategies for Plant Population Viability Modeling

An Overview

MARK A. BURGMAN

Since Lewis (1942) and Leslie (1945), population modeling has been dominated by problems involving large-bodied vertebrates, mostly mammals, birds, and fish. Plants require some different thinking because they don't move, thereby integrating environmental conditions over relatively small areas. In addition, they are frequently susceptible to local disturbance mechanisms such as landslip, fire, and disease, and their reproductive biology and dispersal mechanisms are often episodic and depend on other species.

Werner and Caswell (1977) and Bierzychudek (1982) published population models for the plants *Dipsacus sylvestris* and *Arisaema triphyllum*, using matrix formalism to produce insights into their management. Caswell (1989) provided a complete formal framework, leading Fiedler et al. (1998) to characterize transition matrices as the "modus operandi" for summarizing plant population data. Others (e.g., Burgman and Gerard 1990; Menges 1990, 1998; Damman and Cain 1999) used stochastic population models to argue for the quantification of risk and evaluation of management options by analyzing probability of decline.

Menges (2000a, 2000b) found that most of nearly 100 plant population viability analyses (PVAs) were based on stage- or size-classified matrices. Breakpoints for classes usually were made subjectively. Most studies were based on about 4 years of field data, and very few had more than 10 years of data. Most used deterministic approaches, calculating λ and related measures of elasticity and sensitivity. In contrast to animal studies, only about 10% of the PVAs he reviewed included density dependence and only three included direct assessment of the impact and management of herbivores. Menges (2000a) described several challenges to plant PVAs, including plant and seed dormancy, periodic recruitment and flowering, and clonal growth. This chapter extends Menges's (2000a) list of challenges and outlines some means of dealing with them in population viability models.

Modeling Periodic Recruitment

Plant population dynamics often are driven by disturbance events or recruitment windows—combinations of physical and biotic circumstances that induce flowering, seed production, germination, and establishment. Fire, landslip, storms, smoke, and appropriate mixes of rainfall, light, and temperature—all of these can stimulate a sporadic, mass recruitment event, leading to a series of even-aged cohorts in a population. In contrast, deterministic Leslie matrices presume that recruitment is continuous and that populations are close to the stable age/stage distribution. The solution to this problem is to build population models composed of age or stage classes that can reflect pulses of recruitment. Often it will mean that plant matrix models are composed of subdiagonal elements, with no diagonal elements, except in the oldest stages.

Episodic recruitment is a stochastic process. If the ecological conditions that lead to recruitment are known (e.g., Burgman and Gerard 1990), then these processes may be modeled directly. If the processes are unknown, then the attributes of the process may be modeled with appropriate statistics (in the simplest case, fixed or time-dependent probabilities generating Bournelli trials for an event such as a fire) (e.g., Burgman and Lamont 1992).

Modeling Periodic Mortality

Standard matrix models assume that death is a process of attrition, with (on average) a fixed proportion of each age or stage dying each year. In contrast, many plants die collectively, often as a consequence of the same disturbances that induce mass recruitment (e.g., Burgman and Gerard 1990; Menges 1990; Keith, Chapter 8 in this volume), and are otherwise very unlikely to die at all (e.g., Drechsler, Chapter 6 in this volume; Bekessy et al., Chapter 5 in this volume).

Modeling death as a catastrophe deals adequately with periodic mass mortality, although it is sometimes necessary to model the landscape in which the species' habitat is set. The model may use independent stochastic processes that affect all populations in a metapopulation, all individuals in each population, or parts of populations or metapopulations. For example, Bekessy et al. (Chapter 5 in this volume) modeled fires and habitat loss caused by volcanos, as well as fires lit by people and lightning. The fires affected sets of populations with a likelihood function determined by their distance apart, reflecting the spread of fires from a point source. The modeling framework should allow for more than one kind of event, each represented by a unique and perhaps independent stochastic process.

Ecological Dependencies

The population dynamics of many plant species are driven by dispersal, often by vertebrates that feed on fruits or seeds. Most plants live with diseases. Species interactions may be symbiotic, or the plant may depend on predators with little reciprocal effect. Forecasts for a species may require models of the dynamics of other species that determine its reproductive success and its survival. To develop adequate models, both the dynamics of the species and its interactions with other species need to be reasonably well understood and the modeling framework must be sufficiently flexible to capture all of the important attributes. Yet very few explicitly multispecies studies have been published. More often, multispecies interactions are embedded in model assumptions (e.g., Lesica 1995; Keith, Chapter 8 in this volume). Fiedler et al. (1998), Ettl and Cottone (Chapter 4 in this volume), and Keith (Chapter 8) provide examples of models with explicit representations of dispersal and predation. Albert et al. (Chapter 7 in this volume) provide a model of disease (see also Brigham and Schwartz 2003).

Data Requirements: Making Assumptions about Uncertainty

A substantial field effort is needed to estimate standard matrix model parameters, usually including measurements of the soil seed bank and responses of seeds and adults to disturbances, together with either tagging individual plants or surveying populations repeatedly through time.

Partly in response to the perception that stochastic models need more data than their deterministic counterparts, many ecologists have explored plant population dynamics with analytical models, evaluating the sensitivities and elasticities of matrix elements and life history processes (van Groenendael et al. 1988, 1994; Silvertown et al. 1996). But stochasticity is a dominant feature of the lives of most plant species. In addition, most stochastic and deterministic parameter estimates are very uncertain (Fiedler et al. 1998), making deterministic assessment alone unreliable. Modelers often rely on experts to guess at parameters, and there is no reason why these same experts can't be used to estimate the extent of unreliability in parameters and the structure of a model. The alternative to including uncertainty is to ignore it, thereby hiding the lack of knowledge among model assumptions and losing the opportunity to conduct a stochastic sensitivity analysis that may reveal unexpected possibilities and unanticipated outcomes.

Stage Structure and Dormancy

The choice to model either genets or ramets (genetically identical but physiologically independent individuals) should depend on relevant biological details and management questions (Menges 2000a). For example, Dreschler (Chapter 6 in this volume) modeled ramets because he was interested in the chances of recovery, as measured by probabilities of adult plant populations exceeding target sizes. In other circumstances, it may be better to model ramets and genets in the same population separately, because the vital rates in clonal plants may be different from those in plants that have grown from seed.

Soil and canopy-stored seed banks buffer plant populations against disturbance and environmental perturbations. Dormancy may be modeled relatively easily by using stages for dormancy classes. The modeling environment needs to be sufficiently flexible to provide for appropriate responses to disturbance and periodic recruitment windows. Such features are important because short-term stochastic dynamics often determine the outcomes for populations, long before long-term (asymptotic) properties are expressed. However, seed dormancy creates real problems for parameter estimation and census because field trials to assess transitions are costly (Menges 2000a).

Population Genetics

Genetic variation allows populations to adapt to environmental change. Inbreeding and outbreeding may result in population crashes through loss of fitness. For example, management may seek to retain the existing spatial distribution of genetic variation of a species. Another objective may be to maintain the full complement of genetic variation by isolating different genotypes spatially, thereby ensuring that loss of genetic variation will not be caused by drift. This strategy will minimize within-population diversity and maximize between population diversity. Still another focus may be to maximize diversity within populations, to maximize the chances that each population will adapt to environmental conditions.

Very few plant population models have included any genetic aspects (e.g., Burgman and Lamont 1992, Menges and Dolan 1998). The main reasons are that data on genetic load, purging, and heterozygote advantage are rarely available. Most plant genetic data is for neutral, single-locus variation, whereas variation in important adaptive traits is rarely sampled or evaluated. The relationships between genetic variation and population vital rates have not been explored empirically for most species, so that feedback between levels of genetic variation and survival and reproduction is uncertain.

Density Dependence

Most plant population viability models ignore density dependence (Menges 2000a) because (1) density dependence mechanisms are not known, (2) it is assumed that density dependence has only a modest influence on probabilities of decline (e.g., Menges 1998), (3) interest is in the long-term asymptotic characteristics (e.g., Silvertown et al. 1996), or (4) the populations are so sparse that density independent population growth may be expected for the foreseeable future (Burgman and Lamont 1992; Dreschler, Chapter 6 in this volume). Another factor is that there is no well-developed library of accepted density dependent models for plants, as there is for vertebrates where use of such functions as Ricker and Beverton-Holt density dependence is commonplace and understood.

Regan et al. (2003) illustrated some important issues for plant density dependence. They developed an individual-based model for a shrub that included overshadowing, which affected the growth and mortality of suppressed plants. The model was spatially explicit, and the fates of individual plants and seeds were followed. Their results showed that reduced growth and elevated mortality led to dynamics that were different from the standard scramble, contest, and ceiling density dependence that describe most animal populations. Frequently, competition among plants results in skewed distributions for many parameters such as dry weight, the number of seeds per plant, and individual seed weight (e.g., Obeid et al. 1967). Under reduced light, if growth is reduced but plants don't die, then the population will include a cohort of relatively small, old individuals that wait for a mortality event in the canopy above them. If small plants are intolerant of shade, then elevated mortality will occur when plants are outcompeted for light by their neighbors.

Regan and Auld (Chapter 3 in this volume) summarized individual dynamics in a frequency model in which survival was a function of density. In contrast, Bekessy et al. and Keith (Chapters 5 and 6 in this volume) reduced both survival and growth in different stages. All these models reduced survival as a function of the density of larger plants, simulating overtopping and competition for light, water, and space. Competition occurs on very small spatial scales, and its importance depends on seed dispersal distances and the spatial patterns of disturbance events. These examples show it is possible to abstract these dynamics in a frequency-based model, accounting for the relative importance of shade tolerance, mortality, and reduced growth.

Discussion

Menges (2000a) recommended plant PVA for examining and comparing management alternatives rather than for assessing extinction risks. Useful models must capture the components of a species dynamics that are important in determining responses to management. Many pieces of software have arbitrary limits, such as those on the number of stages or ages, forcing modelers to extend the time steps or to collapse age classes into stages. Because the latter strategy results in unrealistically rapid development of mature or adult forms from seedling or juvenile stages, it can have important effects on the predictions of a model under different management options. The vast array of life histories, interactions, and dependencies in plants means that the successful, routine application of population viability analysis in the management of plant species depends on the availability of sufficiently flexible modeling tools. This has been a substantial impediment to the adoption of PVA for plant management.

Plants are often better described by their physiological stage than by their age. The imposition of discrete classes is artificial and may result in discontinuities in reproduction and survival in the model that are not apparent in the real population (Fiedler et al. 1998). But the demands of periodic recruitment and mortality result in the need to recognize the ages of individuals, so that cohorts may be tracked and treated appropriately in a model. There may be opportunities in the future to develop software that assists the modeler to more easily develop models that account for density dependence and age and physiological stage, in a two-dimensional array.

Acknowledgments I am indebted to Helen Regan, Tracey Regan, Sarah Bekessey, David Keith, Reşit Akçakaya, and Tony Auld for forcing me to think about these issues, most of which they resolved in their modeling work. I am indebted to the ongoing work of Eric Menges and his colleagues. Neil Thomason provided many remarkably helpful suggestions on an earlier draft that substantially improved the clarity of this review.

References

Bierzychudek, P. 1982. The demography of jack-in-the-pulpit, a forest perennial that changes sex. *Ecological Monographs* 52: 335–351.

Brigham, C. A., and Schwartz, M. W. 2003. *Population viability in plants.* Springer, Berlin.

Burgman, M. A., and Gerard, V. A. 1990. A stochastic population model for the giant kelp, *Macrocystis pyrifera. Marine Biology* 105: 15–23.

Burgman, M. A., and Lamont, B. B. 1992. A stochastic model for the viability of *Banksia cuneata* populations: environmental, demographic and genetic effects. *Journal of Applied Ecology* 29: 719–727.

Caswell, H. 1989. *Matrix population models: construction, analysis, and interpretation.* Sinauer, Sunderland, Mass.

Damman, H., and Cain, H.L. 1998. Population growth and viability analysis of the clonal woodland shrub, *Asarum canadense. Journal of Ecology* 86: 13–26.

Fiedler, P. L., Knapp, B. E., and Fredricks, N. 1998. Rare plant demography: lessons from the Mariposa Lilies (*Calochortus*: Liliaceae). Pages 28–48 in P. L. Fiedler and P. M. Kareiva (eds.), *Conservation biology: for the coming decade.* 2nd ed. Chapman and Hall, New York.

Lesica, P. 1995. Demography of *Astragalus scaphoides* and effects of herbivory on population growth. *Great Basin Naturalist* 55: 142–150.

Leslie, P. H . 1945. On the use of matrices in certain population mathematics. *Biometrika* 33: 183–212.

Lewis, E. G. 1942. On the generation and growth of a population. *Sankya* 6: 93–96.

Menges, E. S. 1990. Population viability analysis for an endangered plant. *Conservation Biology* 4: 52–62.

Menges, E. S. 1998. Evaluating extinction risks in plant populations. Pages 49–65 in, P. L. Fiedler and P. M. Kareiva (eds.), *Conservation biology: for the coming decade.* 2nd ed. Chapman and Hall, New York.

Menges, E. S. 2000a. Applications of population viability analysis in plant conservation. *Ecological Bulletins* 48: 73–84.

Menges, E. S. 2000b. Population viability analysis in plants: challenges and opportunities. *Trends in Ecology and Evolution* 15: 51–56.

Menges, E. S., and Dolan, R. W. 1998. Demographic viability of populations of *Silene regia* in midwestern prairies: relationships with fire management, genetic variation, geographic location, population size, and isolation. *Journal of Ecology* 86: 63–78.

Obeid, M., Machin, D., and Harper, J. L. 1967. Influence of density on plant to plant variation in fiber flax, *Linum usitatissimum* L. *Crop Science* 7: 471–473.

Regan, H. M., Auld, T. D., Keith, D. A., and Burgman, M. A. 2003. The effects of fire and predators on the long-term persistence of an endangered shrub, *Grevillea caleyi. Biological Conservation* 109: 73–83.

Silvertown, J., Franco, M., and Menges, E. 1996. Interpretation of elascticity matrices as an aid to the management of plant populations for conservation. *Conservation Biology* 10, 591–597.

van Groenendael, J., de Kroon, H., and Caswell, H. 1988. Projection matrices in population biology. *Trends in Ecology and Evolution* 3: 264–269.

van Groenendael, J., de Kroon, H., Kalisz, S., and Tuljapurkar, S. 1994. Loop analysis: evaluating life history pathways in population projection matrices. *Ecology* 75: 2410–2415.

Werner, P. A., and Caswell, H. 1977. Population growth rates and age vs. size distribution models for teasel (*Dipsacus silvestrus* Huds.) *Ecology* 58: 1103–1111.

3

Australian Shrub *Grevillea caleyi*

Recovery through Management of Fire and Predation

HELEN M. REGAN
TONY D. AULD

Fire plays an important role in the survival and ecology of many plant species. For many of these species, recruitment from a soil or canopy-stored seed bank depends on fire events; these include a number of *Acacia* spp., *Banksia* spp., *Pinus* spp., and *Protea* spp. (Whelan 1995, Bond and van Wilgen 1996). And this is particularly important for species in which all above-ground life history stages are killed by fire. One of the threats species in fire-prone habitats face is the alteration of natural fire regimes, and this threat is exacerbated when it exists in combination with habitat destruction. Such a combination can particularly affect the plants that occur in fragmented populations (a population of genetically indistinct individuals that has been fragmented into small groups) within cleared landscapes or on the edge of urban areas.

Management of species in fire-prone, fragmented landscapes presents a number of challenges, one of which is how to determine an effective fire regime that will ensure the long-term persistence of populations. The optimal fire-management strategy will be one that ensures that seed production, survival of adult plants, and recruitment occur at complementary rates and times. In this chapter we use population viability analysis (PVA) to investigate a range of management strategies for fragmented populations of an endangered plant species, *Grevillea caleyi* (Proteaceae).

Grevillea caleyi is an endangered understory shrub of eucalypt forest in northern Sydney, Australia. Its range is less than 10 km, and its five populations are found on the interface between urban development and remnant bushland, some of which lies within the boundaries of Ku-ring-gai Chase and Garigal National Parks (Figure 3.1). The major threats to populations of *G. caleyi* are habitat destruction, adverse fire regimes, and very high seed predation (Auld and Scott 1997). Road construction and rural and urban development along ridges where the species occurs has destroyed about

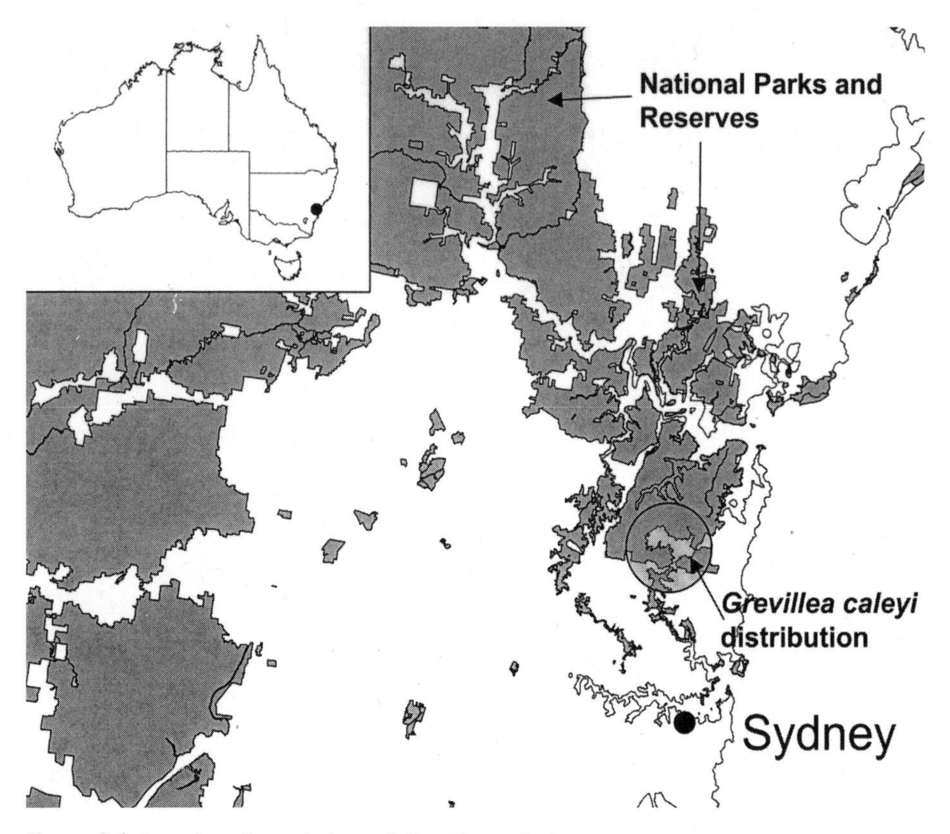

Figure 3.1 Location of populations of *Grevillea caleyi.*

85% of the habitat of *G. caleyi* and has also caused fragmentation of the remaining clusters of plants.

Implementation of a recovery program is currently under way; it aims to "minimise human imposed disturbance to *Grevillea caleyi* populations and to maintain viable wild populations into the foreseeable future" (National Parks and Wildlife Service 2001). A number of recovery actions have been identified. The most relevant for this particular study are research into the population viability of *G. caleyi* and the implementation of appropriate fire management. High fire frequency has been identified as a key threat to this and many other plant species in southeastern New South Wales (NSW Scientific Committee 2000). Population viability analysis is an effective tool that allows the convenient exploration of a range of management options on populations and a comparison of the relative success of each. Regan et al. (2003) used an individual-based model to investigate appropriate fire and predation management for a single small population of *G. caleyi* (70 seeds and 46 four-year-old plants). This model was spatially explicit in that the coordinates of every living plant and all soil-stored seeds were recorded throughout the simulation. It also included the effects of intraspecific shading on plant survival. It was found that the optimal management strategy, out of those investigated, involved a combination of fire and predation management and the prevention of wildfires. The fire-management action that induced the lowest risk of extirpation of the subpopulation depended on the level of seed predation. If seed predation was reduced to 80%, the optimal strategy was to burn 90% to 100% of the subpopulation every 7 to 15 years and to prevent wildfires from occurring. However, if seed predation was reduced to 70%, then the optimal strategy was to burn 20% to 90% of the subpopulation every 5 to 10 years and to prevent wildfires from occurring.

In this study we wish to extend the investigation to the entire population of *G. caleyi*. While the level of detail of individual-based models is relatively easy to incorporate into a model for a small population, it is not practical for the entire population. Computer memory and time constraints make an individual-based model impractical for long-term stochastic simulations of the entire population of *G. caleyi*. Instead, we use the population modeling software package RAMAS Metapop to construct a model for all known populations of the species. The specific management questions of interest here are as follows:

1. What are the management strategies that lead to the highest reduction in extinction and population decline rates for the species?
2. Is it important to prevent wildfires from occurring when implementing a fire-management strategy? If wildfires cannot logistically be prevented, then what are the best management options?
3. What is the effect of seed predation on the risk of extinction and population decline of the species?
4. Do the results obtained for the single small population apply to the entire population?

Here we describe the relevant life history characteristics of *G. caleyi*, the incorporation of those traits into a model, and the range of management strategies investigated with the model. Finally, we present and discuss the results produced from the model simulations and their implications for management.

Life History Characteristics

The relevant life history characteristics of *G. caleyi* for modeling purposes are initial population size, plant survival and growth, seed viability, seed production, seed predation, and response of plants and seeds to fire. Known populations of the plant occur within an area of 8 km × 8 km. Its three main populations are severely fragmented into small patches that vary from one adult (with a soil-stored seed bank) to a few thousand adult plants. Remnant patches of *G. caleyi* range in area from 5 m × 5 m to 350 m × 100 m.

Subpopulations of *G. caleyi* grow as even-age cohorts because all plants are killed by fire and seed germination is predominantly thought to be triggered by one or both (in combination) of smoke and heating of soil. The number of germinants after a fire depends on the heat and smoke outputs of the fire (Auld and Denham 2001a). There are essentially two types of fire: fires that produce high and low soil heating (here called hot and cool fires). Lower seed mortality and germination are predicted to occur for low soil heating fires than for high soil heating fires. How components of a fire may affect smoke as a dormancy-breaking mechanism is currently unknown.

The number of seeds produced by adult plants depends on the age of plants, predominately as a function of increasing plant size with age (cf. Auld and Scott 1997).

This is slightly different from other species in which reproductive output is directly related to the size of the plant (Samson and Werk 1986, Rees and Brown 1992, Klinkhamer et al. 1992). Reproductive maturity occurs when plants are around 3 years of age. Although 2-year-old plants rarely produce seeds, 3-year-old plants produce an average of 3 seeds, and plants 10 years and above can produce up to 15 seeds each flowering season (Regan et al. 2002). Seeds fall directly from the canopy and do not disperse. It is believed that emus may have once acted as key dispersal agents, but emus have disappeared from the area since its settlement by Europeans. Seed predation occurs both in the canopy and in the leaf litter. In the canopy, up to 30% of developing seeds are damaged and destroyed by the weevil *Cydmaea dorsalis* Lea (Coleoptera: Curculionidae) (Auld and Denham 2001b). Once seeds are at the soil surface they are further preyed on by native mammals (bush rats [*Rattus fuscipes*] and swamp wallabies [*Wallabia bicolor*]) at rates of 82% to 100% (Auld and Denham 1999, 2001a). Seed viability is quite high, but it decays exponentially in time when it enters the seed bank, generally at rates of around 5% to 30% per year (Auld et al. 2000). The survival of germinants is also high at around 99%. This rate gradually reduces as plants get older, and it eventually declines to approximately 50% and 80% survival rates (for shaded and nonshaded plants, respectively), for plant populations aged 8 years and higher (for details, see Regan et al. 2003).

Model

An age-based matrix model was constructed for all known populations of *G. caleyi* using RAMAS Metapop. Remnant patches of severely fragmented populations were represented as separate populations because there is no evidence of seed dispersal for this plant and because past fire events occurred independently at remnant sites, resulting in different aged cohorts at the separate sites. Eleven age classes were selected: from age 0 (to represent seeds) to age 10+ (all plants 10 years or older). The initial population sizes of plants were taken from the most recent census records and appear in Table 3.1. The size of the soil-stored seed bank is largely unknown and could only be roughly estimated by accounting for the size of the remnant site, the distribution of *G. caleyi* across the site, the fire history at the site, and limited soil sampling to provide patterns of soil seed abundances. Estimates of soil seed bank size also appear in column 2 of Table 3.1.

Survival rates and fecundity of plants were summarized from field data and assigned according to Regan et al. (2003). Shading is regarded as important for resource uptake for this plant, and shading depends on the size and location of plants, the size of the patch, the density of the plants within the patch, and the number and location of the seeds that germinate. Since populations grow as even-age cohorts, natural variability in the heights of plants is the driver for shading. Individual-based models are well suited to this type of dynamic. Modeling shading in more aggregated models, while not impossible, is difficult and of limited benefit for plant species such as *G. caleyi*, where shading depends on the differential heights within the same age cohort and those heights depend on natural variation within the population. Hence, survival rates were assigned as the average of those for the shaded and nonshaded plants. Ceiling density dependence was selected with a high carrying capacity. Because it is unknown what a likely

Table 3.1 Initial population sizes for all known populations and remnant patches of *Grevillea caleyi*

Site[a]	No. of Seeds	No. of Plants (Age in Years)
Bahai	5,000	1,300 (4)
		50 (10+)
Duffy's a	500	700 (1)
Duffy's b	1,000	350 (10+)
Belrose a	10	3 (1)
		1 (8)
Belrose b	1,200	700 (4)
Belrose c	20	4 (4)
Belrose d	100	8 (10+)
Belrose e	1,200	770 (2)
Terrey Hills a	20	4 (4)
Terrey Hills b	50	1 (10+)
Terrey Hills c	9,500	4,500 (1)
		243 (10+)
Terry Hills d	1,000	16 (10+)
Terrey Hills e	2,500	3,143 (4)
		10 (10+)
Terrey Hills f	200	300 (2)
McCarrs	50	17 (4)

Note: Two sets of plants with two different ages occur when sites were only partially burned at some time in the past.

a. Lowercase letters (a, b, c, etc.) following a site name refer to fragments of a population.

carrying capacity would be for each subpopulation, a high carrying capacity was selected to ensure that subjective estimates would not influence the results of management strategies. It was assumed that surpassing carrying capacity thresholds was not an issue for this species. This is a reasonable assumption since the initial population size is sufficiently low that it is extremely unlikely that carrying capacity would be exceeded.

Fecundity was assigned as the product of the number of seeds produced within the relevant age class and the proportion that survive predation. The number of seeds produced per plant was calculated according to the function in Regan et al. (2003). Predation rates depend on the time since the last fire and were calculated according to the regression line fit to field data appearing in Auld and Denham (2001a):

$$\rho = \begin{cases} 1.0 & \text{if } t \leq 2 \\ -0.01 \times t + 1.01 & \text{if } 2 < t \leq 18 \\ 0.83 & \text{if } t > 18 \end{cases} \tag{1}$$

where ρ is the predation rate (as a proportion) and *t* is the time since the last fire. Seed viability was set at 80% per year, roughly the mean of the seed viability range reported in National Parks and Wildlife Service (2001). The Leslie matrix used in the model for the baseline case of no management intervention appears in Table 3.2. A background germination rate of 1 in 10,000 was chosen in the absence of fires. This was to account for the rare case when germination occurs because of other types of soil disturbance. A non-zero germination rate was also necessary to facilitate recruitment in the model when

Table 3.2 Leslie matrix used in the model for the baseline case of no management intervention

	Seeds	Age 1	Age 2	Age 3	Age 4	Age 5	Age 6	Age 7	Age 8	Age 9	Age 10+
Seeds	0.8	0	0	0	0.14	0.28	0.45	0.62	0.78	0.93	1.54
Age 1	0.00001	0	0	0	0	0	0	0	0	0	0
Age 2	0	0.99	0	0	0	0	0	0	0	0	0
Age 3	0	0	0.99	0	0	0	0	0	0	0	0
Age 4	0	0	0	0.99	0	0	0	0	0	0	0
Age 5	0	0	0	0	0.99	0	0	0	0	0	0
Age 6	0	0	0	0	0	0.97	0	0	0	0	0
Age 7	0	0	0	0	0	0	0.86	0	0	0	0
Age 8	0	0	0	0	0	0	0	0.74	0	0	0
Age 9	0	0	0	0	0	0	0	0	0.68	0	0
Age 10+	0	0	0	0	0	0	0	0	0	0.66	0.65

Note: Survival rates are the average of those for shaded and nonshaded plants. Fecundity includes seed mortality due to predation.

fires occur. Demographic stochasticity was included for the fecundity of plants, whereas both demographic and environmental stochasticity were incorporated for the survival rates of plants. A coefficient of variation of 10% was included for all survival rates. A lognormal distribution was selected for the survival rates. Constraints are imposed within RAMAS Metapop to ensure that all survival rates lie within the bounds of 0 and 1 with minimal truncation of the distribution.

Wildfires are unplanned fires that occur because of natural processes such as lightning but may also occur due to arson or accidental human interference. The probability of a fire occurring depends on the time since the last fire and is represented in the model as a catastrophe. Table 3.3 displays the probability of a fire that gives an expected fire frequency that corresponds with the fire history records in the Sydney region and with prescribed burning practices of the land-management authorities in Sydney (Regan et al. 2003). In the model, only one fire type was represented, with the fire response as the average of that for hot and cool fires: that is, 15% of the seeds die in a fire, and 45% of the remaining seeds germinate. Fires were uncorrelated across the populations. This is a realistic assumption that corresponds with the initial population structure of the plants (Table 3.1). The model description thus far represents the baseline case of no imposed management action.

Table 3.3 Probability of an unplanned fire occurring

Time since last fire (years)	Probability of unplanned fire
1	0.005
2	0.005
3	0.037
4	0.068
5	0.1
≥ 6	0.1

To address the management questions presented above, a number of fire- and predation-management scenarios were tested with the model:

- Seed predation rates set to 90% for all times since last fire
- Seed predation rates reduced to 80% for all times since last fire
- Seed predation rates reduced to 70% for all times since last fire
- Uncorrelated fires every 5 years (wildfires prevented)
- Uncorrelated fires every 10 years (wildfires prevented)
- Uncorrelated fires every 15 years (wildfires prevented)
- Uncorrelated managed fires (every 5, 10, or 15 years) in combination with wildfires
- Seed predation rates reduced to 80% and fires every 5 years (wildfires prevented)
- Seed predation rates reduced to 80% and fires every 10 years (wildfires prevented)
- Seed predation rates reduced to 80% and fires every 15 years (wildfires prevented)
- Status quo–applying current known seed predation rates and wildfire

The timing of managed fires was staggered across populations. For scenarios in which managed fires occurred in addition to wildfires, the managed fire occurred on schedule, regardless of when the wildfire occurred. While this is a somewhat unrealistic assumption, it is useful for the sake of comparison with the results of the individual-based model in Regan et al. (2003). A management time frame of 50 years was selected for simulations, and 1,000 simulations were performed for the status quo case and for every scenario (except the last) in the preceding list. All of the results to be discussed next include the seed bank in the population count.

Results

Results for the baseline case of no imposed management strategy indicate high population declines (Figure 3.2a). While there is a 0.01 probability of extinction of the entire population (Figure 3.2b), by time step 50 an average of approximately three subpopulations remain and population decline is substantial. The population is guaranteed to fall below 25 individuals at least once in the 50-year time frame (Figure 3.2b).

When seed predation rates are managed to 90%, the mean population trajectory is not substantially improved (Figure 3.2a). However, the lowest population that the species is guaranteed to fall below at least once in 50 years is around 298. Population decline rates and risks of decline are somewhat improved for predation rates set at 80% and 70%. Although there is 0% chance of extinction for scenarios that use these predation rates, the mean population declines to low levels within 50 years.

Results are variable when fire is managed to occur every 5, 10, and 15 years. With fires that occur every 5 years, in the absence of wildfires, population declines are greater than the baseline case and the extinction risk is increased from 1% to 96% (Figure 3.3a). But there is noticeable improvement when fires are scheduled to occur every 15 years or 10 years (Figures 3.3a and b). Both the mean population trajectories and the risk curves do not differ substantially between these two management scenarios.

There is a dramatic difference in the risk of population decline when wildfires are permitted to occur in addition to managed fires. Figure 3.4 displays risk curves for the 10-year fire-management scenario, with and without wildfires. Population thresholds corresponding to a probability of 50% are 2,530 when wildfires are prevented and 180 when they are permitted to occur.

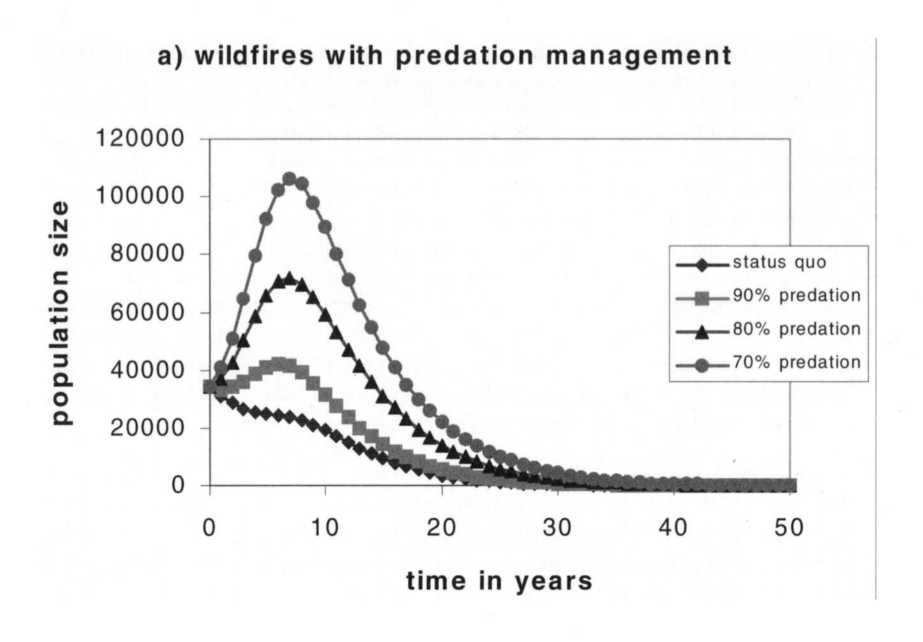

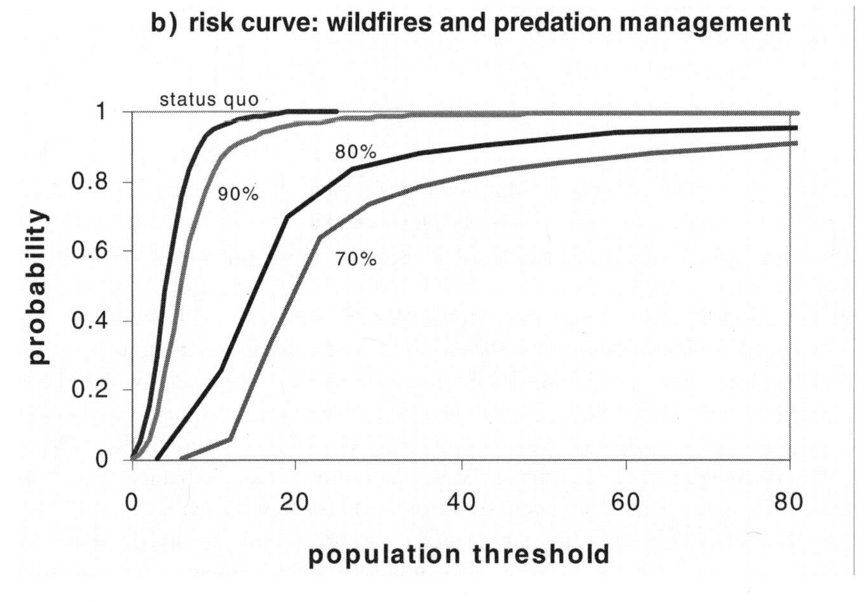

Figure 3.2 (**a**) Mean population trajectories of *Grevillea caleyi* over 50 years. The "status quo" scenario represents the baseline case with no imposed management action. The remaining trajectories represent the mean population trajectories over 50 years with predation rates consistently set to 90%, 80%, and 70% for all times since last fire. (**b**) Risk curves for the four scenarios: baseline and predation set to 90%, 80%, and 70%. The curves represent the chance that the population will fall below the threshold population at least once in the 50-year period.

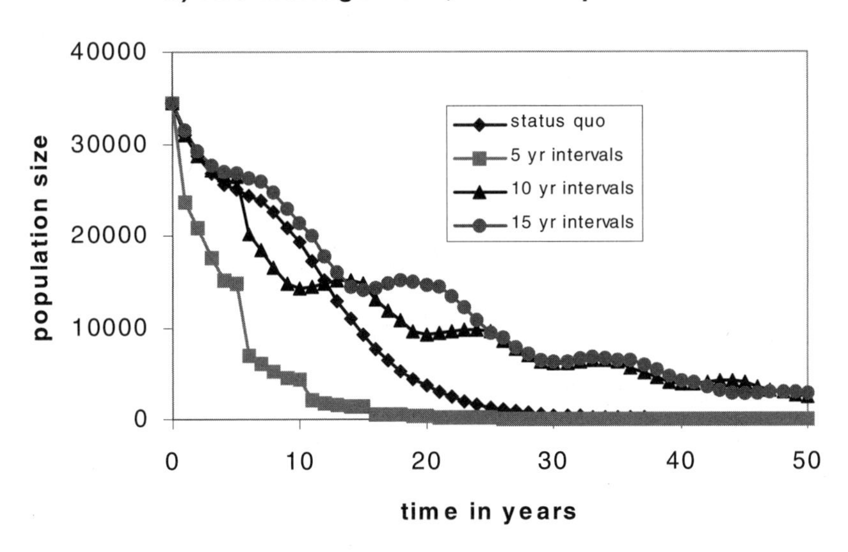

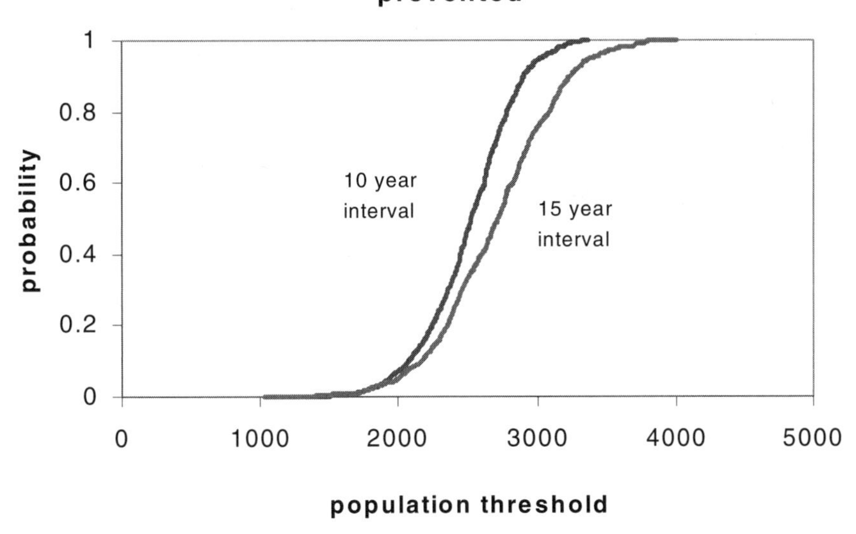

Figure 3.3 **(a)** Mean population trajectories of *Grevillea caleyi* over 50 years. The trajectories represent the mean population trajectories over 50 years with managed fires induced every 10 and 15 years in the absence of wildfires. The management scenario with fires induced every 5 years yielded an extinction risk of 95%. **(b)** Risk curves for two scenarios: fires managed every 10 and 15 years in the absence of wildfires. The curves represent the chance that the population will fall below the threshold population at least once in the 50-year period.

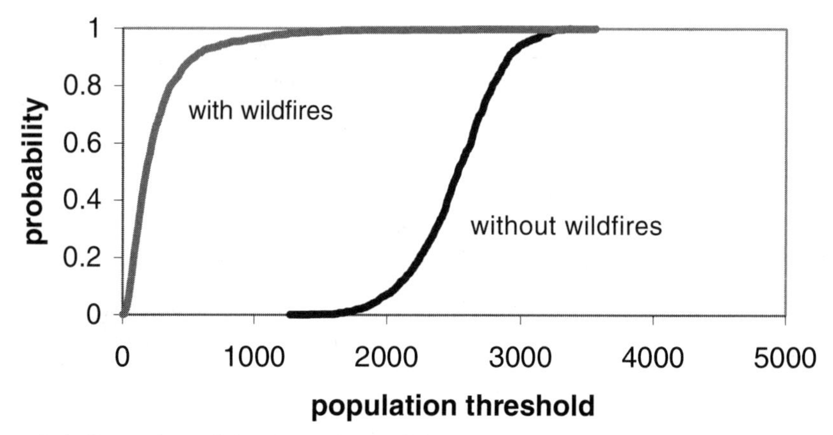

Figure 3.4 Comparison of the 10-year interval fire-management strategy with and without wildfires. When wildfires were permitted to occur, the managed fires occurred according to schedule, irrespective of when the last wildfire occurred.

Finally, Figure 3.5 displays the effect of a combination of predation and fire management. When predation rates are reduced to 80% in combination with managed fires, the impact on population decline can be substantial—the population threshold corresponding to a 50% chance of decline is around 37,000 for fires at 10-year intervals and 38,000 for fires every 15 years. This compares with population thresholds of approximately 2,530 and 2,720, respectively, at the 50% level in the absence of predation management.

Discussion

The results of model simulations show that some type of management action is crucial for the long-term persistence of *G. caleyi* populations: predation management on its own was a fairly ineffective management strategy. This result differs from the single subpopulation assessment in Regan et al. (2002) where predation reduction substantially improved the chances of long-term persistence of small populations. One explanation for this difference is that for a single small population the number of seeds entering the seed bank after predation is extremely low—low enough that there is a greater chance of all seeds being depleted in the seed bank due to loss of both viability and germination. Reducing predation rates for these small populations would therefore reduce their risk of extinction. Larger populations (or in this case, the entire suite of populations) have a seed bank large enough to ensure that there are always seeds remaining in the seed bank (unless there are fires in three consecutive years). Hence, reduction in predation rates for larger populations, or many populations considered simultaneously, does not have a substantial effect on the risk of extinction. For multiple populations, it is more

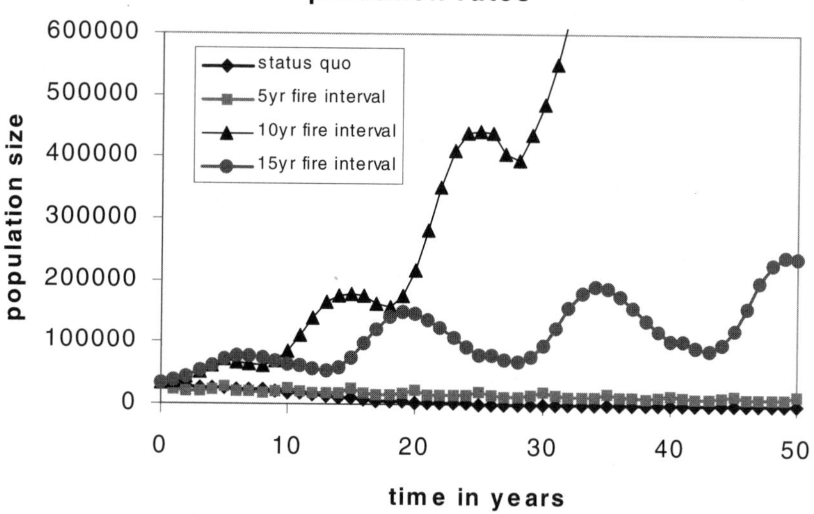

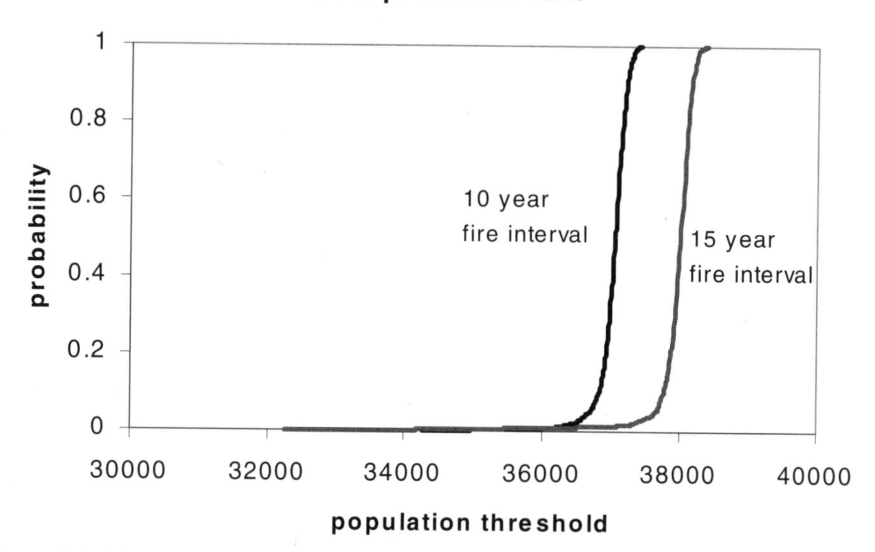

Figure 3.5 (**a**) Mean population trajectories of *Grevillea caleyi* over 50 years. The "status quo" scenario represents the baseline case with no imposed management action. The remaining trajectories represent the mean population trajectories over 50 years with predation rates reduced to 80% throughout and managed fires induced every 5, 10 and 15 years in the absence of wildfires. (**b**) Risk curves for the 2 scenarios: baseline and 80% predation rates with fires managed every 10 and 15 years in the absence of wildfires. The curves represent the chance that the population falls below the threshold population at least once in the 50-year period.

important to ensure that seeds are being produced and that germination is occurring on a regular basis while maximizing the chance of plant survival in years when there are no fire events. Hence, for the entire population of *G. caleyi*, fire management becomes a much more important strategy.

In this study regular fires occurring at constant intervals proved to be an extremely effective management strategy when fires were scheduled every 10 or 15 years. Fires occurring every 5 years were actually detrimental to the overall population, however. The extinction risk rose from about 1% to 96% by inducing fires at short regular intervals. This is not a surprising result. If fires occur before plants are mature enough to replenish the seed bank, then germination will occur too frequently, thereby eventually exhausting the seed bank and leading to extinction of the species. This result was also seen with the individual-based model of one small subpopulation (Regan et al. 2003). When wildfires are permitted to occur along with managed fires, the risk of decline below threshold populations increases substantially. Although prevention of wildfires may be difficult and undesirable, especially when it conflicts with management goals for other species and purposes, results here indicate that it is crucial to prevent wildfires from burning populations of *G. caleyi* when interval fire management is implemented across the entire population. Alternatively, although not explicitly modeled here, substituting a wildfire (when one occurs) for the next planned fire is likely to allow a degree of persistence in the 10- and 15-year managed-fire scenarios and would avoid increasing the impacts of fire frequency by attempting to implement a scheduled managed fire shortly after a wildfire has occurred. However, a number of remnant patches are isolated from surrounding bushland and do not burn in wildfires, so for these patches prevention of wildfires is not an issue.

The greatest improvement in the long-term survival of the entire population occurred when managed fires were combined with predation reduction. When predation rates were reduced to 80%, the 10- and 15-year fire intervals actually invoked an increasing trend in the mean population. Furthermore, the decline thresholds for the risk curves increase by an order of magnitude when fire management is combined with predation reduction. The 5-year fire interval also improved the chances of survival of the population when it was combined with predation reduction, but not to the same extent as the longer fire intervals. A decreasing trend in the mean population size was observed for the 5-year fire interval.

It is clear from the results here that the chances of long-term survival of *G. caleyi* are significantly improved with fire and predation management. Predation management will have the greatest benefit when it is implemented in the years immediately after a fire event, when predation rates are at their highest. This will allow a more rapid replenishment of the soil seed bank and will increase the resilience of the species to frequent fire. Actually implementing predation management, however, may be a challenge, and preventive actions may differ according to which predator is targeted—swamp wallabies or native rats. Further research into predation is necessary to determine which, if any, is the dominant predator. This will assist in guiding management actions for predation. Nevertheless, if fire management is the only action taken, the chances of long-term persistence of *G. caleyi* are considerably improved but to a lesser extent than when in combination with predation.

The results of this study indicate the importance of long-term fire records and wildfire prevention and also highlight the need for a better understanding of the response of seed predators to fire. This type of information may assist in establishing priorities for habitat protection, conservation, and recovery of this endangered plant.

Acknowledgments This work was completed while Helen Regan was a Postdoctoral Associate at the National Center for Ecological Analysis and Synthesis, a center funded by the National Science Foundation (grant no. DEB-0072909), the University of California Santa Barbara.

References

Auld, T. D., and Denham, A. J. 1999. The role of ants and mammals in dispersal and post-dispersal seed predation of the shrubs *Grevillea* (Proteaceae). *Plant Ecology* 144(2): 201–213.

Auld, T. D., and Denham, A. J. 2001a. The impact of seed predation of mammals on post-fire seed accumulation in the endangered *Grevillea caleyi* (Proteaceae). *Biological Conservation* 97(3): 377–385.

Auld, T. D. and Denham, A.J. 2001b. Predispersal seed predation in shrubs of *Grevillea* (Proteaceae) from south-eastern Australia. *Australian Journal of Botany* 49: 17–21.

Auld, T. D., and Scott, J. 1997. Conservation of endangered plants in urban fire-prone habitats. Pages 163–171 in *Proceedings: Fire effects on rare and endangered species and habitats Conference, Coeur D'Alene, Idaho, USA.* International Association of Wildland Fire, Fairfield, Wash.

Auld, T. D., Keith, D. A., and Bradstock, R. A. 2000. Patterns in longevity of soil seedbanks in fire-prone communities of southeastern Australia. *Australian Journal of Botany* 48: 539–548.

Bond, W. J., and van Wilgen, B. W. 1996. *Fire and plants.* Chapman and Hall, London.

Klinkhamer, P. G. L., Meelis, E., de Jong, T. J., and Weiner J. 1992. On the analysis of size-dependent reproductive output in plants. *Functional Ecology* 6: 308–316.

National Parks and Wildlife Service. 2001. Grevillea caleyi *R. Br. (Proteaceae) Draft Recovery Plan.* NSW National Parks and Wildlife Service, Hurstville.

NSW Scientific Committee (2000) Final determination for listing "High frequency fire resulting in the disruption of life cycle processes in plants and animals and loss of vegetation structure and composition" as a key threatening process under the NSW Threatened Species Conservation Act 1995. NSW Scientific Committee, Sydney, Australia.

Rees, M., and Brown, V. K. 1992. Interactions between invertebrate herbivores and plant competition. *Journal of Ecology* 80: 353–360.

Regan, H. M., Auld, T. D., Keith, D. A., and Burgman, M. A. 2003. The effects of fire and predators on the long-term persistence of an endangered shrub, *Grevillea caleyi. Biological Conservation* 109(1): 73–83.

Samson, D. A., and Werk, K. S. 1986. Size-dependent effects in the analysis of reproductive effort in plants. *American Naturalist* 127(5): 667–680.

Whelan, R. J. 1995. *The ecology of fire.* Cambridge University Press, Cambridge.

4

Whitebark Pine (*Pinus albicaulis*) in Mt. Rainier National Park, Washington, USA

Response to Blister Rust Infection

GREGORY J. ETTL
NICHOLAS COTTONE

The decline of whitebark pine (*Pinus albicaulis* Engelm.) across North America is well documented (Tomback et al. 2001) and is attributed to fire exclusion, insect outbreaks, and various fungal blights and rusts (Arno and Hoff 1990, Hoff and Hagle 1990, Keane and Arno 1993, Keane et al. 1994). White pine blister rust is a major source of whitebark pine mortality throughout western North America, with the heaviest mortality rates in Washington, Idaho, and northwestern Montana (Kendall and Keane 2001). In early 1995, blister rust was present on all sampled sites in Mt. Rainier National Park (Rochefort 1995) and was a likely cause of mortality in many trees (DelPrato 1999). Whitebark pine is in danger of becoming locally extinct.

Whitebark pine is a member of the stone pines (family Pinaceae, genus *Pinus*, subgenus *Strobus,* section *Strobus*, subsection *Cembrae*) that contain species with indehiscent cones and require birds for dispersal (McCaughey and Schmidt 2001). Whitebark pine is almost completely dependent on Clark's nutcracker (*Nucifraga columbiana* Wilson) for seed dispersal and successful germination (Tomback and Linhart 1990) through their extensive caching of seeds. Because squirrels cache whitebark pine seeds and bears often raid these caches, the decline of whitebark pine has important implications for many wildlife species. In Yellowstone National Park, whitebark pine cone crops have been linked to grizzly bear (*Ursus arctos* L.) sightings at lower elevations; when seeds are abundant, bears feed on seeds in the high country and avoid encounters with humans (Mattson et al. 1992). There is almost a doubling in mortality rates of grizzly bears in non-mast seed years than in mast years as grizzly bear–human conflicts increase, resulting in declining bear populations (Pease and Mattson 1999).

Whitebark pine is found in Alberta, British Columbia, California, Idaho, Montana, Nevada, Oregon, Washington, and Wyoming (Arno and Hoff 1990) and grows near the tree line (1170 m to 2130 m) throughout the Olympic and Cascade Mountains (Bedwell and Childs 1943, Franklin and Dyrness 1988, Arno and Hammerly 1984). Whitebark pines often exist as isolated populations on steep, rocky, and xeric sites, frequently separated from other whitebark pine populations by stands of subalpine fir (*Abies lasiocarpa* [Hook.] Nutt.), lodgepole pine (*Pinus contorta* Dougl.), and Engelmann spruce (*Picea engelmannii* Parry) (Franklin and Dryness 1988, Arno and Hoff 1990). Whitebark pine is also found on more mesic sites in mixed stands with these subalpine species (Daubenmire and Daubenmire 1968, Pfister et al. 1977). However, it is primarily a seral species in mixed stands in the Cascade Mountains (Franklin and Mitchell 1967, Achuff 1989), eventually being outcompeted in the absence of disturbance (Arno 1986).

The fungus *Cronartium ribicola* (Fischer) causes the disease white pine blister rust (Hunt 1983). White pine blister rust has a complicated life cycle that includes five spore stages and two hosts, both of which are limited to the five-needle pines and members of the genus *Ribes* (currants and gooseberries). Pycniospores and aeciospores are associated with the pine branches and stems, while urediniospores, teliospores, and basidiospores are associated with the leaves of the alternate host *Ribes* species (Hunt 1983).

Basidiospores are the phase that infects pines. Because the basidiospores are thin-walled and small, they rarely travel much farther than 25 m to infect a pine (Hunt 1983), although high precipitation followed by strong winds has been shown to promote basidiospore transport considerable distances (Bedwell and Childs 1943). When the pine needles are moist, basidiospores land and germinate, and mycelia enter the stomata, with chlorotic spots evident 4 to 20 weeks later. The fungus continues through the needle, into the branch at a rate of ~8 cm/year (Sinclair et al. 1987) and can result in tree death ultimately if the fungus girdles the bole.

The fungus is considered native to northcentral and eastern Asia with Siberian and Japanese stone pines serving as the original hosts (McDonald and Hoff 2001). Blister rust spread throughout much of Europe during the 1800s, and eastern white pines (*Pinus strobus* L.), introduced from North America and grown for timber, were very susceptible to white pine blister rust. The fungus was inadvertently introduced to both eastern and western North America on infected eastern white pine seedlings that were shipped from nurseries in Europe (McDonald and Hoff 2001). In western North America, the origin has been traced to 1,000 infected seedlings in Point Grey, British Columbia (near Vancouver) in 1910 (Hoff and Hagle 1990).

Once introduced to western North America, the fungus spread throughout British Columbia and southward into Washington, Oregon, and northern California, in addition to a southeastern expanse into Idaho, Wyoming, and Montana. Lachmund (1926) determined that most regions within a 100-mile radius of Vancouver had been infected by blister rust by 1913. Idaho cankers were dated as early as 1923 (Hoff and Hagle 1990). In 1937, Bedwell and Childs (1943) documented 92% whitebark pine infection on the Olympic Peninsula, Washington; 89% northeast of Mt. Rainier in Hyas Lake, Washington; and stand infection as far south as the White River and Salmon River regions in Oregon. In 1951, 55% infection was recorded in Mt. Rainier National Park (Gynn and Chapman 1951). All sites within Mt. Rainier National Park are currently infected with blister rust, causing a dramatic decline in whitebark pine populations.

To examine the potential of whitebark pine to persist in the face of heavy blister rust infections, we used RAMAS GIS to develop a spatially explicit stage-based model using demographic data from field plots and tree ring data that projects whitebark pine populations in Mt. Rainier National Park.

Modeling Whitebark Pine in Mt. Rainier

Input of Subpopulation Structure

We used an existing geographic information system database of whitebark pine habitat that was created from 1935 field surveys of 67 patches in Mt. Rainier National Park (Darrin Sweeney, unpublished data). RAMAS GIS (Spatial Data subprogram) was used to identify subpopulations (Akçakaya 2002). With a "neighborhood distance" of 10 m, the program identified 46 subpopulations. Color aerial photography was used to estimate the number of whitebark pine adults within each subpopulation (Cottone and Ettl 2001). Photographic counts showed a strong correlation ($r = 0.88$) with ground-truth plots, and correlation analysis was used to estimate populations on the 39% of the sites lacking coverage. An initial population of ~ 38,000 whitebark pine trees was estimated for Mt. Rainier.

We used vegetation plots to establish the proportion of individuals in each life history stage of whitebark pine (Figure 4.1). The diameter of all whitebark pine within 0.1- to 1.0-ha plots were measured at 1.4 m above the ground. Increment cores were extracted from larger trees as part of a dendrochronological study (DelPrato 1999). The age of each relatively healthy whitebark pine (Class 1 and 2) was determined from ring counts when the center of the tree was visible (i.e., pith was present). Correlation analysis was used to establish diameter-age relationships, allowing an estimate of the ages of trees that lacked cores. Because we cored the trees at breast height, we added 30 years to the age determined by annual growth rings; sapling data (see below) indicate that whitebark pine requires approximately 30 years to reach 1.4 m in height. The age estimates allowed us to set initial population structures on a subpopulation basis for model runs, and they were also used in calibrating the model. We created a stage-based model with 13 stages: seed, four seedling stages, saplings, infected saplings, nonreproductive adults, infected nonreproductive adults, healthy adult trees (Class 1), and three stages of blister rust infection (Classes 2–4).

The model incorporates ceiling-type density dependence. Density dependence was applied to nonreproductive adults, infected nonreproductive adults, and Class 1–4 trees. Carrying capacities were set at twice the initial populations. We chose ceiling competition because this allows the catastrophes to affect vital rates. The presence of white pine blister rust results in a rapid decline in all populations, thus making density dependence irrelevant.

Vital Rates

Vital rates for a stage-based matrix were estimated by field sampling, by dendrochronological techniques, and from other studies. One problem with using a simple model for long-lived species is that an individual can move from seed to reproductive adult in seven time steps.

Seed

The amount of seed produced was determined from 18 years of whitebark pine cone crop data in Yellowstone (Chuck Schwartz, unpublished data, USDA Forest Service,

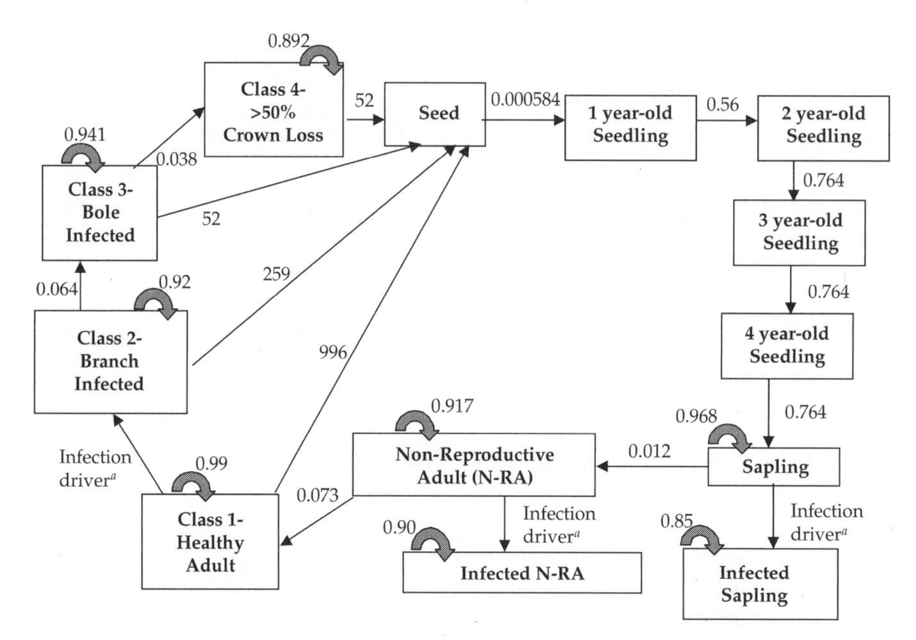

Figure 4.1 Whitebark pine life-history characteristics in the presence of blister rust. Thin arrows represent transition probabilities for moving from one stage to the next. Thick arrows represent the probability of staying in a particular stage. Arrows from Class 1, 2, 3, and 4 to seed represent the average amount of seed produced at each time step. Note a: The catastrophe matrix moves individuals from healthy to infected classes when catastrophe strikes.

Bozeman, Montana), because extended interval cone crop data were not available for the Cascade Mountains. The proportion of cone production from reproductive adults was estimated to be 100% for healthy, Class 1 trees. The seed output is reduced in infected trees: Class 2 adults produce 26.0%, and Classes 3 and 4 produce 5.2% of the cones that Class 1 adults produce annually (DelPrato 1999).

Seedlings

Seed germination rates are difficult to determine due to variation in seed predation, the abundance of nutcrackers, edaphic conditions, and various additional environmental factors that vary on a site-by-site basis. We modified a deterministic equation for the total number of seedlings produced from total number of seeds present in the system based on a study by Keane et al. (1990). Using high and low estimates of various parameters (number of seeds per cache, proportion of seed eaten, cache distance from seed source, and nutcracker and cone density), the equation yielded a range of seedling germination from seed between 1.04×10^{-5} and 3.36×10^{-2}. Seedlings have 56% survival in the first year and 25% survival to the fourth growing season (Tomback 1982). We assigned a survival rate of 0.56 to 1-year-old seedlings and 0.764 for 2-, 3-, and 4-year-old seedlings because 0.56*0.764*0.764*0.764 = 0.25, corresponding to a 25% chance of seedlings becoming saplings.

Sapling Survival and Transition to Nonreproductive Adults

Sapling ages were determined by counting annual whorls in the field. They indicate a residence time of 31 years for saplings (estimated from DelPrato 1999) and a yearly mortality rate of 2% (Kobe and Coates 1997). We used the following equation (Akçakaya et al. 1997) to calculate survival rates from residence time:

$$\text{Residence time in stage } i = 1/(1 - D_i) \quad (1)$$

where D_i is the diagonal element in column i, or the annual probability of an individual in stage i to stay in the same stage. Total survival rate is the sum of D_i and the probability of growth (transition to the next stage).

Assuming a 2% sapling mortality rate, then the survival of saplings is 96.8% with 1.2% transitioning to nonreproductive adults. Regression analysis of cone-bearing trees observed in the field (DelPrato 1999) indicates that 10.3 cm is the maximum diameter at breast height for nonreproductive adults ($R^2 = 0.70$). From tree ring data, a tree of 10.3 cm was determined to be approximately 45 years old; therefore, nonreproductive adults had a resident time of 12 years (between 34 and 45 years old). Assuming a 1% mortality rate for nonreproductive adults (similar to Class 1 below), 7.3% of nonreproductive adults reach maturity and 91.7% remain in each time step.

Class 1, Uninfected Trees

A 1% mortality rate was assumed for all adults due to competition or advanced age (Edmonds et al. 1993, Mast and Veblen 1994, Volney 1998, Monserud and Sterba 1999), and therefore Class 1 trees have a 0.99 probability of surviving each year.

Class 2, Trees with a Branch Canker

Class 1 trees become infected with blister rust through the catastrophe subroutine (see the section titled "Catastrophe"). We presume that blister rust travels horizontally and vertically in branches at a rate of ~8 cm/year (after Sinclair et al. 1987). Therefore fungus entering a branch 1 m from the bole will have a Class 2 residence time of 12.5 years. Infection Class 2 trees progress to infection Class 3 when the cankers reach the bole. The mortality rate of Class 2 trees was estimated to be 1.6%, intermediate between Class 1 and Class 3. From tree ring data, the Class 3 annual mortality rate was estimated to be 2.1%. Therefore, 92% of Class 2 trees survive, while 6.4% transition to Class 3 trees in each time step.

Classes 3 and 4, Severally Infected Trees

Class 3 and 4 vital rates were determined from growth reductions observed in tree rings. DelPrato (1999) used the computer program OUTBREAK (Swetnam 1986) to detect radial growth reductions associated with blister rust infection. OUTBREAK allowed us to compare non-host trees with heavily infected, Class 3 and 4, trees. OUTBREAK corrects for differences in growth that are related to climate by comparing standardized chronologies of non-host trees with standardized chronologies of severally infected trees on a site-by-site basis. We defined sustained growth reduction in severally infected whitebark pine, when growth of infected trees was –1.28 SD of the non-host chronology for 10 consecutive years. We then counted the number of years that each tree showed sustained growth reduction. Class 3 and 4 trees show 26 and 31 years of growth reduction, respectively.

We also used OUTBREAK to determine the number of years of growth reduction experienced in trees where the primary cause of death was blister rust. Survival rates of Class 3 and 4 trees were determined from dead trees (DelPrato 1999). Tree ring time series of all dead trees was set to $t = 1$ in the first year that blister rust caused a decrease in annual growth for that tree (as determined by the computer program OUTBREAK). The series with the longest period of growth reductions is 61 years, and the shortest is 7 years. We determined annual survival rates by counting the number of trees showing growth reduction in one "year" divided by the number of trees showing growth reduction in the previous "year." For example, if 60 cores exhibited reduced growth at year 15 and 57 cores exhibited reduced growth at year 16, then the yearly survival rate is 57/60 = 95%. The average survival rate determined from trees during the first 26 years (based on DelPrato 1999) is assumed to represent the survival rate of Class 3 trees (97.9%). The next 35 years were used to calculate survival rates for Class 4 trees, yielding 89.2%. The 97.9% estimate did not take into account movement of Class 3 trees to Class 4. The mortality rate of Class 4 trees was 10.8%, and the survival rate was 89.2%, which yields a residence time of 9.3 years. Therefore, the time for a Class 3 tree to become a Class 4 tree is 26 – 9.3 = 16.7 years. With a mortality rate of 2.1%, the application of equation 1 yields a Class 3 survival rate of 94.1% and a 3.8% transition to Class 4.

Calibration

We compared field data from whitebark pine stands with stands we simulated (on the same sites) to determine if the model could simulate existing stands. The diameters of all whitebark pine were measured in 35, 100–m^2 vegetation plots on 10 of Mt. Rainier's 67 whitebark pine sites (DelPrato 1999). Tree diameters were correlated with age from tree rings of whitebark pine that were cored ($R^2 = 0.663$, $p < 0.05$). Diameter measurements of whitebark pine from 1997 to 1998 plots were used to reconstruct the stand in 1930. Trees were placed into life-history stages by inferring each tree's age in 1930: (1) seedlings 1–4 years old, (2) saplings 5–34 years old, (3) nonreproductive adults 35–46 years old, and (4) reproductive adults >46 years old.

The frequency of blister rust catastrophe was determined for each subpopulation based on the occurrence of the first major growth reduction observed on each site (DelPrato 1999). For example, a site with a severe growth reduction in 1960 would have greater than 50% of sampled Class 3–4 trees exhibiting bole cankers, which would result in the observed growth reductions. We decided that this site was subjected to its first wave of blister rust in 1948, because Sinclair et al. (1987) determined the lateral spread of blister rust from branch to bole at 8 cm/year and a Class 2 tree transitioned to Class 3 after 12 years. This gave a probability of blister rust catastrophe occurring on this site of 0.056 (1/18 years).

A site was considered calibrated when the model projected 1998 data that were sufficiently similar to actual 1998 data from vegetation plots that they were not significantly different in a χ^2 analysis (after Platt et al. 1988). If the two data sets were significantly different ($p \leq 0.05$ in χ^2 analysis), then the site was calibrated again, this time varying only seedling germination rates or catastrophe transition probability (see "Catastrophe" section). We calibrated sites using only these two parameters because they were the most difficult parameters to estimate. In all, 8 of 10 sites were calibrated successfully by altering the probabilities of the blister rust catastrophe matrix and altering seedling germination probabilities between 5.8×10^{-5} and 11×10^{-4}. The average seed germination probability and catastrophe matrix for the eight sites that were calibrated were applied to the 46 subpopulations in the model.

Environmental and Demographic Stochasticity

We calculated standard deviations for vital rates from data where possible and applied a standard deviation of 10% to either the survival or transition probability (whichever is less) when calculations were not possible. The standard deviation in seed production was calculated from 18 years of cone crop data from Yellowstone. We also calculated the standard deviation in survival rates of Class 3 and Class 4 trees from tree ring data, after subtracting the variance due to demographic stochasticity (as described by Akçakaya 2002). We used the negative correlation between the largest survival rate and other rates in the same column (see "Advanced" button in the Stochasticity dialog under the Model menu), because several columns had sums close to 1. Simulations were run, incorporating demographic stochasticity 10,000 times for 175 years in each trial, thereby allowing us to determine the median, and 95% confidence intervals of different scenarios. Demographic stochasticity was also incorporated into the simulations to more realistically incorporate the local extinction of smaller subpopulations due to chance.

Catastrophe

Blister Rust

The catastrophe routine of RAMAS GIS was used to simulate white pine blister rust entering the system. During the calibration process, the frequency of blister rust catastrophe was initially determined for each site based on the occurrence of the first major growth reduction observed in the tree ring records of Classes 3–4, and dead trees (DelPrato 1999). During simulation years in which the fungus strikes a specific site, a catastrophe matrix is multiplied with the standard whitebark pine stage matrix, thereby reflecting blister rust. When blister rust "strikes," saplings, nonreproductive adults, and healthy adults are transferred to infected saplings, infected nonreproductive adults, and Class 2 stages, respectively. The first two groups continue on to death, while Class 2 trees move through the various levels of infection, with decreasing fecundity. Calibration led to the creation of a catastrophe matrix that transfers 15% of saplings, 58% of nonreproductive adults, and 96% of mature adults to corresponding infected stages at each time step that blister rust catastrophe occurs.

The probability of blister rust catastrophe striking was based on the life cycle and ecology of the fungus, in addition to field observations and tree core samples (Hunt 1983, DelPrato 1999). The average blister rust invasion of a site, given that blister rust is already present on the site, was determined to be 8 years, or a 0.125 probability that the fungus could strike in any time step. The estimates derived from tree ring data (DelPrato 1999) are consistent with the infection rates determined by McDonald and Hoff (2001) using an epidemiological model of blister rust infecting whitebark pine for the Pacific Northwest. The catastrophe probability was simulated locally, so each subpopulation experienced pulses of catastrophe independent of all other subpopulations.

Population Correlation and Interpopulation Dispersal

Populations in the same locale are likely to experience similar biotic and abiotic factors. Population correlations vary as a function of distance between subpopulations. Stochasticity incorporated into the model affects highly correlated subpopulations similarly. The model was run with moderate correlations between subpopulations, based on the correlation distance function (see Chapter 1) with $a = 1$, $b = 0.5$, and $c = 1$, yielding strong correlations for sites within 1 km.

A second parameter of the model that depends on site distance is the seed dispersal function. Previous studies determined a range of maximum dispersal distance for whitebark pine seed from 2.5 to 12.5 km (Tomback 1978, Hutchins and Lanner 1982). Other field studies have documented that only 10% to 20% of harvested seeds are cached closer than 100 m from the seed source, and only a small percentage are transported further than 1000 m (Vander Wall 1988, Hutchins and Lanner 1982). This translates into nutcrackers caching primarily within the same subpopulation in the model. We used dispersal distance function (see Chapter 1) with $a = 1$, $b = 5$, $c = 1$, and D_{max} (maximum dispersal distance) = 2.5 km.

Results

Our model predicts an increasing population in the absence of blister rust, but with blister rust in the system the population declines. The incorporation of density dependence has no effect on model projections because the population is declining. The median time to quasi extinction (population < 100 individuals) is 148 years (Figure 4.2). Because whitebark pines live to 500 years on suitable sites (Arno and Hoff 1990), the decline occurs in less than one generation. The metapopulation is steady for about the first 15 years before trees die from blister rust infection. There is a 94% chance that the population will fall below 100 individuals in 175 years.

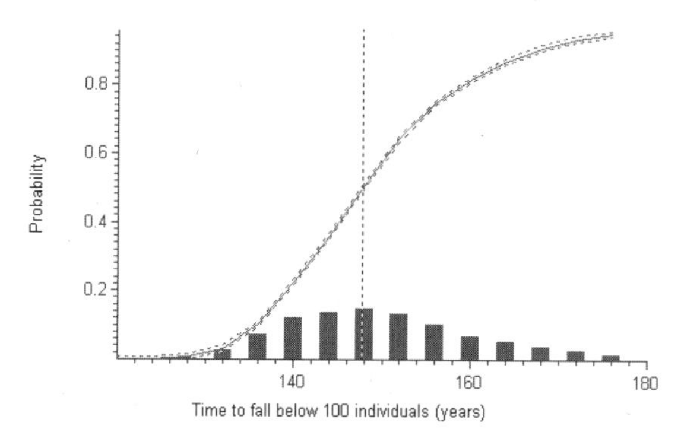

Figure 4.2 Probability of whitebark pine metapopulation declining below 100 individuals. The vertical dotted line represents the median time to quasiextinction at 147.7 years. The bars represent the distribution of time to quasiextinction from 10,000 trials.

Whitebark pine is unequally distributed among 46 subpopulations in Mt. Rainier National Park, with the majority of individuals located at Sunrise (subpopulations 21 and 27), the Palisades (subpopulations 1 and 7), Moraine Park (subpopulation 24), Crystal Mountain (subpopulations 2 and 23), and Summerland (subpopulations 34 and 37) (see the model on the CD-ROM for a map of the metapopulation). The dispersal of seed (maximum dispersal of 2.5 km) allows for interaction among most subpopulations. There is a significant correlation between subpopulation size and the growth rate of the subpopulations ($r = 0.699$, $p < 0.001$). A comparison of growth rate for the first 50 years of simulation demonstrates that smaller subpopulations decline much faster than larger subpopulations and that this relationship is exponential (Figure 4.3; note log-transformation). We interpret deviation from the fitted line to indicate subpopulations that have a growth rate either higher than expected (subpopulations below the line) or lower than expected (subpopulations above the line). The seven largest subpopulations have the highest growth rates, but they also have growth rates that are lower than expected (data points above the trend line), suggesting that these populations are declining faster than expected—possibly serving as sources to adjacent smaller populations. A comparison of other subpopulations that are farther from the trend line also suggests that distant subpopulations decline more rapidly than expected (subpopulations 35 and 44) because there is less chance of immigration from larger subpopulations. Conversely, subpopulations near the larger subpopulations decline more slowly than expected. Subpopulations 9, 11, 20, 4, and 33 benefit from their proximity (< 0.5 km—see distance matrix in Mt.Rainier.ptc) to the larger subpopulations 7, 7, 24, 7, and 34, respectively. In fact, populations 4 and 33 are not visible in the metapopulation map due to the fact that larger

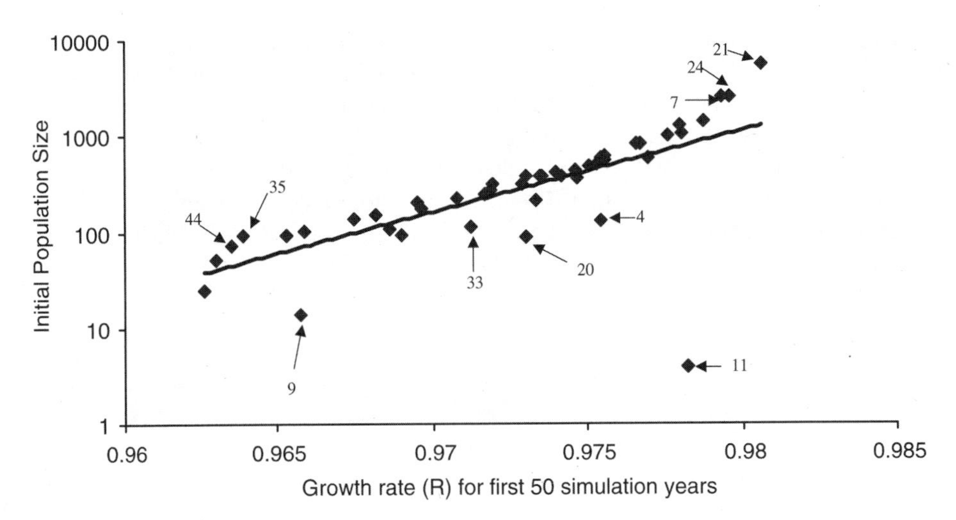

Figure 4.3 Growth rate for the first 50 years of simulation for 46 subpopulations of whitebark pine in Mt. Rainier National Park. There is a relatively strong correlation between growth rate and initial population size ($r = 0.699$). Relatively large subpopulations (7, 21, 24—indicated on the figure) showed lower growth rates than expected, as do two subpopulations that are more distant from other subpopulations (35 and 44). Subpopulations 4, 9, 11, 20, and 33 all have growth rates higher than expected for their population size and are most likely benefiting from their close proximity to larger subpopulations.

subpopulations are plotted as a larger circle overlapping the smaller subpopulations. The metapopulation shows dramatic decline in all subpopulations with only the large subpopulations, on average, able to support small populations after 100 years (Sunrise 436, Moraine Park 149, the Palisades 156, Crystal Mountain 111, and Summerland 58).

The model is more sensitive to changes in the vital rates of healthy adult trees than to changes in the vital rates of infected trees. The model is also more sensitive to changes in the vital rates of less mature than to more mature stages of whitebark pine (Cottone 2001). Furthermore, the model is more sensitive to changes in the probability of blister rust catastrophe occurring than to changes in the blister rust matrix multipliers as we have defined them.

Discussion

Our model predicts a dramatic decline of whitebark pine in Mt. Rainier National Park. The decline occurs in all subpopulations and is driven by mortality related to increases in blister rust. Our model also suggests that source-sink dynamics are important, with seed from larger populations slowing the decline of smaller populations that are more subject to local extinction.

White pine blister rust has dramatically reduced many five-needle pine populations in North America. Considerable effort has been given to preventing the spread of blister rust since 1921 by removing infected pines and eradicating *Ribes*, and yet the outlook for most pine populations is grim (McDonald and Hoff 2001). The decline of whitebark pine has been well documented (Kendall and Keane 2001 provide review). Our model quantifies the speed with which blister rust can devastate whitebark pine populations and demonstrates the potential for local extinction in Mt. Rainier. In the absence of blister rust resistance, the future of whitebark pine throughout most of its range is in question.

The plight of the whitebark pine has not received the attention it may warrant because most sites are inaccessible and its value to the timber industry is limited (Tomback et al. 2001). However, it is clear that the loss of whitebark pine from high-elevation communities will have consequences for a variety of species, including bears, grouse, elk, and a number of predatory birds. Pease and Mattson (1999) estimate that grizzly bear populations decrease by 5% in nonmast whitebark pine cone crop years and increase by 7% in mast years in the Yellowstone ecosystem. This suggests that the spread of blister rust through whitebark pine communities in the Yellowstone ecosystem will dramatically decrease grizzly bear populations. Furthermore, this decrease could occur over a period of one to three decades since cone production decreases well before trees die (DelPrato 1999).

Our model could serve as a starting point for other viability analyses of whitebark pine populations or for other five-needle pines infected with blister rust. Western white pine (*Pinus monticola* Dougl.), sugar pine (*Pinus lambertiana* Dougl.), limber pine (*Pinus flexilis* James), southwestern white pine (*Pinus Strobiformis* Engelm.), bristlecone pines (*Pinus aristata* Enegl. and *Pinus longaeva* Bailey), and foxtail pine (*Pinus Balfouriana* Grev. and Balf.) are all susceptible to white pine blister rust (McDonald and Hoff 2001). It would be possible to modify this model for use with other species using field data and information regarding species-specific vital rates from the literature. However, caution should be used in applying the model without calibration for site and species specific differences in vital rates.

References

Achuff, P. L. 1989. Old-growth forests of the Canadian Rocky Mountain national parks. *Natural Areas Journal* 9: 12–26.

Akçakaya, H. R. 2002. *RAMAS GIS.* Applied Biomathematics, Setauket, N.Y.

Akçakaya, H. R., Burgman, M. A., and Ginzburg, L. R. 1997. *Applied population ecology.* Applied Biomathematics, Setauket, N.Y.

Arno, S. F. 1986. Whitebark pine cone crops: a diminishing source of wildlife food? *Western Journal of Applied Forestry* 1: 92–94.

Arno, S. F., and Hammerly, R. 1984. *Timberline-mountain and arctic forest frontiers.* Mountaineers, Seattle, Wa.

Arno, S. F., and Hoff, R. J. 1990. *Pinus albicaulis* Engelm. Whitebark pine. Pages 268–279 in R. M. Burns and B. H. Honkala (tech. coords.), *Silvics of North America.* Vol. 1. *Conifers.* Handbook 654. U.S. Department of Agriculture, Washington, D.C.

Bedwell, J. L., and Childs, T. W. 1943. Susceptibility of whitebark pine to blister rust in the Pacific Northwest. *Journal of Forestry* 41: 904–912.

Cottone, N. 2001. Modeling whitebark pine infected with blister rust in Mt. Rainier National Park. M.S. thesis, St. Joseph's University, Philadelphia, Pa.

Cottone, N., and Ettl, G. J. (2001). Estimating populations of whitebark pine in Mount Rainier National Park, Washington using aerial photography. *Northwest Science* 75: 397–406.

Daubenmire, R., and Daubenmire, J. B. 1968. *Forest vegetation of eastern Washington and northern Idaho.* Technical Bulletin 60. Washington Agriculture Experiment Station, Pullman.

DelPrato, P. R. 1999. Effect of blister rust on whitebark pine in Mt. Rainier National Park, WA. Master's thesis, St. Joseph's University, Philadelphia, Pa.

Edmonds, R. L., Thomas, T. B., and Maybury, K. P. (1993). Tree population dynamics, growth, and mortality in old-growth forests in the western Olympic Mountains, WA. *Canadian Journal of Forest Research* 23: 512–519.

Franklin, J. F., and Dyrness, C. T. 1988. *Natural vegetation of Washington and Oregon.* Oregon State University Press, Corvallis.

Franklin, J. F., and Mitchell, R. G. 1967. *Successional status of subalpine fir in the Cascade Range.* Res. Pap. PNW-46. U.S. Department of Agriculture, Forest Service, Pacific Northwest Forest and Range Experiment Station, Portland, Ore.

Gynn, J. D., and Chapman, C. M. 1951. Blister rust control, Mt. Rainier National Park, 1951. Pages 57–60 in *White pine blister rust control in the Northwest region.* U.S. Department of Agriculture, Bureau of Entomology and Plant Quarantine, Spokane, Wash.

Hoff, R. J., and Hagle, S. K. 1990. Diseases of whitebark pine with special emphasis on white pine blister rust. Pages 179–190 in W. C. Schmidt and K. J. MacDonald (eds.), *Proceedings of the symposium on whitebark pine ecosystems: ecology and management of a high mountain resource, 29–31 March 1989, Bozeman, Montana.* General Technical Report INT-270. U.S. Department of Agriculture, Forest Service, Ogden, Utah.

Hunt, R. S. 1983. *White pine blister rust in British Columbia.* Pest Leaflet. Canadian Forestry Service, Pacific Forest Research Center, Victoria, B.C.

Hutchins, H. E., and Lanner, R. M. 1982. The central role of Clark's nutcracker in the dispersal and establishment of whitebark pine. *Oecologia* 55: 192–201.

Keane, R. E., and Arno, S. F. 1993. Rapid decline of whitebark pine in Western Montana: evidence from 20-year remeasurements. *Western Journal of Applied Forestry* 8: 44–47.

Keane, R. E., Arno, S. F., Brown, J. K., and Tomback, D. F. 1990. Modeling stand dynamics in whitebark pine (*Pinus albicaulis*) forests. *Ecological Modeling* 51: 73–95.

Keane, R. E., Morgan, P., and Menakis, J. P. 1994. Landscape assessment of the decline of whitebark pine (*Pinus albicaulis*) in the Bob Marshall Wilderness Complex, Montana, USA. *Northwest Science* 68: 213–229.

Kendall, K. C., and Keane, R. E. 2001. Whitebark pine decline: infection, mortality, and population trends. Pages 221–242 in D. F. Tomback, S. F. Arno, and R. E. Keane (eds.), *Whitebark pine communities: ecology and restoration.* Island Press, Washington, D.C.

Kobe, R. K., and Coates, K. D. 1997. Models of sapling mortality as a function of growth to characterize interspecific variation in shade tolerance of eight tree species of northwestern British Columbia. *Canadian Journal of Forest Research* 27: 227–236.

Lachmund, H. G. 1926. Studies of white pine blister rust. *Western Journal of Forestry* 24: 874–884.

Mast, J. N., and Veblen, T. T. 1994. A dendrochronological method of studying tree mortality patterns. *Physical Geography* 15: 529–542.

Mattson, D. J., Blanchard, B. M., and Knight, R. R. 1992. Yellowstone grizzly bear mortality, human habituation, and whitebark pine seed crops. *Journal of Wildlife Management* 56: 432–442.

McCaughey, W. W., and Schmidt, W. C. 2001. Taxonomy, distribution, and history. Pages 105–120 in D. F. Tomback, S. F. Arno, and R. E. Keane (eds.), *Whitebark pine communities: ecology and restoration.* Island Press, Washington, D.C.

McDonald, G. I., and Hoff, R. J. 2001. Blister rust: an introduced plague. Pages 193–220 in D. F. Tomback, S. F. Arno, and R. E. Keane (eds.), *Whitebark pine communities: ecology and restoration.* Island Press, Washington, D.C.

Monserud, R. A., and Sterba, H. 1999. Modeling individual tree mortality for Austrian forest species. *Forest Ecology and Management* 113: 109–123.

Pease, C. M., and Mattson, D. J. 1999. Demography of the Yellowstone grizzly bears. *Ecology* 80: 957–975.

Pfister, R. D., Kovalchik, B. L., Arno, S., and Presby, R. 1977. *Forest habitat types of Montana.* General Technical Report INT-34. U.S. Department of Agriculture and Forest Service, Intermountain Forest and Range Experiment Station, Ogden, Utah.

Platt, W. J., Evans, G. W., and Rathbun, S. L. 1988. The population dynamics of a long-lived conifer (*Pinus palustris*). *American Naturalist* 131: 491–525.

Rochefort, R. 1995. Whitebark pine blister rust survey in Mt. Rainier National Park, 1994. *Nutcracker Notes* 6: 2 pp.

Sinclair, W. A., Lyon, H. H., and Johnson, W. T. 1987. *Diseases of trees and shrubs.* Comstock Publishing Associates, a division of Cornell University Press, Ithaca, N.Y.

Swetnam, T. W. 1986. Western spruce budworm outbreaks in northern New Mexico: tree ring evidence of occurrence and radial growth impacts from 1700 to 1983. Pages 130–141 in G. C. Jacoby and J. W. Hornbeck (eds.), *Proceedings of the international symposium on ecological aspects of tree ring analysis.* U.S. Department of Commerce, Springfield, Va.

Tomback, D. F. 1978. Foraging strategies of Clark's nutcracker. *Living Bird* 16: 123–161.

Tomback, D. F. 1982. Dispersal of whitebark pine seeds by Clark's nutcracker: a mutualism hypothesis. *Journal of Animal Ecology* 51: 451–467.

Tomback, D. F., and Linhart, Y. B. 1990. The evolution of bird-dispersed pines. *Evolutionary Ecology* 4: 185–219.

Tomback, D. F., Arno, S. F., and Keane, R. E. 2001. The compelling case for management intervention. Pages 3–28 in D. F. Tomback, S. F. Arno, and R. E. Keane (eds.), *Whitebark pine communities: ecology and restoration.* Island Press, Washington, D.C.

Vander Wall, S. B. 1988. Foraging of Clark's Nutcracker on rapidly changing pine seed resources. *Condor* 90: 621–631.

Volney, W. J. A. 1998. Ten-year tree mortality following a jack pine budworm outbreak in Saskatchewan. *Canadian Journal of Forest Research* 28: 1784–1793.

5

Monkey Puzzle Tree (*Araucaria araucana*) in Southern Chile

Effects of Timber and Seed Harvest, Volcanic Activity, and Fire

SARAH A. BEKESSY
ADRIAN C. NEWTON
JULIAN C. FOX
ANTONIO LARA
ANDREA PREMOLI
MARCO CORTES
MAURO GONZALEZ
BRUCE BURNS
LEONARDO GALLO
FERNANDA IZQUIERDO
MARK A. BURGMAN

Commonly known as the monkey puzzle tree or pehuén, *Araucaria araucana* (Molina) K. Koch (Araucariaceae) is an impressively large and long-lived conifer, reaching 50 m in height, 2.5 m in girth, and ages of at least 1,300 years (Montaldo 1974). Its current distribution spans only three degrees of latitude and is divided between a main area that straddles both sides of the Andean divide (in Chile and Argentina) and two disjunct populations on the Coastal Range, Chile (Figure 5.1). Its former distribution has been severely diminished by logging, human-set fires, and land clearance since European settlement (Veblen 1982).

Monkey puzzle produces high-quality timber and provides a unique resource for tourism and recreation. The tree figures importantly in the religion of the native Pehuenche people and is valued for its large, edible seeds that are extensively collected for local consumption and distribution to markets across Chile and Argentina (Aagesen 1998a, Tacón 1999). Monkey puzzle has been classified under IUCN criteria as vulnerable (Farjon and Page 1999) and is officially protected in both Chile and Argentina,

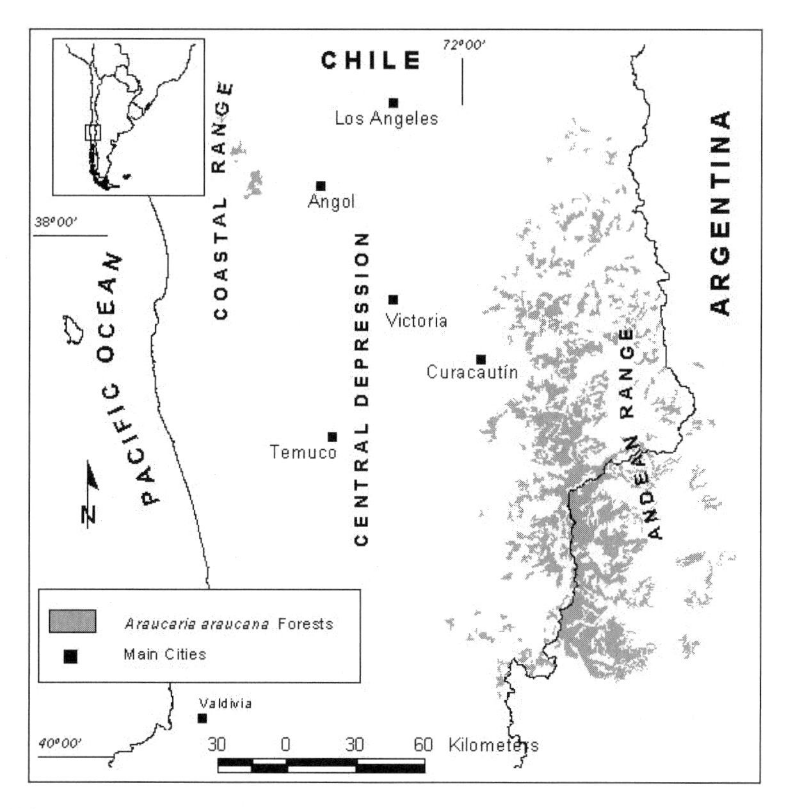

Figure 5.1 Distribution of monkey puzzle in southern Chile and Argentina.

where its logging is illegal. It is listed in Appendix I of the Convention on International Trade in Endangered Species (CITES) of Wild Fauna and Flora. Despite its protected status and exceptional ecological, economic, and cultural significance, this species continues to experience intense human pressures, such as grazing, burning, and exotic species plantation establishment and harvesting, both for seeds and occasionally timber (Aagesen 1998b).

Monkey puzzle persists in a region that experiences landscape level disturbances, such as volcanism, fire, landslides, and wind. The species has effective adaptations to withstand such disturbances, such as thick bark, epicormic buds, and the ability to withstand burial of its lower trunk by volcanic ash tephra layers (Veblen 1982, Burns 1993). Monkey puzzle is generally dioecious but may occasionally be monoecious with predominantly gravity-dispersed seed and wind-dispersed pollen. Both seed and pollen are relatively heavy and generally appear not to disperse over large distances (Muñoz 1984, Heusser et al. 1988). Asexual reproduction by root suckering has been reported (Schilling and Donoso 1976), but it is unknown how important this process is to population maintenance and expansion (Veblen et al. 1995) or for the genetic structuring of populations.

The objective of building a model for monkey puzzle was to explore the following questions:

1. How will the current seed-harvesting regime affect the probability of decline or recovery in the short term (100 years, considered to be a reasonable management planning time horizon) and the long term (500 years)?
2. What is the effect of various fire regimes on population structure in the short term and the long term?
3. Do occasional volcanic eruptions affect population viability in the short term and the long term?
4. What is the effect of small-scale illegal logging, and what would be the effect of large-scale controlled logging on population structure and viability in the short/long term?
5. Which aspects of the population dynamics should be prioritized for further study?

Creating a population model for the monkey puzzle tree posed several interesting challenges. First, the demographic dynamics of monkey puzzle operate on time scales that are much longer than those of most other species; adult trees can be extremely long lived (>1,000 years) and growth rates are often very low (< 1 mm yr^{-1} radial growth). This leads to problems with creating a model that is meaningful in the time frame of typical conservation management. Second, modeling density dependence is a critical aspect to population dynamics yet was complex to incorporate here. Third, the species occurs in different associations and exhibits different ecological characteristics in each. Fourth and finally, the model needed to include the effect of unpredictable catastrophes (fire and volcanic eruptions), as the distribution of monkey puzzle is centered on an area with active volcanism.

Methods

Landscape Data

The locations of populations of monkey puzzle in Chile were recorded during the National Vegetation Mapping Project of Chile, based on interpretation of aerial photographs and ground-truthing (CONAF et al. 1999) and were available in 1:50,000-scale maps integrated to a geographical information system (GIS) database. Because this database did not include the species distribution in Argentina, the model given here is only for the species range in Chile. The GIS records indicate a range of approximately 263,525 ha within Chile, but a further 179,289 ha exist in Argentina (CONAF et al. 1999). The final metapopulation patch structure was determined using RAMAS GIS (Akçakaya 2002), resulting in the identification of 445 populations of monkey puzzle. Monkey puzzle occurs in four main forest types, three defined by the presence of different *Nothofagus* species: monkey puzzle–lenga (*N. pumilio*), monkey puzzle–ñire (*N. antarctica*), monkey puzzle–coihue (*N. dombeyi*), and one representing pure stands of monkey puzzle.

Population Model

There have been several studies of the population dynamics of monkey puzzle, including detailed dendrochronological studies (Burns 1991), long-term surveys of seed production (Muñoz 1984), and studies to determine the ecological dynamics of different forest types (Veblen 1982, Finckh and Paulsch 1995). This information was compiled to give best estimates of stand structure, growth, survival, and response to various disturbance regimes. Additionally, data were collected over a 3-year period (1999 to 2001) from randomly located permanent plots, providing information on the growth and survival of younger stages.

Populations of monkey puzzle were divided into one of the four distinct forest types using spatial data describing associated species. The landscape data were also used to estimate initial abundances and carrying capacities, based on forest type, the level of dominance of monkey puzzle relative to other species, and disturbance history (Table 5.1). As the four monkey puzzle forest types have distinct densities, demographic dynamics, and disturbance regimes, it was necessary to compile four different transition matrices, each varying in survivorship and growth rates.

Stages

The model has 50 stages, including seeds, seedlings, saplings, juveniles, adults, and dominants. These reflect important biological stages in the species life history, with distinct growth rates, survivorship, fecundity, and sensitivity to disturbance (Figure 5.2). It was necessary to include numerous stages in the model to delay transitions, which otherwise may lead to unrealistic times for individuals to move through the various stages. Each time step represented 1 year.

Table 5.1 Classification rules for determining initial abundances and carrying capacities for monkey puzzle populations

Classification Rule	Definition	Initial Abundance	Carrying Capacity
1. Ecosystem type	Lenga	Mean plot density × 0.3[a]	Mean plot density
	Ñire	Mean plot density × 0.3	Mean plot density
	Coihue	Mean plot density × 0.3	Mean plot density
	Pure monkey puzzle	Mean plot density × 0.3	Mean plot density
2. Dominance of monkey puzzle	Dominant	Same as above	Same as above
	2nd dominant	0.6 × above	0.6 × above
	3rd dominant	0.4 × above	0.4 × above
	4th dominant	0.2 × above	0.2 × above
	5th dominant	0.1 × above	0.1 × above
	6th dominant	0.1 × above	0.1 × above
3. Past disturbance regime	No record or no disturbance	Same as above	Same as above
	Minor harvest	0.8 × above	Same as above
	Less than 50% harvested	0.7 × above	Same as above
	50% harvested	0.5 × above	Same as above
	50% to 75% harvested	0.25 × above	Same as above
	Less than 75% burned	0.5 × above	Same as above
	More than 75% burned	0.25 × above	Same as above

Note: "Same as above" refers to the value estimated in the previous line of the table (data from the GIS data base produced by CONAF et al. 1999).

a. It was estimated that 70% of the area mapped as habitat would not contain stands of monkey puzzle (based on field observations).

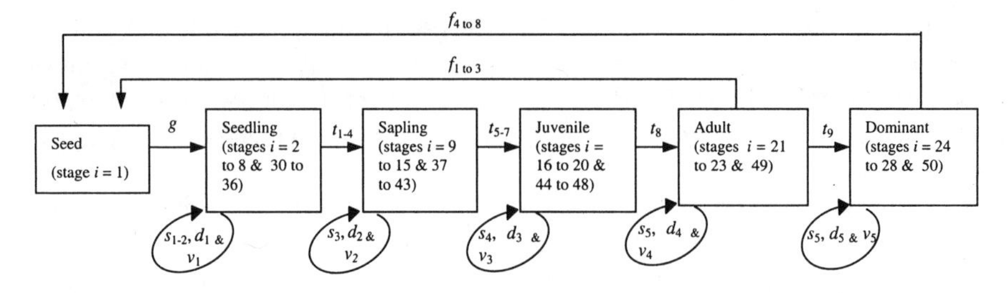

Figure 5.2 Life cycle of monkey puzzle and its parameters: g = germination rate; s = survival; d = fire survival; v = survival following volcanic activity; t = transition; f = fecundity.

Parameter Estimation

A problem associated with the development of transition matrices based on field data is that field measurements are made in the presence of density dependence, whereas model parameters must be estimated in the absence of density dependence. Consequently, vital rates were estimated from data from a glasshouse experiment or from maximum rates cited in the literature (Tables 5.2 and 5.3). Adult and dominant stages were assumed to be immortal, reflecting the extreme longevity of individual trees relative to the time scales evaluated here (Montaldo 1974).

Table 5.2 Vital rates for all female stages (male stages are equivalent) for monkey puzzle–lenga forest (the most extensive type; estimates varied slightly for other forest types)

Stage	Growth Rate Per Year	Probability of Survival (s)	Probability of Transition (t)
Seed	Produced every year, but viable for 1 year only (Montaldo 1974)	60% germination rate (g)[a]	All germinated seed move to seedling 1
Seedling 1	Faster growth than subsequent years due to seed store (6 cm in 1 year)[b]	95% survival (s_1)[a]	All seedlings that survive move to seedling 2 in the first year (0.95) (t_1)
Seedling 2	2 cm in height[a]	95% (s_1)[a]	0.127 (t_2)
Seedling 3 to 6	2 cm in height[b]	98% (s_2)[b]	0.123 (t_3)
Seedling 7	2 cm in height[b]	98% (s_2)[b]	0.093 (t_4)
Sapling 1	2 cm in height[b]	99% (s_3)[b]	0.094 (t_5)
Sapling 2	2 cm in height[b]	99% (s_3)[b]	0.064 (t_6)
Sapling 3 to 7	5 cm in height[b]	99% (s_3)[b]	0.097 (t_7)
Juvenile (1 to 5)	dbh growth at 1 cm[b]	99.5% (s_4)[b]	0.199 (t_8)
Adult (1 to 3)	dbh growth at 0.4 cm (Balzer 1963, Burns 1991, Cavieres 1987)	100% (s_5)	0.04 (t_9)
Dominant (1 to 5)	dbh growth at 0.2 cm (Balzer 1963, Burns 1991, Cavieres 1987)	100% (s_5)	0.02 (t_{10})

Note: Vital rates were estimated using glasshouse data, field data, or previous studies, as indicated.

[a]Based on glasshouse data.
[b]Based on field data.

Table 5.3 Fecundity for female adult and dominant stages

Stage of Female	Proportion of Females Producing Cones (%)	Average No. of Seeds Per Female	Fecundity (f)
Adult 1	6.9	70	4.83 (f_1)
Adult 2	18.5	172	31.82 (f_2)
Adult 3	19.5	274	53.43 (f_3)
Dominant 1	23.5	376	88.36 (f_4)
Dominant 2	36	478	172.08 (f_5)
Dominant 3	50	580	290 (f_6)
Dominant 4	50	682	341 (f_7)
Dominant 5	50	825	412.5 (f_8)

Note: Fecundity is the number of seeds produced per female—that is, the proportion of females producing cones × the number of seeds per female. Based on data from Muñoz 1984 and Schmidt and Caro 1998.

Density Dependence

A user-defined density dependence function (MPT.DLL; to see the source code, open the file MPT.DPR, on the accompanying CD, in Notepad) was developed to model within-species competition, particularly the effect of density dependence between stages; for example, an adult will have a much greater effect on the growth and mortality of other stages than a seedling has. The DLL file works on the basis of space used, which is equal to the number of individuals in each stage multiplied by their respective "weights" (assigned in the Stages dialog box). When the space used by a population is greater than the carrying capacity, individuals are assigned lower growth rates and survivorships according to the user-defined values in the Populations dialog box. This function also ensures that individuals will only be affected by individuals in larger stages. Table 5.4 presents the density-dependent mortality and change in growth for different stages of monkey puzzle in the four forest types.

Mating System

As the species is dioecious, males and females were considered separately. Males were included in the model, as future investigations will focus on the effect of genetic isolation on population viability. Field data indicated that the sex ratio was approximately 1:1 (S. Bekessy, unpublished data). A polygynous mating system was chosen, in that each male can mate with an unlimited number of females at each time step.

Dispersal

The dispersal-distance function of RAMAS GIS was used to specify the dispersal rates between two populations as a function of the distance between them. Distances were measured as the nearest distance between patch edges. The majority of monkey puzzle seeds fall within 13 m of the mother tree, although it is possible that seeds may be dispersed over greater distances by parakeets and rats (Muñoz 1984). The mean dispersal distance was estimated as 10 m and the maximum as 4 km (the maximum estimated distance a parakeet would travel with a seed).

Catastrophes

Two principal types of catastrophes affect populations of monkey puzzle: fire and volcanic eruptions. Volcanic activity occurs on average every 13 years at some place within the natural range of the species (Casertano 1963). However, the main effect of volcanic activity is fire caused by lava and incandescent ejecta and burial in ash. Only rarely are populations completely destroyed by lava flow or mudslides. Therefore, the overall probability of fire included ignition from humans, lightning, and volcanic activity. To model the increasing intensity of fire with time since the last fire, an MCH file was incorporated for each forest type (lenga.mch, coihue.mch, nire.mch, and pure.mch), which increased mortality of each stage with the time since fire (Figure 5.3). To model the change in carrying capacity after a fire or volcanic eruption, a KCH file was written for each population, reflecting the elimination of competing species after a catastrophe (which increases the carrying capacity) and the subsequent return of the competing species (which gradually returns carrying capacity to the original value) (Figure 5.4).

The incidence of a fire or volcanic activity was spread among populations using the "spread by a vector" option in the Advanced settings of the Catastrophes dialog box. It was assumed that fire and volcanic eruptions spread by 2.5 km on average and up to 10 km (the sensitivity of this assumption was tested). An important implication of using the spread by vector function in RAMAS is that the local probability must be estimated as the probability of an ignition event occurring in that population (that is, the

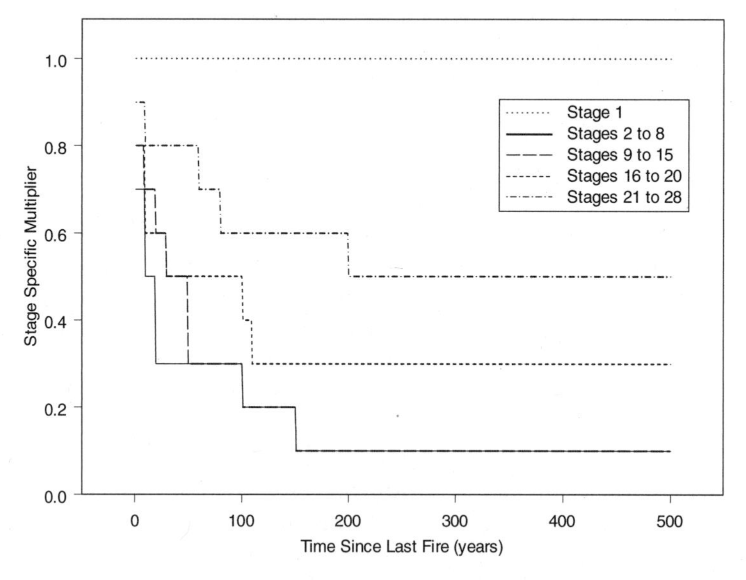

Figure 5.3 Functions used to simulate the effect of fire on different stages of monkey puzzle with time since last fire (data presented are for the monkey puzzle–ñire forest type).

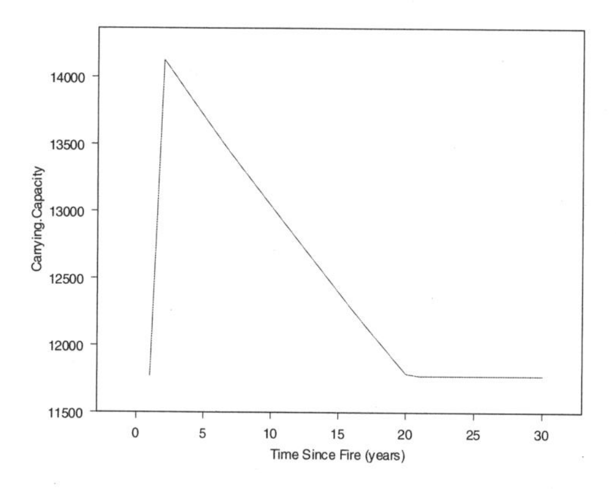

Figure 5.4 Functions used to simulate the increase in carrying capacity after a fire and the subsequent return to original values (data presented are for the monkey puzzle–lenga forest type).

fire actually starting in that population) rather than the overall probability of a fire occurring. Otherwise, fires would occur more regularly than expected. While this approach is spatially realistic, it is difficult to determine the actual number of fires that occur in each population throughout the simulation. The probability of volcanic activity in the Coastal Range was set to 0, as this part of the species range is not affected by volcanic eruptions. Table 5.4 presents the probability and effect of fire and volcanic eruptions for the four forest types.

Stochasticity

The model included environmental and demographic stochasticity. However, standard deviations were only included for fecundity and seed germination (10%) and were not included for growth and survivorship of any stage above seed, because survival is very close to 1, and therefore variation in these parameters is negligible. Additionally, we used "negative correlation for largest survival" (see Model | Stochasticity | Advanced) to ensure values were not truncated arbitrarily in simulations. Because the length and intensity of the summer drought period in Chile varies with latitude (Veblen 1982), a correlation-distance function was estimated to specify the correlation coefficients between two populations as a function of the distance between them. It was assumed that other types of environmental variation would be correlated at approximately the same distances. The sensitivity of the model to the correlation-distance function was tested.

Simulations

A "base" model that represents the best assumptions for environmental conditions and model parameters was developed (monkey-puzzle.mp) and was implemented in RAMAS GIS (Akçakaya 2002). Next, sensitivity analyses were conducted to determine if the results of the model were sensitive to estimates of parameter values (Burgman et al. 1993). Deterministic sensitivities were explored using analytical results, and stochastic sensitivities were evaluated by simulation. The sensitivity of the parameters investigated was calculated as

$$S = \Delta \text{EMP}/\text{EMP} \times 100$$

Table 5.4 Multipliers for density-dependent mortality and reduction in growth for different stages of monkey puzzle and the probability and effect of fire and volcanic eruptions for the four different forest types

	Monkey puzzle–ñire	Monkey puzzle–coihue	Monkey puzzle–lenga	Pure monkey puzzle
Mortality for different stages when N[a] exceeds K[b]	0.05, 0.05, 0.007, 0.007, 0.007[c]	0.07, 0.07, 0.007, 0.007, 0.007 (higher mortality due to the effect of disease)	0.05, 0.05, 0.007, 0.007, 0.007	0.05, 0.05, 0.007, 0.007, 0.007
Reduction in growth for different stages when N exceeds K	0.2, 0.3, 0.8, 0.9, 0.1[c]	0.6, 0.7, 0.8, 0.9, 0.1 (greater reduction in growth due to intense competition with coihue at K)	0.2, 0.3, 0.8, 0.9, 0.1	0.2, 0.3, 0.8, 0.9, 0.1
Probability of fire	Ignition probability in each patch each year was 0.001124	Ignition probability in each patch was 0.000225	Ignition probability in each patch was 0.00049	Ignition probability in each patch was 0.000449
Effect of fire	Higher with time since last fire owing to fuel buildup from both monkey puzzle and ñire (see nire.mch). Change in K is low (*1.3) because competition with ñire is low and occurs in seedling and sapling stages.	Higher with time since last fire (see coihue.mch). Change in K is high (*1.5) because competition with coihue is high and coihue is killed by fire.	Higher with time since last fire (see lenga.mch). Change in K is moderate (*1.2) because competition with lenga mainly occurs in seedling, sapling and juvenile stages.	Higher with time since last fire (see pure.mch). No change in K because there is no or very little competition with other species.
Probability of volcanic eruption	1 in 1,000 years	1 in 1,000 years	1 in 1,000 years	1 in 1,000 years
Effect of volcanic eruption	90% of population affected	90% of population affected	90% of population affected	90% of population affected

[a]Population size.
[b]Carrying capacity
[c]Values are listed in the order of seedlings, saplings, juveniles, adults, and dominants.

where S is sensitivity, EMP is the expected minimum population size of the base model, and ΔEMP is the EMP of the model being investigated minus the EMP of the base model. Sensitivity calculated in this way provides an indication of both the magnitude and the direction (positive or negative) of the change in EMP.

Various scenarios were explored in which parameters including frequency and intensity of fire and volcanic activity, seed harvest, and timber harvest were changed, one at a time, and the response of the model was measured by examining the expected minimum population sizes, the probability of decline, and the probability of a 10% increase. The model simulations were designed to explore the questions outlined in the beginning of this chapter.

When populations were counted in the simulations, all stages except for adults and dominants were excluded, as these are the reproductive stages and the objective of the modeling exercise was to investigate the effect of different scenarios on these stages. Four output statistics were recorded:

- The total number of monkey puzzle trees at the end of 100 or 500 years of simulation
- The expected minimum abundance (McCarthy and Thompson 2001)
- The probability that the population falls below half the initial population size at least once
- The probability that the population will increase by 10% at least once

Attention was paid to the probability of decline over the next 100 years because the International Union for the Conservation of Nature (IUCN) classifies a species as vulnerable when the decline is 30% and endangered when the decline is 50% over 100 years (IUCN 2001). Each scenario was implemented over a projection period of 100 or 500 years, using 100 replicates. The large number of stages and patches, along with the complex regime of catastrophes, resulted in the model being computationally intensive.

Results

The initial population size of monkey puzzle in Chile was estimated as 3.6 million adults and dominants, and the finite rate of increase was estimated as 1.032 (similar to the rate determined for another *Araucaria* species; Enright and Ogden 1979). The results of the simulations representing 35 scenarios for the metapopulation model are given in Table 5.5. The family of risk curves for the model over 100 years and 500 years are given in Figures 5.5 and 5.6, respectively.

The metapopulation model was extremely sensitive to the frequency and effect of volcanic activity, fire, and timber harvesting, particularly in the long term (500 years) (Table 5.5 and Figures 5.5 and 5.6). The effect of seed harvest increased when the frequency of fires was increased or when volcanic activity was included. The probability of a 10% population increase over 100 years was less than 0.01 except for the scenarios with low (up to 50%) or no seed harvest or reduced incidence or spread of fire. Over 500 years, high seed harvest, in combination with fire and volcanic activity, had a substantial effect on population decline. In addition, there was an increase in the probability of 50% population decline over 100 years if vital rates were decreased by 5% or if the probability of volcanic activity was increased.

Table 5.5 Results of 35 parameter combinations and scenarios run for the metapopulation model over 100 or 500 years

Simulations	Initial *N*[a]	*N* at year 100/500	EMP[b]	P of 50% decline[c]	P of 10% increase[c]	Sensitivity
Base (70% seed harvest, fire, no volcanic activity)	3573011	3427114	3395519	<0.01	<0.01	
Low vital rates (–5%)	3573011	163582	163582	1	0	–95.18
Low dispersal (–20%)	3573011	349038	3382837	<0.01	<0.01	–0.37
High dispersal (+20%)	3573011	3401900	3373473	<0.01	<0.01	–0.65
Correlation pessimistic (–20%)	3573011	3436115	3410191	<0.01	0.01	0.43
Correlation optimistic (+20%)	3573011	3406431	3370015	<0.01	0.01	–0.75
Volcanic activity (once every	3573011	3031764	3012490	0.01	<0.01	–11.28
Low volcanic activity (–20%) 50 years)	3573011	3062194	3040724	<0.01	<0.01	–10.45
High volcanic activity (+20%)	3573011	2927638	2913282	0.01	0	–14.2
Low spread of volcanic activity (–20%)	3573011	3240839	3213370	<0.01	0	–5.36
High spread of volcanic activity (+20%)	3573011	2925280	2547090	0.03	0	–24.99
High volcanic activity and high spread	3573011	2447116	2439212	0.05	0	–27.93
Low fire (–20%)	3573011	3448190	3415854	<0.01	0.01	0.60
High fire (+20%)	3573011	3405985	3379008	<0.01	0	–0.49
Low spread of fire by vector function (–20)	3573011	4298085	3640398	0	1	7.21
High spread of fire (+20)	3573011	3183863	3170940	<0.01	0	–6.61
No fire	3573011	4813976	3649441	0	1	7.48
Timber harvest (10% every 20 years)	3573011	3344762	3158613	<0.01	0	–6.98
Low timber harvest (5% every 50 years)	3573011	3367458	3318678	<0.01	0	–2.26
High timber harvest (20% every 50 years)	3573011	3725819	1824744	0.01	0	–46.26
High seed harvest (99%)	3573011	3330779	3313670	<0.01	0	–2.41
50% seed harvest	3573011	3458673	3401839	<0.01	0.03	0.19
High seed harvest (99%) with volcanic activity	3573011	3003291	2993136	<0.01	0	–11.85
High seed harvest (99%) with +20 fire	3573011	3311252	3300800	<0.01	0	–2.79
No seed harvest	3573011	3545318	3442825	<0.01	0.12	1.39
No seed harvest, fire, or volcanic activity	3573011	5009917	3649441	0	1	7.48
High volcanic activity in Pop 406	3573011	1518675	1482478	0.76	0	–56.34
Worst case: volcanic activity, 99% seed harvest, 10% timber every 20 years	3573011	2925280	2831137	<0.01	0	–16.62
Base model run over 500 years	3573011	16089151	3635060	<0.02	1	–100.00
500 years no seed harvest	3573011	6454424	3417206	<0.02	1	77.56
500 years 99% seed harvest with volcanic activity	3573011	1440300	1177774	1	0	–67.60
500 years –20 fire	3573011	15326742	3633497	<0.02	1	–0.04
500 years +20 fire	3573011	6246327	3262331	<0.02	1	–10.25
500 years with volcanic activity	3573011	2894287	2220049	0.4	0.76	–34.62
500 years worst case	3573011	1682190	1187352	1	0	–67.34

[a]Population size.

[b]Expected minimum population size.

[c]P = probability.

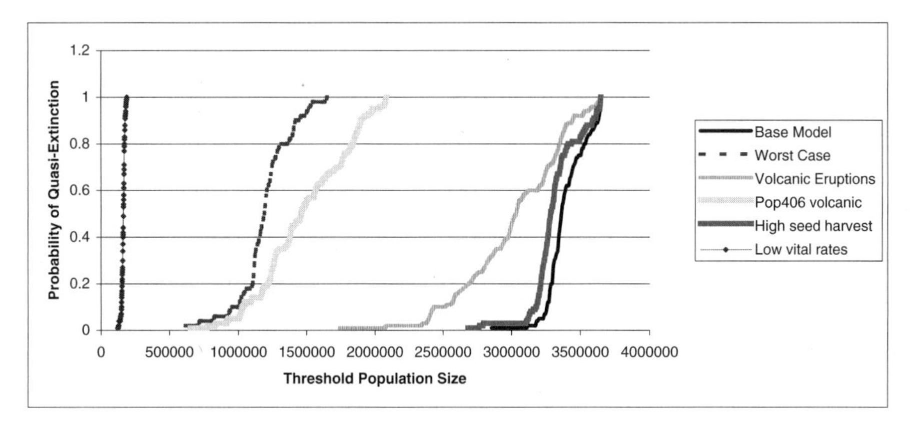

Figure 5.5 Family of risk curves for the monkey puzzle metapopulation model, run for 100 years.

The metapopulation model was very sensitive to the fate of population 406, which includes approximately 1.6 million mature individuals, representing the Conguillio population in the Chilean Andean part of the species range. If the probability of volcanic activity is increased in population 406 by 20%, the likelihood of 50% decline over 100 years increases from <0.01 to 0.76.

Discussion

To model the monkey puzzle tree in southern Chile adequately, several unique aspects of the RAMAS GIS package were required. Four different transition matrices were needed to model the distinct forest types in which monkey puzzle occurs. Additionally, density dependence was customized using a DLL that allowed stage-specific mortality and change in growth. To model the effect of catastrophes, KCH and MCH files were

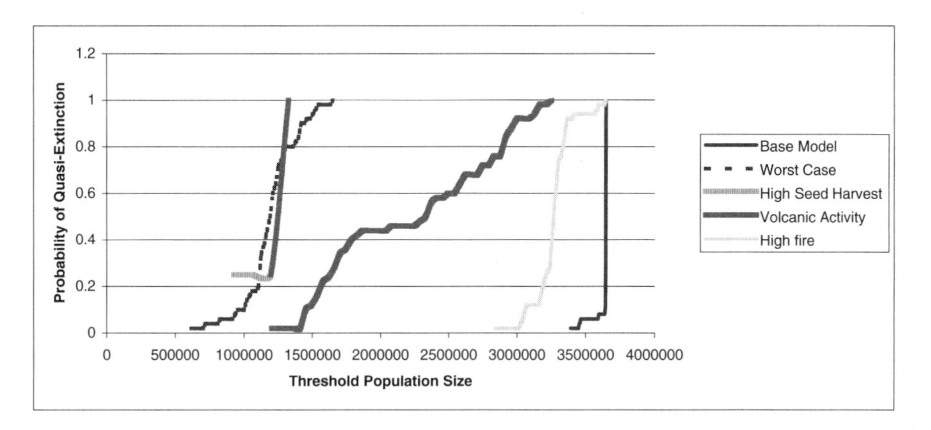

Figure 5.6 Family of risk curves for the monkey puzzle tree metapopulation model, run for 500 years.

used, as well as a spread function to simulate the spatial dynamics of fire and volcanic activity.

The results of the model have several management implications. First, as monkey puzzle has experienced intense historical human disturbance, conservation managers are interested in the likelihood of recovery. The model indicates that the probability of a 10% increase over the next 100 years is very low given the current seed harvesting and burning regime. This is particularly the case in the presence of any level of timber harvesting or a more intensive fire regime, or if seed harvest is increased from the base rate of 70%. To provide a higher probability of recovery over 100 years, regulation of seed harvest is advisable. The current model indicates that 50% seed harvest is unlikely to be detrimental to the regeneration success of the species, although management recommendations must take into account predation by animals. Suggestions that seed predation by introduced species, such as wild boar, has increased substantially in recent years (M. Gonzalez, pers. comm.) are cause for concern. Additionally, the current illegal status of timber harvest should be maintained and actively enforced.

Second, the model indicates that monkey puzzle is very sensitive to the frequency of fire, particularly in combination with other forms of disturbance. The dynamics of monkey puzzle are dominated by fire and volcanic activity, and the species requires a certain level of disturbance to persist (Burns 1991). However, the natural fire regime has been altered dramatically by humans (both indigenous and Western), and it is likely that a greatly increased frequency will be inappropriate, particularly if managers are interested in population recovery. M. Gonzalez (pers. comm.) suggested that since European settlement, fire may have occurred as frequently as every 10 years in some forest types, suggesting that the frequency used in the current exercise may have been too low. The extent to which indigenous people influenced fire frequency in the centuries before Europeans arrived is largely unknown, although it is thought that fire was used to clear undergrowth, to facilitate seed collection and to open routes to the other side of the Andes (M. Gonzalez, pers. comm.). Although the "natural" fire regime is difficult to determine, it can be assumed that it would be less frequent than the present regime (it has been suggested that it may be related to long-term climate-induced patterns due to the El Ñino–Southern Oscillation; Kitzberger et al. 2001). Hence, it is recommended that fire should be managed to occur less frequently, particularly those fires set by humans. Results from this modeling exercise indicate that the fire frequency of the base model (see Table 5.4) is likely to maintain a stable population size over 100 years.

Third, the probability of a 50% reduction in the next 100 years, although low for most scenarios, is substantial if vital rates are lowered by as little as 5% and moderate if the probability of volcanic activity is increased, particularly in the largest population (406). This finding has implications for the current classification of the species (IUCN 2001). Currently, the species is classified as vulnerable (Farjon and Page 1999). However, species that are likely to decline by 50% in the next 100 years fall into the higher risk category of endangered. The risks for monkey puzzle should be considered with reference to the disturbance regime, particularly volcanic activity, as it has the potential to strongly influence population trends.

Fourth, the sensitivity of the model to the fate of the population in Conguillio National Park is an important finding, as this population is likely to be affected by large-scale volcanic activity or fire. This highlights a problem with relying heavily on one large reserve (see, for example, Simberloff and Abele 1976). The risk of volcanic eruptions and fire need to be included in management planning (Bond 1998), particularly as most remaining populations of monkey puzzle lie close to active volcanoes. This situation has arisen partly due to the aesthetic appeal of snow-capped volcanoes as the backdrop to the National Parks and the clearance of surrounding areas for agriculture.

Although many previous studies have examined the ecology of monkey puzzle, several key aspects require further investigation. The influence of fire and volcanic activity was shown to be an important factor for all models, yet the effects of these catastrophic events on the population dynamics of the species are not well understood. Specifically, the spread of fire and volcanic activity into nearby patches should be investigated. Vital rates, initial abundance, density dependence, and carrying capacity were also found to be sensitive parameters and require further investigation to improve the reliability of the model. Finally, the level of seed predation by animals such as rodents, birds, and cattle should be investigated to assist with the development of guidelines for sustainable harvest by humans.

A population model can potentially be of great value in the case of a long-lived, slow-growing species such as the monkey puzzle tree as it is otherwise difficult to foresee the long-term consequences of current management practices. For example, it may appear that seed harvest has no effect in the short term. However, the effect of seed harvesting, in combination with other disturbance events, must be considered over many years. The construction of a population model may also assist in the assessment of the risk that catastrophic events may pose. The results and sensitivity analyses for the monkey puzzle model developed here indicate that the species is likely to remain stable over the next 100 years as long as the occurrence of fire and volcanic activity is low, the intensity of human activities do not change appreciably, and vital rates do not decrease. However, if management of the species includes the objective of population recovery, then seed harvest must be regulated at lower levels than present (that is, no more than 50%) and human-set fires should be controlled. On a cautionary note, the threats posed to the species in Argentina are likely to be very different, and the results of this model should not be simply extrapolated to the management of the species in this part of its distribution.

Postscript

Upon completion of this modeling exercise, a series of severe human-lit forest fires occurred in the ninth region of Chile (February 2002). An area of over 200 km^2 of native forest was burned, of which a large area corresponds to national parks containing monkey puzzle, including Conguillio. The fires have been termed an "environmental tragedy" by the Chilean government and demonstrate the potential for catastrophes to affect populations of this species.

Acknowledgments We are very grateful to CONAF (the forest service of Chile) and the Administratión de Parques Nacionales, Argentina, for permission to establish plots and their assistance in doing so. We are also grateful to Tracey Regan for assistance with developing the model and to Brendan Wintle, Paul Wintle, Lorna Raso, Christian Echeveria, and several extremely helpful students from the Chilean universities for assistance in the field. This research was supported by the European Commission–funded project SUCRE (Sustainable Use, Conservation and Restoration of Native Forest in Southern Chile and Argentina and South-central Mexico), contract no. ERBIC18CT970146, and the BIOCORES project (PL ICA4-2000-10029). Additional funding to Sarah Bekessy was provided by an Australian postgraduate award, the University of Melbourne, and the Sir Robert Menzies Australian Bicentennial Award.

References

Aagesen, D. L. 1998a. Indigenous resource rights and conservation of the Monkey-Puzzle tree (*Araucaria araucana,* Araucariaceae): a case study from southern Chile. *Economic Botany* 52: 146–160.

Aagesen, D. L. 1998b. On the northern fringe of the South American temperate forest: the history and conservation of the Monkey-Puzzle Tree. *Environmental History* 3: 64–85.

Akçakaya, H. 2002. *RAMAS GIS: linking spatial data with population viability analysis. Version 4.0.* Applied Biomathematics, Setauket, N.Y.

Balzer, U. N. 1963. Crecimiento y propiedades de la especie *Araucaria araucana* (Mol.) Koch. Master's thesis, Universidad Austral de Chile.

Bond, W. J. 1998. Ecological and evolutionary importance of disturbance and catastrophes in plant conservation. Pages 87–106 in G. M. Mace, A. Balmford, and J. R. Ginsberg (eds.), *Conservation in a changing world.* Cambridge University Press, Cambridge.

Burgman, M. A., Ferson, S., and Akcakaya, H. R. 1993. *Risk assessment in conservation biology.* Chapman and Hall, London.

Burns, B. R. 1991. Regeneration dynamics of Araucaria. Ph.D. diss., University of Colorado.

Burns, B. R. 1993. Fire-induced dynamics of *Araucaria araucana–Nothofagus antarctica* forest in the southern Andes. *Journal of Biogeography* 20: 669–685.

Casertano, L. 1963. General characteristics of active Andean volcanoes and a summary of their activities during recent centuries. *Seismological Society of America Bulletin* 53: 1415–1433.

Cavieres, A. 1987. Estudio de crecimiento de *Araucaria araucana* (Mol.) C. Koch en un bosque virgen de Araucaria-Lenga. Master's thesis, Universidad de Chile.

CONAF, CONAMA, BIRF, Universidad Austral de Chile, Pontificia Universidad Católica de Chile, and Universidad Católica de Temuco. 1999. *Catastro y evaluación de los recursos vegetacionales nativos de Chile.* Informe Nacional con Variables Ambientales, Santiago.

Enright, N., and Ogden, J. 1979. Applications of transition matrix models in forest dynamics: *Araucaria* in Papua New Guinea and *Nothofagus* in New Zealand. *Australian Journal of Ecology* 4: 3–23.

Farjon, A., and Page, C. N. 1999. *Conifers: status survey and conservation action plan.* IUCN/SSC Conifer Specialist Group, Cambridge, U.K.

Finckh, M., and Paulsch, A. 1995. The ecological strategy of *Araucaria araucana. Flora* 190: 365–382.

Heusser, C. J., Rabassa, J., Brandant, A., and Stuckenrath, R. 1988. Late-Holocene vegetation of the Andean Araucaria region, Province Nequen, Argentina. *Mountain Research and Development* 8: 53–63.

IUCN. 2001. *Redlist categories.* International Union for the Conservation of Nature, Gland, Switzerland.

Kitzberger, T., Swetnam, T. W., and Veblen, T. T. 2001. Inter-hemispheric synchrony of forest fires and the El Niño–Southern Oscillation. *Global Ecology and Biogeography Letters* 10: 315–326.

McCarthy, M. A., and Thompson, C. 2001. Expected minimum population size as a measure of threat. *Animal Conservation* 4: 351–355.

Montaldo, P. R. 1974. La bio-ecologia de *Araucaria araucana* (Mol) Koch. *Instituto Forestal Latino-Americano de Investigación y Capacitación* 46–48: 1–55.

Muñoz, R. I. 1984. Analysis de la productividad de semillas de *Araucaria araucana* (Mol.) C. Koch en el area de Lonquimay–IX region. Master's thesis, Universidad de Chile, Santiago.

Schilling, R., and Donoso, C. 1976. Reproducción vegetativa natural de *Araucaria araucana* (Mol.) Koch. *Investigaciones Agricultura* (Chile) 2: 121–122.

Schmidt, H., and Caro, P. 1998. Producción de semillas en un bosque de Araucaria bajo condiciones silviculturales. Seminar paper, Universidad de Chile, Santiago.

Simberloff, D. S., and Abele, L. G. 1976. Island biogeographic theory and conservation practice. *Science* 191: 285–286.

Tacón, A. C. 1999. Recolección de piñón y conservación de la Araucaria (*Araucaria araucana* (Mol) Koch.): un estudio de caso en la comunidad de Quinquén. Master's thesis, Universidad Austral de Chile.

Veblen, T. T. 1982. Regeneration patterns in *Araucaria araucana* forests in Chile. *Journal of Biogeography* 9: 11–28.

Veblen, T. T., Burns, B. R., Kitzberger, T., Lara, A., and Villalba, R. 1995. The ecology of the conifers of southern South America. Pages 120–155 in N. J. Enright and R. S. Hill (eds.), *Ecology of the southern conifers.* Melbourne University Press, Melbourne.

6

Banksia goodii in Western Australia

Interacting Effects of Fire, Reproduction, and Plant Growth on Viability

MARTIN DRECHSLER

This study demonstrates the use of RAMAS Metapop to simulate the very slow dynamics of *Banksia goodii,* a species of shrub with almost negligibly small vital rates, which strongly depend on disturbance, namely fire. A special feature of the model is the interaction of several processes: the random occurrence of fires, the development of fecundity after a fire, the aging of plants, and the growth of the population. All these processes can be considered explicitly in the software.

B. gooddii is a rare rhizomatous shrub that is endemic to southwestern Australia (Lamont and Markey 1995). Despite frequent fires and summer drought, adult mortality is very low and the main mortality factor is the loss of adults through land clearing (Witkowski and Lamont 1997, Lamont et al. 1993). Reproduction in the species is also very low. There is a significant Allee effect (Allee 1949), such that plants in populations below eight individuals do not produce any seeds at all. Seeds are only produced by adults (plants above 20 years age) and only between 3 and 30 years after a fire. Seeds die after 10 years (Witkowski and Lamont 1997). They are stored on the canopies of the adults and are released only after a fire. Only few of the seeds germinate, and the mortality of seedlings and young plants, especially in a fire, is very high.

Due to its low adult mortality, land users perceive the species as very robust and safe, but they do not realize that if an adult is lost there is only a small chance of recovery. This perception problem is aggravated by the fact that even if recovery takes place, it will take decades or even centuries—time spans that exceed those of human perception and planning. In this context a model is helpful to discuss the vulnerability of the species and highlight the problem of population recovery.

Life History of *Banksia goodii*

The life history of *B. goodii* may be divided into four stages: seedling, juvenile, subadult, and adult. A plant is called juvenile when it is more than 1 year old and has an area of less than 1.15 m^2, which is the threshold size for flowering and corresponds to an age of about 20 years (B. Lamont and E. Witkowski, unpublished data). Plants younger than 20 years do not produce any seeds and are characterized by a comparatively high mortality. Mortality decreases gradually with age (B. Lamont, unpublished data). A subadult is more than 20 years old and has negligible mortality. Its seed production increases as the plant grows and reaches a maximum when net plant growth stops due to competition for light, water, and nutrients. It is not clear at what age plant growth stops; this will certainly depend on the environment and the density of plants. Observations and extrapolations (see Drechsler et al. 1999) show that plant growth stops at an age of between 100 and 500 years. Seed production is assumed to increase linearly with plant age. After plant size and seed production have reached their maximum, a plant is termed an adult.

Immediately after a fire, seed production drops to zero, regardless of population size. Following the third year after the last fire, it increases linearly with time, reaching a maximum at 5 years. From then on, seed production remains constant (unless there is another fire) until 15 years after a fire. Between 15 and 30 years after a fire, seed production decreases linearly, a consequence of litter accumulation in the canopy until 30 years after the last fire, when seed production ceases (Figure 6.1, solid line). Seed viability drops to zero at an age of 10 years. Consequently, 40 years after the last fire there will be no viable seeds left on the plant.

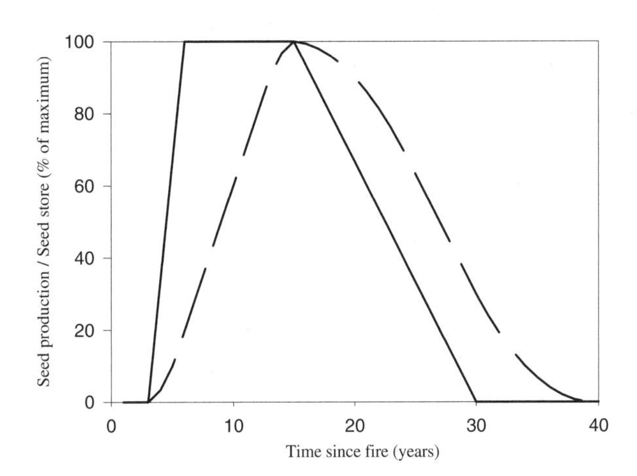

Figure 6.1 Seed production and seed store versus time since fire. *Solid line*: seed production; *dashed line*: seeds stored. Each line is scaled to its respective maximum (see text) (from Drechsler et al. 1999).

Model Structure

Stage and Spatial Structure

The plants are divided into 50 age classes (stages). The time to become an adult is assumed to be 98 years, which means that each of the first 49 stages encompasses 2 years of age: that is, the first stage contains plants 1 or 2 years old, the second stage contains 3- and 4-year-olds, and so on. In this way, the first 10 stages contain seedlings and juveniles, stages 11–49 contain subadults, and stage 50 contains the adults.

The 16 known populations are all isolated. Therefore there is no metapopulation structure, and one representative isolated population is considered.

Density Dependence: Allee Effect

Seed production decreases when the population size becomes small. Following Drechsler et al. (1999), this Allee effect is modeled by

$$f = \alpha - \frac{\beta}{N} \tag{1}$$

where f is the number of seeds per adult, N the total population size, and $\alpha = 30.55$ and $\beta = 277.4$ are constants, estimated by fitting the function to observations of population size and seed production (see Drechsler et al. 1999).

Effect of Fire

Seed production depends on the time since fire. Seeds die 10 years after they have been produced. Figure 6.1 shows the seed storage curve (dashed line) relative to its maximum, which occurs 15 years after the last fire and is given by eq. (1). In this way, for any year t since the last fire, seed store is given by eq. (1) multiplied by a fire factor $c(t)$, which is given by Figure 6.1 (dashed line).

Age Dependence and Canopy Seed Store

The canopy seed store depends on plant age, in addition to population density and fire. As eq. (1) refers to adults, to obtain the seed store in subadults it has to be multiplied by an age factor. This is zero for subadults of age 20 and increases linearly with age to a value of 1, reached when the plant becomes an adult. Thus the mean seed store of a plant in stage j ($j = 11, \ldots, 50$) is given by

$$C_j = f * c(t) * (j - 10)/40 \qquad (j = 11, \ldots, 50) \tag{2}$$

where t is the time (measured in years) since the last fire.

Reproduction

The number of seedlings produced by a subadult or adult in stage i is given by the seed store, C_j, multiplied by a seed viability factor, V, a release factor, R, and a germination rate, G. The viability factor accounts for granivory and other sources of seed loss or death before the end of their 10-year life span. Plants will release seeds only in response to a fire, so the release factor is non-zero only in a fire year. The release and germination rates take into account that only a certain proportion of seeds held in the canopy is released and germinates after a fire. Each year a number (N_0) of seedlings produced by the N_i subadults and adults in stages $j = 11, \ldots, 50$ is given by

$$N_0 = \sum_{j=11}^{50} F_j * N_j \tag{3}$$

with stage-specific fecundity

$$F_j = C_j \cdot V \cdot R \cdot G. \tag{4}$$

The parameters V, R, and G are all assumed to be normally distributed with means $m(V)$, $m(R)$, $m(G)$ and standard deviations $\sigma(V)$, $\sigma(R)$, $\sigma(G)$. Correlation among V, R, and G is assumed to be zero.

Survival

The survival from seedling stage to the age of 1 year is described by the establishment rate E which has mean $m(E)$ and standard deviation $s(E)$. The number of survivors from one year to the next is modeled by annual survival rates s_j where j is the plant age. Survival rates of juveniles increase with plant age. Plant growth is limited by herbivory and the abundance of water and nutrients in the soil. However, intraspecific competition for water and nutrients is likely to be weak, because there is plenty of room for populations to spread. Therefore in the model, survival of seedlings and older plants is assumed to be density-independent. However, it is affected by rainfall and varies annually. We assume that the survival rates, s_i, are normally distributed random numbers with means $m(s_j)$ and standard deviations, $s(s_j)$. In addition, survival of juveniles is affected by fire. If there is a fire, all juvenile survival rates are multiplied by a fire survival multiplier, which depends on the age of the plants. In the stage-based model, the product Es_2 is the transition rate from the first to the second stage. The transition from the second to the third stage is given by the product $s_3 s_4$, and so on.

Model Parameterization

The model parameterization is based on the field data given in Table 6.1. In the following discussion, all parameter inputs for RAMAS Metapop are provided in the order they are entered into the program.

General Information

The simulation duration was set at 500 time steps, with one time step equal to 2 years (note that each stage contains two years).

Density Dependence

Density dependence affects fecundities and is based on the abundance of all stages. All populations have the same density dependence type, which is user-defined in the file

Table 6.1 Field data on vital rates of *Banksia goodii*

	Minimum	Average	SD
Seed viability (V)	0.200	0.362	0.05
Seed release (R) (fire years only)	0.500	0.75	0.08
Germination (G)	0.000	0.013	0.004
Establishment in year 1 (E)	0.072	0.400	0.11
Annual survival in year 2 (s_2)	0.167	0.350	0.06
Annual survival in year 3 (s_3)	0.250	0.500	0.08
Annual survival in year 4 (s_4)	0.500	0.750	0.08
Annual survival in year 5 (s_5)	0.750	0.875	0.04
Annual survival in year 6 (s_6)	0.950	0.975	0.01
Annual survival in years 7 to 20 (s_7–s_{20})	0.98	0.99	0.00
Annual survival in years 20+ (s_{21}–s_{50})	0.999	0.9999	0.0000

Note: The difference between the observed average and minimum is assumed to be due to environmental variation, such that the minimum values mark the 1% quantile of environmental variation. This means that the difference between average and minimum values corresponds to 3 SD. All numbers are rates (proportion of individuals) so carry no dimension.

Allee2.DLL (on the CD accompanying this volume) based on Drechsler et al. (1999). To see the source code, open the file Allee2.DPR in Notepad. According to eq. (1), fecundity under density dependence (Allee effects) is $f = \alpha - \beta/N$, which can be written as fecundity without density dependence ($f = \alpha$) multiplied by the factor $1 - \beta/(\alpha N)$. Therefore in the source code, the Allee effect is modeled by multiplying the transition (stage) matrix elements $T\text{matrix}_{1,j}$ by the factor (1 – Allee/PopSize), where

$$\text{Allee} = \beta/\alpha = 277.4/30.55 = 9.08 \tag{5}$$

Stages

A total of 50 stages are defined. The other entries of this input window are irrelevant.

Sex Structure

The model includes all individuals (mixed).

Stage Matrix

The first 10 elements of the first row (fecundity of juveniles) are zero. The last entry of the first row (fecundity of adults without Allee effect) is given by

$$T\text{matrix}_{1,50} = V{\cdot}R{\cdot}G{\cdot}\alpha = 0.107 \tag{6}$$

Compare eqs. (1), (2), and (4).

The fecundity entries for the subadults ($j = 11, \ldots, 49$) are given by

$$T\text{matrix}_{1,j} = T\text{matrix}_{1,50}{\cdot}(j\text{–}10)/40 = 0.107{\cdot}(j\text{–}10)/40 \tag{7}$$

Compare eq. (2).

The survival from the first to the second stage is

$$T\text{matrix}_{2,1} = s_2 \cdot E = 0.35 \cdot 0.4 = 0.14 \tag{8}$$

See Table 6.1, average values.

Accordingly, the other elements of the subdiagonal are

$$T\text{matrix}_{j,j-1} = s_{2(j-1)} \cdot s_{2j-1} \qquad (j = 3, \ldots, 50) \tag{9}$$

with the survival rates s_j given by the corresponding average numbers in Table 6.1. In the constraints matrix, the elements of the first row are set to 0 and all other elements are 1.

Standard Deviations Matrix ("*S*matrix")

For the calculation of the elements of the standard deviations matrix, $S\text{matrix}_{i,j}$, a formula (eq. (10)) is needed for the standard deviation of products of random numbers. Let a and b be two normally distributed uncorrelated random numbers with means m_a and m_b and standard deviations σ_a and σ_b. Let further $<x>$ and $\sigma(x)$ denote the mean and the variance of a quantity x. Then

$$\begin{aligned}[\sigma(ab)]^2 &= \langle(ab - m_a m_b)^2\rangle = \langle(a^2b^2\rangle - \langle(m_a m_b\rangle^2 = \langle a^2\rangle\langle b^2\rangle - (m_a m_b)^2 \\ &= (\sigma a^2 + m_a^2)\cdot(\sigma_b^2 + m_b^2) - (m_a m_b)2 = \sigma_a^2\sigma_b^2 + \sigma_a^2 m_b^2 + \sigma_b^2 m_a^2\end{aligned} \tag{10}$$

With eq. (10), the standard deviation of the product (VR) is $\sigma(VR) = [0.05^2(0.08)^2 + 0.36^2(0.08)^2 + 0.75^2(0.05)^2]^{1/2} = 0.047$ (see Table 6.1). Similarly, the standard deviation of $(G\alpha)$ is $\sigma(G\alpha) = 30.55(0.004) = 0.12$. The means are $<VR> = 0.27$ and $<G\alpha> = 0.40$. Applying eq. (10) to the product $[(VR)(G\alpha)]$ leads to the standard deviation $\sigma(VRG\alpha) = [0.047^2(0.12)^2 + 0.27^2(0.12)^2 + 0.40^2(0.047)^2]^{1/2} = 0.038$.

This is the standard deviation of the fecundity of adults (upper right element of the *S*matrix):

$$S\text{matrix}_{1,50} = \sigma(V \cdot R \cdot G \cdot \alpha) = 0.038 \tag{11}$$

Analogous to eq. (7), the standard deviation of the fecundity of subadults ($j = 11, \ldots, 49$) is

$$S\text{matrix}_{1,j} = S\text{matrix}_{1,50} \cdot (j-10)/40 = 0.107 \cdot (j-10)/40 \tag{12}$$

The standard deviations of the survival rates are calculated in the same way. The standard deviation of (Es_2) is $\sigma(Es_2) = 0.11^2(0.06)^2 + 0.40^2(0.06)^2 + 0.35^2(0.11)^2 = 0.046$. This is the first element of the subdiagonal of the *S*matrix (cf. eq. (8)):

$$S\text{matrix}_{2,1} = \sigma(s_2 \cdot E) = 0.046$$

The following elements of the subdiagonal of the *S*matrix are {0.072, 0.04, 0, . . . , 0}(cf. eq. (9)).

Populations/General

Initial abundance is set to 80, reflecting the size of an important remnant population. Relative fecundity varies with time since fire and is described by a fecundity change (fch) file.

As described in Figure 6.1 (dashed line), the abundance of seeds in the canopy that can be released in case of another fire varies with the time since the last fire. These seeds are released only in the case of another fire. This means that fecundity is zero in non-fire years and given by eq. (4) in fire years with $c(t)$ given in Figure 6.1 (dashed line). With the fecundity and catastrophe multipliers provided by RAMAS Metapop, this cannot be modeled exactly but is well approximated by multiplying the fecundity of eq. (6) with $\varphi c(t)/\varphi_0$, where $\varphi \ll \varphi_0$ in non-fire years and $\varphi = \varphi_0$ in fire years. This means that in non-fire years, fecundity of eq. (6) is multiplied by a very small factor $\varphi c(t)/\varphi_0 \ll 1$—and thus is negligible—and in fire years is multiplied by $\varphi c(t)/\varphi_0 = c(t)$ as required by eq. (2).

To model this with RAMAS Metapop, relative fecundity has to be set to $c(t)/\varphi_0$ with $\varphi_0 \gg 1$ and the catastrophe multiplier (which by definition is 1 in non-fire years and $\neq 1$ in fire years; see "Catastrophes") to φ_0. In the study reported here, a value of $\varphi_0 = 10^5$ was chosen, and so the (.FCH) file contains the $c(t)$ of Figure 6.1 (dashed line) divided by 10^5. The particular values are given in Table 6.2 (note that the model has a time step of two years). If there is no fire within 40 years after the previous one, $c(t)$ stays at the value of year 40, which is zero (see "Catastrophes").

Populations/Catastrophes

The local fire probability and the local multiplier are set to 0.1 (i.e., one fire on average in 20 years) and 1.0, respectively.

Initial Abundance

All stages start with zero individuals, except stage 50, which starts with 80.

Stochasticity

Demographic stochasticity is used, environmental stochasticity is lognormal, and fecundity and survival are perfectly correlated.

Catastrophes

Vital rates are affected. The catastrophe multipliers for fecundity (first row of the multiplier matrix, $M_{0,j}$, $j = 1, \ldots, 50$) are all set to φ_0, as described in the section "Populations/General." The chance of surviving fires increases linearly from 0 to 1 as age increases from 0 to 20. The fire survival rate in the first year (seedlings) is 0, that in the second year is 0.05, and so on. The fire survival in stage 1 then is the geometric mean of

Table 6.2 Values of the fecundity change (.FCH) file

Year	2	4	6	8	10	12	14	16	18	20
Value	0	10^{-6}	$3*10^{-6}$	$5*10^{-6}$	$7*10^{-6}$	$9*10^{-6}$	10^{-5}	$99*10^{-7}$	$95*10^{-7}$	$88*10^{-7}$
Year	22	24	26	28	30	32	34	36	38	40
Value	$79*10^{-7}$	$67*10^{-7}$	$53*10^{-7}$	$40*10^{-7}$	$27*10^{-7}$	$16*10^{-7}$	$8*10^{-7}$	$3*10^{-7}$	0	0

Note: In the file the values are ordered one after the other in a single column, without the year numbers (see RAMAS Metapop help file).

0 and 0.05, which is 0. Similarly, the fire survival in stage 2 is the geometric mean of 0.1 and 0.15, which is $[0.1(0.15)]^{1/2} = 0.12$. The fire survival of stages 3–10 are 0.22, 0.32, 0.42, 0.52, 0.62, 0.72, 0.82, and 0.92, respectively. The following 39 elements of the subdiagonal (fire survival of subadults and adults) are all set to 1.

In a fire, all seeds are released and canopy seed store starts to build up again according to Figure 6.1. This is achieved by setting the fch file back to the beginning, as specified in the "Advanced" window.

Other

Menu items "Dispersal" and "Correlation" are irrelevant in this study because it applies to a single population, and no management actions are considered.

Results

A typical population trajectory is shown in Figure 6.2, which shows the seedlings produced immediately after stochastic fire events. Mortality in young plants is extremely high, so that a few years after a fire the population size is more or less the same as just before the fire. Consequently, the population size is almost constant (with a slight decreasing long-term trend). The weak and negative population trend is confirmed statistically when looking at the trajectory summary obtained from 500 simulation runs (Figure 6.3). Both population increase and decline are unlikely, as can be seen also from the interval quasi-extinction and quasi-explosion risk curves (Figures 6.4 and 6.5) (cf. Ginzburg et al. 1982).

A characteristic of this study is the interaction of different time scales: the frequency of fires, the development of the canopy seed store, the growth of the individual plants, and the growth of the population all operate on a variety of scales. In contrast to the

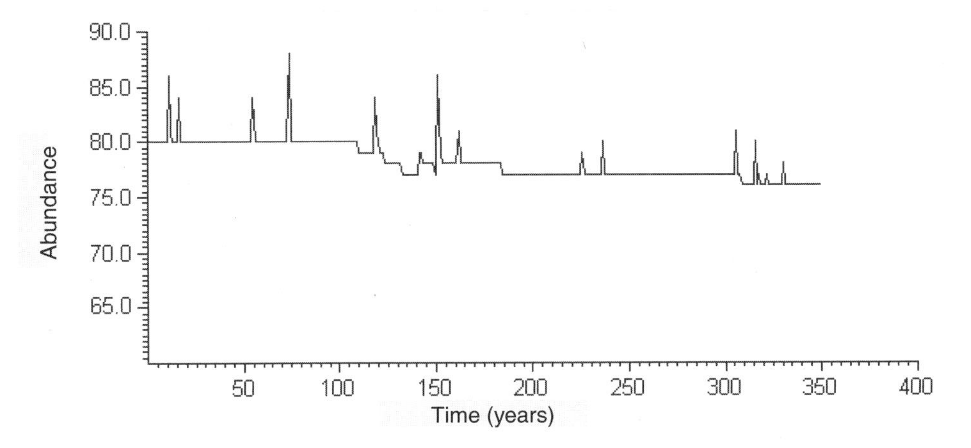

Figure 6.2 Typical population trajectory, showing the total number of living plants (including seedlings) over time. Spikes in abundance are caused by germination of seedlings after fire, followed by the mortality of most during the first decade after the fire.

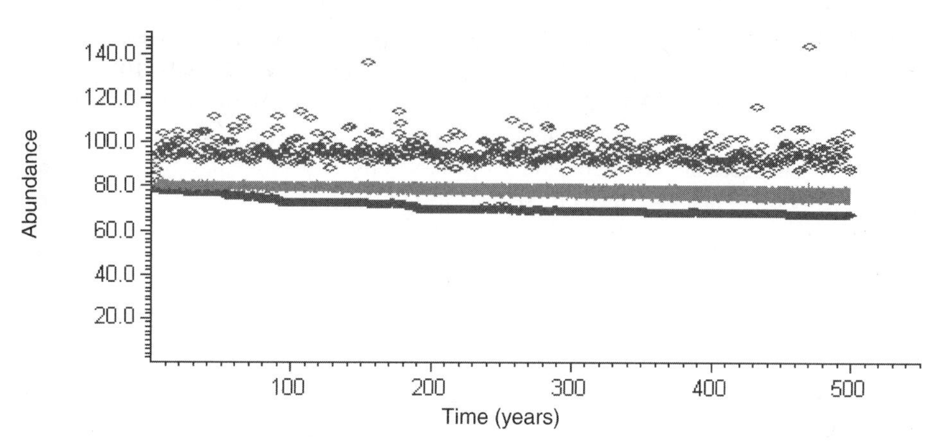

Figure 6.3 Mean population size with bounds (plus or minus 1 SD; middle—light gray) and population maxima and minima (dark gray diamonds) as a function of time, based on 500 replicates.

study reported here, in Drechsler et al. (1999) the development of the canopy seed store—and thus its interaction with the fires—could only be described in a statistical manner. RAMAS Metapop allows a more explicit treatment of this topic, and it is therefore useful to have another look at this issue and to analyze how the population dynamics change when fire frequency is varied. Table 6.3 shows the effect of fire frequency on the chance of population recovery.

It is evident that the optimal fire frequency that leads to maximum population increase (50% chance of a population increase to 97 individuals within 500 years) is about 0.05/year to 0.1/year: that is, one fire in about 10 to 20 years. Moreover, the

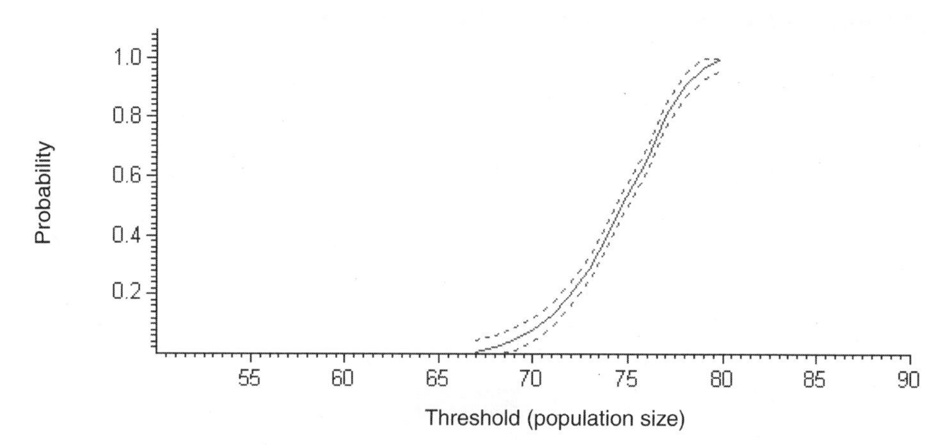

Figure 6.4 Interval quasi-extinction risk curve: the probability of falling below a particular population size within 500 time steps (from 500 replicates). Dotted lines show the 95% confidence intervals.

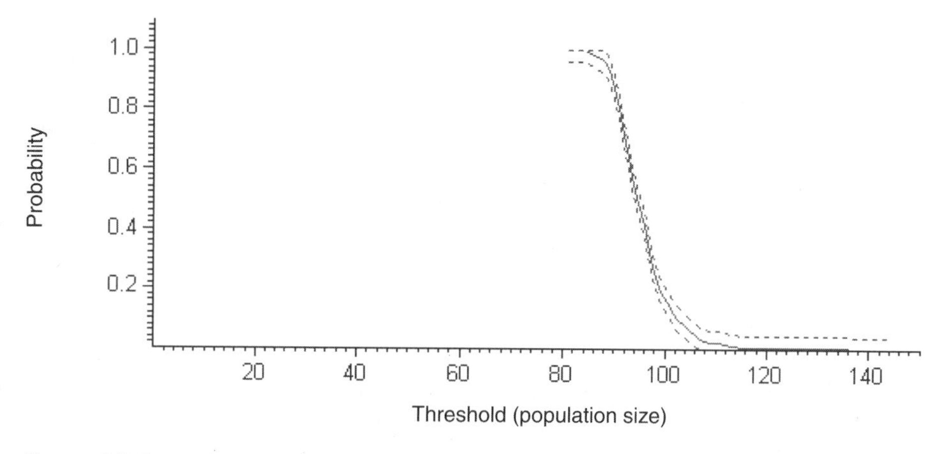

Figure 6.5 Interval quasi-explosion risk curve: the probability of topping a particular population size within 500 time steps (from 500 replicates). Dotted lines show the 95% confidence intervals.

optimum is relatively flat: all values within this range are about equally good for the population.

Discussion

RAMAS Metapop was used to model the slow dynamics of the rare serotinous shrub *Banksia goodii* in southwestern Australia. The results show that the population slowly declines and that significant population changes will occur only on a very long time scale of centuries. This agrees well with results of Drechsler et al. (1999), despite the changes in the model structure from 1-year to 2-year time steps and 50 instead of 10 stages, along with the parameter transformations necessary for these changes. Obviously, the model results are rather stable to minor changes in the model structure, which is encouraging for the use of simulation models in population viability analysis (PVA).

In this context, one may ask how the model results would change if the parameters *V*, *R*, and *G* were assumed to be correlated. Clearly, this would increase the temporal variation in the population size, especially in the height of the peaks in Figure 6.2. However, Figure 6.2 also shows that changes in the population size are very slow, and

Table 6.3 The population size that is exceeded within 500 years with probability 50% for different fire frequencies (measured in units of 1/year)

Fire probability (1/yr)	0.2	0.1	0.075	0.05	0.025
Population size	94	97	97	97	94

Note: Due to the 2-year time step of the model, if a fire frequency of *x*/year is desired, a value of 2*x* has to be specified in Populations/Catastrophes.

it is likely that it does not matter (in the long run) whether after one fire fecundity is a little higher than average and after another fire a little lower. This suggests that the statistics of the population dynamics, including the explosion and extinction risks of the population, are probably not very sensitive to the degree of temporal variation in fecundity (cf. table 5 in Drechsler et al. 1999). Within plausible bounds, even the fire frequency does not seem to have a decisive impact on the population dynamics, as indicated here by Table 6.3. Therefore, it should not matter very much whether or not *V, R,* and *G* are correlated.

An interesting facet of the model is the interaction of various processes, which usually would require an individual-based modeling approach but are well supported by the RAMAS Metapop software. These processes include the random occurrence of fires, the development of fecundity after a fire, the aging of plants, and the growth of the population.

This allowed a relatively thorough analysis of the effects of different fire frequencies. The optimum fire frequency found is identical to that obtained by Drechsler et al. (1999), which confirms the statistical description of the canopy store used in that study. In all, RAMAS Metapop is a tool that is well suited for the analysis of long-term plant dynamics with interacting processes, such as those in *Banksia goodii.*

References

Allee, W. C. 1949. Group survival value for *Philodina roseola,* a rotifer. *Ecology* 30: 395–397.

Drechsler, M., Lamont, B. B., Burgman, M. A., Akcakaya, H. R., Witkowski, E. T. F., and Supriyadi. 1999. Modelling the persistence of an apparently immortal *Banksia* species after fire and land clearing. *Biological Conservation* 88: 249–259.

Ginzburg, L. R., Slobodkin, L. B., Johnson, K., and Bindman, A. G. 1982. Quasiextinction probabilities as a measure of impact on population growth. *Risk Analysis* 2: 171–181.

Lamont, B. B., and Markey, A. 1995. Biogeography of fire-killed and resprouting *Banksia* species in south-western Australia. *Australian Journal of Botany* 43: 283–303.

Lamont, B. B., Klinkhamer, P. G. L., and Witkowski, E. T. F. 1993. Population fragmentation may reduce fertility to zero in *Banksia goodii*: a demonstration of the Allee effect. *Oecologia* 94: 446–450.

Witkowski, E. T. F., and Lamont, B. B. 1997. Does the rare *Banksia goodii* have inferior vegetative, reproductive or ecological attributes compared with its widespread co-occurring relative *B. gardneri? Journal of Biogeography* 24: 469–482.

7

Erodium paularense in Spain

Relevance of Microhabitats in Population Dynamics

MARÍA J. ALBERT
DAVID DRAPER
JOSÉ M. IRIONDO

Erodium paularense Fern. Gonz. and Izco (Geraniaceae) is a woody rosulate chamaephyte that is endemic to central Spain. It is confined to a single locality in the Lozoya Valley (Madrid), where there are three small populations that grow on isolated dolomitic outcrops in a widely extended siliceous landscape. This species occupies two different microhabitats: the crevices and cavities of rocks, and the shallow soils (lithosols) of the nearby grassy communities (González-Benito et al. 1995, Albert et al. 2001a). New populations recently discovered in the province of Guadalajara (200 km from the Lozoya Valley), growing on andesite outcrops, have been ascribed to this species, although comparative genetic studies show great differences between the populations at the two locations (Martín et al. 1999).

This species has been classified as endangered (EN) (VV.AA. [Various Authors] 2000) according to the World Conservation Union (IUCN) criteria. In addition to its narrow distribution and the small size of the populations, plants have very low reproductive success (González-Benito et al. 1995, Albert et al. 2001b). Fruit set is about 16%, and mean number of viable seeds per fruit is only 0.31 (out of five possible seeds per fruit). Adult plants produce, on average, four seeds per plant in one reproductive season (Albert et al. 2001b), and there is not a permanent soil seed bank. Available data also show evidence of seed predation by ants and low seedling recruitment (M. J. Albert, J. M. Iriondo, and A. Escudero, unpublished data). Furthermore, populations are also subject to human impact, such as cattle herbivory (Albert et al., in press) and the effects of recreational activities and plant collection.

The model described in this chapter was built with data from the smallest population of the Lozoya Valley (Population I of González-Benito et al. 1995), with an occupancy area of 443 m^2. In this study we analyzed landscape data of this population; the resulting patch structure of five subpopulations, four located in rock microhabitat and one in lithosol, was exported to a metapopulation model. We built the metapopulation model using both spatial and demographic information gathered from the two microhabitats since 1993. Different simulations were run to estimate extinction risk and population decline under present and possible future scenarios, to assess human impact, and to evaluate the effectiveness of different conservation actions.

Methods: The Metapopulation Model

Landscape Data

The whole occupancy area of the population and its immediate surroundings (a ring-buffer of 2 m) were surveyed to determine the relationships between *E. paularense* plants and their habitat features. The survey was performed by establishing a 1-m^2 grid throughout the population. The total sampled area was 2329 m^2. Data of microhabitat features were obtained from each 1-m^2 plot (see appendix to this chapter). The presence or absence of *E. paularense* plants was used as a dependent variable to generate a habitat suitability model based on a logistic regression (Hosmer and Lemeshow 1989). No subsets were used to validate the model, as the whole occupancy area was sampled and all *E. paularense* presences were used to construct the model.

Habitat Suitability Model

Two models, one for rock and another for lithosol microhabitat, were built due to the different demographic and reproductive behaviors of the species on these microhabitats (Albert et al. 2001b). The same variables were used in both models. Habitat suitability was calculated using the total pool of presences on rock ($n = 229$) and an equivalent set of absences of *E. paularense* on rock ($n = 285$) selected by a randomized process. The same process was carried out for the lithosol (presences $n = 128$; absences $n = 193$). Only significant variables were used. The resulting habitat suitability models had the following expression:

For lithosol habitat:

$$y_{\text{lithosol}} = 3.0736 - 0.0180*[\%\text{cvpern}] - 0.0019*[\%\text{cvrock}] - 0.1455*[\text{soil\_dep}] \\ - 2.2061*[\text{erod\_dis}] + 0.2194*[\text{ant\_dis}] + 0.0001*[\text{tran\_alt}] + 0.0001*[\text{slope}]$$

For rocky habitat:

$$y_{\text{rocky}} = 48.2553 + 0.0082*[\%\text{cvpern}] + 0.0316*[\%\text{cvrock}] - 0.0061*[\text{soil\_dep}] \\ - 2.4059*[\text{erod\_dis}] + 0.0076*[\text{ant\_dis}] - 0.4980*[\text{tran\_alt}] - 0.0197*[\text{slope}]$$

In both cases the logit transformation was applied:

$$p = (\exp(y)/(\exp(y) + 1))$$

where y = presence or absence of *E. paularense* and p = probability of occurrence of *E. paularense.* The GIS environment used was Idrisi v.2 (Clark University, Worcester, Mass.) complemented with Statistica v.5.5 (StatSoft, Tulsa, Okla.).

The resulting models make biological sense because the slopes of the coefficients ratify many of the features we perceive about the plant: a rupicolous plant with an aggregated spatial pattern that grows on the lower part of rock outcrops or on nearby shallow soils.

Habitat suitability threshold values were obtained from the minimum value of habitat suitability where *Erodium* was present. Thus, 0.5681 was considered for lithosol and 0.5718 for rock.

Neighborhood distances for patches of two cells for lithosol and four cells for rock were used to obtain subpopulations that were compatible with our previous knowledge. Distances were considered from edge to edge. Five patches or subpopulations were identified: one from the lithosol microhabitat, and four from the rocky microhabitat (rock 1, rock 2, rock 3, and rock 4).

The Mann-Whitney test was used in the two models to evaluate its fitness based on examination of the residuals and considering the residuals and the predicted absolute values. In the lithosol, the model correctly classified 91.19% of absences ($U_{0.001\,(2),\,32,\,161} = 885.5$) and 92.97% of presences ($U_{0.005\,(2),\,36,\,92} = 1794$). In the rocky habitat, the model correctly classified 87.72% of absences ($U_{0.001\,(2),\,74,\,211} = 4{,}009$) and 91.70% of presences ($U_{0.001\,(2),\,89,\,140} = 7{,}140$).

Stage Structure

As *E. paularense* flowers are hermaphroditic, subpopulations were not structured according to sex and all individuals were considered for the model. Taking into account plant size and the ability to produce flowers, plants were grouped into four stages, one vegetative and three reproductive. These classes were obtained from field data by cluster classification and comprised the following categories:

Vegetative	< 6 cm
Adult 1	6–12 cm
Adult 2	13–21 cm
Adult 3	> 21 cm

Plant size was estimated by the maximum diameter of the rosette cluster (González-Benito et al. 1995). Since established plants do not migrate, all plants were assigned a relative dispersal of 0. The vegetative category was mainly represented by seedlings coming from seeds of the previous year and, to a lesser extent, by juvenile plants that have not reached the capacity of reproduction. We did not consider a seed stage because field data showed that there was no permanent soil seed bank (M. J. Albert, J. M. Iriondo, and A. Escudero, unpublished data). Thus, seeds are immediately incorporated into the population as vegetative individuals.

Two different stage matrices were constructed, one for plants from the rock microhabitat and the other for plants from the lithosol microhabitat (Table 7.1). The transitions from the stages adult 1, adult 2, and adult 3 to the vegetative stage include both survival and fecundity elements. Therefore, we imposed constraints on matrix elements to ensure realistic simulated survival rates (survival rates varying between 0 and 1, and the sum of all survival transitions from a given stage being less than 1). The constraints matrix elements were calculated by $S/(S + F)$, where S is survival rate and F is fecundity.

Table 7.1 Stage matrices (coefficients) for lithosol and rock microhabitats of Population I of *Erodium paularense*

	Lithosol				Rock			
	Vegetative	Adult 1	Adult 2	Adult 3	Vegetative	Adult 1	Adult 2	Adult 3
Vegetative	0.279	0.098	0.101	0.178	0.622	0.038	0.043	0.065
Adult 1	0.275	0.678	0.120	0	0.241	0.822	0.118	0
Adult 2	0.003	0.191	0.760	0.190	0	0.148	0.833	0.374
Adult 3	0	0	0.102	0.804	0	0	0.043	0.625

Fecundity rates were obtained from field reproductive data. Linear regressions were made each year to estimate fruit and seed production as dependent variables of plant size. Each category was assigned a reproductive value considering these estimations, and the number of seedlings was recorded in the following year. Growth and mortality data from all monitored plants from each microhabitat were used to estimate survival rates.

Stochasticity

Demographic stochasticity was taken into account in the model by checking the appropriate option in RAMAS Metapop. We estimated the variability in vital rates (survival and fecundity) due to demographic stochasticity, following the method proposed by Akçakaya (2002). A weighted average of demographic variance was calculated by

$$\frac{\sum_{t=1}^{Y} p_t(1-p_t)}{\sum_{t=1}^{Y} N_t}$$

where Y = number of years, p_t = survival rate, and N_t = number of individuals at year t. This component was subtracted from the total observed variance to obtain an estimate of environmental stochasticity and to avoid overestimating this parameter (Akçakaya 2002). This estimate was then used to construct standard deviation matrices for both rock and lithosol microhabitats.

Environmental stochasticity was set to follow a lognormal distribution. We considered that within-population F, S, and K were uncorrelated because fecundities essentially depend on environmental conditions during the reproductive process (March–June), whereas survival depends on the environmental conditions experienced throughout the entire year. Moreover, the basic environmental limitation for survival is water availability, whereas fecundities also depend on many other factors, such as temperature or number of sunny days for pollination and ant activity for seed predation.

Initial Abundances

Data from Population I of the Lozoya Valley were used in the model. Lithosol plants were monitored for 9 years for demographic and reproductive data, and rock plants were studied for 3 years. Total census and population structure were obtained from field data. These were used to input initial abundances at each stage.

Spatial Structure

There are five subpopulations, four located in rock microhabitat (rock 1, rock 2, rock 3, and rock 4) and one in a lithosol. Dispersal among subpopulations was not considered because observed seed-dispersal distances are always less than 2 m.

We calculated the correlation between the vital rates in the average stage matrices recorded in the two microhabitats for the monitored years. This correlation coefficient was used in the interpopulation correlation matrix for the elements corresponding to correlations between subpopulations of different microhabitats. A value of 0.99999 was given to the elements corresponding to correlations between subpopulations of the same microhabitat. Moreover, relative fecundity and survival values were assigned to each subpopulation based on their habitat suitability values. These values were estimated for each subpopulation by dividing the average habitat suitability where *E. paularense* occurs in the subpopulation by the average habitat suitability where *E. paularense* occurs in the whole microhabitat.

Populations

We considered that the most appropriate density dependence type to use in all subpopulations was the ceiling model: that is, the subpopulations grow exponentially until all microhabitats are occupied and then remain at that level. Since we have evidence of decreased plant size and fecundity due to intraspecific competition (Albert et al. 2001b), we set density dependence to affect all vital rates and considered it to be based on the abundance of all stages.

Carrying capacity for each subpopulation was estimated in the RAMAS GIS Landscape data module using the following expression:

$$K = \text{ahs}*\text{noc}*(X/Y)$$

where ahs = average habitat suitability for the subpopulation, noc = area as the number of cells in the subpopulation, X = maximum number of *E. paularense* plants found in the 10 cells with the maximum habitat suitability values, and Y = maximum value of habitat suitability where *E. paularense* is present for each microhabitat.

The values considered for each subpopulation were

$$K_{\text{lithosol}} = \text{ahs}*\text{noc}*(9/0.9779)$$
$$K_{\text{rock}} = \text{ahs}*\text{noc}*(9/0.9999)$$

To model the dynamics of the subpopulations and project them into the future, the simulation was repeated 1,000 times and considered a period of 30 years.

Diagnosis of the Present Situation

According to the model, *E. paularense* shows a declining trend (Figure 7.1). Model projections point out that total population size has a 100% probability of a 10% decline in a

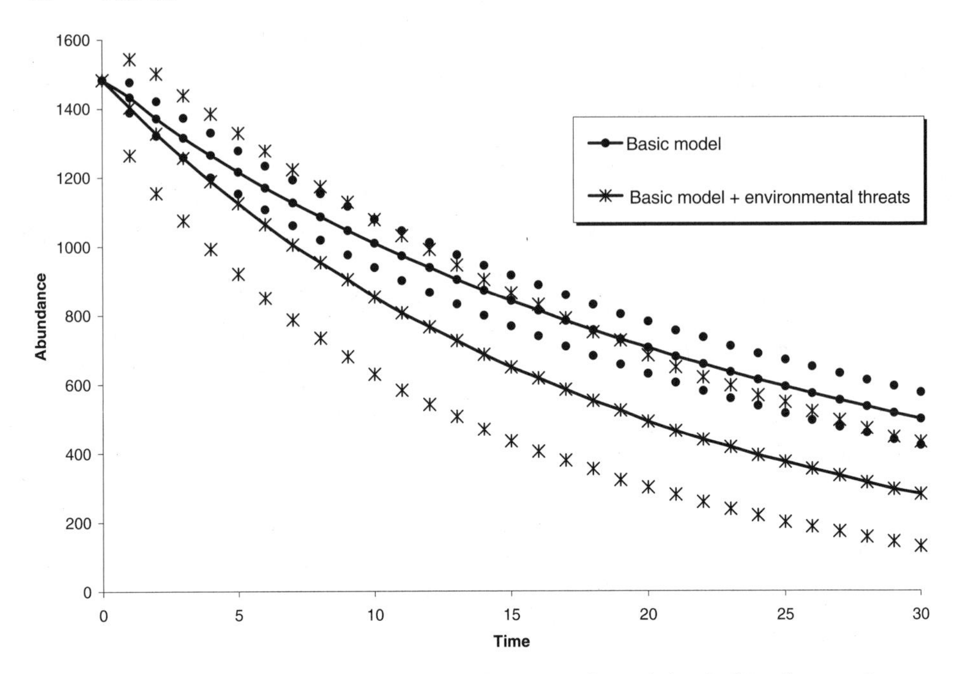

Figure 7.1 Projected abundance over a period of 30 years of Population I of *Erodium paularense* in the Lozoya Valley, as a sum of the six identified subpopulations, for the basic PVA model and a modified model with a 100% increase in environmental variability and the inclusion of regional catastrophic droughts. Solid lines represent the average value of 1,000 simulations; the corresponding marks below and above each line represent the mean ±1 SD.

period of 10 years. The model suggests that the population studied, if considered all by itself, would be classified as vulnerable (VU) under type C criteria, according to IUCN Red List Categories (IUCN 2001): an estimated population size below 10,000 mature individuals and an estimated continuing decline of at least 10% within 10 years. This classification of threat is not in agreement with that given for the species entire range (VV.AA. [Various Authors] 2000), because the latter was assigned using type B criteria.

Finite rate of increase (λ) varied among subpopulations (0.9333–1.0096) and was slightly higher in lithosol (λ for lithosol = 0.9955; mean λ for rock 1 to rock 4 = 0.9713). Greater differences between microhabitats were observed in the elasticity matrices. In lithosol, the highest elasticity values corresponded to the survival of the largest individuals (adult 3 and adult 4) in their corresponding stages. However, in the rock populations the highest elasticity values corresponded to the survival of adult 2 and adult 3 individuals in their own stages. This is probably due to the existing differences in average plant size between microhabitats: plants growing on rock are smaller than those growing on lithosol (Albert et al. 2001b), so the contribution of each transition to population growth differs between microhabitats.

Total reproductive value of individuals in the rock populations was smaller than in the lithosol (6.29 vs.11.965). As previously shown (Albert et al. 2001b), such differences in fecundity between habitats may be caused by the smaller mean plant size of rocky plants. Another important implication of microhabitat variability is that the average residence time of rocky plants was highest at intermediate stages and strongly decreased in the largest plants, whereas the average residence time of individuals growing on lithosol progressively increased along with the successive stages of the life cycle (Table 7.2). Growth capacity of rocky plants may be affected by intense resource limitation due to the reduced soil pocket of rock cavities (Matthes-Sears and Larson 1999). As a chasmophyte, *E. paularense* is able to colonize bare rock crevices, but plants face more intense competition for nutrients and water (Fowler 1986, Tilman 1987, Goldberg and Novoplansky 1997).

In spite of smaller plants and lower seed production, a higher stability of reproductive parameters in time has been observed at the rock microhabitat (Albert et al. 2001b). This can also be observed by comparing the values of the standard deviation matrices. Moreover, according to the average residence times (Table 7.2), rock microhabitat plants would live a little longer (14.96 years) than lithosol microhabitat plants (13.77 years). Thus, the ability to grow on rock microhabitat may have an adaptive value. Plants growing in rock microhabitats show lower but more stable seed production, which may ensure the long-term persistence of the populations by complementing the irregular reproductive output in the lithosol (Fiedler 1987, Albert et al. 2001b).

The analysis of subpopulation trends revealed that the greatest subpopulations (rock 1 and rock 4) are the most vulnerable to environmental and demographic stochasticity, and they show declines in their initial abundances of 47% and 57%, respectively, over a period of 10 years. These subpopulations have the lowest values of relative fecundity and survival. In contrast, the lithosol presented the lowest decline in 10 years (8%). Although the lithosol's habitat suitability value is lower than those in rock microhabitat, its vital rates provide more favorable projections in time. These differences in reproductive and demographic behavior between the two microhabitats may benefit the viability of the population, especially under the unpredictable climatic conditions of the Mediterranean region.

Simulation Scenarios

Assessment of Potential Future Threats

E. paularense populations are potentially vulnerable to the occurrence of extreme environmental fluctuations that may affect either their fecundities or their survival rates, or both. Taking into account that current climate change models predict higher environmental fluctuations and a higher incidence of catastrophic extreme droughts for this area (Butterfield et al. 1997), we modeled three different scenarios:

Table 7.2 Average residence time (years) for each life stage in both microhabitats

	Stage			
Microhabitat	Vegetative	Adult 1	Adult 2	Adult 3
Lithosol	1.39	3.11	4.17	5.10
Rock	2.45	4.86	5.13	2.52

1. An increase in environmental variability, modeled with a 100% increase in the elements of the standard deviation matrix.
2 Introduction of catastrophic regional droughts in the model with an average of one incidence every 25 years. We assumed that these events would drastically affect all vital rates, causing death to all vegetative individuals, preventing all reproduction, and decreasing adult survival by 30%.
3. A combination of the above-mentioned scenarios.

Sensitivity analyses were made to assess the impact of such situations on population viability. We observed that all different modeled scenarios would strongly increase extinction risk, cause additional declines in population abundance, and lead the population to the endangered (EN) status according to IUCN categories (Table 7.3 and Figure 7.1). Interval extinction risk and percent decline curves of the modeled scenarios were significantly different from those of the basic population viability analysis (PVA) model used as the control (Table 7.3). The worst projection was obtained with the concurrence of increased environmental variability and catastrophic events.

Model results suggest that *E. paularense* is sensitive to environmental alterations of the present situation induced by climate change. Taking into account the low water-retention capability of the substrate (dolomitic rock and shallow lithosol), it is foreseeable that occurrence of extreme regional droughts will increase the extinction risk of the population. A simple increase in environmental variability can also change the synchrony of the flowering process with the life cycle of their pollinators (Albert et al. 2001a) and adversely affect reproductive success.

Population Management Strategies

The main purpose of population management strategies is to regenerate the population and to guarantee its viability in the future. Given the present environmental and demographic conditions, the population can be classified as vulnerable (VU) according to IUCN categories. Since we aim to decrease its danger of extinction in an objective and quantifiable manner, we are interested in addressing actions that will place the population in the near threatened (NT) category. To achieve this goal, we evaluated popula-

Table 7.3 Effect of potential future threats, based on an increase in environmental variability and/or the inclusion of catastrophic droughts, on quasi-extinction risk, expected abundance, and probability of a 10% decline in population abundance

Scenario	Quasi-extinction Risk (%)	Expected Abundance (No. of Plants)	Probability of a 10% Decline (%)
Present situation with basic PVA model (control)	57.0	493.1	100
Increase in environmental variability	80.8	444.3	99.8
Catastrophic regional droughts	88.3	305.2	100
Increase in environmental variability and catastrophic regional droughts	92.7	279.0	99.9

Note: Quasi-extinction risk (considering a threshold of 500 individuals) within a 30-year interval. Abundance for which there is a 50% probability that the population will fall below this value at least once within a 30-year interval. Probability of a 10% decline in population abundance within a 10-year interval.

tion trends under different potential management actions that would enhance both plant survival and plant reproductive success.

Management must provide suitable conditions for both seed production and establishment of new plants (Lennartsson and Oostermeijer 2001). Since the rock surface and other environmental characteristics do not allow the creation of new habitat for the species, we can only manage the species life cycle.

Low seed production may be a major cause of the narrow distribution and small population sizes of *E. paularense*. We have observed that plant size strongly conditions the value of all absolute reproductive components (flowers, fruits, and seeds), as seen in many other plants (Samson and Werk 1986, Herrera 1991, Mitchell 1994, Ollerton and Lack 1998). Thus, strategies that favor adult plant survival may encourage greater reproductive values in the population. Although pollen limitation is common in plants and may be a causing agent of low fruit set (Schemske 1980, Bawa and Webb 1984, Campbell and Motten 1985, Garwood and Horvitz 1985, Horvitz and Schemske 1988, Burd 1994), available data showed that there was no pollinator limitation (i.e., quantity of pollen) on reproductive success of *E. paularense* in natural populations (Albert et al. 2001b). Thus, we cannot increase reproductive output by increasing pollination rates (e.g., by manual pollination of flowers or by the establishment of beehives in the surrounding area). However, the number of available seeds may be increased by limiting seed predation by ants. Experimental assays carried out at the Lozoya Valley populations (M. J. Albert, J. M. Iriondo, and A. Escudero, unpublished data) showed that seed predation by the ant *Messor capitatus* reached 30% of total seed production.

Seed production may also be increased by limiting the effect of herbivory on plant growth and reproduction. Preliminary studies on the impact of cattle herbivory on *E. paularense* populations (Albert et al., in press) showed that total seed production of adult plants was reduced by 50% in those areas where cows grazed. In addition, survival of the smallest adult plants was slightly affected by herbivory. Other human effects include recreational activities of people in nearby areas and plant collection. Based on our field experience, we estimate that this collection eliminates 0.3% of adult plants each year. All these aspects will be taken into account in the modeling of population viability under different management options (Figure 7.2).

Actions on Human Activities

The first step of a management program for the studied population would be the limitation of human access. The effect of this action can be modeled by adding 0.003 to the survival rates of adult plants in the stage matrices (Figure 7.2).

In addition to this, if we do not allow cattle to graze at the population, we may also increase plant reproductive values (Figure 7.2). The effect of limiting herbivory must be studied with caution, however. Although herbivory has generally been assumed to reduce plant reproduction and fitness (Harper 1977), both negative (Louda 1984, Parker and Salzman 1985, Karban and Strauss 1993) and positive (McNaughton 1985) effects of herbivores on population dynamics have been documented. Further monitoring is needed to assess how grazing may be affecting population dynamics by differential elimination of competitors and creation of suitable conditions for germination, seedling emergence, and growth. In consequence, the 50% increase in reproductive output de-

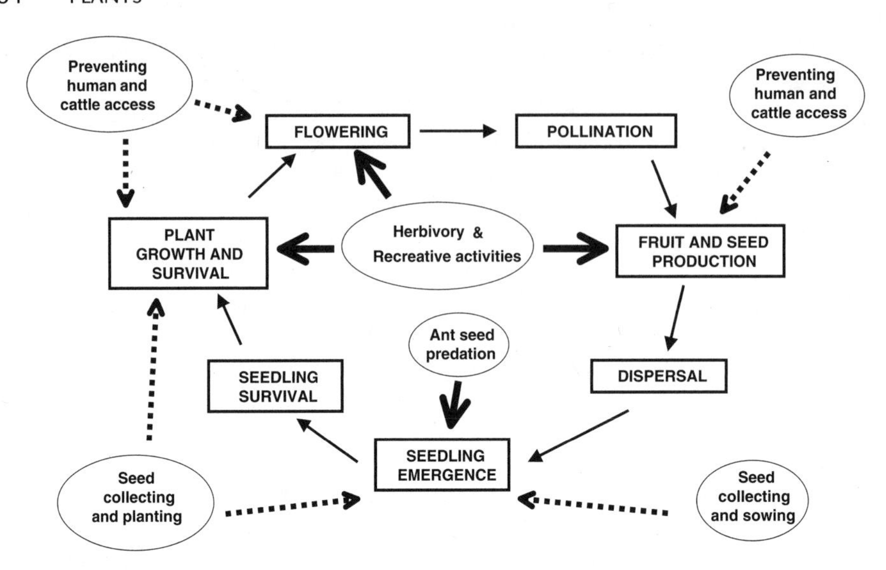

Figure 7.2 Life cycle of *Erodium paularense*. Thick bold arrows indicate the stages that are negatively affected by human impact (cattle herbivory and recreational activities) and ant seed predation. Dotted arrows point to the stages that would be affected by the proposed management actions.

tected in short-term cattle exclusion experiments may be an overestimation if it is applied to models for long-term projections.

Model projections that include the action of preventing human and cattle access to the population reduce the 100% probability of a 10% population decline within 10 years to 98.4% and set quasi-extinction risk (threshold: 500 individuals) within 30 years at 0%. Mean λ values increased in both microhabitats (lithosol $\lambda = 1.0210$; rock $\lambda = 0.9801$). Since prevention of cattle access may have medium-term or long-term negative effects, which are difficult to quantify at this moment, a second set of temporary management actions will be evaluated in order to assure long-term population viability.

Actions on Plant Reproductive Cycle

An increase in the number of available seeds for germination and establishment of new plants may also be obtained by partially preventing seed predation by ants (Figure 7.2). One management action to increase recruitment at the vegetative stage may be to collect some seeds immediately after dispersal, before ant seed predation takes place, and to sow them in a safe site. Total seed production in the population was estimated to ensure that the collection would not surpass 10% of total seed production. Seed production was calculated considering the present initial abundances for each stage. Total seed production per plant for each stage was estimated using reproductive data gathered since 1995 in the studied population, and we assumed 70% seed germination (M. Gris, unpublished data) to estimate the number of seeds we would need to recruit the desired number of vegetative plants.

We considered two types of management actions. The first action consisted of sowing a number of seeds in available safe sites each year for a period of 3 years, once access to humans and cattle is limited. Seeds would be sown in autumn, and both seeds and seedlings would be periodically watered to facilitate their establishment and survival for this period (Figure 7.2). Seeds to be sown would be distributed over the subpopulations according to their carrying capacities and their current abundances. Total seed availability from seed collection and the results of several simulations run for different alternatives were also taken into consideration. Thus, in the simulations, 50, 175, 15, 20, and 120 seeds were sown each year for three consecutive years in populations lithosol, rock 1, rock 2, rock 3, and rock 4, respectively. The second action consisted of collecting a similar number of seeds, sowing and growing the plants in greenhouse conditions, and planting the resulting adult 1 stage plants in available sites of the populations according to the previous distribution (Figure 7.2). In both cases, we used the Population Management option of RAMAS Metapop, and since we wanted to perform a specific management action for each subpopulation, we added as many introduction actions as existing subpopulations and assigned a specific treatment to each one of them.

We compared the results of these population reinforcements with those of the basic PVA model. Both types of reinforcement provided a 0% quasi-extinction risk in a 30-year interval (considering a threshold of 500 individuals) and significantly reduced the extinction curve and the probability of a 10% decline (Table 7.4). The sowing alternative provided somewhat better results than planting (Figure 7.3), although only the lithosol and rock 3 subpopulations presented a positive trend.

Concluding Remarks

Microhabitat characterization of the population with the use of a geographical information system (GIS) provides useful information on optimum microenvironmental conditions for the species, facilitates the estimation of habitat suitability throughout the population's area of occupancy, and allows for a quality and quantity assessment of the available sites. Relevant demographic information such as identification of subpopula-

Table 7.4 Effect of population management actions on quasi-extinction risk, expected abundance, and probability of a 10% decline in population abundance

Scenario	Quasi-extinction risk (%)	Expected abundance (no. of plants)	Probability of a 10% decline (%)
Present situation with basic PVA model (control)	57.0%	493.1	100%
Preventing human and cattle access + sowing	0%	1078.6	0%
Preventing human and cattle access + planting	0%	965.6	9%

Note: Quasi-extinction risk (considering a threshold of 500 individuals) within a 30-year interval. Abundance for which there is a 50% probability that the population will fall below this value at least once within a 30-year interval. Probability of a 10% decline in population abundance within a 10-year interval.

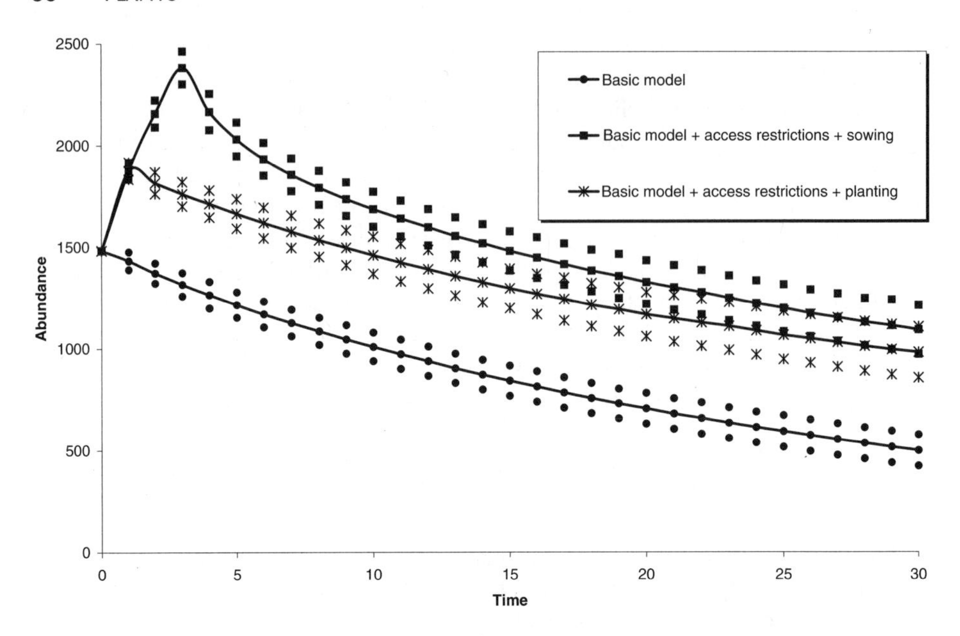

Figure 7.3 Projected abundance over a period of 30 years of Population I of *Erodium paularense* in the Lozoya Valley for the basic PVA model and the models including the access restrictions to humans and cattle plus seed sowing or planting of Adult 1 stage individuals. Solid lines represent the average value of 1,000 simulations; the corresponding marks below and above each line represent the mean ±1SD.

tions and estimation of their carrying capacity can also be estimated from these data. Nevertheless, the logistic approach used to model habitat suitability has several limitations derived from intrinsic assumptions. As there is no single best modeling approach, the use of other response variables (such as abundance) and other statistical approaches could provide new perspectives on *E. paularense* habitat suitability.

The population as a whole showed a decreasing population trend in spite of its subpopulations having λ values near 1. This suggests that environmental stochasticity is the main agent originating this trend. Furthermore, the survival of the population will be negatively affected by any future change that may increase environmental variability or catastrophic regional droughts.

This subpopulation-structured model shows that microhabitat heterogeneity found in populations of some species can have a relevant effect on the dynamics and behavior of the population. The particular differential features of the subpopulations in the rock and the lithosol microhabitats may be beneficial for the survival of the whole population and may buffer the effects of environmental stochasticity. Therefore, maintaining habitat heterogeneity may be crucial to species conservation (Potter 1996), and management strategies should consider population dynamics at the microhabitat scale (Albert et al. 2001b).

Given the population management alternatives studied, it seems advisable to limit the access of humans and cattle to the population. Similarly, a 3-year program of selective reinforcements with seeds obtained from the same population, and distributed ac-

Table 7.5 Significant variables used to model habitat suitability

Variable	Units	Layer Code
Distance to *E. paularense* plants	m	[erod_dis]
Coverage of bare soil	%	[%cvsoil]
Coverage of perennial plants	%	[%cvpern]
Coverage of rocks	%	[%cvrock]
Soil depth	cm	[soil_dep]
Distance to the nearest ant nest	m	[ant_dis]
Altitude – 1000	m	[tran_alt]
Slope	°	[slope]

cording to the needs, carrying capacity, and trends of each subpopulation, should maintain current population size levels for the next 10 years, thereby downgrading the threat of the population from a vulnerable status (VU) to a near threatened status (NT) according to IUCN categories. Nevertheless, we are aware that some of the assumptions adopted in the model are based on fragmentary knowledge and, therefore, consider that the model and the management alternatives should be periodically revised as new information and new monitoring data become available.

Appendix: Variable Selection in the Logistic Model

As exploratory methods, the Mann-Whitney test ($p < 0.001$) and the Kolmogorov-Smirnov test ($p < 0.025$) were used to select the independent variables. Several transformations were tested on some variables to improve their significance, although only a transformation on altitude was finally adopted (Table 7.5). To avoid multicolinearity problems, correlation among independent variables was also studied. When correlation between two variables was greater than 0.7, one of the variables was eliminated.

Univariate logistic regressions were used to test parameter significance. The Log-Likelihood ratio goodness-of-fit test (G) and the Wald $p(t)$ test were applied in the univariate logistic regression. The significance of $p(G)$ and $p(t)$ for each variable was evaluated, and variables with $p > 0.25$ in at least one of the tests were rejected (Hosmer and Lemeshow 1989).

A backward stepwise method ($p < 0.001$) was used to maximize R^2. All possible combinations between the significant variables were considered, and the model with the lowest value of deviance (D) (Legendre and Legendre 1998) was selected for each microhabitat.

References

Akçakaya, H. R. 2002. Estimating the variance of survival rates and fecundities. *Animal Conservation* 5: 333–336.

Albert, M. J., Escudero, A., and Iriondo, J. M. 2001a. Environmental range of narrow endemic *Erodium paularense* and its vulnerability to changing climatic conditions. *Bocconea* 13: 549–556.

Albert, M. J., Escudero, A., and Iriondo, J. M. 2001b. Female reproductive success of narrow endemic *Erodium paularense* in contrasting microhabitats. *Ecology* 82: 1734–1747.

Albert, M. J., Iriondo, J. M., and Escudero, A. In press. Reproductive costs of herbivory in small-sized populations. *Bocconea.*

Bawa, K. S., and Webb, C. J. 1984. Flower, fruit and seed abortion in tropical forest trees: implications for the evolution of paternal and maternal reproductive patterns. *American Journal of Botany* 71: 736–751.

Burd, M. 1994. A probabilistic analysis of pollinator behavior and seed production in *Lobelia deckenii. Ecology* 75: 1635–1646.

Butterfield, R. E., Lonsdale, K. G., and Downing, T. E. 1997. *Climatic change, climatic variability and agriculture in Europe (CLIVARA).* European Commission's Environment Programme, Oxford.

Campbell, D. R., and Motten, A. F. 1985. The mechanism of competition for pollination between two forest herbs. *Ecology* 66: 554–563.

Fiedler, P. L. 1987. Life history and population dynamics of rare and common mariposa lilies (*Calochortus* Pursh: Liliaceae). *Journal of Ecology* 75: 977–995.

Fowler, N. 1986. The role of competition in plant communities in arid and semiarid regions. *Annual Review of Ecology and Systematics* 17: 89–110.

Garwood, N. C., and Horvitz, C. C. 1985. Factors limiting fruit and seed production of a temperate shrub, *Staphylea trifolia* L. (Staphyleaceae). *American Journal of Botany* 72: 453–466.

Goldberg, D., and Novoplansky, A. 1997. On the relative importance of competition in unproductive environments. *Journal of Ecology* 85: 409–418.

González-Benito, E., Martín, C., and Iriondo, J. M. 1995. Autoecology and conservation of *Erodium paularense* Fdez. Glez. & Izco. *Biological Conservation* 72: 55–60.

Harper, J. L. 1977. *Population biology of plants.* Academic Press, London.

Herrera, C. M. 1991. Dissecting factors responsible for individual variation in plant fecundity. *Ecology* 72: 1436–1448.

Horvitz, C. C., and Schemske, D. W. 1988. A test of the pollinator limitation hypothesis for a neotropical herb. *Ecology* 69: 200–206.

Hosmer D., and Lemeshow, S. (Eds.). 1989. *Applied logistic regression.* J. Wiley and Sons, New York.

IUCN. 2001. *Red List categories: version 3.1.* Prepared by the IUCN Species Survival Commission. IUCN Gland, Switzerland.

Karban, R., and Strauss, S. Y. 1993. Effects of herbivores on growth and reproduction of their perennial host, *Erigeron glaucus*. *Ecology* 74: 39–46.

Legendre, P., and Legendre, L. 1998. *Numerical ecology.* 2nd English ed. Elsevier Science BV, Amsterdam.

Lennartsson, T., and Oostermeijer, J. G. B. 2001. Demographic variation and population viability in *Gentianella campestris*: effects of grassland management and environmental stochasticity. *Journal of Ecology* 89: 451–463.

Louda, S. M. 1984. Herbivore effect on stature, fruiting and leaf dynamics of a native crucifer. *Ecology* 65: 1379–1386.

Martín, C., González-Benito, M. E., and Iriondo, J. M. 1999. The use of genetic markers in the identification and characterization of three recently discovered populations of a threatened species. *Molecular Ecology* 8 (S12): 31–40.

Matthes-Sears, U., and Larson, W. 1999. Limitations to seedling growth and survival by the quantity and quality of rooting space: implications for the establishment of *Thuja occidentalis* on cliff faces. *International Journal of Plant Sciences* 160: 122–128.

McNaughton, S. J. 1985. Ecology of a grazing ecosytem: the Serengeti. *Ecological Monographs* 55: 259–294.

Mitchell, R. J. 1994. Effects of floral traits, pollinator visitation, and plant size on *Ipomopsis aggregata* fruit production. *American Naturalist* 143: 870–889.

Ollerton, J., and Lack, A. 1998. Relationships between flowering phenology, plant size and reproductive success in *Lotus corniculatus* (Fabaceae). *Plant Ecology* 139: 35–47.

Parker, M. A., and Salzman, A. G. 1985. Herbivore exclosure and competitor removal: effects

on juvenile survivorship and growth in the shrub *Gutierrezia microcephala*. *Journal of Ecology* 73: 903–913.
Potter, T. L. 1996. Population ecology of a winter annual (*Lesquerella filiformis* Rollins) in a patchy environment. *Natural Areas Journal* 16: 216–226.
Samson, D.A., and Werk, K. S. 1986. Size-dependent effects in the analysis of reproductive effort in plants. *American Naturalist* 127:667–680.
Schemske, D. W. 1980. Evolution of floral display in the orchid *Brassavola nodosa*. *Evolution* 34: 489–493.
Tilman, D. 1987. On the meaning of competition and the mechanisms of competitive superiority. *Functional Ecology* 1: 304–315.
Various Authors. 2000. Lista roja de flora vascular española (valoración según categorías UICN). *Conservación Vegetal* 6 (extra):11–38.

8

Australian Heath Shrub (*Epacris barbata*)

Viability under Management Options for Fire and Disease

DAVID KEITH

Southern hemisphere heathlands are characterized by high levels of plant diversity and endemism (Cowling 1983, Hopper et al. 1996). Many of the plant species found in these regions are rare and have small geographic ranges and narrow habitat specificity, although local populations may be large (Rabinowitz 1981). As well as these common patterns of rarity, many southern heathland plants also share a number of life history traits, such as seed dormancy mechanisms, poor propagule dispersal, and the ability of established plants to resprout when burned. With very few exceptions, their life history processes are episodic, driven by recurring wildland fires (Keith et al. 2002). *Epacris barbata*, an endangered shrub restricted to fire-prone heathland in eastern Tasmania (southern Australia), typifies these patterns of rarity and fire-driven life history. Even though the species' entire 30-km range is within a conservation reserve and its life history traits are thought to buffer populations against rapid decline under some fire regimes, appropriate management of fire regimes represents a major challenge for its conservation. In addition, diseases caused by alien pathogens pose a new threat to the persistence of *E. barbata* and many other species in temperate Australia. This chapter describes a model of the impacts of disease on a threatened plant species in a fire-prone environment. The aims are to understand which life history processes have the greatest influence over population dynamics, to quantify the effect of the disease on population viability, and to explore the relative value of alternative management options.

Life History of the Study Species

Epacris barbata is an erect shrub, typically with multiple basal stems that may grow to 1.5 m long. The species is found only in heathland on granite on the Freycinet Peninsula and nearby Schouten Island, a total range of 30 km (Figure 8.1). Populations may be found in either of two habitat types: shallow sandy loams ("soil" populations) or skeletal sands among rocky outcrops ("rock" populations). All known populations are within Freycinet National Park, a popular bushwalking destination.

The life cycle may be divided into five phases: seeds, seedlings, juveniles, mature, and senescent. Seeds are dispersed regularly each summer, only a few meters from the parent plants, and they remain dormant until released by heat shock and smoke-related cues associated with the passage of fire (Keith 1997 and unpublished data). Seedlings emerge exclusively in the post-fire environment and suffer high rates of mortality that decline steadily in the first 5 years of life. Juvenile plants are at least 6 years old, have lower rates of mortality, and may occasionally survive fire, but they are not capable of reproduction. Mature plants are at least 8 years old; they have low rates of mortality, and both their fruit production and post-fire survival are positively related to plant size. Senescent plants have elevated rates of mortality and reduced fecundity, and they cannot survive fire. Senescence generally occurs after 30 years, although some plants may live for up to twice that age (Keith 2002).

Seedling recruitment from the persistent soil seed bank occurs en masse in years after fires, but otherwise it is negligible. Fire kills all seedlings and most juveniles, as well as

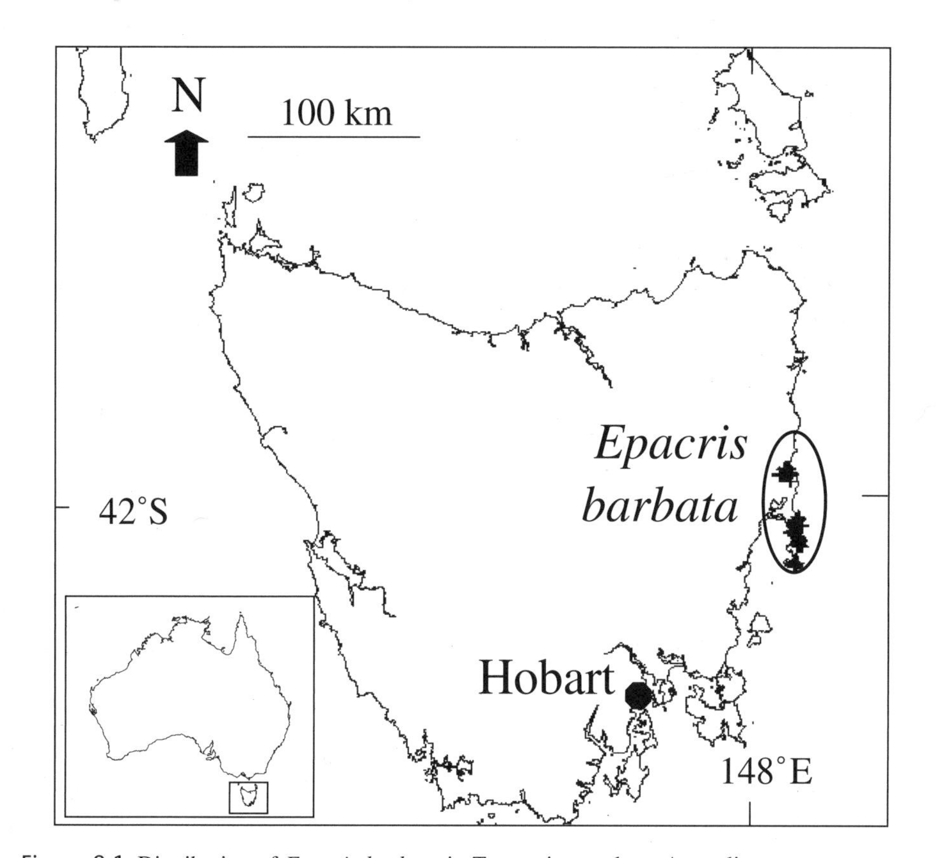

Figure 8.1 Distribution of *Epacris barbata* in Tasmania, southern Australia.

some of the mature plants. The survivors re-sprout new foliage from the bases of stems. High-frequency fire regimes are likely to cause population declines through attrition of mature plants, while recruits are killed before they reach maturity (Keith 1996). Declines are also likely under low frequency fire regimes because recruitment may be insufficient to compensate deaths due to senescence (Keith 1996).

Standing plants of *E. barbata* are susceptible to an introduced soil-borne fungal pathogen, *Phytophthora cinnamomi*, which infects roots, thereby causing necrosis and plant death (Barker and Wardlaw 1995). Death may occur suddenly or gradually as plant size—and, hence, fecundity—is reduced by sequential death of branches. The soil seed bank is apparently unaffected by the disease. The disease may be recognized in the field where susceptible taxa (e.g., Fabaceae, Proteaeaceae, Epacridaceae, Xanthorrhoeaceae) display necrotic symptoms, and the diagnosis may be confirmed with laboratory tests of soil or root samples (Shearer and Dillon 1995). The pathogen spreads locally with movement of soil water and hyphal growth. It may disperse between sites in mud transported by walkers, vehicles, and probably native mammals.

Methods

Model Structure

A stage-based matrix model was implemented with annual time steps in RAMAS Metapop version 4 (Akçakaya 2002). The model comprised 12 stages, including a seed stage, five seedling stages (ages 1 to 5 years), two juvenile stages (young and established), three adult stages (small, medium, and large), and a senescent stage (Figure 8.2). Individuals in seedling stages either grow into the next stage or die. Individuals in the juvenile and mature stages may grow to the next stage and either survive in their current stage or else die. In addition, mature plants may regress to a previous stage, but senescent and juvenile individuals may not. Fecundity transitions from each of the mature stages represent the product of the mean number of seeds produced per plant, the proportional viability of newly released seeds, and the proportion that escape postdispersal predators.

Density dependence was modeled using a threshold carrying capacity (K) specified as total habitat space (see Bekessy et al., Chapter 5 in this volume). Each stage was assigned an average weight, which is the minimum amount of space required by a plant such that its survival and growth are independent of conspecific neighbors. The habitat space occupied by a population is calculated from the product of the average weight and the number of individuals in a stage, summed over all stages. Seeds were excluded since they are not subject to density-dependent processes. When a population exceeds the total habitat space (K), stage-specific multipliers (<1) are applied to the matrix elements for survival and growth of respective stages. The multipliers are applied in successive time steps until the population falls below K. The function ensures that individuals will only be affected (i.e., with reduced survival or growth) by individuals in larger stages. This function is implemented in the user-defined density dependence file EB.DLL (available in the CD that accompanies this volume; to see the source code, open the file EB.DPR in Notepad).

The two catastrophes included in the model (Figure 8.2), fires and disease, not only are different kinds of disturbance regimes but also affect population dynamics of *E.*

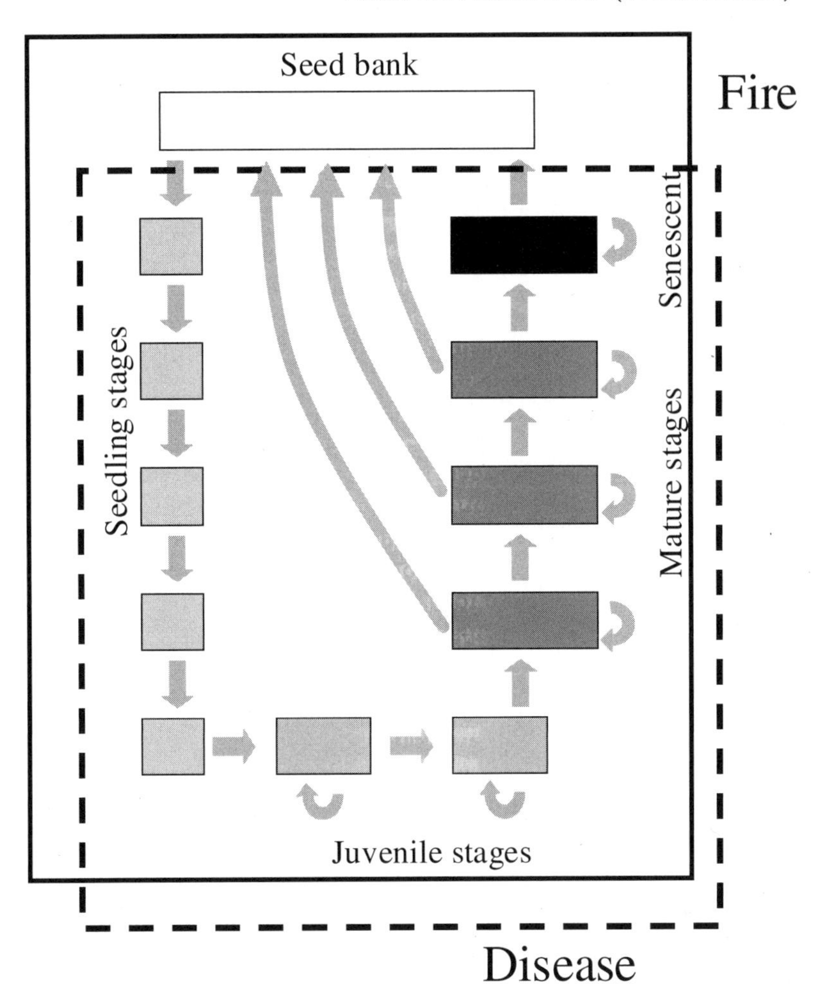

Figure 8.2 Model structure. Fire affected all vital rates represented by arrows within the box with solid lines. Disease affected the carrying capacity for all stages represented by rectangles within the box with broken lines (i.e., excluding the seed bank).

barbata in fundamentally different ways. Fires temporarily interrupt growth and seed production and cause some immediate mortality in all standing plant stages. More important, they activate the transition from seeds to seedlings for 1 to 2 years (Keith 2002). These effects were modeled using stage-specific multipliers.

In contrast, disease has a lasting negative effect on survival and growth of all standing plant stages but no effect on seeds. Moreover, the effects on mean vital rates are gradually amplified as disease spreads through a population several years after initial infection. Again, this contrasts with fires, which spread within populations instantaneously (i.e., within a single annual time step).

The effects of disease were simulated by reducing the carrying capacity of infected populations. A temporal multiplier, representing the proportion of the population remaining uninfected at time *t*, was applied to carrying capacity to simulate spread of the disease throughout the population. It was assumed that initial infection occurred in a 10-m^2 patch and had a mean local spread rate of 2 $m.yr^{-1}$ in all directions. This compares with rates of 1.0–1.6 $m.yr^{-1}$ estimated for heathlands on sandy soils in southwestern Australia (Hill et al. 1994). In each time step, the radius of the infected area was increased by 2 m until the infected area equaled the total area of the population. The temporal multiplier therefore declines to a minimum value of 0 when the whole population area is infected (Figure 8.3). Survival and growth rates of infected populations therefore decline with time since infection to a minimum specified by the product of the mean values specified in the transition matrix and the stage-specific density dependence multiplier specified in the density dependence function, and they continue at those levels for the remainder of the simulation. For simplicity, it was assumed that disease could only spread from one infection point within each population and that populations did not recover following infection. Therefore, after an initial infection occurred, the probability of further infections was set to 0 for the remainder of the simulation.

Fires were modeled as a probability of ignition within or near a population. Perusal of fire history records for Freycinet National Park suggested that ignition probabilities could be adequately modeled by assuming an approximately logistic fire model that reaches a maximum fire probability of 10% some 7 years after a previous fire (Figure 8.4). This results in a mean interval between successive fires of 13.4 (SD 9.6) years (McCarthy et al. 2001 and pers. comm.). Fire spread between populations was ignored because, in RAMAS Metapop, correlated occurrences of catastrophes could not be modeled within a single time step. Thus, adjacent populations were only burned in the same year according to the product of their ignition probabilities. In contrast, the likelihood of disease infection is governed by the proximity of *E. barbata* populations to existing infections. The probability of infection was therefore set to a value $I = \min(1, 0.01 + 10/D)$, where D is the distance in meters to the nearest known infection. Thus, spread of disease from an infection 10 m away is certain, and the probability diminishes rapidly to a minimum of 0.01. The assumed rates of spread from infected sites to *E. barbata* populations are potentially faster than rates of spread within populations. This reflects the likely greater roles of vectors, such as walkers, vehicles, and macropods in disease spread across a landscape, compared with spread at local scales, which is likely to be governed largely by flows of surface and groundwater.

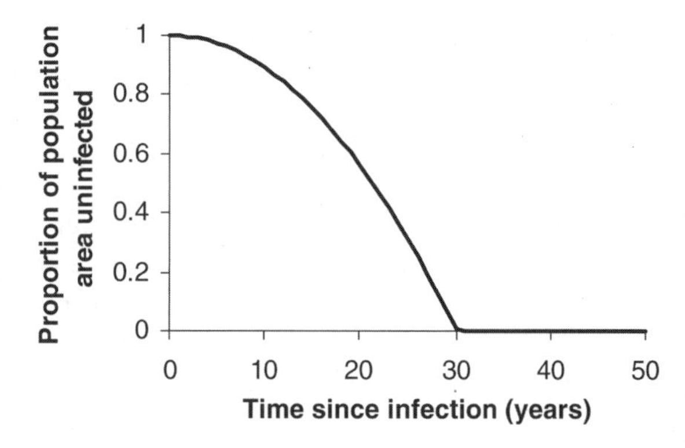

Figure 8.3 Temporal change in the multiplier for carrying capacity (proportion of population area uninfected) representing the spread of disease throughout a population at the rate of 2 m/yr^{-1} from a single point of infection. Example from population 7, area 11,300m^2.

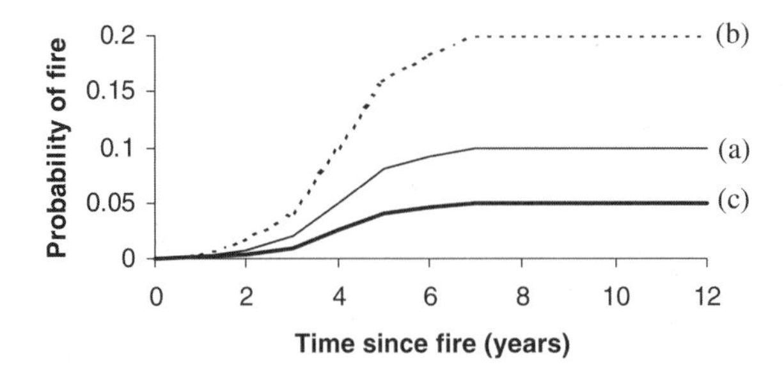

Figure 8.4 Fire probability curves for (a) the base model, mean (SD) fire interval of 13 (10) years; (b) management scenario for doubled fire probability, mean (SD) fire interval of 6 (5) years; and (c) management scenario for halved fire probability, mean (SD) fire interval of 24 (19) years.

Both demographic and environmental stochasticity were incorporated into the model. RAMAS implements demographic stochasticity by sampling the number of survivors from binomial distributions, as well as the number of seeds produced from a Poisson distribution (Akçakaya 2002). Environmental stochasticity was sampled from lognormal distributions. To reduce likely truncations due to high survival rates, a negative correlation was imposed between the largest survival rate and other survival rates for each stage (Akçakaya 2002).

The spatial structure of the model did not include dispersal of seeds between populations because this was assumed to be 0 over the time frames investigated. Environmental correlations between populations *i* and *j* were assumed to be inversely proportional to the distance *D* between populations $C_{ij} = e^{-0.00005*Dij}$. The function was based on comparatively low levels of variation in annual rainfall between weather stations spanning the range of the species.

Model Parameterization

A base model was parameterized to represent the status of *E. barbata* populations in 1996. A systematic survey of *E. barbata* habitat was carried out in 1996 using methods described by Keith (2000). New populations discovered and surveyed in 2001 were added to the data (P. Black, unpublished data). A total of 17 populations were located, each separated from the others by distances of at least 0.5 km. For each one, the following were recorded: grid location, habitat type (soil or rock), an estimate of the area occupied (Keith 2000), an estimate of the proportion of the occupied area infected with *P. cinnamomi* (if any), and an estimate of population size derived from density subsamples (Keith 2000). Grid locations were used to parameterize environmental correlations and the probabilities of disease infection. Separate transition matrices were developed for each habitat type (see below). The estimated areas of occupancy were used to set carrying capacities and calculate sizes of each population from density estimates. The proportion of the occupied area infected was used to estimate the time elapsed since initial infection using the disease spread model for each population (e.g., Figure 8.3).

The three mature stages in the model were based on the total length of basal stems, which was estimated by multiplying the number of stems by the mean length of the longest and shortest stems. These measurements were taken from samples of 30 plants at four soil populations and five rock populations. The structure of the remaining populations was estimated using data from the nearest sampled population of the same habitat type. Plants were assigned to one of the three mature stages based on these values (small <100 cm, medium 100–300 cm, and large >300 cm). The sample proportions ($N = 30$) were multiplied by the estimated population size to parameterize the initial number of individuals in each mature stage of the model. No seedlings were observed during any of the surveys, as none of the populations had been burned within the 5 years prior to the survey. Because most populations had been unburned for >10 years, all small nonreproductive plants were assigned to the small mature stage, so that none of the populations had any seedlings or juveniles in the initial population modeled.

Data on vital demographic rates were gathered from two census sites (one in each habitat type) that were monitored annually during 1994–2001, using techniques similar to those described by Keith (2002). As no seedling or juvenile plants were censused, survival rates for these were estimated from a census of the closely related *E. stuartii*, also an endangered heath plant (Keith 2002). The survival and growth rates for each non-seed stage were taken to be the mean rates across all years observed. Because survival was high and observed variation between years was low, the effects of temporal autocorrelation and sampling error on variation between years were assumed to be negligible. Demographic stochasticity was also assumed to be negligible in the censused populations, which each exceeded several hundred thousand individuals (including seeds). Therefore, raw estimates of standard deviations in vital rates were used to parameterize environmental stochasticity in the model.

Fecundity transitions for the three mature stages represented the number of seeds per plant incorporated into the seed bank per year. Estimates of fecundity for stage i (F_i) were calculated from $F_i = f_i * s * v * p$, where f_i is the mean number of fruit produced per plant per year in stage i across all years of the census, s is the mean number of seeds per fruit (24) obtained from a sample of 25 fruits, and v is mean seed viability (0.9) obtained in germination trials for *E. barbata* (D. Keith, unpublished data; see Keith 1997 for details of the viability test). There were no data on postdispersal seed-predation survival rates (p), but this was estimated at 0.1 based on data from similar Australian sclerophyllous plant species (T. D. Auld, unpublished data; D. Keith, unpublished data). Survival of the soil seed bank was estimated from seed burial trials on *E. stuartii* (M. Ilowski, unpublished data), which indicate that approximately 10% of a buried seed cache may remain viable for 12 months. Seed bank dynamics assumed exponential decay (Auld 1987) based on this rate. It was assumed that a fire would germinate half of the viable seed bank, leaving half as a residual seed bank available for a subsequent fire (Auld 1987).

Multipliers for survival and growth during fire years were estimated from census data for *E. stuartii* (Keith 2002). Thus, growth rates were set to 0 for those years, while survival multipliers were set to 85% of total background survival for small mature plants and 90% for medium and large stages, to allow for the fire-caused death of some individuals (Keith 2002). Census data for infected populations of *E. barbata* were used to estimate the density-dependent multipliers (for survival and growth), which govern modeled rates of decline for infected populations. Thus, mortality of all standing plant stages was increased by 30% on soil and 25% on rock, while growth was reduced by 80%, 70%, and 60% for transitions from seedling, juvenile, and mature stages, respectively, in both habitat types. Transitions from large mature to senescent were unaffected.

Simulations

To simplify the sensitivity analyses, the stages were lumped into five groups (seeds, seedlings, juveniles, matures, and senescent). Mean estimates and standard deviations for survival, growth, and fecundity were increased and reduced by 10%, in turn, for each group of stages. Each simulation was run for 50 years 1,000 times. The effects of these perturbations on model behavior were assessed using the proportional change in expected minimum population (EMP) size (McCarthy and Thompson 2001) relative to the value obtained from the base model.

The model was used to examine the effects of fires and disease on population viability by alternately setting the probabilities of these catastrophes to 0 and comparing outcomes to the base model.

Several management scenarios were tested to determine the most effective means of reducing the extinction risk of *E. barbata.* The first scenario examined the effect of increased survey effort and the consequent discovery of a new population. Two simulations were implemented: one in which the new population was infected with disease and the other in which it was not. In both cases the new population was located in a large area of unexplored rock habitat. The new populations were configured to be equivalent to the existing south Wine Glass Bay (c. 4,000 standing plants) and Mt. Amos (c. 11,000 standing plants) populations for the infected and uninfected simulations, respectively.

The second scenario entailed varying fire frequency by doubling and halving the probability of ignition at all populations. These probabilities resulted in mean fire intervals of 8.4 (SD 4.7) and 23.4 (SD. 19.6) years, respectively.

The third scenario examined the effect of disease quarantine by varying the probability of infection. Two quarantine scenarios were examined: one in which probabilities of infection for all currently uninfected populations were halved and one in which they were reduced to 0. In both cases, disease was allowed to spread through populations that were currently infected, as in the base model.

Finally, the effect of on-site disease treatment was examined by reducing the rate of mortality of standing plants in infected populations. In the base model, disease increased mortality by 25% in rock populations and 30% in soil populations. In two simulations, these effects were reduced by half (i.e., 12.5% increase for rock and 15% for soil) and by 90% (i.e., 2.5% and 3%, respectively). These simulations represented on-site disease treatments that were 50% and 90% effective in preventing plant death.

The impacts of all management scenarios were assessed by examining the proportional changes in effective minimum population size, shifts in extinction risk curves, and changes in metapopulation occupancy relative to the base model.

Results

Sensitivity

There was almost a 10-fold change in EMP in response to 10 % reduction in survival of mature stages, which was more than double the sensitivity of any other parameter (see Figure 8.5 for all data in this paragraph). An increase in survival did not produce an equivalent response because values were already close to unity in the base model. Seedling growth rates were the second most sensitive parameter, producing a 3- to 4.5-fold change in EMP. Variation in growth rates of mature stages produced modest changes in EMP (16%). The model was largely insensitive to the remaining survival, growth, and fecundity parameters, since variation of these by 10% only produced commensurate or smaller variation in EMP. The model was insensitive to small variations in the standard deviations of all vital rates, with variations of 10% producing less than 2% changes to EMP. Similarly, there was little response (<2%) to changes in survival and growth multipliers in the density dependence function.

Management Scenarios

The base model scenario forecasted a very substantial decline in the total population of *E. barbata* from 178,000 standing plants in 1996 to 8,800 ± 3,100 (mean, SD) over the next 50 years. Although a 90% decline in population size was virtually certain, the model predicted that there was less than 0.1% chance of the species becoming extinct in the next 50 years. However, local extinctions were predicted for 5.9 ± 1.1 of the 17 existing populations. Extension of the simulation to 100 years indicated a continuing decline to 2500 ± 1200 individuals, including the loss of 9.9 ± 0.9 populations.

The model predicted that the discovery of a new, large, uninfected population similar to the existing population on Mt. Amos would increase the expected minimum population size by 48% and reduce quasi-extinction risks (Figure 8.6). However, a decline in the total population of at least 90% would still be almost certain. The discovery of a moderately large infected population similar to the existing population at south Wine Glass Bay would not appreciably alter quasi-extinction risks.

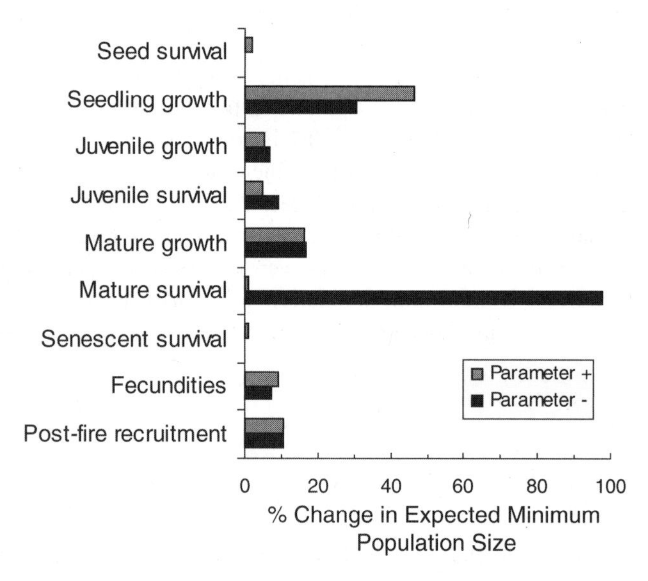

Figure 8.5 Sensitivity of expected minimum population size to variations of +10% (*hatched bars*) and –10% (*filled bars*) in parameter estimates.

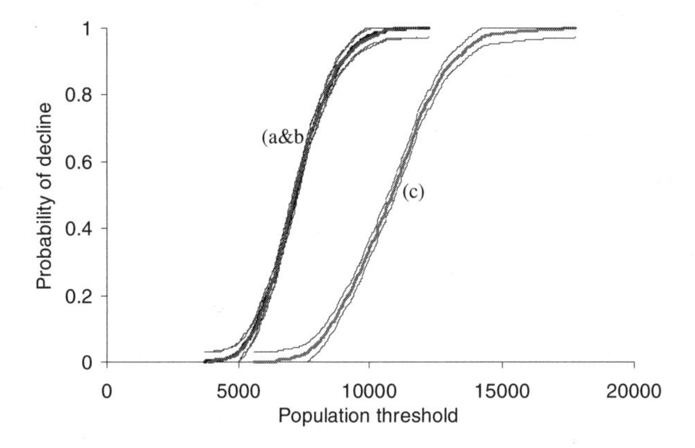

Figure 8.6 Risk curves for survey scenarios (a) base model, (b) discovery of large infected population, and (c) discovery of large uninfected population. Thin lines represent 95% confidence limits.

Both alternative fire-management scenarios slightly reduced population viability (Figure 8.7). A doubling of fire probability reduced EMP by 7.2%, while halving the fire probability reduced EMP by 15.7%.

A disease quarantine scenario that reduced the chance of disease infection by 50% had no appreciable effect on population viability (Figure 8.8). Even a quarantine scenario that was fully effective in preventing infection had marginal influence on quasi-extinction risks (<1% increase in EMP); this was because the largest populations were already infected when simulations began. A fully effective quarantine scenario also failed to reduce the number of local extinctions.

An on-site disease treatment that reduced disease-related mortality by 50% also had only a marginal effect on population viability (Figure 8.9; 3.1% increase in EMP), but reduced the predicted number of local extinctions to 3.2 ± 1. If such a treatment could reduce disease-related mortality by 90%, this would substantially improve population viability (141% increase in EMP) and reduce the predicted number of local extinctions to 2 ± 0.8. It also would reduce the chance of a 90% decline to 54 ± 3% (Figure 8.9). If disease could be eliminated immediately, this would increase EMP by 493% and all but eliminate the chance of a 90% decline over the next 50 years.

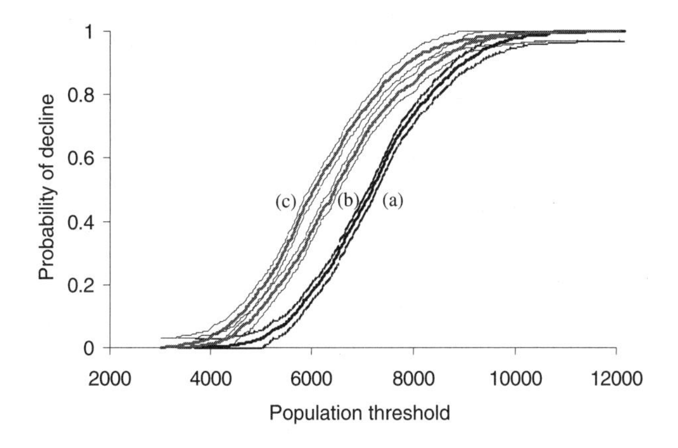

Figure 8.7 Risk curves for fire-management scenarios: (a) base model, (b) doubled fire probability, and (c) halved fire probability (see Figure 8.4). Thin lines represent 95% confidence limits.

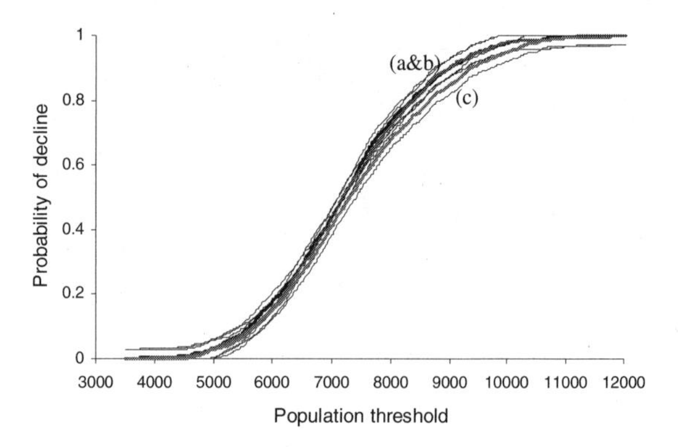

Figure 8.8 Risk curves for disease quarantine scenarios: (a) base model, (b) 50% effective quarantine, and (c) 100% effective quarantine. Thin lines represent 95% confidence limits.

Discussion

The sensitivity analysis underscored the importance of survival of established plants in *E. barbata* populations. Even small losses produced a marked reduction in population viability. By reducing survival of mature stages by 25% to 30%, disease presents a major threat to persistence of the species. The buffering effect of a persistent soil seed bank extends an inevitable decline over several decades when it otherwise might be more immediate. The development of an on-site treatment that reduces the susceptibility of established plants to disease is by far the most effective management option for conservation of the species. Even so, such a treatment needs to be 90% effective to confer appreciable gains in the viability of the species. It should be noted, however, that a more extensive sensitivity analysis that examines the interactions between parameters in a factorial design (Swartzman 1980) may yield new insights into which factors are crucial to population viability in *E. barbata.*

Although quarantine would improve the viability of small populations that remain uninfected, on-site treatment is needed to arrest continuing declines and eventual elimination of the largest populations. Complete quarantine of any population may be logistically difficult, given the popularity of bushwalking and the possibility of disease spread by native mammals. Nonetheless, declines in *E. barbata* may be slowed by the implementation of hygiene procedures and access restrictions, depending on the extent to which these measures reduce the probability of infections.

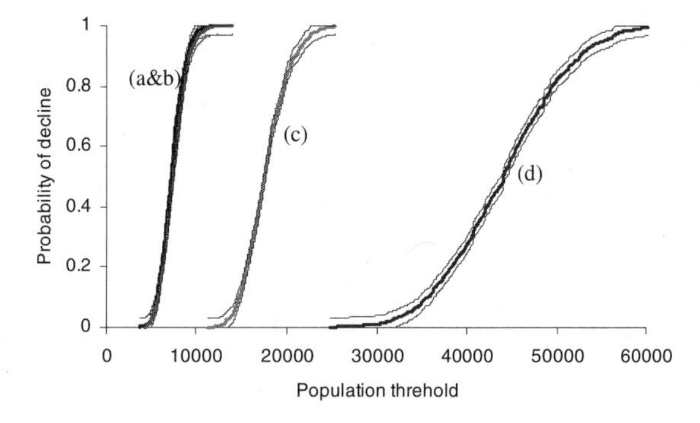

Figure 8.9 Risk curves for disease treatment scenarios: (a) base model, (b) 50% effective treatment, and (c) 90% effective treatment. Line (d) shows effect of immediate disease cure at all sites. Thin lines represent 95% confidence limits.

If surveys of uninfected habitat yield hitherto undiscovered populations, the viability of *E. barbata* would be higher than forecast here, depending on the size of any new populations. However, surveys of infected habitat are unlikely to yield such benefits.

Variation in fire frequency, the most practical management tool, produced surprisingly little response in population viability, given the importance of fire for reproduction. It seems likely that the effects of fire frequency are masked by the effects of disease. Nonetheless, the current fire regime seems more favorable to *E. barbata* than the alternatives explored here, and this frequency (mean interval 13 years) is similar to fire frequencies found to be compatible with the persistence of a wide range of Australian sclerophyll shrub species (Keith 1996, Dreschler et al. 1999, Regan et al. 2002).

Unfortunately, the most effective management options appear to be the least tractable. Trials in Freycinet National Park with phosphonite, a chemical spray that increases the resistance of standing plants to *P. cinnamomi* under controlled conditions, failed to slow rates of infection and mortality in *E. barbata* and co-occurring susceptible species in the field (P. Black, pers. comm.). The large number of Australian heath plants susceptible to the disease (Barker and Wardlaw 1995, Shearer and Dillon 1995) clearly justifies a more extensive research effort to explore alternative chemicals, as well as varying dosages and timings of application.

The modeling of two independent catastrophes is novel in plant population viability analysis (Menges 2000). Dreschler et al. (1999) modeled a *Banksia* species affected by fire regimes and habitat loss, however the latter was not explicitly included in their model. A limitation of the current model is that disease effects were modeled indirectly by reducing carrying capacity (K), which, in turn, caused reduced survival and growth once K had fallen below population size (N). Declines may be more immediate in reality than predicted if population sizes are already well below carrying capacity because it may take some years after infection for modeled K to fall below N. The model also assumes a radial pattern of disease spread from a single infection point within populations. The reality is more complex, and probability involves multiple infection points with spatial spread influenced by irregular microtopography and infill of gaps. The resulting declines could be slower or more rapid than those modeled here.

Another simplification is that fires were assumed to occur independently in different populations. While this seems more plausible than perfect correlation of all populations, nearby populations are likely to experience fires in the same year more often than random because of fire spread. Fire correlations are not simply a function of distance, however, because prevailing wind patterns and topography create nonrandom fire paths. In addition, one population is isolated from the others on an island. The complex nature of fire spread demands a more complex approach to modeling correlations of catastrophes than is currently available in RAMAS Metapop.

The correlation between fire and disease infection was not explored in the study reported here, but it poses an interesting issue for management. The occurrence of fire soon after infection simultaneously depletes the seed bank and exposes new individuals to higher risk of mortality than in the absence of disease. The exclusion of fire after infection may preserve a capacity for recovery by maintaining a higher proportion of the population in the seed stage, which is unaffected by disease. This could be important if the voracity of disease at a site declines with time since infection, as suggested by data from long-term studies of infected sites (Weste and Ashton 1994, Weste and

Kennedy 1997). Further development of the model will allow these complex life history interactions and trade-offs to be explored.

Acknowledgments I am indebted to Resit Akçakaya for his assistance in diagnosing and resolving problems at various stages of model development. Sarah Bekessy kindly allowed me to use the density dependence function that she developed for monkey puzzle trees. Mick McCarthy calculated mean fire intervals from the hazard functions. Thanks to Mark Burgman, Mick McCarthy, and Helen Regan for many productive discussions that stimulated this venture into PVA. Finally, I am very grateful to Paul Black and Mick Ilowski (Nature Conservation Branch, Primary Industries, Tasmania) for their assistance in the field and willingness to gather and share additional data.

References

Akçakaya, H. R. 2002. *RAMAS Metapop: viability analysis for stage-structured metapopulations (version 4.0).* Applied Biomathematics, Setauket, N.Y.

Auld, T. D. 1987. Population dynamics of the shrub *Acacia suaveolens* (Sm.) Willd.: survivorship throughout the life cycle, a synthesis. *Australian Journal of Ecology* 12: 139–151.

Barker, P. J. C., and Wardlaw, T. J. 1995. Susceptibility of selected Tasmanian rare plants to *Phytophthora cinnamomi. Australian Journal of Botany* 43: 379–386.

Cowling, R. M. 1983. Diversity relations in Cape shrublands and other vegetation in the southeastern Cape, South Africa. *Vegetatio* 45: 103–127.

Dreschler, M., Lamont, B. B., Burgman, M. A., Akçakaya, H. R., Witkowski, E. T. F., and Supriyadi. 1999. Modelling persistence of an apparently immortal *Banksia* species after fire and land clearing. *Biological Conservation* 88: 249–259.

Hill, T. C. J., Tippett, J. T., and Shearer, B. L. 1994. Invasion of Bassendean dune *Banksia* woodland by *Phytophthora cinnamomi. Australian Journal of Botany* 42: 725–738.

Hopper, S. D., Harvey, M. S., Chappill, J. A., Main, A. R., and York Main, B. 1996. The Western Australian biota as Gondwana heritage: a review. Pages 1–46 in S. D. Hopper, J. Chappill, M. Harvey, and A. George (eds.), *Gondwana heritage: past, present and future of the Western Australian biota.* Surrey Beatty, Sydney.

Keith, D. A. 1996. Fire-driven mechanisms of extinction in vascular plants: a review of empirical and theoretical evidence in Australian vegetation. *Proceedings of the Linnean Society of New South Wales* 116: 37–78.

Keith, D. A. 1997. Combined effects of heat shock, smoke and darkness on germination of *Epacris stuartii* Stapf, an endangered fire-prone Australian shrub. *Oecologia* 112: 340–344.

Keith, D. A. 2000. Sampling designs, field techniques and analytical methods for systematic plant population surveys. *Ecological Management and Restoration* 1: 125–139.

Keith, D. A. 2002. Population dynamics of an endangered heathland shrub, *Epacris stuartii* (Epacridaceae): recruitment, establishment and survival. *Austral Ecology* 27: 67–76.

Keith, D. A., McCaw, W. L., and Whelan, R. J. 2002. Fire regimes in Australian heathlands and their effects on plants and animals. Pages 199–237 in R. A. Bradstock, J. E. Williams, and A. M. Gill (eds.), *Flammable Australia: the fire regimes and biodiversity of a continent.* Cambridge University Press, Cambridge.

McCarthy, M. A., and Thompson, C. 2001. Expected minimum population size as a measure of threat. *Animal Conservation* 4: 351–355.

McCarthy, M. A., Gill, A. M., and Bradstock, R. A. 2001. Theoretical fire-interval distributions. *International Journal of Wildland Fire* 10: 73–77.

Menges, E. S. 2000. Applications of population viability analysis in plant conservation. Pages 73–84 in P. Sjogren-Gulve and T. Ebenhard (eds.), *The use of population viability analysis in conservation planning.* Ecological Bulletin 48. Munksgaard, Copenhagen.

Rabinowitz, D. 1981. Seven forms of rarity. Pages 205–217 in H. Synge (ed.), *The biological aspects of rare plant conservation.* Wiley, Chichester.

Regan, H. M., Auld, T. D., Keith, D. A., and Burgman, M. A. 2002. Fire, predation and population viability: management of the endangered shrub *Grevillea caleyi*. *Biological Conservation* 109: 73–83.

Shearer, B., and Dillon, M. 1995. Susceptibility of plant species in *Banksia* woodlands on the Swan coastal plain, Western Australia, to infection by *Phytophthora cinnamomi*. *Australian Journal of Botany* 43: 113–134.

Swartzman, G. 1980. Evaluation of simulation models. In W. M. Getz (ed.), *Mathematical modelling in biology and ecology*. Springer-Verlag, Berlin.

Weste, G., and Ashton, D. H. 1994. Regeneration and survival of indigenous dry sclerophyll species in the Brisbane Ranges, Victoria, after a *Phytophthora cinnamomi* epidemic. *Australian Journal of Botany* 42: 239–253.

Weste, G., and Kennedy, J. 1997. Regeneration of susceptible native species following a decline of *Phytophthora cinnamomi* over a period of 20 years on defined plots in the Grampians, western Victoria. *Australian Journal of Botany* 45: 167–190.

PART II

INVERTEBRATES

9

Modeling Invertebrates

An Overview

OSKAR KINDVALL

Invertebrates are by the far most-species rich group of organisms. Consequently, it is hard to point out how to model population dynamics for this group. The variety of life histories and mating systems that are represented by this group is enormous. Actually it is fully possible to find examples of demographic configurations among invertebrate species that resembles what is described as typical for other groups of organisms. Plants can have seeds that rest for considerable periods of times, thus providing a seed bank that enables future recovery of a population without any reproduction during several years (Burgman, Chapter 2, this volume). The same phenomenon occurs with diapausing eggs in several insect groups or by particular resting stages among crustaceans. Invertebrates are often considered to be relatively short-lived with prevalence for annual or even shorter life cycles. Still, there exist several examples of quite long-lived invertebrates with life cycles that correspond well to many vertebrate species.

Modeling Approaches

RAMAS GIS provides good opportunities to adjust the model to the variety of demographic structures and life cycles found among invertebrates. Depending on how field data has been sampled and whether or not the species reproduce in distinct intervals, one can choose either to construct a simple model based on only population growth rate and carrying capacity or to specify fecundates and transition rates in a stage- or aged-structured matrix. As maximum growth rates of invertebrates may often be high, it is usually needed to define some kind of density dependence on population growth. Which one to choose depends on the particular system and amount of knowledge. A simpler solution (e.g., the exponential growth with ceiling) may be recommended when data are scarce because it really requires much data to discriminate between different types of density dependence. Alternatively, it may sometimes be possible to suggest a certain type based on knowledge about existing predators or parasitoids. Sometimes a model that includes Allee effects is recommended for the sake of reducing the risk of making too optimistic predictions of population persistence. Allee effects, among other mechanisms, may arise as a consequence of reduced mating success at low population densities. However, as most invertebrates are quite capable of moving, and adjust their movement behavior to current circumstances, it may be expected that Allee effects of this kind are less prevalent among this group of organism than among more sedentary life forms, especially plants (Kindvall et al. 1998).

The possibility of modeling temporal changes in parameters associated with habitat quality and environmental impact is a useful feature of RAMAS GIS. Many invertebrates, such as insects inhabiting seminatural grasslands, are today threatened partly due to gradual detoriation of the habitats required for reproduction and survival. It is of great importance to consider these changes when trying to assess extinction risks or when making preservation plans for species of this kind. Chapter 15 in this volume clearly illustrates the significance of environmental changes due to natural succession as a consequence of ceased former management. Ignoring these processes would heavily overestimate the possibilities of long-term survival of the species.

When it is necessary to simulate huge numbers of local population, with on average rather many reproductive individuals, demographic models such as RAMAS GIS may be too cumbersome. There are examples of invertebrates—for example, the butterfly *Melitaea cinxia*—that experience high levels of environmental stochasticity, where local extinctions thereby are common, and long-term persistence merely depends on large-scale configuration of habitat patches (Hanski et al. 1995). In such cases the number of local populations that need to be simulated may often significantly exceed the limits of what is feasible using a demographic approach. Of course, one solution to this problem is to divide the whole system into smaller subsets of local populations, which was used by Fox et al. (Chapter 13 in this volume).

Another solution for the problem of handling a large population with many demographic subunits, or local populations, may be to apply some kind of occupancy model, whereby local population dynamics are summarized as merely changes in transient states of occupancy. The application of such metapopulation models, which may be suitable for many invertebrate systems (Hanski 1999, Sjögren-Gulve and Hanski 2000), is beyond the scope of this volume.

Environmental Stochasticity

As many invertebrates have the potential to produce large numbers of eggs, the population growth rate can occasionally become very high, as a result of only minor changes in the survival rate of juvenile stages. These changes may often be generated by fluctuating environmental factors. The combination of generally high fecundity and sensitivity to environmental factors such as weather conditions often results in high temporal variability in invertebrate populations. Several species do not fluctuate much in numbers, due to either rather low fecundity or high environmental stability (e.g.,

Ranius and Hedin, Chapter 14, and Fox et al., Chapter 13 in this volume). Nevertheless, it is difficult to obtain information that is relevant for prediction of long-term variability. For example, the grasshopper, *Stauroderus scalaris* (Orthoptera: Acrididae), was described as being on the fringe of extinction, with no capacity for high reproduction or dispersal based on the information sampled between 1984 and 1989. In 1992 the behavior of the species' population dynamics changed dramatically, with a great expansion of its spatial range as a result (Carlsson and Kindvall 2001), and the parameter values obtained from the years 1993–1998 differ markedly from the earlier investigation. An attempt to model the population dynamics using an occupancy model failed to predict the expansion based on parameters representative for the 1980s, while the stagnation observed before 1992 could not be predicted based on the parameter combination of the 1990s. Furthermore, it would not have helped to just increase the environmental stochasticity associated with the various parameters of the model. The parameter values must be shifted completely and only at certain intervals to obtain the observed pattern. This example clearly illustrates the difficult task of choosing the right parameter values and model structure when attempting to predict future population dynamics.

The performance of invertebrates is often to a significant extent determined by weather conditions. Because, weather conditions are typically rather similar within large proportions of a species distribution area, it is expected that the temporal variability of different demographic parameters will be spatially synchronized. This phenomenon has also been demonstrated for many insects (e.g., Hanski and Woiwod 1993). The outcome of simulations using spatially structured models is extremely sensitive to the degree of environmental correlation. When modeling invertebrates, therefore, it is important to try to estimate this parameter, which can be easily incorporated into the model using RAMAS GIS. If one decides to set the correlation parameter to 0, due to lacking information about the simulated system, it is wise to emphasize that the result will most likely underestimate the expected extinction risk (Harrison and Quinn 1989).

Dispersal

When constructing spatially structured models, one of the most critical aspects is how to model interpatch transitions of individuals. Both movement behavior and the capacities to disperse long distances across hostile or unsuitable habitat vary among different invertebrate species. Also, the mechanism for dispersal varies. While many invertebrates are capable of active flight, which is the case for most insects, others are forced to move by walking or crawling on the ground or in the vegetation. Spiders and caterpillars of many butterflies and moths may also transport themselves long distances through the air by the help of wind. Yet other species may use different kinds of animal vectors such as migrating birds in order to hitchhike to distant places.

Despite the underlining mechanisms affecting invertebrate dispersal, it is only possible to incorporate this process into a RAMAS GIS model by parameterizing a phenomenological function relating interpatch dispersal rate to distance. It is generally difficult to estimate representative parameter values of this function, and the problems associated with this procedure are not unique for invertebrates. Some of these problems are discussed in more detail in Kindvall, Chapter 11 in this volume.

To obtain empirical information on dispersal distance functions, one would like to monitor individual movements through the landscape. This can be done directly by using some kind of transmitter attached onto individuals, but because most invertebrates are rather small, this type of direct observation on movements is not feasible. Recently, microtransmitters have been used for this purpose on some types of insects (Riley and Osborne 2001), and this technique will probably become even more sophisticated and useful in the future. Nevertheless, for most invertebrates it is extremely hard to make observations of dispersal, and in many cases we have to rely on rough guesses.

Applications

The possibility of making both qualitative and quantitative predictions of species population dynamics in different scenarios is useful for many applied ecological problems. The number of population variability analyses of different invertebrate species that have been conducted so far is probably quite large, but the exact figure is hard to determine. In many cases, these applications seem to be reliable enough to generate important insights that give guidance for those implementing various preservation efforts (Akcakaya and Baur 1996, Kindvall 2000, Regan et al. 2001). However, conservation biologists have more often focused on mammals and birds, and the interest of working with preservation plans for insects and other invertebrates has often been rather small. In most countries, invertebrates are not even considered when compiling Red Lists. One reason the number of good examples on invertebrate applications of population viability analyses is rather high is that this group of organisms provide many practical model organisms for ecologists. It may often be much easier to obtain relevant data on a short-lived insect than on, for example, a large carnivore mammal species. By applying demographic models on an insect, ecologists try to understand the processes that affect population extinction risks generally and how population dynamics should be modeled in order to generate reliable predictions.

It is not only within the field of conservation biology where population models, constructed using RAMAS GIS, can be useful. However, in this book only one chapter deals with an invertebrate that is not studied because of conservation purposes; this is the chapter describing how to model a parasite population (Milner-Gulland et al., Chapter 16 in this volume). Another subject of particular importance for invertebrates is the issue of how to manage species that are pests in agricultural or forest environments. RAMAS GIS modeling gives good opportunities to analyze important aspects of population dynamics that are related to pest management problems as well. I definitively encourage ecologists active within the field of pest management to consider RAMAS GIS as a practical tool for developing predictive models also for pests like insects and molluscs.

References

Akcakaya, H. R., and Baur, B. 1996. Effects of population subdivision and catastrophes on the persistence of a land snail metapopulation. *Oecologia* 105: 475–483.

Carlsson, A., and Kindvall, O. 2001. Spatial dynamics in a metapopulation network: recovery of a rare grasshopper *Stauroderus scalaris* from population refuges. *Ecography* 24: 452–460.

Hanski, I. 1999. *Metapopulation ecology*. Oxford University Press, Oxford.

Hanski, I., and Woiwod, I. P. 1993. Spatial synchrony in the dynamics of moth and aphid populations. *Journal of Animal Ecology* 62: 656–668.

Hanski, I., Pakkala, T., Kuussaari, M., and Lei, H. 1995. Metapopulation persistence of an endangered butterfly in a fragmented landscape. *Oikos* 72: 21–28.

Harrison, S., and Quinn, J. F. 1989. Correlated environments and the persistence of metapopulations. *Oikos* 56: 293–298.

Kindvall, O. 2000. Comparative precision of three spatially realistic simulation models of metapopulation dynamics. *Ecological Bulletins* 48: 101–110.

Kindvall, O., Vessby, K., Berggren, Å., and Hartman, G. 1998. Individual mobility prevents an Allee effect in sparse populations of the bush cricket *Metrioptera roeseli:* an experimental study. *Oikos* 81: 449–457.

Regan, T. J., Regan, H. M., Bonham, K., Taylor, R. J., and Burgman, M. A. 2001. Modelling the impact of timber harvesting on a rare carnivorous land snail (*Tasmaphena lamproides*) in northwest Tasmania, Australia. *Ecological Modeling* 139: 253–264.

Riley, J., and Osborne, J. L. 2001. Flight trajectories of foraging insects: observations using harmonic radar. Chapter 7 in I. P. Woiwod, D. R. Reynolds, and C. D. Thomas (eds.), *Insect movement: mechanisms and consequences.* CABI Publishing, New York.

Sjögren-Gulve, P., and Hanski, I. 2000. Metapopulation viability analysis using occupancy models. *Ecological Bulletins* 48: 53–71.

10

Carnivorous Land Snail *Tasmaphena lamproides* in Tasmania

Effects of Forest Harvesting

TRACEY J. REGAN
KEVIN J. BONHAM

Tasmaphena lamproides (Cox 1868) is a carnivorous land snail that inhabits native forests in northwest Tasmania, Australia. Suitable habitat for the species encompasses approximately 380 km^2 of forest areas in the northwest corner of Tasmania, with the total range extending to the main land in Victoria. Due to its small range, the species is currently listed as rare under the Tasmanian Threatened Species Protection Act (1995). Much of the species habitat is within timber production forest, an area administered by Forestry Tasmania. Future use of the area involves converting native forest to eucalypt plantations or harvesting native forest followed by burning to promote regeneration. Forestry Tasmania's management decisions require consideration of the competing demands for timber production and conservation of this sensitive species to ensure that the forest activities are sustainable.

The purpose of this study was to develop a model of *T. lamproides* in order to provide managers with a decision support tool with which they can explore the trade-offs between timber production requirements and conservation of the species under various management scenarios.

Life History

Tasmaphena lamproides is found in wet forests and requires deep leaf litter to survive. It is usually found under logs or in leaf litter at the base of trees or in ditches and depressions. A full-size adult can have a shell width of up to 21 mm, while juveniles have shell widths of less than 14 mm. The basic survival requirements for *T. lamproides* are adequate food supplies, including calcium for shell growth (often obtained from the shells of other snails), moisture, shelter from predators, and protection from adverse weather. *T. lamproides* feeds on other abundant snail species, including *Stenacapha cf. ducani.* (approx. 60% of its diet), *Helicarion cuvieri* (20%), and other less abundant snail species and worms (20%) (Bonham and Taylor 1997).

T. lamproides is thought to live for around 4 to 8 years. Although it is hermaphroditic, it requires a mate to reproduce. Generally, individuals first begin to reproduce at 2 years of age, but reproduction can begin as early as age 1 year. Studies of lowland snail populations suggest that there is a propensity for rapid juvenile growth (see Baur 1994), which may account for the lower bound of 1 year to reach maturity. Although reproductive adults produce a clutch of 5 to 10 eggs once a year, overall fecundity is thought to be less than this due to egg infertility and egg cannibalism by hatchlings (Baur 1992, 1993; Chen and Baur 1993; Desbuquois and Madec 1998).

T. lamproides are relatively fast movers, traveling at maximum speeds of 7 mph. However individuals can take several decades to reinvade an area 1 km away from a starting point. Studies of more populous land snail species indicate that passive dispersal (e.g., mediated by birds, grazing animals, or human activities) is negligible (Baur and Baur 1989, Pfenninger et al. 1996). The likelihood of an individual dispersing depends on the hostility of the intervening environment. Roads, fast-flowing streams, and areas of approximately 50 m of alien habitat are likely to form barriers for snail dispersal. Adults and subadults tend to have the same capacity for dispersal (Baur and Baur 1993).

Forestry Tasmania manages a portion of the total distribution of *T. lamproides*, and future use of the forest will involve converting some native forest into eucalypt plantations, while other areas will be harvested then burned to promote regenerative regrowth. Burning after harvesting of native forest is thought to eliminate populations of *T. lamproides*, but they will reinvade native forest areas once the necessary niche attributes such as adequate level of litter and food sources have returned. The time taken for an area to recover and be able to support snail populations after a disturbance is estimated at 30 to 60 years (Bonham and Taylor 1997). Snails probably survive in streamside reserves, wildlife habitat strips, and adjacent undisturbed habitat. Areas converted to eucalypt plantations eliminate snail populations. It is unlikely that individuals will reinvade plantation forests mainly due to changes in soil and understory structure, nutrient composition, and reduction of the leaf litter and the abundance of rotting logs.

Landscape Data

Geographical information system (GIS) layers of landscape data of areas administered by Forestry Tasmania in the northwest provide details of areas and locations for *T. lamproides*. The total extent of the range investigated in this study is approximately 290 km^2, with an occupied area of approximately 132 km^2. The total distribution of *T. lamproides* lies further afield, but these areas are not administered by Forestry Tasmania and have not been considered in our study. The available habitat of the study area is approximately 35% of the total area of occupancy for *T. lamproides*.

Three main geographic areas provide habitat for snail populations: Bond Tier Reserve (18 km^2), the Togari Forest Block (85 km^2), and the Christmas Hills Forest block (38 km^2). These areas are a combination of production forest and formal and informal reserves, but they differ in the quality of habitat and the proportion designated for wood production (Figure 10.1).

Areas of production forest are subdivided into management units called coupes each of about 40 to 80 ha. Production forest occurs predominately in the Togari Forest Block and to a lesser extent in the Christmas Hills Forest Block. Formal reserves include the Bond Tier Reserve, which is mostly isolated from production forests and will not be logged, and others in the Christmas Hills Forest Block. Informal reserves, such as streamside reserves and wildlife habitat strips are present in both the Togari and Christmas Hills Forest Blocks. Although informal reserves are usually consolidated entirely around the edges of coupes, some are fragmented within a coupe and include areas such as swamps and buffers around small intermittent streams. The Christmas Hills Forest Block includes many gullies that are not accessible to harvesting but are able to support snail populations.

Snail dispersal between the three main areas is unlikely. The Togari Forest Block is surrounded mostly by unsuitable habitat and road boundaries. The Christmas Hills Forest Block is surrounded by a road boundary and unsuitable habitat. In the south, it borders

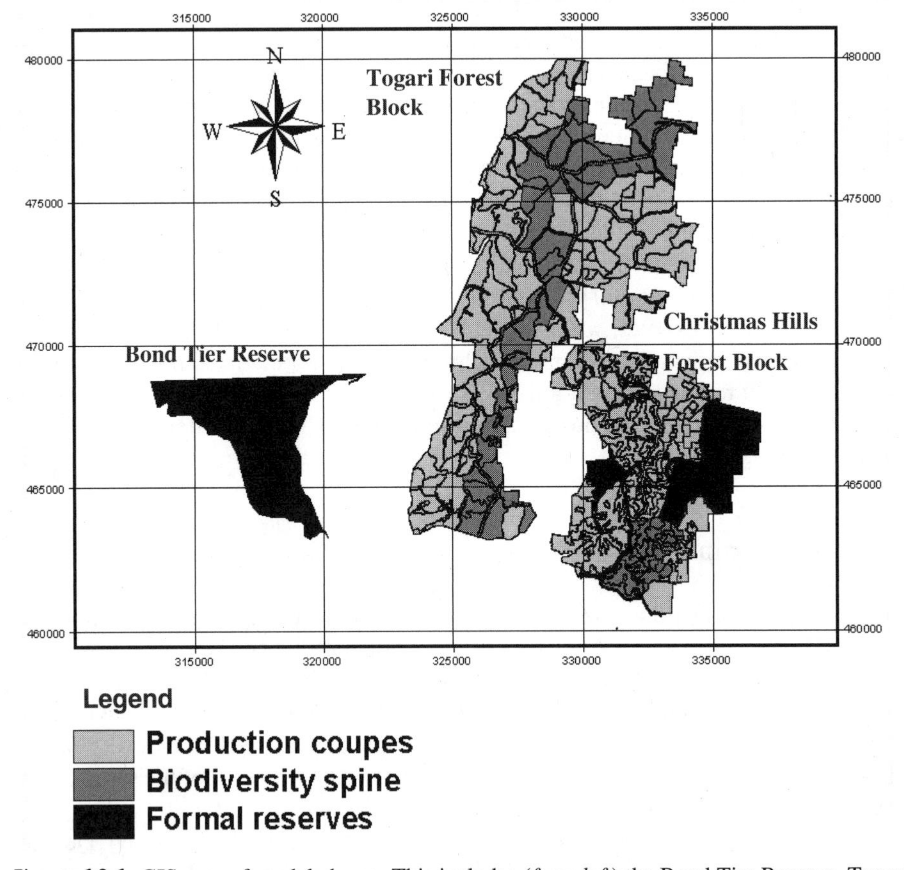

Figure 10.1 GIS map of modeled area. This includes (*from left*) the Bond Tier Reserve, Togari Forest Block, and Christmas Hills Forest Block. Each polygon represents a patch.

suitable habitat (State Forest managed by Forestry Tasmania), but this potential source has been ignored in this model. The Bond Tier reserve is surrounded by suitable habitat to the north and west, unsuitable habitat to the east and a road boundary to the south.

Each polygon in Figure 10.1 represents a patch, which can be either a coupe, a formal reserve, or an informal reserve. Each patch is interpreted as a separate subpopulation within a metapopulation structure in the RAMAS GIS Spatial Data subprogram. There are a total of 317 polygons in the study area. The Togari Forest Block has 184 polygons (130 coupes, 54 informal reserves), Christmas Hills Forest Block has 132 polygons (79 coupes, 53 formal and informal reserves), and the Bond Tier Reserve is made up of 1 polygon (Table 10.1).

Approximately 13.6 km^2 of area in the northern part of the Togari Forest Block contains high-quality habitat for the species. The remaining occupied habitat is thought to be medium-quality habitat for the species. Population densities are approximately 25 individuals per hectare in high-quality habitat and 12 individuals per hectare in medium-quality habitat (Bonham and Taylor 1997). It is assumed that the initial population structure is close to an equilibrium population—that is, a population structure that might be expected after very long periods without disturbance. A habitat suitability map was created incorporating information on habitat quality to set the initial conditions for carrying capacity and initial abundances for the model.

Population Model

A spatially explicit stage-structured metapopulation was developed using RAMAS Metapop (Akçakaya 2002). The model is an adaptation of the model developed by Regan et al. (2001), which focused only on the high-quality habitat in the northern section of the Togari Forest Block.

Vital Rates

A pre-reproductive census is assumed with five age classes, since individuals can reproduce at a lower bound of 1 year and an upper bound of 5 years. The last age class consists of individuals 5 years and older. Estimates for survival and fecundity were taken from Regan et al. (2001). The estimates on survivorship and fecundity are based on the observed population structure and live/dead ratios of specimens.

Stochasticity

Both environmental and demographic stochasticity were incorporated into the model. Measurements of standard deviations were unavailable. A coefficient of variation (CV)

Table 10.1 Three management areas with number of polygons designated for harvesting

	No. of Harvested Polygons	No. of Nonharvested Polygons
Bond Tier Reserve	0	1
Togari Forest Block	130	54
Christmas Hills Forest Block	79	53

of 50% was used for the fecundities, F_i. S_1 and S_2 were assumed to be from a lognormal distribution with a mean of 0.45 and a range of [0.4, 0.5]. It was assumed that this translated to a confidence interval (CI) of 50%, resulting in a CV of 16%. Similarly, lognormal distributions was assumed for S_3, S_4, and $S_{5+,}$ with means of 0.8, 0.75, and 0.7, respectively, and a CV of 10% for each distribution.

The survival rates and fecundities were sampled from a lognormal distribution. Spatial correlations represent the degree of similarity in environmental conditions experienced by spatially separate populations. It was assumed that all coupes are sufficiently close to experience the same ambient environmental variation; hence, fecundities and survival rates were assumed to be perfectly correlated between coupes. To evaluate the sensitivity of results to this assumption, simulations with uncorrelated vital rates between populations were also carried out. It was also assumed that fecundities, survivorship, and carrying capacities within a subpopulation are all correlated.

Density Dependence

T. lamproides is carnivorous and depends on a broadly dispersed prey population. There is no evidence of clumping or of overdispersion, which suggests that snails move in pursuit of their prey unimpeded by spatial or behavioral constraints. Such circumstances are very likely to result in scramble competition in which some of the resources (i.e., some of the prey) are consumed by individuals that are not successful. The model assumes scramble competition implemented with a Ricker function (Burgman et al. 1993). The population is likely to grow exponentially at low population sizes, as long as it is large enough to escape Allee effects (Allee et al. 1949). Thus, a maximum growth rate greater than 1 was assumed, which is limited at moderate to large sizes by the carrying capacity that is derived from the area and by the habitat quality of each patch. The standard simulations use a maximum growth rate of $\lambda = 1.1$. The growth rate implied by the stage matrix is $\lambda = 1.08$.

Dispersal

Dispersal rates for *T. lamproides* are based on expert opinion. It is estimated that the dispersal rates for *T. lamproides* into neighboring populations is at a maximum rate of 5% per year. It is assumed that within a year, snails will only disperse into an adjacent coupe with which they share a common boundary. The number of individuals that disperse is proportional to the length of the common boundary. Studies suggest that snails are more likely to disperse in wetter years than in drier years (Baur and Baur 1993), and a CV of 10% is used for dispersal rates to account for this.

The three areas in the study area are isolated from one another, and the model assumes that there is no effective dispersal between them. It is also assumed that effective dispersal only occurs within the boundaries of each area and not into neighboring habitats. To incorporate barriers to dispersal, stream reserves are modeled as separate populations. This means that individuals can disperse into a streamside reserve within a year but cannot disperse to a patch on the other side of the stream reserve until at least the following year.

In the model, it is assumed that snails cannot distinguish the suitability of the intervening environment or of the patch into which they disperse. Dispersing individuals will migrate to another patch despite that patch having been recently logged. In the absence of a moisture gradient, snails trapped on an open flat surface like a road will simply walk randomly, often dying when exposed to the sunlight (Baur and Baur 1993).

Snails are responsive to a number of cues such as moisture gradients and light levels. A moisture gradient between a mature forest and an area that has been recently harvested is possible, but this will only occur when the weather is dry. Snails are also sensitive to light levels, so a snail dispersing during the daytime (which they will do in relatively gloomy and rainy conditions) is likely to avoid a wide open space. Another factor is that *T. lamproides* often hunt other snails by following their trails for some distance. If the population of the prey is lower in the adjacent coupe, then it is more likely that the snail following a prey trail would remain within its own coupe than disperse. These factors suggest that when a forest is young a snail will be less likely to disperse into it.

To model the possibility of snails responding to cues such as light and moisture gradients, simulations were compared by incorporating the dispersal rates as a function of the habitat quality of the target patch. When the carrying capacity of the target patch reduces to 50% of the maximum carrying capacity, then the dispersal rate is altered according to the function in Figure 10.2. (See RAMAS GIS Metapop subprogram, Populations dialog box).

Using this function, it is assumed that when the carrying capacity is 0 in the adjacent patch, then there will be no dispersal at all into the patch. This is unlikely for snails in the specific case of a logged coupe in the way that it may be for other more mobile species. Dispersal is an uncertain parameter, and the simulations presented here compare two somewhat extreme cases: one in which the dispersal rate is kept constant at a maximum of 5% per year and one in which dispersal rate changes as a function of the carrying capacity of the target population.

Habitat Dynamics (Change in Carrying Capacity)

Each coupe has an associated management option, which has a direct impact on snails. The options include either logging of native forest followed by regenerative burning, or clearing native forest for the development of eucalypt plantations that will eventually be logged and replanted. There is also the option of not logging at all, which is only relevant to patches of habitat that make up formal and informal reserves. Although logging and fire activities eliminate individuals that experience direct disturbance, many snails will survive in streamside reserves; for this reason, 100-m wide wildlife habitat strips are

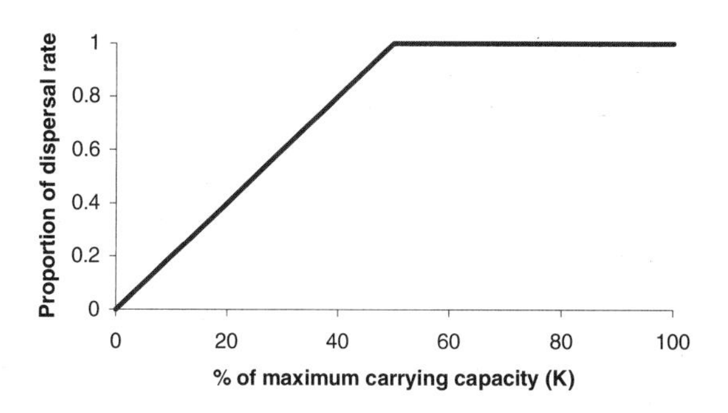

Figure 10.2 Dispersal as a function of the carrying capacity of the target population (adapted from Akçakaya 2002).

maintained between coupes. The recovery of the carrying capacity for a harvested coupe was determined by the following equation and is illustrated graphically in Figure 10.3:

$$K(t)=\begin{cases} K_{\text{init}} & \text{for } t < t_{\text{harv1}} \quad \text{(2a)} \\ 0 & \text{for } t_{\text{harv1}} \leq t \leq t_{\text{harv1}}+20 \quad \text{(2b)} \\ K_{\text{init}}\left(1-\exp\left(-\left(\frac{t-20}{20}\right)^2\right)\right) & \text{for } t_{\text{harv1}}+20 < t \leq t_{\text{harv2}} \quad \text{(2c)} \\ 0 & \text{for } t_{\text{harv2}} < t < t_{\text{harv2}}+20 \quad \text{(2d)} \\ K\left(t_{\text{harv2}}-1\right)\left(1-\exp\left(-\left(\frac{t-\left(20+t_{\text{harv2}}\right)}{20}\right)^2\right)\right) & \text{for } t \geq t_{\text{harv2}}+20 \quad \text{(2e)} \end{cases}$$

where $K(t)$ = carrying capacity at time t, K_{init} = initial carrying capacity, t_{harv1} = time at which first harvesting event occurs, t_{harv2} = time at which second harvesting event occurs, and $K(t_{\text{harv2}} - 1)$ = carrying capacity in the year before second harvest.

The initial carrying capacity is K_{init} until a harvest year where it reduces to 0. It remains at 0 for 20 years and then increases at a recovery rate according to equation (2c) until it is harvested again 65 years later. Once a coupe is logged and converted to plantation, its carrying capacity remains at 0 for the duration of the simulation.

Simulations

Various scenarios were implemented to investigate the impacts of different levels of disturbance on the viability of the species. Model assumptions were also tested to determine the sensitivity of the model to uncertain parameter estimates. The scenarios were run over two time frames: 100-year and 10-year time horizons, each using 1,000 replications. The 10-year time frame was implemented to investigate the impact of harvesting activities in the short term. The 100-year time horizon was assessed to investigate the effects of different management scenarios and the ability of snails to reinvade harvested areas in the long term.

Four main scenarios were investigated (Table 10.2). Scenario 1 involves no harvesting; this enables an analysis of the dynamics of snail populations in the absence of

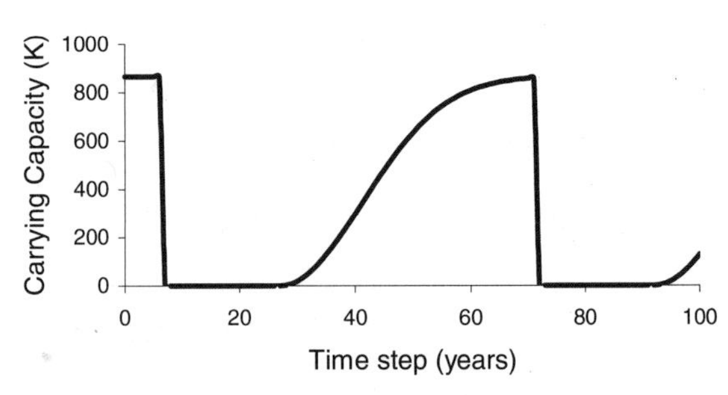

Figure 10.3 Example of the carrying capacity function used in the simulations. Harvesting and regeneration occur at year 7 and then again at year 72 (adapted from Regan et al. 2001).

Table 10.2 Harvesting scenarios

1. No harvesting
2. Harvesting for native forest regeneration only
3. Inclusion of a biodiversity spine
4. Harvesting for plantation conversion only

anthroprogenic disturbance. Three other scenarios introduce disturbance at varying levels of intensity. Scenario 2 involves logging coupes followed by burning to promote regeneration of native forest. This treatment involves the creation of large gaps through the complete removal of the overstory. Burning of the site is then conducted to promote regeneration across the site.

Scenario 3 is the implementation of a biodiversity spine, a connected set of coupes that are at least one coupe wide that runs roughly down the center of both Togari Forest Block and part of the Christmas Hills Forest Block (refer to Figure 10.1). Coupes within the biodiversity spine are only logged for native forest regeneration purposes, whereas coupes outside the spine are converted to plantation if the soils are suitable. Plantation treatment involves clearing native forest patches for plantations that eventually will be harvested and replanted or otherwise regenerated. In this case, harvesting may not be followed by broadcast burning but by windrowing and furrowing instead.

The fourth scenario is somewhat extreme in that all coupes are converted to plantation. Only coupes in which the soil structure is inappropriate for plantation are logged for regeneration purposes.

The simulation period begins in the year 2001. It was assumed that the initial population size was at carrying capacity and that production coupes follow a 65-year harvesting rotation. The year in which each coupe is logged was determined by one of many possible proposed logging strategies set out by Forestry Tasmania. Initial population size under the no-harvesting scenario is 176,540 individuals. The initial population abundance under the three harvesting scenarios is 148,127 individuals. This smaller initial abundance for the harvesting scenarios is due to harvesting prior to the beginning of the simulation period (i.e., before 2001).

Scenarios are compared in terms of either the risk of quasi extinction and expected minimum population (EMP) size. Quasi-extinction risk is the probability that the metapopulation will fall below a specified population threshold within the simulated time horizon (Ginzburg et al. 1982). The EMP is the smallest population that is expected to occur within a particular time period; it is calculated by averaging all the minimum values for the 1,000 trajectories in a simulation (McCarthy and Thompson 2001).

Results

Effects of Harvesting

Under all scenarios and disturbance intensities, the model predicts that there is no chance that *T. lamproides* will become extinct from any of the three areas modeled within the next 100 years. Even in the extreme case where all coupes are converted to plantation, the risk of extinction was 0. This is mostly due to retention of habitat within formal reserves such as the Bond Tier reserves, as well as informal reserves such as streamside buffers and wildlife habitat strips.

Figure 10.4 provides a comparison of the population dynamics of *T. lamproides* in terms of quasi-extinction risk in the presence and absence of harvesting. The probability that the population will reduce to half its size (88,270 individuals) at least once over the next 100 years, even in the absence of harvesting, is almost 1.0. Once harvesting is introduced, the risk increases. The EMP under no harvesting is 37,345 individuals, or 21% of the initial population size. When harvesting is introduced, the EMP reduces to 10,773 individuals, or approximately 7.3% of the initial population size.

A more useful comparison for forest managers is the difference in the risk curves under various harvesting intensities. Figure 10.5 illustrates the risk curves for the three harvesting scenarios. The plantation scenario and the biodiversity spine scenario are similar in their quasi-extinction risks, with risks slightly lower in the biodiversity spine scenario. Under all disturbance intensities, the EMP as a percentage of the initial population size is 6.7%, 7.3%, and 8.3% for the plantation, biodiversity spine, and regeneration scenarios, respectively.

Sensitivity Analysis

Model sensitivities were investigated using the Sensitivity Analysis in RAMAS GIS. Sensitivities to maximum growth rates, mean vital rates, standard deviations in vital rates, dispersal rates, and correlations were all investigated. This aspect of the modeling exercise provides the most useful insight into the processes affecting the persistence of *T. lamproides* at both time scales.

Vital rates, standard deviations of vital rates, and dispersal rates were all altered by ±10%, while the maximum growth rate was altered by ±5%. These were done over both the 100-year and 10-year time horizons. Sensitivities are expressed as the proportional change in the EMP resulting from a proportional change (indicated in Table 10.3 in parentheses) in the model parameters. The results of the sensitivity analysis for the 100-year time frame appear in Table 10.3. For the 100-year time horizon, the most sensitive parameters that were tested were the maximum growth rate and the standard deviations of the vital rates. A 5% change in the maximum growth rate produced a proportional change in the EMP of 0.25 and 0.37, depending on the direction of the change. The model was also sensitive to changes in the standard deviations of the vital rates as a 10% change produced a proportional change in the EMP of 0.19 and 0.17 in each direction. The model was not sensitive to changes in dispersal rates and estimates of mean vital rates.

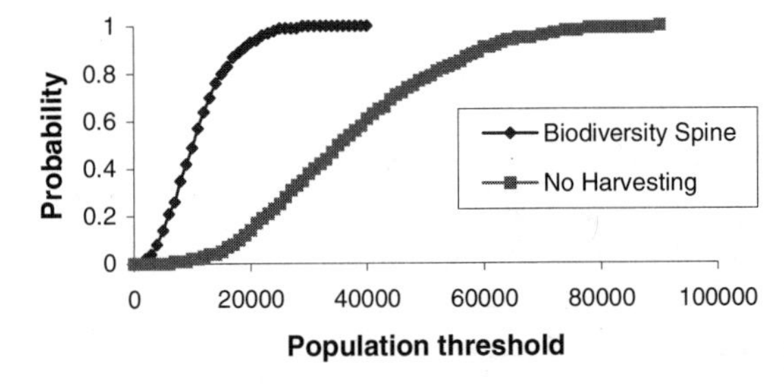

Figure 10.4 Risk curves for 100-year simulations for two scenarios: no harvesting and biodiversity spine.

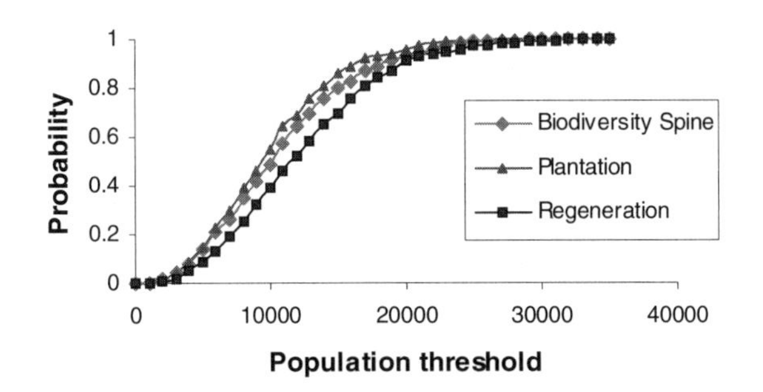

Figure 10.5 Risk curves for 100-year simulations for three scenarios: biodiversity spine, plantation, and regeneration. Initial population size was 148,127 individuals.

Correlations of vital rates were unknown and assumed to be perfectly correlated due to the close proximity of populations. It is likely that the correlations are positive, but the actual figures are uncertain. The base model has been compared with the situation when the correlations are 0. The EMP, when the vital rates are uncorrelated, is 112,877 individuals, or 64% of the initial abundance compared to 21% when the vital rates are perfectly correlated. The actual correlations between the populations (assuming they are positive) will produce results somewhere within these bounds.

Simulations comparing two different assumptions about dispersal appear in Figure 10.6. The scenario is the biodiversity spine scenario with dispersal rates constant at 5% per year. This is compared to the situation in which habitat quality is reduced and individuals respond to cues such as light and moisture gradients. In this latter situation, the dispersal rate into the target patch is reduced linearly once the carrying capacity reaches 50% of the maximum carrying capacity of the target patch. The EMP as a percentage of the initial abundance is 7.3% and 8.5% for the two scenarios, respectively. This comparison represents two somewhat extreme cases of dispersal, and the actual dispersal dynamic for *T. lamproides* is likely to be somewhere within these bounds.

Sensitivity analyses of various assumptions over a 10-year time horizon appear in Table 10.4. It indicates the EMP as a proportion of the initial abundance for both the no

Table 10.3 Sensitivity analysis for the 100-year time frame using the minimum expected population (EMP) size as a proportion of the initial population size

	No Harvesting	Sensitivity
Base model	0.21	
Changes relative to the base model		
Max growth rate (+5%)	0.26	0.25
Max growth rate (–5%)	0.13	0.37
SD of vital rates (+10%)	0.17	0.19
SD of vital rates (–10%)	0.25	0.17
Vital rates (+10%)	0.20	0.04
Vital rates (–10%)	0.21	0.01
Dispersal rates (+10%)	0.21	0.02
Dispersal rates (–10%)	0.21	0.02
Correlations = 0	0.64	2.02

Note: Sensitivity is calculated as the proportional change in the EMP resulting from a proportional change in model parameters (indicated in parentheses).

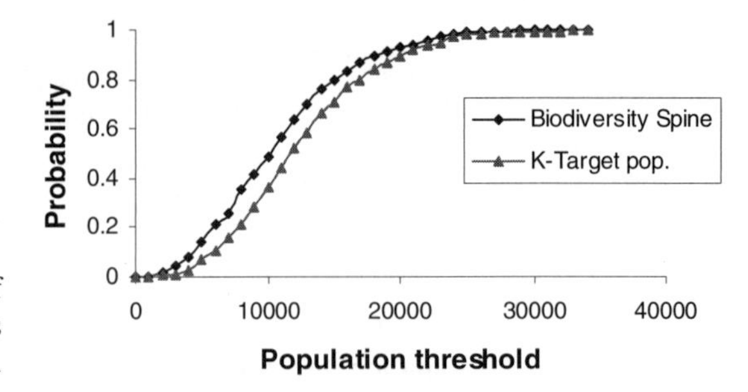

Figure 10.6 Comparison of risk curve for two assumptions about dispersal.

harvesting and the biodiversity spine scenarios. Due to background variation alone under the assumptions of the base model, the EMP at some stage within the next 10 years is 56% of the initial population size even in the absence of disturbance. Once harvesting is introduced the declines increase with the EMP decreasing to 48% of the initial abundance under the biodiversity spine scenario.

In the short term, the model is insensitive to changes in model parameters for both the no-harvesting and the biodiversity spine scenarios. Even though growth rate and standard deviations of vital rates have large effects over the long-term simulations, there is little variation in the outcomes over the short term. Only when there was an extreme change in the parameter (i.e., correlations reduced from 1 to 0), was there a significant change in the EMP. This suggests that the intensity of harvesting in the short term is having the greatest effect on the population dynamics of *T. lamproides*.

Discussion

Current knowledge of the species under model assumptions suggests that there is no risk of extinction of *T. lamproides* in the foreseeable future. The ability of the snail to

Table 10.4 Sensitivity analysis for the 10-year time frame using the minimum expected population (EMP) size as a proportion of the initial population size

	No Harvesting	Sensitivity	Biodiversity Spine	Sensitivity
Base model	0.56		0.48	
Changes relative to the base model				
Max Growth rate (+5%)	0.55	0.01	0.48	0.01
Max Growth rate (–5%)	0.56	0.01	0.49	0.02
SD* of vital rates (+10%)	0.50	0.10	0.44	0.07
SD of vital rates (–10%)	0.58	0.05	0.50	0.06
Vital rates (+10%)	0.55	0.01	0.49	0.02
Vital rates (–10%)	0.56	0.01	0.47	0.02
Dispersal rates (+10%)	0.55	0.02	0.48	0.01
Dispersal rates (–10%)	0.56	0.01	0.48	0.01
Correlations (–10%)	0.57	0.03	0.49	0.02
Correlations = 0	0.85	0.53	0.68	0.44

Note: Sensitivity is calculated as the proportional change in the EMP resulting from a proportional change in model parameters (indicated in parentheses).

reinvade previously disturbed areas and to survive in streamside reserves and wildlife habitat strips allows the snail to persist in the presence of disturbance. However, the expected minimum population size at some stage within the next 10 years is only 56% of the initial population size even in the absence of disturbance, due to the natural tendency of the population to fluctuate. Once harvesting scenarios are implemented, the expected minimum population size will be further reduced.

Sensitivity analyses were done at two time scales. This highlighted some interesting aspects regarding the factors affecting the persistence of the snail at these two scales. In the long term, the model is sensitive to changes in the maximum growth rate and the variation of the vital rates, yet changes in dispersal rates and mean vital rates were less important. Additionally, assumptions about dispersal and the correlations between populations need to be investigated further as there are large differences depending on the assumption. In the short term, however, these aspects become less significant. The magnitude of harvesting over a 10–year period had the most effect on the population dynamics in the short term. This suggests that the persistence of *T. lamproides* is largely dependent on the availability and quality of suitable habitat, in particular, the magnitude of harvesting within 10-year intervals.

The area modeled accounts for approximately 35% of the total available habitat for *T. lamproides*, and the results suggest that decisions about the intensity of logging over the short term are crucial in the management and conservation of the species. This is especially significant since the model is not sensitive to other parameter changes within this time frame.

Various assumptions were made about *T. lamproides* populations. It is uncertain whether individuals can successfully reinvade plantation areas. The model assumes that they cannot. In addition, coupe boundaries are provisional, and the actual harvested areas can often be much smaller than those modeled due to operational and environmental constraints. This was not accounted for in the model as these areas are quite variable, and it is uncertain whether the remaining area provides suitable habitat for the species. Additionally, various assumptions regarding the dynamics of *T. lamproides* populations are based on studies from other snail species. Use of such findings is provisional and has generally only been used when they apply to *T. lamproides* or when the differences between the species are unlikely to affect model outcomes.

The benefit of this model is that forest managers can now use it to investigate alternative harvesting strategies, especially spatial and temporal options, in order to meet similar production needs over the long term but to reduce declines in the short term. The model provides a benchmark for further development and discussion of management of *T. lamproides*. It provides a transparent and logical tool for managers to make more informed decisions about conservation of the species and wood production objectives. Once research into snail population dynamics comes to hand, the model can be updated to reflect the new knowledge. Key research areas for *T. lamproides* should be directed toward information regarding the growth rate and the variation of the vital rates. Other research areas include investigating the assumptions used for dispersal and correlations of vital rates between populations in order to narrow the bounds of uncertainty.

Acknowledgments We would like to thank Julian Fox from the University of Melbourne for his help with the GIS data and map. We furthermore wish to acknowledge Mark Burgman and

Mick McCarthy from the University of Melbourne for their comments on earlier drafts, as well as Penny Wells and Mick Brown from Forestry Tasmania for providing the GIS information and for their support of this study.

References

Akçakaya, H. R. 2002. *RAMAS GIS: linking landscape data with population viability analysis.* Applied Biomathematics, Setauket, N.Y.

Allee, W. C., Emerson, A. E., Park, O., Park, T., and Schmidt, K. P. 1949. *Principles of animal ecology.* Saunders, Philadelphia.

Baur, B. 1992. *Cannibalism in gastropods.* Oxford University Press, Oxford.

Baur, B. 1993. Intraclutch egg cannibalism by hatchlings on the land snail *Arianta arbustorum*: non random consumption of eggs. *Ethology Ecology and Evolution.* 5(3): 329–336.

Baur, B. 1994. Inter-population differences in propensity for egg cannibalism in hatchlings of the land snail *Arianta arbustorum. Animal Behaviour* 48(4): 851–860.

Baur, A., and Baur, B. 1989. Are roads barriers to dispersal in the land snail *Arianta arbustorum*? *Canadian Journal of Zoology* 68: 613–617.

Baur, A., and Baur, B. 1993. Daily movement patterns and dispersal in the land snail *Arianta arbustorum. Malacologia* 35(1): 89–98.

Bonham, K., and Taylor, R. J. 1997. Distribution and habitat of the land snail *Tasmaphena lamproides* (Pulmonata: Rhytididae) in Tasmania. *Molluscan Research* 18: 1–10.

Burgman, M. A., Ferson, S., and Akcakaya, H. R. 1993. *Risk assessment in conservation biology.* Chapman and Hall, London.

Chen, X. F., and Baur, B. 1993. The effect of multiple mating on female reproductive success in the simultaneously hermaphroditic land snail *Arianta Arbustorum. Canadian Journal of Zoology* 71(12): 2431–2436.

Desbuquois, C., and Madec, L. 1998. Within-clutch egg cannibalism variabilty in hatchlings of the land snail *Helix aspersa* (Pulmonata, Stylommatophora): influence of two proximate factors. *Malacologia* 39(1–2): 167–173.

Ginzburg, L. R., Slobodkin, L. B., Johnson, K., and Bindman, A. G. 1982. Quasiextinction probabilities as a measure of impact on population growth. *Risk Analysis* 21: 171–181.

McCarthy, M. A., and Thompson, C. 2001. Expected minimum population size as a measure of threat. *Animal Conservation* 4: 351–355.

Pfenninger, M., Bahl, A., and Streit, B. 1996. Isolation by distance in a population of a small land snail *Trochoidea geyeri*: evidence from direct and indirect methods. *Proceedings of the Royal Society of London Series B: Biological Sciences* 263(1374): 1211–1217.

Regan, T. J., Regan, H. M., Bonham, K., Taylor, R. J., and Burgman, M. A. 2001. Modelling the impact of timber harvesting on a rare carnivorous land snail (*Tasmaphena lamproides*) in northwest Tasmania, Australia. *Ecological Modelling* 139: 253–264.

11

Bush Cricket *Metrioptera bicolor* in Sweden

Estimating Interpatch Dispersal Rates

OSKAR KINDVALL

Interpatch migration is a process of great importance for large-scale population dynamics and survival of species. It is therefore necessary to be careful about how this process is implemented into any spatially structured model that is aimed for population viability analysis (PVA). In most cases, interpatch migration is modeled by simply applying the negative exponential equation, which can be parameterized by fitting it to the inverse cumulative distribution of dispersal distances observed in mark and recapture studies (e.g., Chapter 15 in this volume). Usually the fit is very good, but occasionally other functions have been suggested to give better descriptions of the probability of migration in relation to distance. For example, the mark and recapture data of the butterfly *Hesperia comma* fitted better to the inverse-power function than to the negative exponential (Hill et al. 1996). The tradition of fitting different equations to data on migration distances has a long history, and the number of examples is tremendous. (For a vast array of examples, see Wolfenbarger 1946.)

Despite the fact that it is often possible to obtain some kind of mathematical expression that accurately summarizes the observed migration pattern in relation to interpatch distance, it is far from safe to implement them into predictive models of population dynamics. There are several aspects to consider when constructing the model: (1) What determines the probability of emigration from source patches? (2) How is immigration influenced by target patch features? (3) Are the parameters and the shape of migration distance functions severely dependent on landscape structure, rather than on actual migration ability of the organism? (4) Do the migration parameters change temporally?

In this chapter, I describe how to use RAMAS GIS to construct a spatially structured model of a bush cricket, *Metrioptera bicolor* Philippi, which occurs on seminatural grassland patches within a restricted area in southernmost Sweden. The species is classified as vulnerable (VU) in the Swedish Red List (Gärdenfors 2000). Hence, a model of this kind is desirable for conservation planning. As the interpatch migration of this species has been carefully examined (Kindvall 1995, 1999) and the aspects listed here are fairly well understood, this chapter focuses on how to implement the migration part of the model. I compare simulations based on four alternative approaches of how to parameterize the migration section of the model. The predictive power of these alternatives are interpreted here in relation to empirical observations of local population sizes. Then the consequences of choosing wrong models are discussed in relation to long-term predictions of metapopulation occupancy.

Methods

Species

Metrioptera bicolor (Orthoptera: Tettigoniidae) is a medium-sized (12–19 mm) bush cricket (katydid), typically with short wings that do not permit flight. Long-winged specimens with flight ability occasionally occur in the populations, but to what extent these individuals influence large-scale population dynamics is not known. During the years intensive studies have been conducted, the numbers have always been too low to be significant. However, there is a possibility that certain weather conditions may trigger the frequency of long-winged morphs, and a large-scale expansion of the species distribution may occur in the future.

The species reproduce once a year (July–September), and during this period males continuously stridulate to attract females acoustically. Females then insert their eggs in grass stems, and the eggs hibernate. Eggs hatch either the next spring (early May) or, alternatively, the spring one year later (Ingrisch 1986). In this way, the species has overlapping generations, although the adults only occur at discrete annual intervals. The development from egg to adult includes six larval (nymphs) stages.

In Sweden, the species is restricted to sun-exposed seminatural grassland on sandy soils (Kindvall and Ahlén 1992). The grassland should not be too heavily grazed; if it is, preferred substrates for oviposition and feeding (mainly pollen and seeds) disappear, and the habitat becomes unsuitable. The main threat to the species is forestation of the grassland remnants that are still occurring in the landscape today. In the eighteenth century the habitat of the distribution area of *M. bicolor* was completely open; today, the remaining open grasslands are fragmented into 116 separate patches, separated by various dispersal barriers, such as pine forest, roads, and arable land (Figure 11.1). The species avoids moving into these matrix habitats (Kindvall 1999). Actually, interpatch migration occurs so rarely, even between patches separated by just a 10-m-wide road, that direct observation of it is hard to obtain empirically. However, indirectly, migration of this kind has been documented by observing colonizations on previously vacant habitat patches.

The large-scale dynamics of *M. bicolor* can be described as a metapopulation, where the number of recolonizations compensates for the extinctions that mainly occur locally due to apparently stochastic reasons (Kindvall and Ahlén 1992). The colonization probability is related to isolation from source patches, and the local extinction probability correlates with both patch size and habitat heterogeneity (Kindvall 1996)—that is, patches

Figure 11.1 Distribution area of the bush cricket *Metrioptera bicolor* in Sweden during 1989–1994 (116 patches). The black areas show the distribution of available habitat patches.

with more types of vegetation structures harbor more persistent populations than do patches with only a single type of vegetation.

Data

The Swedish population of *M. bicolor* has been studied intensively during the years 1989 to 1994 (Kindvall 1995). Local population size has been estimated annually on 40 separate habitat patches. These estimates are total counts of stridulating males observed within a patch. The method, which is described in more detail elsewhere (e.g., Kindvall and Ahlén 1992), is expected to give rather accurate approximations of population size.

In 1989, the local population size was estimated on all habitat patches (n = 116) occurring in the distribution area of *M. bicolor*. Occupancy data were sampled for all patches in 1985 and every year between 1989 and 1994. From these data it is possible to quantify turnover events—local extinctions and colonizations. Observations of colonization are used here to obtain information on interpatch migration distances.

Movement behavior of *M. bicolor* has been studied in both natural populations and experimentally constructed populations (Kindvall 1995). An individual-based model of movements in a heterogeneous landscape has been developed and successfully corroborated by means of independent field experiments (Kindvall 1999).

Parameterization

A demographic metapopulation model of *M. bicolor* was constructed using RAMAS GIS. To incorporate spatial structure, a raster map of the distribution area was used.

Using the Landscape Data program, patch configuration and different types of interpatch distance measures were obtained from the raster map.

Local population dynamics were modeled according to the exponential equation with a ceiling (K). No stage structure was assumed, although overlapping generations may occur via the egg stage. However, as the transition probabilities between different stages in the life cycle of *M. bicolor* are uncertain, due to limited amount of data, I preferred to keep the model simple. Besides, an analysis of possible density dependence on growth rate suggested that the annual life cycle was dominating the pattern (Kindvall 1995).

To obtain an estimate of the average population growth rate (R) and its standard deviation, a lognormal distribution was fitted (Kolmogorov: Max. deviation = 0.042; $p = 0.88$; $n = 194$) to the frequency distribution of all R_t-values ($= N_{t+1}/N_t$) obtained from the 40 population trajectories (Table 11.1). The mean ± SD was 2.12 ± 3.10. These values were inserted into the stage matrix and the standard deviation matrix, respectively.

Patch-specific estimates of population ceiling (K) was estimated as the median of maximum population density observed in the 40 investigated local populations (85 ha^{-1}) multiplied by the patch area, which was obtained from the raster map using the RAMAS Spatial Data subprogram. The number of males observed in 1989 was used as the initial population size (N_0).

Most local populations of *M. bicolor* appear to fluctuate independently of each other. Cross-correlation analyses between trajectories of local populations vary from –1 to 1. There was no distance effect within the distribution area on the degree of synchrony (Kindvall 1996). The median correlation coefficient was 0.19 ($n = 780$). To include synchrony in the model of *M. bicolor,* I used the correlation distance function and set parameter a to 0.19, b to 107, and c to 0. Thus I assumed that all populations were equally correlated with any other population over time, independent of its location.

Interpatch migration was modeled in four alternative ways: (1) according to an individual-based model; (2) as a function of center-to-center distances; (3) as a function of edge-to-edge distances; and (4) as a function of center-to-edge distances. An individual-based simulation model of movements was developed to enable realistic predictions of per capita transition rates in heterogeneous landscapes, defined as a raster map (Kindvall 1999). This model, which has been parameterized for *M. bicolor,* was able to accurately predict net displacements of individuals, released as young nymphs and recaptured as adults, in four independent field experiments. As the investigated landscapes differed considerably in terms of spatial arrangement of habitat patches, it is evident that the model enables predictions of interpatch migration rates, which are not obviously dependent on the landscape where parameterization was conducted.

The individual-based model of animal movements consists of two equations that can be iterated stochastically to obtain time-step-specific spatial coordinates, X_t and Y_t, describing an individual's movement path, starting from any initial location (X_0, Y_0):

$$X_t = X_0 + \sum_{t=1}^{t} d_t \cdot \cos(\nu_t)$$
$$Y_t = Y_0 + \sum_{t=1}^{t} d_t \cdot \sin(\nu_t) \qquad (1)$$

The model contains two stochastic parameters, d_t and v_t, which reflect an individual's variation in movement step length (d) and turning angles (v) between successive moves.

Table 11.1 Number of *Metriotera bicolor* males on 40 separate habitat patches observed annually between 1989 and 1994

ID	Area (ha)	N_{89}	N_{90}	N_{91}	N_{92}	N_{93}	N_{94}
7	2.0	28	31	79	5	101	136
22	0.7	13	30	13	0	34	41
31	1.4	7	5	7	2	4	4
39	0.3	6	15	15	15	13	32
46	0.4	10	25	19	39	4	13
57	2.2	102	246	532	122	167	463
58	3.1	11	75	126	77	35	357
59	0.2	0	7	1	1	2	7
60	1.1	22	85	90	32	27	51
64	0.1	2	11	25	8	10	10
65	0.9	14	70	10	12	22	108
66	0.7	3	119	75	69	32	35
67	3.2	66	176	95	108	0	0
71	3.7	83	133	241	142	150	324
73	1.7	34	264	139	30	36	23
74	2.1	27	60	126	64	34	144
76	0.5	4	3	9	8	26	23
80	1.6	25	79	128	46	41	154
81	0.3	1	2	2	9	5	0
82	0.4	9	9	18	7	2	2
83	0.5	7	9	14	7	2	0
84	0.7	10	6	3	1	5	25
87	0.5	2	0	19	0	36	2
88	0.3	29	47	20	25	16	42
90	0.8	0	2	13	9	4	0
91	0.8	20	41	17	12	7	73
92	1.3	42	246	245	45	78	320
93	5.3	292	210	497	133	470	116
94	0.5	14	20	35	26	14	14
95	2.6	6	27	18	55	51	75
96	1.0	40	44	101	36	46	153
97	0.3	8	9	15	10	4	3
101	0.9	19	13	48	4	46	14
103	0.1	1	5	2	0	0	0
105	0.7	13	49	94	49	49	125
108	0.6	8	45	81	65	69	131
109	0.1	7	0	0	5	16	7
110	0.3	13	15	32	5	4	22
111	2.1	40	254	123	170	109	222
113	0.1	1	0	1	2	3	0

Note: ID refers to the population index in the RAMAS input files. Patch sizes (hectares) were obtained from raster maps using the RAMAS Landscape Data program.

In the model for *M. bicolor,* turning angles were obtained from a uniform distribution (0°–360°), assuming no directionality or correlation in the movement path. Movement step lengths were obtained from a lognormal distribution. The mean and standard deviation of this distribution were dependent on habitat type: that is, within habitat patches, *M bicolor* moves on average 3.2 ± 13.7 m per day; within matrix types of habitat, the corresponding value is 49.3 ± 31.1 m (Kindvall 1999). Thus, for each time step, the model has to use the raster map to check current habitat type in order to properly adjust movement step length. The model also includes the effects of habitat edges. If the simulated individual happens to cross an edge toward an unsuitable matrix habitat, it is not certain that the individual will proceed. On the contrary, individuals of *M. bicolor* do only rarely proceed across such an edge ($P_{out} = 0.11$).

To obtain values for all interpatch migration rates in the Swedish distribution area of *M. bicolor,* according to the individual-based model of animal movements, 100 or more specimens were released uniformly within each separate habitat patch ($n = 116$). The model was iterated 150 time steps (corresponds roughly to the life span of an individual). The per capita migration rate (m_{ij}) from patch i to j was estimated directly from the simulation results. All emigrating individuals did not enter any new recipient patch but ended up somewhere in the matrix at the end of the simulation. In the RAMAS model, this fraction of individuals was directed into an imaginary "population," which I have called "Sink." This population was constructed so that the population ceiling was set to 0 and was excluded from the summary statistics.

The reminding three alternatives to model migration were mainly based on the observed colonization events and associated frequency distribution of interpatch distances. For each of the colonizations ($n = 22$), the distance (edge-to-edge) to the nearest patch occupied by *M. bicolor* was measured. The negative exponential equation was fitted to the inverse cumulative frequency distribution of observed distances using a nonlinear regression analysis (Figure 11.2). The best-fitted model ($R^2 = 92.7\%$) suggested that the average migration distance in the current distribution area of *M. bicolor* was 0.11 km.

The RAMAS Spatial Data subprogram was used to generate three different migration models, setting the migration distance function parameter a to 1, b to 0.11, and c to 1; D_{max} was set to 10^7. Then, this function was applied for all interpatch distances, measured by the program as "center-to-center," "edge-to-edge," and "center-to-edge" distances.

The resulting emigration rate from all patches was calculated for the three migration matrixes. In all cases, the maximum emigration rate was much higher than possible—that is, >1.0. For the center-to-center model, maximum emigration was 1.8; for the edge-

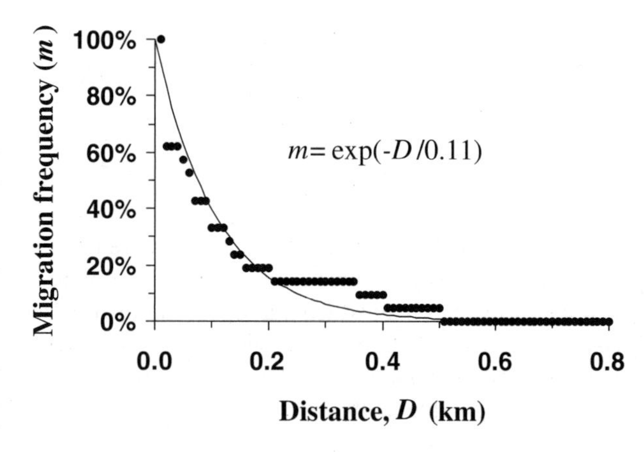

Figure 11.2 Inverse cumulative distribution function (*dots*) of interpatch migration distances inferred from observed colonization events. A negative exponential function describes the observed pattern apparently well ($R^2 = 92.7\%$).

to-edge model, it was 5.92; and for the center-to-edge model, it was 3.04. As the highest emigration rate observed in the individual-based simulations of *M. bicolor* movements was 0.82, I recalculated all the interpatch migration distances by changing the value of parameter *a* for the tree models to 0.456, 0.138, and 0.270, respectively. These figures were calculated as the maximum observed divided by the maximum achieved: for example, the center-to-center model was 0.82/1.8 (0.456.

Simulations

Two series of simulations (1,000 replicates) were run with the four alternative models of *M. bicolor*. In the first set of simulations, each model was simulated for 5 years, starting from the initial conditions observed in 1989. Thus the metapopulation program could calculate the average local population size expected in 1994, which could be compared with observed sizes in 1994. In the other set of simulations, the long-term (100 years) changes in metapopulation occupancy were recorded.

Results

Short-Term Predictions and Observed Patterns

There is a clear positive relationship between local population size (number of males) and habitat patch area, according to what is observed on a subset of the habitat patches suitable for *Metrioptera bicolor* in southern Sweden in 1994 (Figure 11.3). The linear regression model, with a slope of 1.07, explains 34% of the variation (field data; Table 11.2).

Simulations were conducted in order to predict local population sizes of the 40 patches included in the field data set shown here. The four alternative models generated different predictions of the average local population size after 5 years based on 1,000 replicates. These differences become apparent by investigating the relationship between the population size and the area of the habitat patch. Results from linear regres-

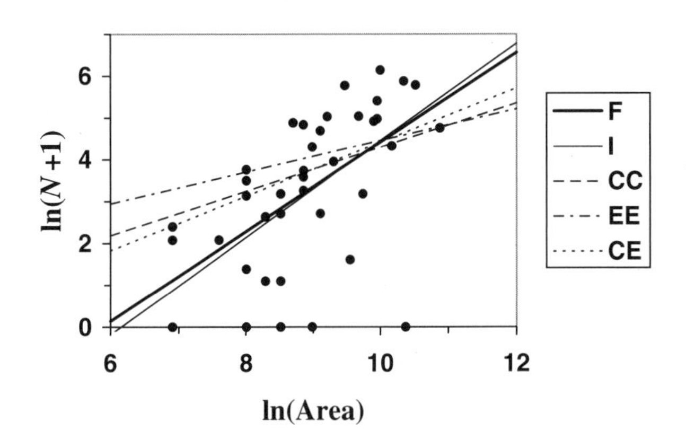

Figure 11.3 Local population size (number of adult males) in relation to patch area (m²). Data observed in 1994 (Table 11.1) are represented by dots. Regression lines from five separate analyses (Table 11.2) are shown for comparison: F = field data; I = individual based model; CC = center-to-center; EE = edge-to-edge; CE = center-to-edge.

Table 11.2 Results from linear regression analysis of the relationship between local population size (N_{1994}+1) and patch area (m^2) according to four alternative simulation models and field data (see Table 11.1)

Data	Slope	P_s	Intercept	P_i	R^2 (%)
Individual-based model	1.16	0.70	−7.17	0.93	92.3
Center-to-center	0.53	<0.05	−1.00	<0.05	66.9
Edge-to-edge	0.38	<0.01	0.66	<0.01	39.1
Center-to-edge	0.65	0.094	-2.08	0.067	78.6
Field data	1.07	—	-6.28	—	34.0

Note: The analyses are based on the same set of 40 habitat patches. Both variables were log-transformed. All regression models were highly significant ($p < 0.0001$). I tested whether estimated slopes and intercepts differed from the field data regression model. A low value of P_s or P_i suggests that estimates of slopes or intercepts are significantly different.

sion analyses are presented in Table 11.2. The prediction made by the simulation model that was based on an interpatch migration matrix, parameterized from the individual-based model of movements, was most similar to field data: both the slope and the intercept of the regression model were statistically indistinguishable from the field data model (Table 11.2).

Visually, all other simulation models differed from the observed data set with regard to the relationship between local population size and patch area (Figure 11.3). The general difference was that the intercepts were higher and the slopes lower than actually observed. Two of these models (center-to-center and edge-to-edge) differed significantly, while the simulation model based on center-to-edge estimates of interpatch distances generated a regression model that was not significantly different from the field observations (Table 11.2).

In 1994, the number of occupied patches was 89 according to the field survey of the whole distribution area of *M. bicolor* (Kindvall 1995). The corresponding average number of occupied patches was predicted by the four alternative simulation models (Table 11.3). It was only the prediction from the model using the individual-based model of

Table 11.3 The number of occupied patches predicted after five time steps, thus corresponding to the year 1994, by four alternative models of *Metrioptera bicolor*

Model	Minimum	−1 SD	Average	+1 SD	Maximum
Observed 1989			83		
Observed 1994			89		
Individual-based model	45	84.9	91.1	97.2	106
Center-to-center	84	104.3	107.9	111.6	114
Edge-to-edge	105	111.4	112.9	114.5	116
Center-to-edge	92	105.7	109.3	112.9	116

Note: The initial number of occupied patches (observed in 1989) and the number observed in 1994 are presented for comparison. The total number of patches was 116.

animal movements that agreed with the observed value. The other models overestimated the number of occupied patches.

Long-Term Predictions

Among the four alternative simulation models of *M. bicolor,* one generated a distinctly different occupancy pattern within the forthcoming century compared to the others (Figure 11.4). The model, in which interpatch migration rates were determined from an individual-based model of movements, only predicts a slight increase in occupancy compared to initial conditions. All the other models predict that the species should occupy almost every available patch. On average, the three models, which all use some kind of interpatch migration function, predicted 1.23 times more occupied patches than the one using the individual-based model of movements.

Predictions about the probability of future declines in occupancy differed between the alternative models in a similar manner, as was the case with average occupancy. The probability that the occupancy will decrease to 50% within 100 years was 0.863, according to the individual-based model. Corresponding values for the models based on center-to-center, edge-to-edge, and center-to-edge estimates of interpatch distances were equal to 0.795, 0.794, and 0.648, respectively.

Discussion

In the study reported here, four alternative spatially structured models of *Metrioptera bicolor* were investigated. The only difference between these models was the way interpatch migration was implemented. Three of the models were parameterized simply by applying the negative exponential function of migration distances. The other model was parameterized more specifically by estimating each interpatch migration rate from an individual-based model of animal movements (i.e., Kindvall 1999). It was apparent from the results that the latter model, which is by far the most complicated and requires much more work to parameterize, generated the best prediction in terms of similarity with field observations (e.g., Figure 11.3). Although the rest of the models used the

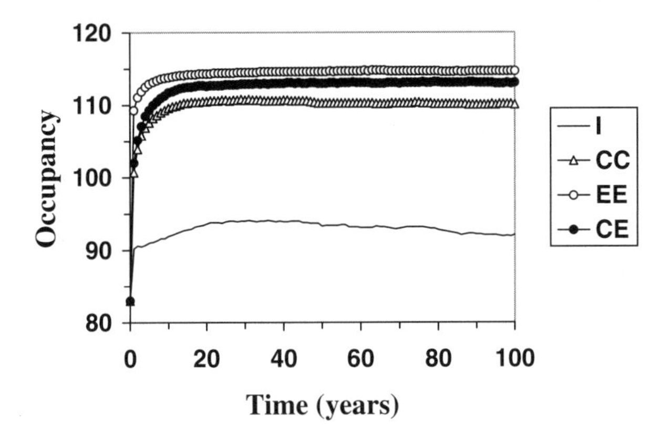

Figure 11.4. Number of occupied habitat patches within the Swedish distribution area of *Metrioptera bicolor* during 100 years, according to four alternative simulation models: I = individual-based model; CC = center-to-center; EE = edge-to-edge; and CE = center-to-edge.

negative exponential function differently, they made similar predictions, both in a short-term and a long-term perspective. These models overestimated the number of occupied patches in relation to what was observed in 1994 (Table 11.3). Furthermore, they over-estimated the number of males on relatively small habitat patches and simultaneously underestimated the number of males on larger patches (see Figure 11.3). The most complicated model provided acceptable predictions of all these summary statistics.

What makes the most complicated model a better predictor than the others? The individual-based modeling approach definitely handles a broader array of environmental factors that influence the flow of individuals between separate patches in a complex landscape than can be dealt with using only a simple function of distance. For example, the distance a migrating individual will move before it settles on a new patch can be influenced by patch configuration. In an area where patches occur densely, the average migration distance will probably be shorter than in an area where patches are situated further apart from each other. A migration model, which only uses a simple function of distance, will predict the same migration rate between two patches independently of whether other patches occur in between. The individual-based model will often predict a lower migration rate between two patches if other patches are located along the individual's movement path. Another feature that was incorporated by using the individual-based model was the effect of patch configuration on the fraction of emigrating individuals that actually succeed in migrating to a new patch. For example, emigrants that leave a patch in the periphery of the distribution area will have smaller opportunities to find a suitable new patch than will emigrants from a patch that is completely surrounded by neighboring patches. By implementing the "sink patch," it was possible to collect and omit all migrants that fail to find new patches.

The emigration rate is expected to vary in relation to patch size and shape. Several empirical studies have reported a negative relationship between the emigration rate and patch area (e.g., Pokki 1981, Kareiva 1985, Turchin 1986, Hill et al. 1996, Kuussaari et al. 1996). This effect is mechanistically generated by the individual-based model of animal movements (Kindvall 1999). When using an interpatch migration function in combination with center-to-center or edge-to-edge distances, emigration rates become independent on patch geometry. However, if center-to-edge estimates of distances are used instead, emigration rates become negatively correlated with patch size. The distance measured from the center of a larger patch to the edge of a smaller patch will be longer than the estimate from the center of the small patch to the edge of the large one. Applying the migration function to these distances will generate relatively lower migration rates from the larger patches to the smaller ones, and vice versa. Interestingly, the simulation model of *M. bicolor* that used center-to-edge distances made a better prediction of the relationship between population size and patch size than did the other two models that also used the negative exponential distance function (Figure 11.3; Table 11.2).

When assuming that migration rates are independent on patch size and that patch size varies in the simulated landscape, it is generally expected that the distribution of individuals among patches becomes increasingly shifted toward smaller patches as the average migration rate increases. Thus, smaller patches will be crowded while larger patches will be underpopulated. Another related effect of this phenomenon is that more patches become occupied (see, e.g., Figure 11.4), as the local extinction risk on the relatively smaller patches decreases. When migration is adjusted to patch geometry, as with the most complex model investigated here, the local population size will track the local carrying capacity more closely, hence producing the distribution pattern expected from ideal free distribution.

Whether or not patch geometry is taken into account when trying to predict the prospects of a species to survive may be crucial. In another simulation study (Kindvall and Petersson 2000), the extinction risk of a species, experiencing a decreasing habitat scenario, was investigated for different assumptions about interpatch migration. The extinction risk increased much faster with decreasing total amount of suitable habitat when migration was adjusted to patch geometry than if migration rate was independent of patch size.

Why use simple interpatch migration functions when modeling large-scale population dynamics of species? Historically, the use of, for example, the negative exponential distance function must have been motivated by its generally good agreement with observed interpatch migration distances. However, any model that ignores the geometric effects of migration will often be too poor when predictions of quantities are really needed. Therefore, the only acceptable reason for merely using information from frequency distributions of migration distances is that high precision of the prediction is not needed and the efforts of applying an individual-based model of animal movements is far too demanding to be justified. Obviously, it requires much more time and resources to be able to parameterize migration from individual-based modeling than to use simple functions. One possibility to gain the main features of the individual-based modeling approach without performing exhausting calculations would be to combine some kind of phenomenological model (e.g., Kindvall and Petersson 2000) describing the relationship between emigration and patch geometry with an interpatch distance function.

References

Gärdenfors, U. 2000. *The 2000 Red List of Swedish species.* ArtDatabanken, Swedish University of Agricultural Sciences, Uppsala, Sweden.

Hill, J. K., Thomas, C. D., and Lewis, O. T. 1996. Effects of habitat patch size and isolation on dispersal by *Hesperia comma* butterflies: implications for metapopulation structure. *Journal of Animal Ecology* 65: 725–735.

Ingrisch, S. 1986. The plurennial life cycles of European Tettigoniidae (Insecta: Orthoptera). 2. The effects of photoperiod on the induction of an initial diapause. *Oecologia* 70: 617–623.

Kareiva, P. 1985. Finding and losing host plants by *Phyllotreta:* patch size and surrounding habitat. *Ecology* 66: 1809–1816.

Kindvall, O. 1995. *Ecology of the bush cricket* Metrioptera bicolor *with implications for metapopulation theory and conservation.* Report 29, Department of Wildlife Ecology, Swedish University of Agricultural Sciences, Uppsala, Sweden.

Kindvall, O. 1996. Habitat heterogeneity and survival in a bush cricket metapopulation. *Ecology* 77: 207–214.

Kindvall, O. 1999. Dispersal in a metapopulation of the bush cricket, *Metrioptera bicolor* (Orthoptera: Tettigoniidae). *Journal of Animal Ecology* 68: 172–185.

Kindvall, O. 2000. Comparative precision of three spatially realistic simulation models of metapopulation dynamics. *Ecological Bulletins* 48: 101–110.

Kindvall, O., and Ahlén, I. 1992. Geometrical factors and metapopulation dynamics of the bush cricket, *Metriopotera bicolor* Philippi (Orthoptera: Tettigoniidae). *Conservation Biology* 6: 520–529.

Kindvall, O., and Petersson, A. 2000. Consequences of modelling interpatch migration as a function of patch geometry when predicting metapopulation extinction risk. *Ecological Modelling* 129: 101–109.

Kuussaari, M., Nieminen, M., and Hanski, I. 1996. An experimental study of migration in the butterfly *Melitaea cincia. Journal of Animal Ecology* 65: 791–801.

Pokki, J. 1981. Distribution, demography and dispersal of the field vole *Microtus agrestis* (L.) in the Tvärminne archipelago, Finland. *Acta Zologica Fennica* 164: 1–48.

Turchin, P. 1986. Modelling the effect of host patch size on Mexican bean beetle emigration. *Ecology* 67: 124–132.

Wolfenbarger, D. O. 1946. Dispersion of small organisms: distance dispersion rates of bacteria, spores, seeds, pollen, and insects; incidence rates of diseases and injuries. *American Midland Naturalist* 35: 1–152.

12

Puritan Tiger Beetle (*Cicindela puritana*) on the Connecticut River

Habitat Management and Translocation Alternatives

KRISTIAN SHAWN OMLAND

Puritan tiger beetles (*Cicindela puritana*) dwell on sandy beaches along large bodies of slow-moving water. As adults, they are fast and furious diurnal hunters of other arthropods, which makes them exciting insects to watch and popular (as are all tiger beetles) among entomologists; as larvae, they are treacherous sit-and-wait predators, and they also prey on other arthropods (Knisley and Schultz 1997, Leonard and Bell 1999). The species is endemic to the Connecticut River and Chesapeake Bay of eastern North America (U.S. Fish and Wildlife Service 1993); furthermore, molecular research has shown that populations in those two areas are diagnosably distinct, and therefore each is an evolutionarily significant unit in its own right (Vogler et al. 1993).

The number of populations of Puritan tiger beetles on the Connecticut River declined precipitously from no fewer than 12 around the turn of the twentieth century to exactly 2 by the late 1980s (U.S. Fish and Wildlife Service 1993). As a result, the species was listed as threatened by the U. S. federal government (U.S. Fish and Wildlife Service 1990).

Conservationists in the Connecticut River region aim to foster the recovery of the Puritan tiger beetle on the river to the point where we can view the species as a viable member of the region's biota. We seek both to sustain the numbers of individuals within the lone thriving metapopulation and to increase the number of metapopulations along the river. The recovery plan established the criterion that there must be at least three metapopulations, two of which should be large (> 500 adults), within the species' historical range on the Connecticut River before the species can be removed from threatened status (U.S. Fish and Wildlife Service 1993). While that may seem an arbitrary criterion for recovery, it should suffice to observe that the species will be more secure when that criterion is met than it is now, when there are only two metapopulations, one large (~600 adults, at river km 46–50) and one small (~30 adults, at river km 146–149). (Throughout this chapter I identify beaches [or habitat patches] and clusters of beaches [or metapopulations] by their distance from the mouth of the river, measured in kilometers.)

A number of management tactics have been proposed with the intent of approaching the goal of security. In this study, I used population viability analysis (PVA) implemented in RAMAS Metapop to prioritize those tactics by addressing a series of questions. I first considered habitat management within the extant large metapopulation at river km 46–50:

1. Which among four habitat management tactics minimizes the risk of decline in the river km 46–50 metapopulation?

I then considered translocation:

2. Can we remove individuals from river km 50 to supply a translocation program without exacerbating the risk of decline in that metapopulation?
3. How would the prospects for (a) metapopulation persistence and (b) patch occupancy respond to translocation at each of two potential recipient areas?

Finally, I considered one overarching question:

4. Which tactic minimizes the risk of regional extinction?

Metapopulation Model

Puritan tiger beetles predominantly have a biennial life cycle (Knisley and Schultz 1997, Leonard and Bell 1999, Omland 2001), thus the population is structured into larvae and adults at the time of a prebreeding census. Because both cohorts of larvae are active in the fall, the potential for interaction between larvae of the year and off-year larvae (i.e., competition, parasitoid-mediated apparent competition, or cannibalism) precludes modeling the cohorts independently. The Leslie matrix representing that biennial life cycle is as follows:

$$\mathbf{L} = \begin{bmatrix} 0 & \rho F p_1 \\ p_2 & 0 \end{bmatrix}$$

where ρ is the sex ratio, F is fecundity, and p_1 and p_2 are annual survival probabilities. In the study reported here, I assumed that the Leslie matrix applied to all subpopulations in all metapopulations. Parameter estimates were obtained by maximum-likelihood fitting of that model to annual census data.

Annual census data include only adults, so initial abundance of larvae was set at $\rho F p_1$ times observed adult abundances from 1999; initial abundance of adults was set at observed abundance from 2000.

The beach habitat of Puritan tiger beetles is intrinsically patchy. Furthermore, while adult beetles possess wings and are capable of moving among subpopulations, such movement is rare (Omland 2001). Therefore, I view the system as a metapopulation. Movement in the metapopulation is modeled with a redistribution matrix with parameters M_{ij}, the net annual probability that a beetle moves from beach i to beach j. I conducted a 2-year mark-recapture study to estimate those parameters for the river km 46–50 metapopulation. The study yielded over 400 resightings and resulted in precise (i.e., narrow confidence intervals) estimates of net annual movement probabilities among three habitat patches in that metapopulation (Omland 2001). Here I assumed that a negative exponential model fitted to those data predicts reasonably accurate redistribution matrices for other metapopulations; the model fitted to both years of data was $M_{ij} = -0.15\, e^{-0.98D_{ij}}$ (RAMAS Metapop dispersal distance function parameters: $a = 0.15$, $b = 1/{-0.98} = -1.02$, $c = 1$).

Along the Connecticut River, distances between clusters of apparently suitable habitat patches (i.e., potential metapopulations) are typically far greater than a Puritan tiger beetle is likely to fly. For example, the distance between the two extant metapopulations is 80 km, and the distance from the larger of those metapopulations to the nearest vacant but apparently suitable patch is 12 km (Omland 2001). Recognizing the dangers of extrapolating the preceding fitted model beyond the range of observed distances (maximum distance between patches in the mark-recapture study was 2.7 km), it is apparent that colonization of vacant habitat patches is unlikely, at least on the time scale of decades. By implication, augmentation of a dwindling metapopulation to take advantage of the rescue effect (Burkey 1989) or establishment of new populations (akin to recolonization in classic metapopulation dynamics; Levins 1970) will require intervention. Thus, movement of individuals among metapopulations (as opposed to within metapopulations) is shifted in this case into the realm of management (see later in this chapter).

Tiger beetle populations are typically thought to be limited by the availability of food (Pearson and Knisley 1985, Knisley and Juliano 1988, Knisley and Schultz 1997), and there is some support for inferring that density dependence has affected population growth in the river km 46–50 metapopulation during the period of population monitoring (Omland 2001). There are no data, however, that are adequate for estimating the parameters of a density-dependent growth function. Consequently, I modeled exponential growth both with and without population ceilings. At one extreme, I ran simulations without population ceilings, an approach that is incautious in the sense that some trajectories may climb to implausibly high densities from which they are unlikely to fall back to 0. An intermediate approach was to estimate the area of potential larval habitat at each beach, and then multiply it by 20 larvae m^{-2}, which corresponds to high larval density (Table 12.1). Finally, the most cautious approach—the one that confined simulated trajectories to the lowest densities—was to use population ceilings estimated from observed densities. For two of the three beaches in the river km 46–50 metapopulation, maximum likelihood fitting of a population model to the monitoring data yielded such estimates (Omland 2001). For the other beaches, including ones where Puritan tiger beetles have not been observed in recent years, I entered my best guesses based on observations about the area and quality of habitat (Table 12.1). Estimating the variability of those ceilings from empirical observations is even farther beyond reach; I ran simulations with the coefficient of variation of K equal to 0.10.

July or August floods could cause catastrophic mortality of eggs and young larvae, which are buried shallowly in the sand. At other times of the year, only larger larvae are present, and they are burrowed deeply enough to avoid being washed away and are otherwise adapted to tolerate prolonged submersion (Knisley and Schultz 1997, Hoback et al. 1998). Because only two major floods in the months of July and August were re-

Table 12.1 Estimated area and population ceilings of beaches in three metapopulations (or potential ones) designated by river km

Beach (River km)	Area (m^2)	High Ceiling	Low Ceiling
149	50	1,000	300
147	100	2,000	300
146	200	4,000	1,000
91	200	4,000	1,000
89	100	2,000	300
87	50	1,000	300
50	250	5,000	5,000*
49	50	1,000	300
47	150	3,000	2,000*
46	100	2,000	1,000*

Note: High ceilings = area × 20 larvae m^{-2}; low ceilings estimated from observed abundances (indicated by *) or best guesses (see text for explanation).

corded in a 60-year hydrograph from Middletown, Connecticut (U.S. Geological Survey 2001), I set the probability of such a flood at 0.03. No data exist about the actual impact of such a flood; here I modeled 50% mortality of first-year larvae. The frequency at which a flood was recorded at one gauge given that a flood was recorded at another gauge within 25 miles upstream or downstream was 0.95 (unpublished analysis of data from U.S. Geological Survey 2001). Therefore, in the simulation model I assumed that floods catastrophically affected all populations when they occurred.

Simulation Scenarios

I view the available tactics (Figure 12.1) as candidates to be prioritized based on predicted ecological payoff. Although resources may be available to implement more than one of the tactics simultaneously, I treated the candidates as mutually exclusive. Ultimately, one or more tactics will be selected based also partly on other criteria such as cost. The simulation scenarios are organized under the four questions given in the introduction.

1. Which among Four Habitat Management Tactics Minimizes the Risk of Decline in the River km 46–50 Metapopulation?

The beach at river km 46 is adjacent to a cultivated field, which is treated with insecticide. Insects, including the protected tiger beetles, have been observed dead or exhibiting symptoms of insecticide exposure on two occasions over the last 10 years. Following one such incident in 1998, I observed an 80% decline overnight in the number of marked Puritan tiger beetles (unpublished data). Furthermore, the number of adult beetles seen 2 years later was one-third what we expected based on abundance of the even-year cohort in 1996 and 1998. Thus under the status quo I modeled insecticide drift as a catas-

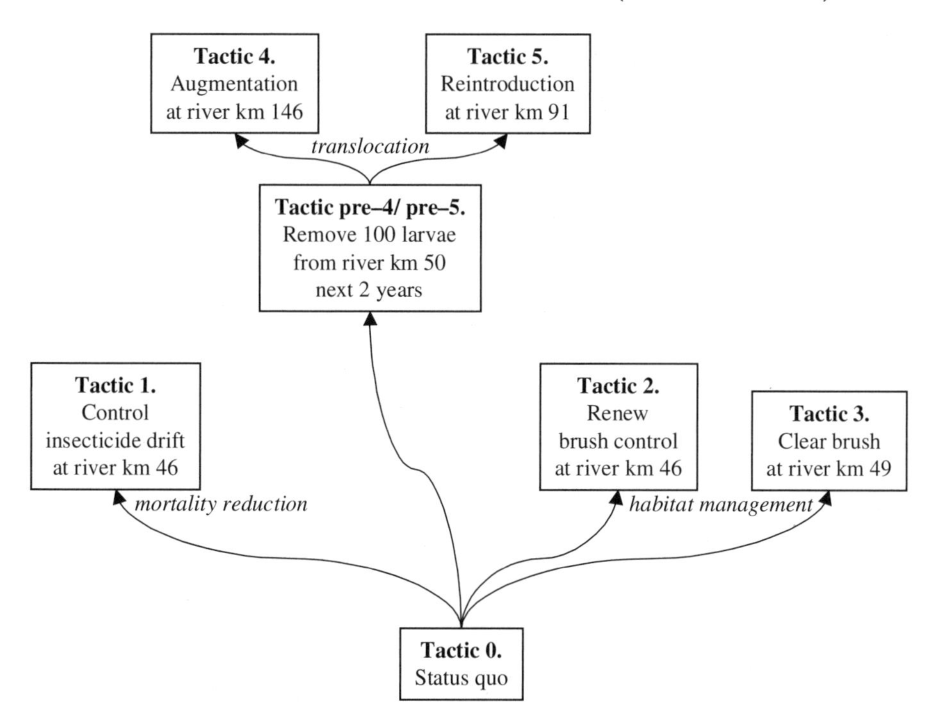

Figure 12.1 Decision tree for choosing among six candidate management tactics, including no action (tactic 0). My goal in this study was to prioritize the tactics according to the predicted chances of achieving the goals stated in the text, assuming that costs of implementation are acceptable. In question 1, I compared tactics 0–3. In question 2, I addressed whether tactic pre-4/pre-5 would exacerbate the risk of decline in the river km 46–50 metapopulation: then in question 3 whether tactic 4 or 5 would yield a greater payoff from a translocation program. Finally, in question 4, I asked which tactic among all six would maximize the chance of having two or three large metapopulations in the region.

trophe that occurs at that beach with an annual probability of 0.2 and causes 50% mortality of adult beetles (not 80% because emergence and activity are spread across the approximately 6-week flight season). Tactic 1 differs from the status quo in that catastrophic mortality due to insecticide drift would be eliminated—for example, by having an observer monitor insecticide application.

Populations of tiger beetles that live in sparsely vegetated habitats decline as vegetative succession encroaches on their habitat (Knisley and Hill 1992). Realizing that, a brush removal and control program was implemented at river km 46. Brush was cleared in 1993, and sprouting stumps were treated with herbicide again in 1994 (Lapin and Nothnagle 1995). The population appeared to respond favorably: geometric mean population size since 1995 (the first cohort to have begun as eggs after clearing) has been 2.3 times what it was prior to the brush clearing. Unfortunately, that program was not sustained after 1994, and we can expect the carrying capacity of that beach to decline as brush again encroaches. Under the status quo, I modeled a change in carrying capacity at that beach such that it will be halved over the 30 years (linear change in K, slope = ½ $K/30$). Alternatively, if we renew brush control at that beach (tactic 2), carrying capacity will remain at its current level.

Another beach in the river km 46–50 metapopulation, that at river km 49, is currently overgrown with vegetation. Small numbers of adults have been seen there over the last several summers, but, more importantly, I have also seen concentrations of larvae that are as dense as at the most populous beaches. From that observation, I infer that conditions are otherwise appropriate at that beach for Puritan tiger beetle recruitment. Vegetation removal could be implemented at the river km 49 beach with the intent of increasing its carrying capacity (tactic 3). Here I assumed that clearing brush (and maintaining brush control, as in the preceding tactic) could double carrying capacity.

It should be noted that human use and edaphic conditions make brush encroachment less problematic for tiger beetles at other beaches, so I did not include change in K at any other beaches.

The model simulated here, particularly under the assumption of exponential population growth, yields a vanishingly small risk of extinction in the river km 46–50 metapopulation over a 30-year time horizon. For comparisons to report, I selected the level of decline that occurred with about 50% probability under status quo (i.e., where the sigmoid terminal percent decline curves were steepest).

2. Can We Remove Individuals from River km 50 to Supply a Translocation Program without Exacerbating the Risk of Decline in that Metapopulation?

An apparently reliable translocation tactic has been developed for tiger beetles that entails digging up prepupal larvae (i.e., individuals of the cohort that will emerge as adults that summer) and placing them on a different beach early enough in the season that they can reestablish burrows, pupate, and emerge to breed that summer. Here I assess the impact of removing 100 prepupal larvae from river km 50 each of the first two years of the simulations (tactics pre-4 and pre-5). I compared the predicted risk of decline in the river km 46–50 metapopulation with and without removals for status quo management, as well as for the best tactic that emerged from question 1.

3. How Would the Prospects for (a) Metapopulation Persistence and (b) Patch Occupancy Respond to Translocation at Each of Two Potential Recipient Areas?

If removing 100 larvae each of the next 2 years would not exacerbate the risk of decline in the river km 46–50 metapopulation, I would address the question of how those individuals could best be used. They could be translocated to river km 146, where there is currently a small population, in hopes of pushing that population onto a steeper part of its growth curve (tactic 4). They could also be translocated to a currently unoccupied beach in hopes of establishing a new population (or metapopulation); I previously identified a cluster of beaches at river km 87–91 as the best possibility for reintroduction on the Connecticut River within the state of Connecticut (Omland 2002); tactic 5). Specifically, I compared the prospects for (a) metapopulation persistence above a threshold and (b) patch occupancy between the two metapopulations given the introduction of 100 prepupal larvae each of the first 2 years of the simulations.

4. Which Tactic Minimizes the Risk of Regional Extinction?

Finally, I asked which among all the tactics depicted in Figure 12.1 minimizes the risk of regional extinction. I compared the tactic that minimized risk of decline in the river km 46–50 metapopulation to both of the tactics involving translocation of larvae. Since risk of extinction over a 30-year time horizon is vanishingly small for the river km 46–50 metapopulation, risk of true regional extinction is inappropriate for use as a response variable. Instead, I computed the probability that each metapopulation would exceed a given threshold (i.e., akin to a quasi-explosion probability). The question then became one of comparing the predicted probability that one, two, or three metapopulations in the region would exceed a given threshold in 30 years. Maximizing those quantities, particularly the probability of having three large metapopulations, is equivalent to minimizing the risk of regional extinction.

Results

1. Which among Four Habitat Management Tactics Minimizes the Risk of Decline in the River km 46–50 Metapopulation?

Tactic 3, clearing brush at the river km 49 beach, emerged as the one that minimized predicted risk of decline in the river km 46–50 metapopulation. With exponential growth, there is very little risk of decline, even under status quo management: the predicted probability of a 1% decline was only 0.02 (95% CI, 0.00–0.05). If populations were confined by a ceiling, however, any of the three candidate tactics significantly reduced the risk of decline (Figure 12.2). With the high estimates of population ceilings, the level of decline that occurred with about 50% probability under the status quo was 10%; with

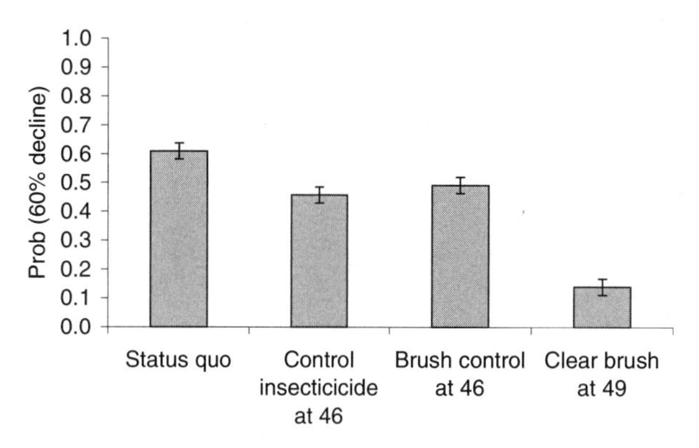

Figure 12.2 Probability of observing a 60% decline over 30 years (± 95% CI) with status quo or implementation of three alternative management tactics. Results are from the model with low ceilings.

the low estimates, it was 60%. In either case, doubling carrying capacity at the river km 49 beach by clearing brush resulted in a significantly lower predicted risk of decline than any other candidate. Tactic 1, eliminating insecticide-caused catastrophes at the river km 46 beach, ranked next, regardless of the ceilings used, followed by tactic 2, preventing brush encroachment at the river km 46 beach.

2. Can We Remove Individuals from River km 50 to Supply a Translocation Program without Exacerbating the Risk of Decline in that Metapopulation?

Simulation results suggested that we could remove 100 larvae from river km 50 for 2 years without compromising metapopulation viability, thus opening the door for tactic 4 or 5. As in question 1, there was negligible risk of decline under the assumption of exponential growth even with status quo management plus removal (probability of a 1% decline 0.03; 95% CI, 0.00–0.05). With density dependence, however, there was measurable risk of decline, but that risk was not significantly greater with the removal than without, regardless of whether ceilings were set to the high or low alternatives and regardless of whether the baseline management was status quo or clearing brush at river km 49. In all cases, the estimated probability of a 60% decline was slightly higher, but with broad overlap of 95% confidence intervals (Figure 12.3).

3. How Would the Prospects for (a) Metapopulation Persistence and (b) Patch Occupancy Respond to Translocation at Each of Two Potential Recipient Areas?

Prospects of metapopulation persistence were practically indistinguishable for the two cases of augmenting the existing metapopulation at river km 146–149 (tactic 4) or reintroducing the species at river km 87–91 (tactic 5; Figure 12.4). That was true of results from simulations with no density dependence or high ceilings; simulations with low ceilings suggested that the river km 87–91 metapopulation was more likely to exceed 500 individuals by the end of the 30-year interval.

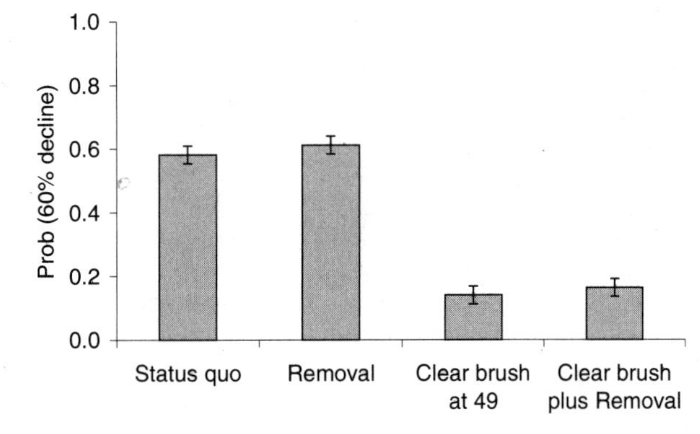

Figure 12.3 Probability of observing a 60% decline over 30 years (± 95% CI) without and with removal of 100 prepupal larvae each of the first 2 years. Removal is compared to two baselines, status quo (*left*) and clearing brush at the river km 49 beach (*right*). Results are from the model with low ceilings.

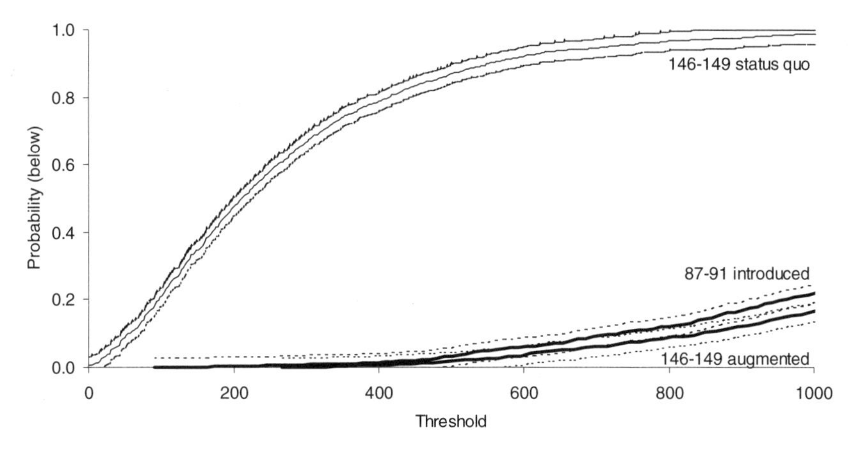

Figure 12.4 Predicted risk that metapopulation abundance will fall below given thresholds over a 30-year time horizon. Results are from the exponential model. The thin top line shows the risk without augmentation for the river km 146–149 metapopulation (no such line is shown for river km 87–91, which is 1.0 for any threshold), while the heavy lower lines show risk with augmentation or reintroduction of 100 larvae each of the first 2 years of the simulations. Dashed lines show 95% confidence intervals.

The simulations yielded a prediction of higher patch occupancy in the river km 87–91 metapopulation. Reintroducing a population at the river km 91 beach was predicted to lead to colonization of two other patches nearby over the course of about 15 years (Figure 12.5c). In contrast, the simulations did not suggest that augmenting the river km 146 population would appreciably increase the chance that two adjacent patches would be colonized (Figure 12.5a and b). Thus the somewhat nuanced answer to question 3 was that predicted payoff from translocation would be greater by adopting tactic 5 (translocate larvae to river km 91) than tactic 4 (translocate them to river km 146).

I investigated a conditional strategy of translocating 100 individuals when abundance at the recipient beach fell below 100 individuals rather than only in the first two years of the simulation, but the results were qualitatively similar in terms of predicted patch occupancy in either metapopulation.

4. Which Tactic Minimizes the Risk of Regional Extinction?

Simulations of reintroducing Puritan tiger beetles to river km 87–91 (tactic 5) generally yielded the highest predicted probability that three metapopulations in the region would exceed a given threshold (Figure 12.6), thus the lowest predicted risk of regional extinction. Simulation of any tactic yielded the prediction that the metapopulation at river km 46–50 would still be large (^{3}1,000 individuals) in 30 years with virtual certainty (Figure 12.6a). Under the status quo management, or mere improvement of habitat within river km 46–50, though, there is appreciable risk that river km 46–50 would be the only metapopulation larger than 100 individuals. There is but a slender chance that river km 146–149 will exceed 1,000 individuals in 30 years and essentially no chance that a third area on the river will be colonized naturally. Translocating 100 prepupal larvae to the river km 146 beach each of the next 2 years greatly improves the chance that it will also exceed 100 individuals or even 1,000 individuals (Figure 12.6b) but still leaves essen-

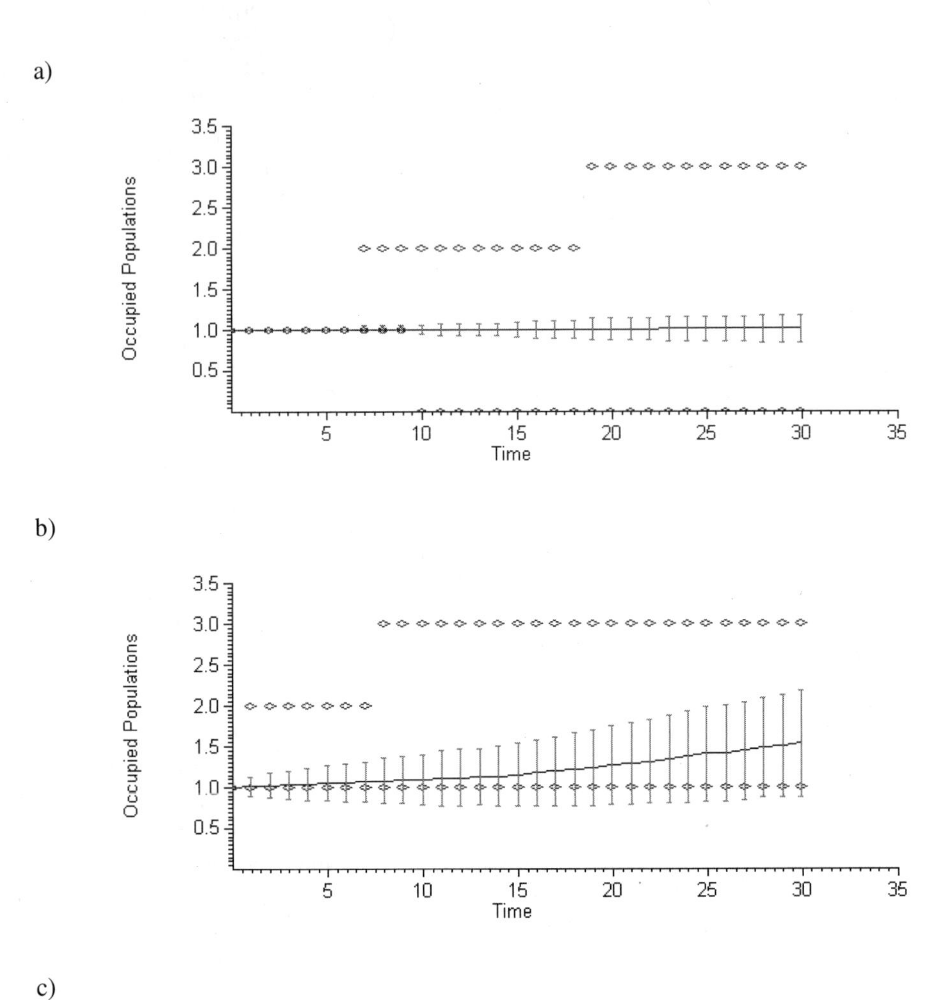

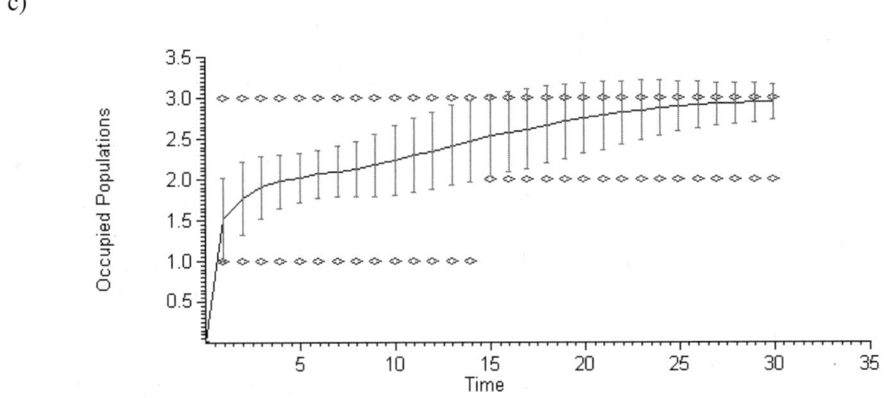

Figure 12.5 Predicted patch occupancy in two 3-patch habitat configurations: river km 146–149 **(a)** without augmentation and **(b)** with augmentation; river km 87–91 **(c)** with reintroduction. Error bars are 1 SD; ovals are minima and maxima.

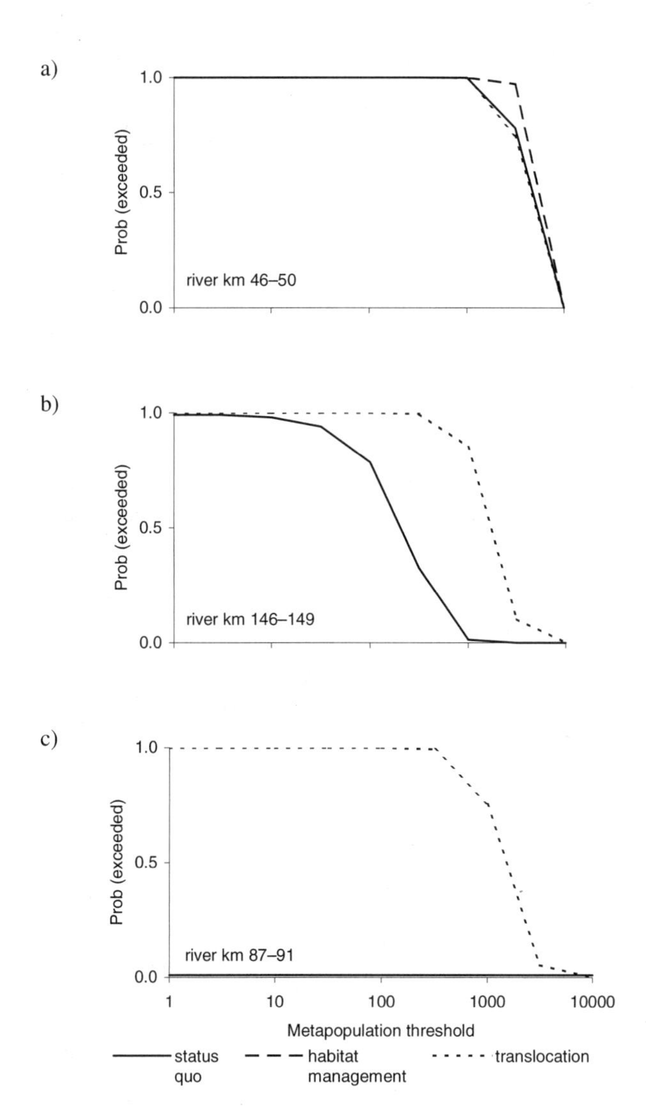

Figure 12.6 Predicted probability that three metapopulations (identified in each panel) will exceed the threshold abundance given on the *x*-axis in 30 years; results of simulations with high ceilings. Solid lines show status quo management or, in **(b)** and **(c)**, any scenario other than translocating larvae to that metapopulation; note that in (c) the solid line lies directly on the x-axis. The coarsely dashed line in **(a)** shows the predicted improvement resulting from the best habitat management tactic within the river km 146–149 metapopulation—namely, clearing vegetation from river km 49. The finely dashed lines show the predicted effect of translocating individuals from river km 46–50 to river km 146–149 or river km 87–91; note the minor shift from the solid line in (a).

tially no chance that there will be a third metapopulation. Only reintroducing beetles to river km 87–91 yields a reasonable chance that there will be three metapopulations larger than 100 individuals in the region 30 years hence (Figure 12.6a–c).

Conclusion

Population viability analysis is here shown to be an effective tool for prioritizing a set of candidate management tactics. In this chapter, I used PVA to predict that reintroducing Puritan tiger beetles to an area that is unlikely to be recolonized naturally is the tactic among the candidates I considered that would minimize the risk of regional extinction. To some extent, that is an intuitive result. Unless we succeed with a reintroduction, there is virtually no chance that there will be three viable metapopulations in the region. The results depicted in Figure 12.6, however, take understanding beyond intuition by showing in quantitative terms the chance that three metapopulations will exceed a given threshold 30 years hence.

I also used PVA to determine that eliminating catastrophic mortality associated with insecticide drift at one patch is more likely to avert decline in the river km 46–50 metapopulation than is controlling encroachment of brush in the same patch. In that case, the simulation results lead us to favor one management tactic over another in a manner that could not be approached by intuition alone.

Employing PVA effectively in the context of adaptive management requires a number of elements. Precise objectives must be articulated (Possingham et al. 2002), a population model must be selected and its parameters estimated (Foley 2000), population consequences of candidate management tactics must be represented in quantitative terms, and standards for measuring ecological payoff must be adopted. In addition, a complete decision-making process ought to entail assessment of the economic costs associated with each management tactic. In the study reported here, I assumed that the economic cost of any one tactic is acceptable, thus glossing over that aspect of decision-making.

An important point is to recognize that after the selected tactic is implemented, the decision should be updated. Any element (e.g., the objectives, the parameter estimates) may change; most importantly, we are likely to gain new information about the consequences of the adopted tactic. In the case presented here, I used PVA as the basis for recommending reintroducing Puritan tiger beetles to an area of the Connecticut River that they are unlikely to recolonize on their own. Embedded in the simulations was the assumption that translocating prepupal larvae would be followed by emergence, reproduction, and recruitment. The validity of that assumption should be assessed through a specific monitoring program at the reintroduction site.

References

Burkey, T. 1989. Extinction in nature reserves: the effect of fragmentation and the importance of migration between reserve fragments. *Oikos* 55: 75–81.

Foley, P. 2000. Problems in extinction model selection and parameter estimation. *Environmental Management* 26: S55–S73.

Hoback, W. W., Stanley, D. W., and Higley, L. G. 1998. Survival of immersion and anoxia by larval tiger beetles, *Cicindela togata. American Midland Naturalist* 140: 27–33.

Knisley, C. B., and Hill, J. M. 1992. Effects of habitat change from ecological succession and human impact on tiger beetles. *Virginia Journal of Science* 43: 133–142.

Knisley, C. B., and Juliano, S. A. 1988. Survival, development and size of larval tiger beetles: effects of food and water. *Ecology* 69: 1983–1992.

Knisley, C. B., and Schultz, T. D. 1997. The biology of tiger beetles and a guide to the species of the South Atlantic States. Special Publication no. 5. Virginia Museum of Natural History, Martinsville.

Lapin, B., and Nothnagle, P. 1995. Control of false indigo (*Amorpha fruticosa*), a non-native plant, in riparian areas in Connecticut. *Natural Areas Journal* 15: 279.

Leonard, J. G., and Bell, R. T. 1999. *Northeastern tiger beetles: a field guide to tiger beetles of New England and Eastern Canada*. CRC Press, Boca Raton, Fla.

Levins, R. 1970. Extinction. Pages 75–107 in M. Gerstenhaber (ed.), *Some mathematical questions in biology*. Lecture notes on mathematics in the life sciences. American Mathematical Society, Providence, R.I.

Omland, K. S. 2001. Population management modeling for the Puritan tiger beetle. Ph.D. diss., University of Connecticut, Storrs.

Omland, K. S. 2002. Larval habitat and reintroduction site selection for *Cicindela puritana* in Connecticut. *Northeastern Naturalist* 9: 433–450.

Pearson, D. L., and Knisley, C. B. 1985. Evidence for food as a limiting resource in the life cycle of tiger beetles (Coleoptera: Cicindelidae). *Oikos* 45: 161–168.

Possingham, H. P., Lindenmayer, D. B., and Tuck, G. N. 2002. Decision theory for population viability analysis. Pages 470–489 in S. R. Beissinger and D. R. McCullough (eds.), *Population viability analysis*. University of Chicago Press, Chicago.

U.S. Fish and Wildlife Service. 1990. Endangered and threatened wildlife and plants: determination of threatened status for the Puritan tiger beetle and the northeastern beach tiger beetle. *Federal Register* 55: 32088–32094.

U.S. Fish and Wildlife Service. 1993. *Puritan tiger beetle* (Cicindela puritana *G. Horn*) *recovery plan*. Hadley, Mass.

U.S. Geological Survey. 2001. *Surface water for USA: peak streamflow*. Available at http://water.usgs.gov/nwis/peak/?site_no=01193000. Retrieved on 5/26/1999 (URL updated 9/21/2001).

Vogler, A. P., Knisley, C. B., Glueck, S. B., Hill, J. M., and DeSalle, R. 1993. Using molecular and ecological data to diagnose endangered populations of the Puritan tiger beetle, *Cicindela puritana*. *Molecular Ecology* 2: 375–385.

13

Giant Velvet Worm (*Tasmanipatus barretti*) in Tasmania, Australia

Effects of Planned Conversion of Native Forests to Plantations

JULIAN C. FOX
ROBERT MESIBOV
MICHAEL A. McCARTHY
MARK A. BURGMAN

The giant velvet worm belongs to the phylogenetically significant phylum Onychophora, which has a close but as yet undetermined relationship to the largest animal phylum, Arthropoda (Moore 2001). Onychophora are believed to have changed little over the last hundred million years (Poinar 1996), and the phylum is of great interest to evolutionary biologists for its phylogenetic significance. Globally, approximately 140 species have been described, and there are phylum-wide tendencies toward local endemism, disjunct populations, and low densities. Such characteristics resulted in the entire phylum Onychophora being cited as vulnerable in the IUCN Invertebrate Red List (Wells et al. 1983). The giant velvet worm shares these attributes, being endemic to northeast Tasmania and having a low population density across a largely unprotected environment subject to disturbance from forestry and agriculture. The giant velvet worm is listed as rare under the Tasmanian Threatened Species Protection Act of 1995.

Onychophora are cryptic animals and often live deep within decaying logs. The giant velvet worm and its sister species the blind velvet worm were not recognized until 1984 and 1987, respectively. Today the greatest threats to the giant velvet worm are inappropriate fire management and the planned conversion of large tracts of wet eucalypt forest to plantation. Conservation management planning permits further plantation establishment within the range of *T. barretti* with the caveat that special actions may be required to reduce the chance of local extinction and fragmentation of populations. The framework for plantation expansion prescribed in the Tasmanian Regional Forest Agreement (RFA) is operational in Forestry Tasmania's Tactical Plan that outlines plans for native forest harvesting and plantation conversion over the next 10 years and beyond. In this chapter we compare management scenarios based on this tactical plan in terms of their impact on the average population sizes and the risk of decline. In this context the development of population models can assist in quantifying the magnitude of this threat and its consequences for the predicted population size and the risk of decline of *T. barretti*.

Methodology

Habitat Model

The known range of the giant velvet worm was estimated using presence records, expert judgments (R. Mesibov, unpublished data), and known parapatric boundaries. Records of presence and absence (collated since the species was discovered in 1984) indicate a range of approximately 600 km^2 in northeast Tasmania. The primary habitat requirement is the presence of moist rotting eucalypt logs within which *T. barretti* shelters and lives. It was formerly believed that the species was restricted to flow lines in dry sclerophyll forest (Mesibov and Ruhberg 1991), but more recently it has been found to occur more or less continuously through dry forest subject to the availability of a decaying log habitat (Mesibov 1997). Given that considerable effort has been expended in searching for *T. barretti* over the past 14 years, presence records should reasonably reflect the actual range of the species; for this reason a minimum convex polygon enclosing presence records was chosen to determine range extent. The distribution also has a parapatric boundary on its southern edge with its congener the blind velvet worm, and the northwest range boundary lies along the Goulds Country Break (Mesibov 1996: 22). Within this range, areas of dry and wet eucalypt forest identified from Forestry Tasmania's Forest Class 2001 GIS layer were included as suitable habitat. Areas of existing plantation (see the following section on disturbance), cleared land, and unsuitable forest types were removed as potential habitat.

Areas of suitable habitat were demarcated according to Forestry Tasmania's coupe structure (an area of forest of variable size and shape subject to disturbances such as plantation conversion and native forest harvesting) to facilitate a consistent disturbance regime within each patch. The final metapopulation patch structure was determined using RAMAS GIS (Akçakaya 2002), with patches being considered distinct when separated by more than 35 m (see the following section on dispersal). This process resulted in the identification of 882 habitat patches, covering a total of 56,078 ha of suitable habitat. Because RAMAS can only simulate up to 500 patches, two separate population models were built: one for the dry eucalypt part of the range (451 patches), and one for the wet eucalypt part (431 patches). The dispersal model described later in this chapter indicated that there would be negligible dispersal between the two parts of the range. The two sections of the range were modeled separately, implying that environmental stochasticity in the two regions was uncorrelated. This should be a safe assumption, given the different forest types and the small magnitude of stochasticity (coefficient of variation [CV] of 5%) affecting vital rates (see the following section on life history attributes). This subdivision is useful as it facilitates a separate examination of viability for populations in the wet eucalypt forests, which are scheduled for a majority of the plantation conversion.

Contrary to early findings that the largest populations comprised those in continuous tracts of wet eucalypt forest in the western portion of the range (Mesibov and Ruhberg

1991), *T. barretti* were assumed to be equally abundant in wet and dry sclerophyll forest. This follows more recent findings that *T. barretti* are equally abundant in the two forest types (Mesibov 1997; R. Mesibov, unpublished data). Therefore, a uniform density of 80/ha was applied to suitable habitat.

The current range of *T. barretti* consists of a mosaic of forest ages due to previous disturbance from forestry activities. The structure of this mosaic was extracted from Forestry Tasmania's Forest Class 2001 GIS layer, and an approximate forest age was associated with each of the 882 habitat patches. The age of each patch was used to adjust initial density according to the rate of habitat recovery after disturbance from native forest harvest (see the following description of disturbance dynamics). With a predicted habitat extent as described, and taking into account previous anthropogenic disturbance, the initial, current population size of *T. barretti* was estimated at approximately 4.5 million individuals. This is within the bounds of what has previously been estimated: between 2 and 10 million individuals (R. Mesibov, unpublished data).

Life History Attributes

Given its recent discovery, there has been little opportunity for detailed studies of *T. barretti*'s life history. Therefore, the life history attributes required for model development have been largely inferred from other Onychophora and the field observations of R. Mesibov.

Reproductive rates were inferred from closely related Onychophora, in particular *Euperipatoides rowelli*, a well-studied (e.g. Sunnucks et al. 2000) velvet worm from New South Wales, which should have similar life history attributes (H. Ruhberg, pers. comm.). The reproductive attributes of Onychophora are diverse and often very unusual (Sunnucks and Tait 2001). The birth of young has not been observed for *T. barretti*, and a brood size of 10–15 young was inferred from that for *T. anophthalmus*, which has been observed producing 12 and 16 young (Mesibov and Ruhberg 1991), and from reproductive trends for other Onychophora (Monge-Najera 1994). Seasonal reproduction has been observed in other Onychophora (Sunnucks et al. 2000), so a single seasonal brood is expected for *T. barretti*. Following Sunnucks et al. (2000), male and female *T. barretti* are assumed to be sexually mature beyond 1 year of age, and will reproduce in their second year. Onychophoran longevity has been estimated to range between 1 and 7 years (Monge-Najera and Lourenco 1995). A mean longevity of 5 years is assumed to be appropriate for *T. barretti*.

T. barretti emerge from their decaying log habitat to mate, disperse, and feed on insects and other forest litter invertebrates (Mesibov and Ruhberg 1991). It is while they are outside of their log habitat that they may be most at risk of predation or desiccation. Mortality rates are assumed to be low in the absence of disturbance, with neither predation nor desiccation ever being witnessed for *T. barretti* (Mesibov and Ruhberg 1991).

A stage-structured population model was built, based on the life history attributes of juvenile and adult life stages. The stage matrix was

$$\begin{bmatrix} 0.0 & 2.0 \\ 0.5 & 0.75 \end{bmatrix}$$

The finite rate of increase for this stage matrix is 1.44 and average stage occupancy for juveniles and adults is 1 and 4 years, respectively. A fecundity rate of two offspring per adult reflects a high mortality rate immediately after birth, and first-year survivorship was estimated at 50% (Monge-Najera and Lourenco 1995). At each time step of the simulation, the fecundity and survival rates were sampled from lognormal distributions with mean values as shown in the stage matrix and standard deviations of 5% of the mean for both fecundity and survival rates. These low levels of environmental stochasticity reflect the minor levels of stochasticity likely to be affecting the decaying log habitat in the absence of major disturbance. Effects of fires were modeled separately by changing the carrying capacity of patches (see below). Density dependence was modeled using the "scramble" (Ricker) function (see Chapter 1 in this volume).

Dispersal

Adult *T. barretti* have been found to disperse at least 20 m from their decaying-log habitat (R. Mesibov, unpublished data). Barclay et al. (2000), in a rare study of onychophoran dispersal, found quite high rates of dispersal among logs in close proximity; Onychophora dispersed between logs 10 m apart in 4 months. They hypothesize that onychophoran dispersal may be pheromonally mediated. Males are the predominant dispersers and females disperse to new log habitat when levels of male pheromone reach a certain level (Barclay et al. 2000). Based on these observations, *T. barretti* is assumed to be capable of approximately 20 m net dispersal per year.

To characterize the unusual dispersal dynamics of *T. barretti*, a dispersal model was constructed based on that developed for spatially structured populations by Possingham and Davies (1995). The model was constructed assuming that only individuals located within 20 m of the patch edge were capable of dispersing into adjoining habitat patches and that individuals had a maximum dispersal of 40 m per year. The number of potentially dispersing individuals was calculated as the area within 20 m of the patch edge as a proportion of the total patch area. The probability of arriving at an adjoining patch was calculated as a function of the edge-to-edge distance and the width of the destination patch. The larger the edge-to-edge distance and the smaller the width, the lower the chance of successful dispersal. Individuals dispersing and failing to reach an adjoining patch were subject to dispersal mortality. This was implemented in RAMAS Metapop by establishing a "dispersal mortality" patch, where unsuccessful dispersers were removed from the simulations.

Disturbance

The current body of evidence suggests that the clearfelling of coupes followed by a low-intensity regeneration burn has little effect on populations of *T. barretti*. In the study of Horner (1998), *T. barretti* populations were maintained following clearfall logging and regeneration burning. The limited effect of forestry practices is due to the fact that the decaying log habitat is generally left undisturbed (Horner 1998: 8). There is also field evidence of previous fires in a majority of known *T. barretti* sites, and *T. barretti* have been found sheltering in the charcoal deposits of severely burnt logs (Mesibov and Ruhberg 1991). Despite this, frequent or very hot fires may potentially destroy decaying log habitats and desiccate local populations. Current forest practices prohibit the use of high-intensity regeneration burns within the range of *T. barretti*. The effect of clearfall logging followed by a low-intensity regeneration burn is assumed to result in approximately 25% mortality from incidental desiccation, with populations fully recovering within 2 to 3 years. This is described in the recovery function

$$R = 0.75 + \left(0.25 * \left(1 - e^{-0.8t}\right)\right)$$

where R is the relative quality of the habitat (i.e., $R = 1$ is fully recovered), and t is the time since harvesting. The impact of harvesting was modeled by multiplying the usual carrying capacity (80 velvet worms per ha) by R.

It has previously been assumed that conversion of native forest to softwood and hardwood plantations will result in habitat that is unsuitable for *T. barretti* (Mesibov 1988: 19; Mesibov 1990: 54). However, eucalypt logs will survive several plantation rotations and may provide suitable habitat. Furthermore, thriving populations of other Onychophora (*Ooperipatellus* spp.) have been found in Tasmanian softwood and hardwood plantations (Mesibov 1998, Bonham et al. 2002). Given the uncertainty surrounding this assumption, R. Mesibov was commissioned, as part of this study, to search for *T. barretti* in plantations within the range of the species. *T. barretti* were absent from first-rotation plantations but were readily found in adjoining native forest (Mesibov 2001). Eucalypt logs remaining in plantations appeared to be unsuitable habitat for *T. barretti*. It was hypothesized that the impoverished invertebrate litter fauna of plantations may limit prey availability (Mesibov 2001). Given these findings, conversion of native forest to plantation was assumed to render the habitat permanently unsuitable for *T. barretti*.

Wildfires are very common in dry sclerophyll forests and will generally be of a similar intensity to low-intensity regeneration burns. Given this, 25% mortality from incidental desiccation, with populations fully recovering within 2 to 3 years should also apply to wildfires and the same habitat recovery function as used for timber harvesting was used to simulate the habitat response to wildfires. A stochastic wildfire module was designed to correctly model the influence of wildfires and is described in Appendix 1 to this chapter (available on the CD that accompanies this volume).

Six management scenarios were identified based on Forestry Tasmania's Tactical Plan, ranging from minimum to maximum levels of habitat disturbance. Scenarios 1 through 6 were subject to stochastic wildfires, in addition to increasing levels of anthropogenic disturbance, as described in Appendix 2 to this chapter (available on the accompanying CD).

Results

Mean Tendency

The mean population trajectories of 100 replications for each management scenario are shown for the entire extent of *T. barretti* in Figure 13.1, together with box and whisker plots representing 1 SD either side of the mean and extreme values.

The slight population decline in the no-disturbance scenario over the first 10 years can be attributed to the introduction of stochastic fire events. Frequent small fires in the dry eucalypt part of the range may be responsible for this decline. For this reason, trends in the mean tendency for the five disturbance scenarios will be quantified relative to the mean tendency for the no-disturbance scenario rather than the initial population size.

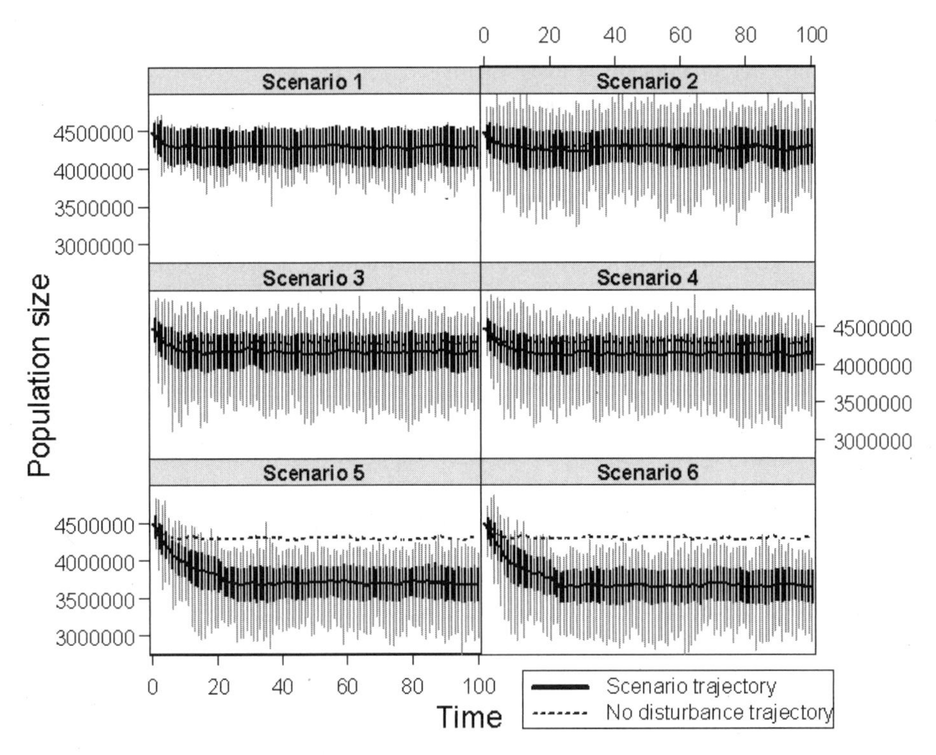

Figure 13.1 Trajectory summaries for each scenario over 100 years. Mean population trajectories of 100 replications for each management scenario are shown for the entire extent of *T. barretti*, together with box and whisker plots representing 1 SD either side of the mean and extreme values.

Because stochastic wildfires will have affected populations previously, initial population estimates may be unrealistic and are perhaps better reflected by the stable population size achieved after 10 years in the no-disturbance scenario. Therefore, a revised current population estimate can be inferred from this stable population size of approximately 4.3 million individuals.

The scenario involving only native forest harvesting results in very little net decline over the projection period. This is because populations within coupes only suffer 25% mortality as a result of low-intensity regeneration burns. The harvesting of standing timber has a negligible effect on the decaying-log habitat. The surviving 75% is capable of population growth rates in the vicinity of 1.4, which allow the population to effectively regain predisturbance population levels within 5 years.

Plantation conversion results in habitat becoming permanently unsuitable, effectively causing local extinctions at the coupe level. Mean tendencies for disturbance scenarios reflect these varying levels of habitat loss and local extinction. Because population density is assumed to be uniform, change in the mean tendency relative to the no-disturbance scenario should reflect the amount of habitat lost to plantation conversion. Any additional loss in total population size will reflect flow-on effects resulting from disruption of dispersal dynamics. A comparison of percent total habitat lost to plantation conversion and percent reduction in total population size is shown in Table 13.1. Percent reduction in total population size is estimated from the more or less stable tendency achieved after plantation conversion ceased in 2020.

There is almost exact correspondence between percent habitat lost and percent population decline. The disruption of dispersal dynamics appears to have only a minor effect for *T. barretti* resulting in an additional decline of only 0.4%. Conversion to plantation will impede dispersal, slowing recovery, and will increase mortality in individuals dispersing from adjacent native forest. Despite this, the additional decline due to the breakdown of dispersal dynamics is minor, and this suggests that *T. barretti* is relatively insensitive to disruption of dispersal at the scale of coupes.

Forestry Tasmania's current Tactical Plan is represented by scenarios 3 and 4; scenario 3 includes tactical plantation conversion scheduled within the next 10 years, while scenario 4 includes tactical and strategic plantation conversion scheduled within the next 20 years. Relative to the entire range of *T. barretti*, tactical and strategic conversions affect only a small area (2.5% and 3.2%, respectively). This is due to the large, dry eucalypt portion of *T. barretti*'s range being unsuitable for plantation conversion. Corresponding to this, the observed population declines are small: 2.9% and 3.6% decline for scenarios 3 and 4, respectively. Given this small projected decline and its relative insensitivity to disturbance from native forest harvesting, *T. barretti* appears robust to the disturbance regimes described by scenarios 3 and 4.

Scheduled plantation conversion associated with the current tactical plan is concentrated in the wet eucalypt forests of the northern portion of *T. barretti*'s range. When we examine mean tendencies for the northern patch, population declines resulting from plantation conversion become more dramatic. Relative declines for scenarios 3 through 6 are detailed in Table 13.2.

For the northern section of *T. barretti*'s range, projected population declines of 8.4% and 10.7% for scenarios 3 and 4 indicate substantial departures from the no-disturbance scenario. Such declines were not observed when trajectories were examined for the entire range. Despite these declines, populations in the northern region suffer no discernable loss of viability, with trajectories remaining stable after 2020. A projected decline of 10.7% after 100 years also satisfies Forestry Tasmania's threatened species caveat that projected population declines (attributable to forestry activity) should not exceed 20% over the next 10 years.

Risk Analysis: Quasi-extinction Risk

Quasi-extinction risk curves for a 100-year time horizon for the entire range of *T. barretti* are shown in Figure 13.2. For risk curves, the extent to which the curve is moved to the

Table 13.1 Effects of plantation conversion on available habitat and total population size, northern and central sections combined

Management Scenario	Habitat Lost (%)	Decline in Total Population (%)
Scenario 3	2.5	2.9
Scenario 4	3.2	3.6
Scenario 5	12.5	13.8
Scenario 6	13.2	14.8

Table 13.2 Effects of plantation conversion on available habitat and total population size, northern section

Management Scenario	Habitat Lost (%)	Decline in Total Population (%)
Scenario 3	7.9	8.4
Scenario 4	10.3	10.7
Scenario 5	22.6	26.1
Scenario 6	24.8	28.3

left represents the increase in risk of falling to small population sizes at some point in the projection period. It can be observed that increasing levels of disturbance move risk curves further to the left and thus increase the risk of smaller populations. This is most noticeable for the extreme disturbance scenarios, scenarios 5 and 6.

To offset the decline to stable population size affecting all scenarios, the magnitude of decline was also calculated as a percentage of the average no-disturbance (scenario 1) population size at 3,001. These results for the five disturbances are examined in Table 13.3.

Table 13.3 reveals that all the scenarios involving some degree of plantation conversion (scenarios 3–6) will always (probability of 1.0) result in a decline of more than 10% at some point over the next 100 years. The risk of a 20% decline is also inflated by

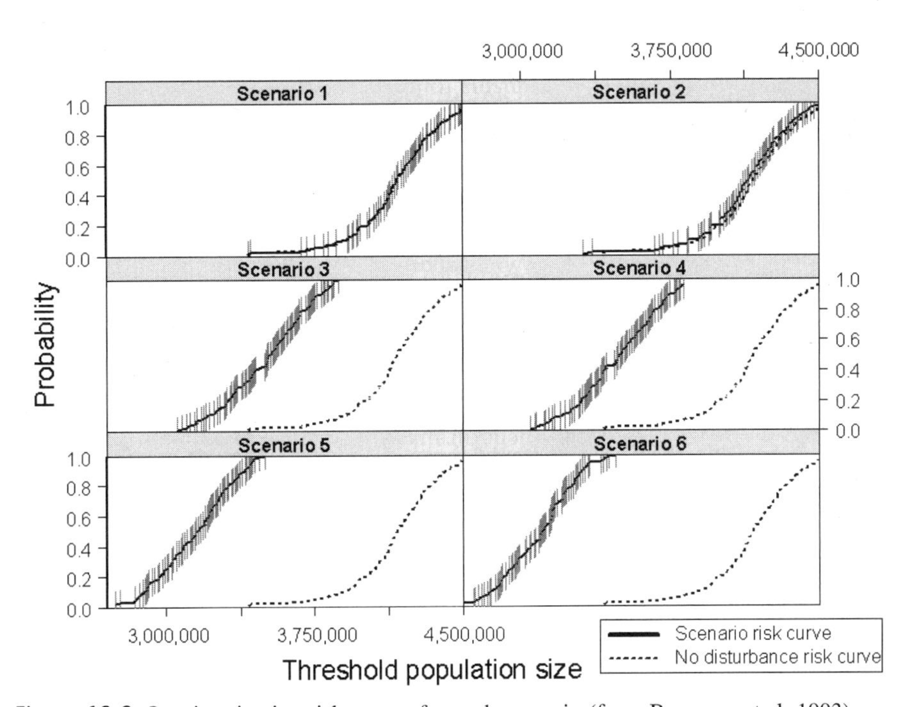

Figure 13.2 Quasi-extinction risk curves for each scenario (from Burgman et al. 1993).

Table 13.3 Probability of a 5%, 10%, 20%, 30%, and 50% interval decline in total population size relative to the no-disturbance scenario at least once in the next 100 years for the entire range of *Tasmanipatus barretti*

	5%	10%	20%	30%	50%
Scenario 2	0.34	0.10	0.02	0.00	0.00
Scenario 3	1.00	1.00	0.38	0.00	0.00
Scenario 4	1.00	1.00	0.42	0.00	0.00
Scenario 5	1.00	1.00	0.97	0.33	0.00
Scenario 6	1.00	1.00	1.00	0.34	0.00

Note: A stable population size for the no-disturbance scenario was calculated as 4,292,311.

scenarios involving plantation conversion. For scenarios 3 and 4, respectively, there is a 0.38 and 0.42 probability that the projected population will decline by more than 20% at least once over the next 100 years. Quasi-extinction risks reflect the consequences of extreme events, and the risks detailed in Table 13.3 quantify the probability of these events occurring through the projection period. For *T. barretti*, a quasi-extinction event is likely to be the result of coincident disturbance from a very large wildfire and anthropogenic disturbances. This could be further exacerbated by demographic and environmental stochasticity, which controls fecundity and mortality rates. In the context of Forestry Tasmania's threatened species caveat, there is an approximate probability of 0.4 of such a quasi-extinction event causing a projected population decline of more than 20% at least once over the next 100 years. This risk is not trivial, but may be acceptable, given the apparently negligible risk of extinction.

Although the change in risk of decline due to the plantation conversion prescribed in Forestry Tasmania's Tactical Plan is apparent, there appears to be negligible risks of declines exceeding 30%. This examination of risk coupled with the stable trend of the mean tendencies indicates that the disturbance prescribed in the current tactical plan does not appreciably reduce total population size nor threaten the viability of *T. barretti*.

Assumptions and Limitations

Despite renewed interest in Onychophora, many of their life history and dispersal traits remain a mystery, and the development of models in this instance was only possible given the field observations of Robert Mesibov and the recent work of Sunnucks et al. (2000). Parameterizing the life history component of the model involved some speculation. However, selected parameters were always estimated within the bounds of qualitative observations on *T. barretti* and quantitative results from other Onychophora. A sensitivity analysis revealed that model outcomes were not sensitive to any component of the life history parameterization. Model outcomes consistently reflected the extent of plantation conversion associated with each scenario. This eases our reliance on the accuracy of the life history parameters. If the results depended sensitively on the life history parameterization, then further study of the life history part of the model would be required. For *T. barretti*, other parts of the model should be validated before life history parameters are refined. Furthermore, such a recommendation would entail a large expenditure of time and effort and would require the destruction of the log habitat (e.g., Sunnucks et al. 2000). The conservation effort required to further refine life history parameters may be better directed at lesser-known endemic invertebrates.

Estimating the dispersal dynamics was also based on the qualitative observations of Robert Mesibov. A sensitivity analysis revealed that outcomes were insensitive to dispersal dynamics. This is because dispersal rates were very small and made a negligible contribution to population maintenance. Dispersal is also relatively unimportant when 75% of the population survives native forest harvesting, and a population growth rate of 1.44 allows it to quickly regain carrying capacity.

Scale considerations may be particularly important for *T. barretti* and other invertebrates, as each coupe may comprise an individually functioning population. Results from analysis of mean tendencies indicated that dispersal dynamics are only making a minor contribution to population maintenance. This is consistent with observations that *T. barretti* has an ability to maintain itself in isolated pockets of habitat (R. Mesibov, unpublished data) and is consistent with its opportunistic use of an often sporadically distributed decaying-log habitat. Given this, an area the size of a coupe may support a self-sufficient population of *T. barretti*, capable of maintaining itself in perpetuity. Evidence of this has been observed for *Euperipatoides rowelli* in New South Wales. The application of DNA markers by Sunnucks and Tait (2001) revealed that Onychophora of the same species were genetically distinct in two populations separated by less than 10 kilometers on a transect through the Great Dividing Range. Remarkably, genetic divergence was as great as that between a bee and a fly. What must be ensured is that such intra-species diversity in *T. barretti* is not reduced by the local extinctions resulting from plantation conversion. Sunnucks and Tait (2001) suggest that the actual species diversity of Australian velvet worms may be 100 times greater than the number currently identified using traditional morphological traits. Such considerations make it imperative that populations remain viable in the northern, wet eucalypt portion of the range. Although this area is subject to quite intensive plantation conversion (approximately 10% of habitat area), a separate analysis for the northern section indicated that the *T. barretti* populations were capable of maintaining themselves and suffered no appreciable loss of viability.

An informal sensitivity analysis revealed that model outcomes were sensitive to disturbance response parameters. The resilience of *T. barretti* in the face of native forest harvesting reflects our current understanding. An estimate of 25% incident mortality from desiccation followed by a rapid recovery within three years is consistent with the qualitative observations of R. Mesibov. Any refinement on these estimates would entail census before, during, and after native forest harvest, which is impossible as an initial census requires the destruction of habitat and therefore precludes further census. The qualitative observations of R. Mesibov were collected by observing populations at different stages of recovery, and clearly, this is the best information available. In the current model it is assumed that the availability of a suitable decaying log habitat is maintained over successive native forest harvesting events. In the absence of improved models for coarse woody debris accumulation over successive rotations, which may become available in 2002 (S. Grove, pers. comm.), this assumption reflects our current understanding. The assumption that plantations are unsuitable habitat for *T. barretti* was confirmed with findings from Mesibov (2001) indicating their absence from first rotation plantations, despite their presence in adjoining native forest.

The assumption that regeneration burns are low intensity may also be optimistic (P. Wells, pers. comm.), but it is a management directive for native forest harvesting within the range of *T. barretti*. High-intensity regeneration burns are believed to result in much higher mortality and much slower recovery rates. The results detailed in this report are dependent on high-intensity regeneration burns being averted. Therefore, it is important that the current directive is implemented and maintained. If high-intensity regeneration burns eventuate then projected declines will be much larger, as almost half of *T. barretti*'s range is subject to this disturbance on an ongoing basis.

In the absence of further quantitative information on model parameters, and the associated investment of time and effort, the current model and results can be used to indicate the appropriateness of anthropogenic disturbances within the range of *T. barretti*.

Appendix 1. Stochastic Fire Regime

This appendix is available as VelvetWorm-App1.PDF in the "Invertebrates\Giant Velvet Worm" folder on the accompanying CD.

Appendix 2. Management Scenarios

This appendix is available as VelvetWorm-App2.PDF in the "Invertebrates\Giant Velvet Worm" folder on the accompanying CD.

Acknowledgments We thank the staff of Forestry Tasmania for their collaboration, in particular Jeff Meggs, Luke Ellis, Penny Wells, and Mick Brown.

References

Akçakaya, H. R. 2002. *RAMAS GIS: linking landscape data with population viability analysis* (version 4.0). Applied Biomathematics, Setauket, N.Y.

Barclay, S. D., Rowell, D. M., and Ash, J. E. 2000. Pheromonally mediated colonization patterns in the velvet worm *Euperipatoides rowelli* (Onychophora). *Journal of Zoology* (London) 250: 437–446.

Bonham, K. J., Mesibov, R., and Bashford, R. 2002. Diversity and abundance of some ground-dwelling invertebrates in plantation vs. native forests in Tasmania, Australia. *Forest Ecology and Management* 158: 237–247.

Burgman, M. A., Ferson, S., and Akçakaya, H. R. 1993. *Risk assessment in conservation biology*. Chapman and Hall, London.

Horner, D. 1998. Comparative study of the effects of logging operations on the GVW populations of GC171A. Unpublished report to Forestry Tasmania, Hobart.

Mesibov, R. 1988. Tasmanian Onychophora. Unpublished report to the Department of Lands, Parks and Wildlife, Hobart.

Mesibov, R. 1990. Velvet worms: a special case of invertebrate fauna conservation. *Tasforests* 2: 53–55.

Mesibov, R. 1996. *Invertebrate bioregions in Tasmania*. Report to Tasmanian RFA Environment and Heritage Technical Committee, Hobart.

Mesibov, R. 1997. The blind velvet worm (*Tasmanipatus anophthalmus*): a background report for conservation planners. Unpublished report for the Theatened Species Unit, Department of Environment and Land Management, Tasmania, Hobart.

Mesibov, R. 1998. Velvet worms in plantations. Internal report. Forestry Tasmania, Hobart.

Mesibov, R. 2001. Giant velvet worms (*Tasmanipatus barretti*) and plantations. Unpublished report for the project "Linking Landscape Ecology and Management to Population Viability Analysis." Hobart, Tasmania.

Mesibov, R., and Ruhberg, H. 1991. Ecology and conservation of *Tasmanipatus barretti* and *T. anophthalmus*, parapatric onychophorans (Onychophora: Peripatopsidae) from north eastern Tasmania. *Papers and Proceedings of the Royal Society of Tasmania* 125: 11–16.

Monge-Najera, J. 1994. Reproductive trends, habitat type and body characteristics in velvet worms (Onychophora). *Revista de Biologia Tropical* 42: 611–622.

Monge-Najera, J., and Lourenco, W. R. 1995. Biogeographic implications of evolutionary trends in Onychophorans and Scorpions. *Biogeographica* 71: 179–185.

Moore, J. 2001. *An introduction to the invertebrates*. Cambridge University Press, Cambridge.

Poinar, G. 1996. Fossil velvet worms in Baltic and Dominican amber: Onychophoran evolution and biogeography. *Science* 273: 1370–1371.

Possingham, H. P., and Davies, I. 1995. ALEX: a model for the viability analysis of spatially structured populations. *Biological Conservation* 73: 143–150.

Sunnucks, P., and Tait, N. 2001. What's so interesting about velvet-worm sex? Tales of the unexpected. *Nature Australia* Winter: 60–69.

Sunnucks, P., Curach, N. C., Young, A., French, J., Cameron, R., Briscoe, D. A., and Tait, N. N. 2000. Reproductive biology of the Onychophoran *Euperipatoides rowelli. Journal of Zoology* (London) 250: 447–460.

Wells, S. M., Pule, R. M., and Collins, N. M. 1983. *The IUCN Invertebrate Red Data Book*. IUCN, Gland, Switzerland.

14

Hermit Beetle (*Osmoderma eremita*) in a Fragmented Landscape

Predicting Occupancy Patterns

THOMAS RANIUS
JONAS HEDIN

Trunk hollows in old, deciduous trees harbor a specialized insect fauna that consists mainly of beetles and flies (Dajoz 2000). Today, this fauna is threatened in Europe due to habitat loss (Speight 1989). To enable efficient conservation work in order to preserve this fauna, there is a need for detailed knowledge about habitat requirements of the species. Field inventories of presence or absence suggest that the spatial structure of the habitat is important for several beetle species, as they are absent or occur in lower frequencies in individual hollow trees or in small stands (Ranius 2000, 2002a). The decrease of old, hollow trees has taken place during the past 200 years, and perhaps for that reason the metapopulations of inhabiting species are not in equilibrium; in many landscapes we expect the extinctions from small stands to be more frequent than the colonizations. To understand how the present occupancy patterns have arisen, detailed studies on the population dynamics are needed. Therefore, we have carried out more thorough studies on one species, *Osmoderma eremita* (Coleoptera: Scarabaeidae), within the framework set by metapopulation ecology (Ranius and Nilsson 1997; Ranius 2000, 2001, 2002a, 2002b; Ranius and Hedin 2001; Hedin and Ranius 2002).

In the study reported here, we used empirical data on the population dynamics and life history of *O. eremita* to parameterize a metapopulation model with use of RAMAS. We simulated populations that inhabit a landscape in which we assume the stands of hollow oaks of different sizes to have been isolated from each other 150 years ago. The outcome from the simulation was compared with presence/absence data from an inventory in this landscape.

Study Species

Osmoderma eremita lives in tree hollows, in Sweden mainly in oaks *Quercus robur* L. with large amounts of wood mold (Ranius and Nilsson 1997). The adults occur in July to September and never hibernate. They lay eggs that develop to larvae. After 2 years of larval development a pupa is formed from which an adult beetle is hatched the following year (Tauzin 1994). The larvae live deep in the wood mold and are found only occasionally. Therefore, only adults can be studied without destroying the habitat. In contrast to many other saproxylic beetles, the *O. eremita* adults rarely visit flowers or sap flows, but remain mainly in the tree hollows (Martin 1993; pers. obs.). We have estimated population sizes and dispersal rates from capture-recapture data obtained from pitfall trapping in trunk hollows of oaks (Ranius 2001, Ranius and Hedin 2001). We have also studied dispersal by using telemetry (Hedin and Ranius 2002). At the same trees, but also at other sites, we have collected presence/absence data by taking wood mold samples (8 liters/tree) and assessed the presence of fragments (pronotum, elytras end heads) of adult *O. eremita* beetles (e.g., Ranius 2000).

Study Areas

All data used to parameterize the model were collected from two sites—Bjärka-Säby and Brokind—both of which have high densities of hollow oaks. They are cores in an area south from Linköping (province of Östergötland, southeastern Sweden) with one of the largest concentrations of old oaks in northern Europe (Ranius et al. 2001).

Simulation results were compared with the result from an inventory of a 14×15 km^2 wide area around Kättilstad (mapped in Ranius 2000), which is situated 10 and 20 km south from Brokind and Bjärka-Säby, respectively. Today, this area contains many small stands of hollow oaks (Ranius 2000), but historical documents reveal that the density of hollow oaks in Kättilstad was much higher earlier (Anon., 1749). Currently, about 90% of the hollow oaks in Kättilstad are situated on land previously owned by the nobility (pers. obs.). In an inventory of land owned by farmers and the church in 1749, several thousand oaks in Kättilstad were classified as "old and unusable" (Anon. 1749), but now very few old oaks can be found there. The majority of the old oaks were removed in the nineteenth century (Eliasson and Nilsson 2002). Therefore, we assume the stands of hollow oaks to have occurred in such high densities that *O. eremita* populations were connected with dispersal up until 1850 but since that time they have been isolated, and the stands have been of a constant size, equal to the size today.

Parameterization in RAMAS

Landscape Data

We treated every single tree as a habitat patch, and all trees clustered in a wooded pasture or forest stand sustain together a metapopulation. This is because most *O. eremita* individuals remain in the same tree throughout their entire lifetime (Ranius and Hedin 2001). An individual tree may contain a trunk hollow during at least 200–300 years. It is reasonable to believe that the carrying capacity changes over that time, but we have no data on this. Therefore we assumed each habitat patch to have a constant carrying capacity and that no hollow trees are lost or generated over time.

Our objective was to predict the occupancy per tree and stand in a landscape that has been fragmented during the last 150 years. A stand was defined as a cluster of hollow oaks with a distance of <250 m from one tree to another, based on flight distances of up to 190 m found for *O. eremita* (Ranius and Hedin 2001). We assumed the dispersal rate between stands to be 0, while the dispersal rate within a stand was equal between all pairs of trees. For that reason there was no need to define any spatial structure of the trees, and thus no coordinates of the trees were used. The size of the stands were adjusted to facilitate comparisons with the classes presented in figures 2 and 3 in Ranius (2000): that is, we simulated five different metapopulations, inhabiting stands consisting of 2, 5, 9, 21, or 97 hollow oaks.

Migration Function

We have studied the migration between trees with use of capture-mark-recapture over 5 years in Bjärka-Säby and Brokind. Among those 377 individuals that were captured at least twice, 2.4% moved from one tree to another (Ranius and Hedin 2001). However, this figure is an underestimate of the dispersal, because many individuals disperse to trees without traps or they disperse before the first or after the last capture. Therefore we constructed and used a computer simulation program which suggested that 15% of the individuals moved between trees, while the remaining 85% stay in the same tree throughout their entire lifetime (Ranius and Hedin 2001).

For each pair of oaks we estimated the dispersal rate by dividing the total dispersal rate by the number of other hollow oaks in the stand (i.e., dispersal rate = 0.15/(number of hollow oaks – 1)). The dispersal rate to trees with low K ($K < 50$) was assumed to be lower. This was based on the observation that in some trees there have never been any beetles captured during 7 years, even though the trees are near other trees with large populations (Ranius 2001). This suggests that very few beetles move to trees that are not suitable for the species.

Age and Stage Structure

Experience from breeding in the laboratory suggests that the larvae usually construct a cocoon in the autumn after 2 years of development, and metamorphosis takes place in the spring of the following year (Tauzin 1994). Thus, development normally takes 3 years, but this may vary due to habitat quality. Our own observations support this view, as among larvae in a single hollow tree it is usually possible to identify two or three size classes, probably representing different cohorts. For many saproxylic beetles the development time may vary due to, for instance, weather and habitat quality. Therefore, we have parameterized our model based on the assumption that the development time is 3 years for 50% of the individuals and 4 years for the remaining 50%.

The rate of population increase may be estimated as the ratio between the offspring population size and the parent population size. From a time series of population size censuses (Table 14.1), we estimated the mean rate of population increase to 1.64, assuming a lognormal distribution. These population size censuses have been performed by capture-mark-recapture in 26 trees in the core of the Bjärka-Säby area. Here we only used data from those 7 trees with the largest populations. This is because the sampling error becomes larger as the population size becomes smaller. Therefore, in small populations the censused population sizes fluctuate more widely between years due simply to large sampling errors. If these populations were included, the average rate of population increase would be overestimated. When the rate of population increase was estimated, the population size in 1998 was divided by the size in 1995, and so on, as the development time is 3 years. Thus we have data for four different time periods: 1995–1998, 1996–1999, 1997–2000, and 1998–2001. In many cases this 3-year time period is too short for achieving reliable estimates of the population increase rate; for instance, extreme weather during these few years may cause the rate of population increase to be much lower or higher than the long-term average. However, regarding *O. eremita* this problem is smaller than for most other insects, as the population fluctuations have been found to be to a large extent asynchronous between trees (Ranius 2001).

In the RAMAS model, we calculated parameters in the Stage Matrix dialog (fecundity, mortality, and transition between age classes) from the rate of population increase. Fecundity data have been reported from a laboratory study, which suggested that each *O. eremita* female lays some tens of eggs (Luce 1996). Thus, we assumed that there are on average 20 females in the offspring of one female (as the sex ratio has been found to be 1:1; Ranius 2001). The ratio between rate of population increase and fecundity (i.e., 1.64/20 = 0.082) was used as an estimate of the survival rate from the first to the latest stage. We assumed the survival rate to be equal between the stages and between individuals with 3 and 4 years of development. It is only one combination of survival rates that is consistent with these assumptions, and this combination was used in the parameterization of the stage matrix.

Obviously, it is only the adults that reproduce and are able to migrate between trees but not the larvae. We assumed that the larvae compete with each other and are restricted by a ceiling but that the adults are not affected by this ceiling.

Because Luce (1996) has reported a variation in fecundity, the standard deviation of the fecundity was set to 5. In the RAMAS model, a value of the standard deviation of the survival rate is also required. We obtained that by setting the standard deviation to different values and searching for a case where the population size variability in the simulation output was similar to our empirical data. This resulted in a standard deviation that was 52% of the mean survival rate. The method to estimate survival rate means and variabilities that we used generates one possible combination that is consistent with empirical data. However, there are also other combinations (for instance, we can decrease the variability and increase the mean of the survival rates at the same time) that generate similar population size fluctuations. Because we have no direct data on the survival rates of different stages, it is impossible to determine which of these combinations represents the most correct description of the population dynamics.

The population size fluctuations also have been found to be asynchronous in trees within stands (Ranius 2001). Therefore, we assumed the correlation between local populations to be 0.

Density Dependence

We assumed that there was a ceiling in the population size. The ceiling was estimated from capture-recapture data from 43 trees situated in the core of the Bjärka-Säby area by using population estimates from the year when each tree had its maximum population size. These population estimates were the same as those given in Table 14.1, but they were multiplied by 1.25 to compensate for the fact that males and females were pooled when the population sizes were estimated (see further Ranius 2001). The value of the maximum population size (of adults, both males and females) was multiplied with a constant (11.4) to provide a measure of *K*, which in this case is the number of female larvae that could possibly be harbored in the tree. The constant was calculated from the survival rates given in the Stage Matrix dialog, and the assumption was that there was a stable distribution between cohorts. Then the number of 1-, 2-, and 3-year-old female larvae would be 6.09, 4.68, and 1.30, respectively, when the number of adults (both males and females) was 1. The constant (11.4) was calculated by summing the number of larvae of all ages. To be able to also use these *K* values in stands with <43 trees, we stratified all trees into equally large categories based on their *K* value. For each category a mean *K* value was estimated. This implies that in our models, the mean *K* per tree was equal between stands differing in size. However, the maximum value was higher and the minimum value was lower in the largest stands. Easily measured characteristics of the oaks, which may reflect their suitablility as habitat for *O. eremita,* have been compared between Bjärka-Säby and Kättilstad, but no significant difference was found (Ranius 2000, 2002a). This indicates that there is no difference in the average *K* between oaks in these areas.

Tree hollows are a stable habitat that does not seem to change much from one year to another. Therefore we assumed *K* to be constant over time.

Use of Output Data

The simulations were run with 1,000 replicates, as this was found to be sufficient to give robust results between simulation runs. The output data were compared with an

Table 14.1 Population size of *Osmoderma eremita* in those seven oaks with the highest population size in Bjärka-Säby, Sweden

Oak No.	1995	1996	1997	1998	1999	2000	2001
1	38	18	90	101	94	100	56
2	79	55	36	90	80	32	8
3	41	59	42	11	29	37	27
4	19	16	11	10	21	9	13
5	38	22	0	3	8	6	5
6	11	4	9	22	10	12	4
7	6	0	2	4	6	12	8

Note: Population size is the number of adults per tree and year estimated by Craig's model, without correction for difference in catchability between males and females; see also Ranius 2001.

inventory (presence/absence) of remains of dead adults. Because the fragments may remain in the tree hollows a few years, their presence or absence was compared with the presence or absence of *O. eremita* in any stage, and not only the adults. The "metapopulation occupancy" in RAMAS corresponds to the occupancy per tree (figure 2 in Ranius 2000), while the "time to quasi-extinction" provides data that can be used to calculate the occupancy per stand (figure 3 in Ranius 2000). We simulated the metapopulation over 250 years. Because we assumed the stands to become isolated in 1850, we compared the situation after 150 years with the present occupancy and used the situation after 250 years to forecast the occupancy pattern in 100 years.

Results and Discussion

Metapopulation Survival of *Osmoderma eremita*

The simulation results show that what we know about the population ecology of *O. eremita* is consistent with the observed occupancy pattern (Figures 14.1 and 14.2). It has been suggested that migration between stands occurs today so rarely that the occupancy pattern reflects mainly the extinction risk in stands that were colonized before the landscape was fragmented. The lower occupancy in smaller stands would then be a result from a higher extinction risk in these stands (Ranius 2000). This study confirms that this is a possible explanation to the observed occupancy pattern; no catastrophic events, breaks of the historical continuity, or extensive dispersal between stands are needed to explain the present occupancy of *O. eremita* in Kättilstad.

Osmoderma eremita populations in smaller stands will not survive in the long run, but it may take several decades or centuries before they go extinct (Figure 14.3). In contrast, a stand with at least 20 suitable hollow oaks seems to be sufficient to maintain a viable population of *O. eremita*. This is a result that is robust quantitively; it is clear that in larger stands the populations have a much lower extinction risk. There-

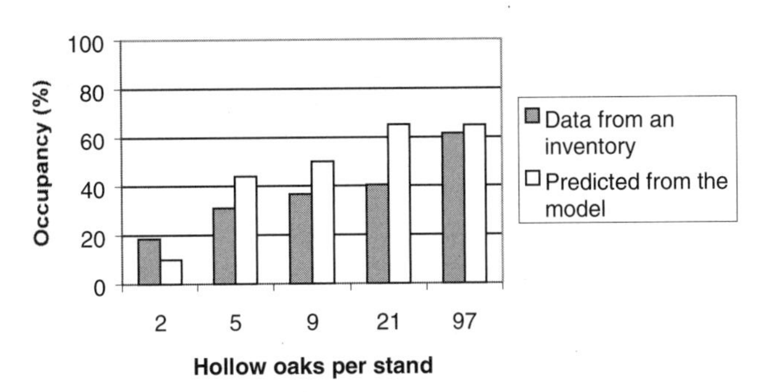

Figure 14.1 Frequency of occurrence per tree of *Osmoderma eremita* in relation to stand size. The prediction is from the model described in this chapter. The inventory is described in Ranius (2000) and was carried out in Kättilstad (stands with <50 hollow oaks) and Bjärka-Säby (one stand with 97 hollow oaks), Sweden (*n* (1–3) = 27, *n* (4–6) = 29, *n* (7–10) = 30, *n* (11–32) = 42), *n* (97) = 43).

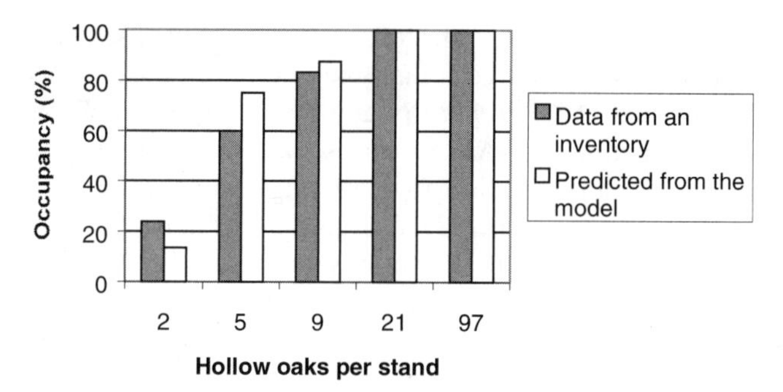

Figure 14.2 Frequency of occurrence per stand of *Osmoderma eremita* in relation to stand size. The prediction is from the model described in this essay. The inventory is described in Ranius (2000) and was carried out in Kättilstad (stands with <50 hollow oaks) and Bjärka-Säby (one stand with 97 hollow oaks), Sweden (n (1–3) = 21, n (4–6) = 11, n (7–10) = 5, n (11–32) = 4, n (97) = 1).

fore, survival of *O. eremita* in a landscape such as the study area around Kättilstad relies solely on the survival of the populations in the largest stands. As resources to nature conservation are limited, they should not be spread to all localities where *O. eremita* is present, but priority should be given to the largest stands, as these sustain the most viable populations, both of *O. eremita* and probably also of other species living in tree hollows.

Uncertainties of the Model

We checked that the values of rate of population increase and the standard deviations that we used were realistic by estimating rates of population increase from model output data in the same way as we did from the empirical data. We found that the population size fluctuations in the model were consistent with empirical data, but there were large deviations between simulation runs. Furthermore, we compared the population sizes predicted from the model and found that this fitted fairly well with empirical data; in the largest stand (Bjärka-Säby, 97 oaks), the average adult population size after 150 years was 54% of K, given a stable stage distribution, whereas from empirical data we

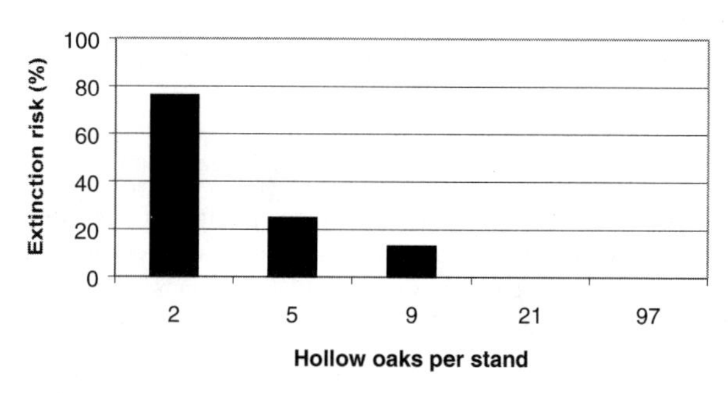

Figure 14.3 Expected extinction risk (%) during the following 100 years according to the outcome from the model. This was calculated by the following equation: extinction risk = 1 – (proportion of populations surviving 250 years) / (proportion of populations surviving 150 years).

should expect this proportion to be 62%. Because even small changes in the standard deviation of survival rates have a large impact on the estimated extinction risk, the confidence limits around estimated extinction risks (Figure 14.3) and occupancy per stand (Figure 14.2) would probably have been wide, if they had been possible to estimate. A much longer time series from more trees would be required to estimate extinction risks with higher certainty.

A principal problem with the model construction is that it assumes the number of trees and the carrying capacity to be constant over time. In a stand consisting of, for instance, 20 trees, only a few trees comprise the main part of the population. Thus, there is a risk that there will be periods when trees with high carrying capacities are completely absent or at least fewer than on average. Stochasticity in the formation and loss of suitable hollow trees may increase the extinction risk considerably in comparison with the present model, which does not take the dynamics of the habitat itself into account.

Acknowledgments Oskar Kindvall helped us with the modeling and offered valuable comments on the manuscript.

References

Anonymous. 1749. Östergötlands landskanslis arkiv. EIII:1 Förteckning över tillgången av ekar i Östergötland [Handwritten in Swedish. Available at Landsarkivet i Vadstena].

Dajoz, R. 2000. *Insects and forests: the role and diversity of insects in the forest environment.* Intercept, London.

Eliasson, P., and Nilsson, S. G. 2002. "You should hate young oaks and young noblemen": the environmental history of oaks in 18th and 19th century Sweden. *Environmental History* 7: 657–675.

Hedin, J., and Ranius, T. 2002. Using radio telemetry to study dispersal of the beetle *Osmoderma eremita,* an inhabitant of tree hollows. *Computers and Electronics in Agriculture* 35: 171–180.

Luce, J.-M. 1996. *Osmoderma eremita* (Scopoli, 1763). Pages 64–69 in P. J. van Helsdingen, L. Willemse, and M. C. D. Speight (eds.), *Background information on invertebrates of the Habitats Directive and the Bern Convention.* Part I: *Crustacea, Coleoptera and Lepidoptera.* Council of Europe, Strasbourg.

Martin, O. 1993. Fredede insekter i Danmark. Del 2: Biller knyttet til skov. *Entomologiske Meddelelser* 57: 63–76.

Ranius, T. 2000. Minimum viable metapopulation size of a beetle, *Osmoderma eremita,* living in tree hollows. *Animal Conservation* 3: 37–43.

Ranius, T. 2001. Constancy and asynchrony of populations of a beetle, *Osmoderma eremita* living in tree hollows. *Oecologia* 126: 208–215.

Ranius, T. 2002a. Influence of stand size and quality of tree hollows on saproxylic beetles in Sweden. *Biological Conservation* 103: 85–91.

Ranius, T. 2002b. *Osmoderma eremita* as an indicator of species richness of beetles in tree hollows. *Biodiversity and Conservation* 11: 931–941.

Ranius, T., and Hedin, J. 2001. The dispersal rate of a beetle, *Osmoderma eremita,* living in tree hollows. *Oecologia* 126: 363–370.

Ranius, T., and Nilsson, S. G. 1997. Habitat of *Osmoderma eremita* Scop. (Coleoptera: Scarabaeidae), a beetle living in hollow trees. *Journal of Insect Conservation* 1: 193–204.

Ranius, T., Antonsson, K., Jansson, N., and Johannesson, J. 2001. Inventering of skötsel av gamla ekar i Eklandskapet söder om Linköping. *Fauna och flora* 96: 97–107 [Inventories and management of old oaks in an area south from Linköping, Sweden. In Swedish, Engl. summary].

Speight, M. C. D. 1989. *Saproxylic invertebrates and their conservation.* Council of Europe, Strasbourg.

Tauzin, P. 1994. Le genre Osmoderma le Peletier et Audinet-Serville 1828 (Coleopt., Cetoniidae, Trichiinae, Osmodermatini). Systématique, biologie et distribution (Deuxième Partie). *Entomologiste* 50: 217–242.

15

Woodland Brown Butterfly (*Lopinga achine*) in Sweden

Viability in a Dynamic Landscape Maintained by Grazing

OSKAR KINDVALL
KARL-OLOF BERGMAN

Landscape changes are one of the most important factors affecting extinction risks among species (Caughley 1994). Despite this, most studies using population viability analysis (PVA) assume constant landscape scenarios in which the quality, size, and existence of habitat patches remain the same throughout the simulated time frame. In fact, there are very few examples of PVA where landscape changes have been considered (Gyllenberg and Hanski 1997, Akçakaya and Raphael 1998, Carlsson and Kindvall 2001, Wahlberg et al. 2002).

Many species are associated with certain successional vegetation stages—for example, many butterflies (Thomas 1991, Bergman 2001). Grazing by cattle, sheep, or horses maintains grassland habitats in a successional stage that favors a majority of the butterfly species in Sweden. However, in many European countries the grazing becomes increasingly concentrated within certain areas, leaving huge areas to become overgrown with trees and bushes. Obviously, this landscape change will reduce the amounts of suitable habitat for many species living in pastures.

One of the butterfly species that are threatened in many parts of western Europe due to changes in agricultural practices is *Lopinga achine* Scopoli (Lepidoptera: Nymphalidae). The habitat quality for this species is related to the amount of bush and tree cover within the pastures and the occurrence of its host plant *Carex montana* L. (Cyperaceae) (Bergman 1999). As grazing ceases, the essential open glades with host plants become overgrown, and the habitat deteriorates. In this chapter, using RAMAS GIS we parameterize a model for *L. achine* for one of the two distribution areas in Sweden. The model includes realistic descriptions of the landscape and its probable changes that are expected within the next 100 years. We simulate three contrasting scenarios: (1) a constant landscape in which all available patches are maintained by management; (2) a changing landscape in which most patches become overgrown due to lack of management, but where all currently grazed patches are maintained; (3) a changing landscape in which no patches are grazed. The importance of implementing landscape changes into models aimed for a PVA is the main focus of this chapter.

Methods

Species

Lopinga achine is a univoltine butterfly with only one generation per year. Adults fly, disperse, and mate in June and July. Females lay their eggs at the edges of glades of the partly open oak (*Quercus robur* L.) woodland pastures where its host plant *Carex montana* grows (Bergman 1999). Eggs hatch after 18–22 days, and larvae grow during approximately 10 weeks until it is time for hibernation. After hibernation the surviving larva grow during another 10–12 weeks before pupating. The pupae stage lasts for approximately 16 days.

In Sweden, *L. achine* occurs in two separate areas: in Östergötland, which is a province on the southeastern part of the mainland, and on the island Gotland in the Baltic Sea (Figure 15.1). The species is considered as being near threatened (NT) according to the Swedish national Red List (Gärdenfors 2000). This chapter deals only with the population in Östergötland, which has been thoroughly studied during the last decade by Bergman (1999, 2000a, 2000b, 2001). The system currently contains 158 discrete habitat patches, among which 79 were occupied by at least one male or female of *L. achine* in 1998 (Bergman and Landin 2001). The local population sizes are correlated both to the total area and the current cover of bushes and trees. The temporal changes in the numbers of adult males and females were studied by mark-and-recapture at eight localities between 1989 and 2001 (Bergman 2001; Bergman, unpublished data). Methodological details are presented in Bergman (2001). Mark-and-recapture surveys also revealed information on lifetime interpatch migration of adult females (Bergman 2000a).

The structure of the population of *L. achine* fits to the general idea of a metapopulation where occupancy is related to both the patch quality (size and tree cover) and connectivity in a compensatory way (Bergman and Landin 2001). Thus, the probability of occupancy increases with the size and optimality of tree and bush cover (intermediate values: i.e., 65%–90% is the best amount of cover) and decreases with increasing isolation from other occupied patches. Both local extinctions and colonizations have been documented, but the main factor driving a local population toward extinction is habitat deterioration due to vegetation succession (Bergman 2001).

The rate of landscape changes due to succession of habitat was investigated by comparing aerial photographs from different years between 1941 and 1997 (Bergman 2001). The area not covered by trees or bushes was estimated for all sites using a measuring program (NIH Image 1.52; Rasband 1993) on scanned versions of the photographs. The percentage cover at different time periods were then calculated for each site.

Parameterization

We used RAMAS GIS to construct a demographic metapopulation model for *L. achine* in Östergötland. As the data were too limited for determine the proper mode of density

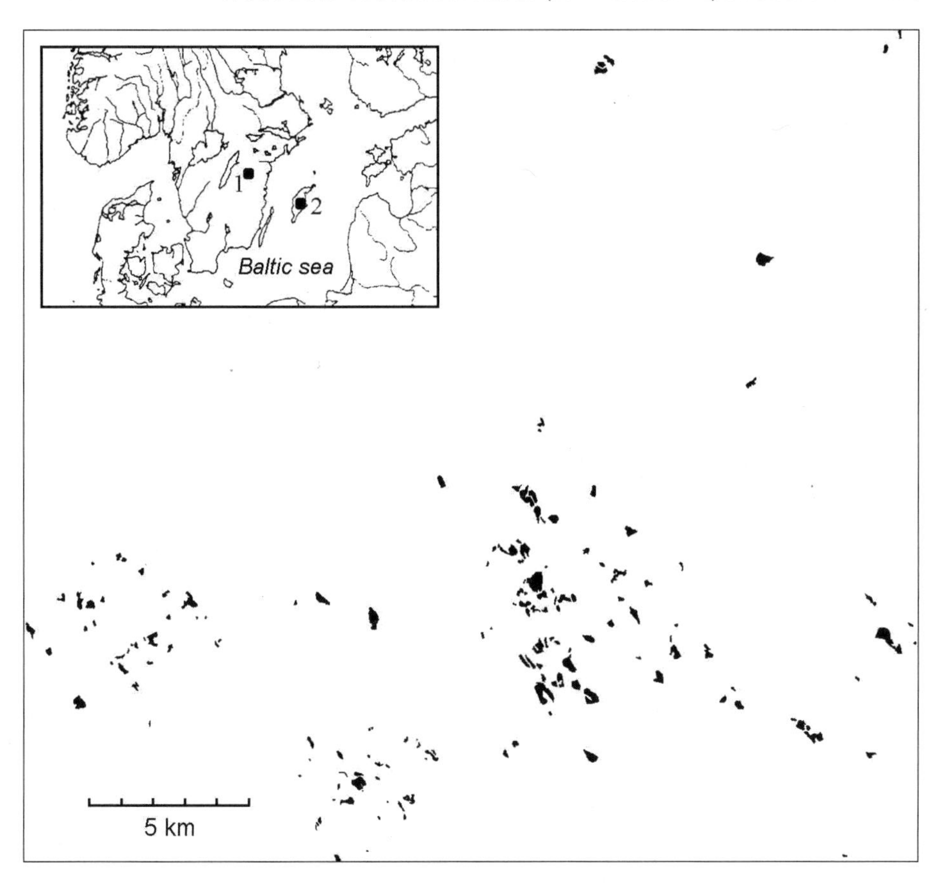

Figure 15.1 Distribution area of the butterfly *Lopinga achine* in Sweden. The species occur in two separate regions: (1) Östergötland and (2) Gotland. The spatial structure of the Östergötland population is shown. Black areas indicate the distribution of available habitat patches.

dependence, we chose to apply the simplest form of population regulation of local populations—that is, a combination of the exponential growth model with a ceiling (K). Because this species does not have overlapping generations, we used a scalar model (i.e., with no stage structure). The observed trajectories of adult population sizes (both sexes) from the eight investigated sites were used to estimate the average population growth rate (R) and its temporal variability (standard deviations of R). We did this by first calculating all current values of R_t according to the following equation:

$$R_t = \frac{N_{t+1}}{N_t} \qquad (1)$$

We obtained 53 values of R_t, which were, of course, not completely independent statistically. Despite this, we fitted all data to a lognormal distribution (Statgraphics Plus 3.1). According to the best fitted function ($c^2 = 3.79$; d.f. = 2; $p = 0.15$), the average of R was 1.16 and the standard deviation was 0.82 (see Stage Matrix and Standard Deviations

Matrix dialogs). As the local populations of *L. achine* fluctuate synchronously to some extent (see later in this chapter), and as the trajectories are expected to include effects of demographic stochasticity, it is possible that the estimated standard deviation represents an overestimation of the actual variation of local populations over time.

The type of distribution for environmental stochasticity was set to lognormal (in the Stochasticity dialog box). We also included demographic stochasticity in the model, although this source of uncertainty is expected to have influenced the estimated standard deviation of *R*. However, it is necessary to add demographic stochasticity mechanistically in order to give realistic migration between patches. With low values of migration rates between pairs of populations, in relation to expected population sizes, no individuals will actually move between patches even though single individuals are expected to migrate in the long term. When we use the demographic stochasticity algorithm in RAMAS, the number of migrating individuals will vary stochastically between time steps, thereby resulting in a long-term average migration rate that is equal to the value estimated by the migration distance function.

Patch-specific estimates of population ceiling (K) were determined according to the following equation:

$$K = A \cdot P_{\text{cover}} \tag{2}$$

where A is the area of the habitat patch in ha. (estimated from the raster map) and P_{cover} is the maximum population density expected in different tree and bush cover classes (Table 15.1). Based on an analysis of aerial photographs, the annual rate of change (Δ) of percentage cover (cover) was estimated. This rate varied approximately linearly in relation to current cover ($n = 22$; $R^2 = 0.86$; Bergman 2001): that is, as succession proceeds, the rate of succession decreases:

$$\Delta = -0.0297 \cdot \text{cover} + 2.799 \tag{3}$$

In addition to the expected changes in maximum population densities of *L. achine* in relation to the cover of trees and bushes, the extinction risk of the host plant will also increase as it becomes overgrown (Bergman 2001). On average, the host plant abundance will become reduced by 80% about 12 years after the cover has exceeded 90%.

Table 15.1 Estimates of maximum density (P_{cover}) of adult female *Lopinga achine*

Cover (%)	P_{cover} (females/ha)	No. of Sites (n)
<60	21	—[a]
60–65	101	—[a]
65–90	182	34
90–95	101	6
>95	21	6

Note: The estimates were calculated by doubling the average density that was observed on patches with different percentage cover of trees and bushes (Cover). Empirical data from Bergman (unpublished).

[a]Insufficient data: the data for 90%–95% and >95% are used in the model.

As the population ceiling (K) will be practically equal to 0 at sites where the host plant has declined to that low levels, we deterministically set K to 1 for all patches 20 years after the 90% cover class was entered.

To implement a changing landscape scenario, six successive raster maps were produced, each representing the spatial structure at the onset of a 20–year period. In these maps, a pixel corresponds to an area of 20 × 20 m. Based on eq. (3) and Table 15.1, it was possible to set the pixel values (integers) to the appropriate value of P_{cover}. In case the host plant was lost, P_{cover} was set to 1. Then we used the RAMAS GIS Spatial Data program to calculate all values of K according to eq. (2). In case a patch was assumed to be grazed continuously also in the future (scenario 2), P_{cover} and consequently K remained the same for all habitat maps. For the static scenario (scenario 1), this was true for all patches.

To incorporate the initial population size (initial number of females) we also used the Spatial Data program by loading an extra raster map where the color value of each pixel was set to the population size observed in 1998.

To parameterize dispersal, we used the results obtained from the mark-and-recapture study and calculated the inverse cumulative distribution function of dispersal distances on a relative scale (Figure 15.2). This function was then fitted to the negative exponential function with a nonlinear regression analysis (Statgraphics Plus 3.1). This one-parameter function fitted better than any other function with either two or three parameters. The parameter b of the migration distance function in RAMAS was set to the coefficient (0.17) obtained from the nonlinear regression analysis (Figure 15.2). The other two parameters were set to 1 to correspond to the negative exponential function. D_{max}, which sets the maximum migration distance, was set to some huge value (1 million km). Then, we calculated all the interpatch migration rates with RAMAS based on center-to-center distances.

By summing all dispersal rates from all patches—that is, each column in the Dispersal matrix—we found that the maximum emigration rate (0.98) in the system was too high in relation to the maximum emigration observed (0.44; Bergman and Landin 2001). Therefore, we adjusted the dispersal-distance function by a factor (0.44/0.98) implemented as parameter a of the migration distance function in RAMAS. This

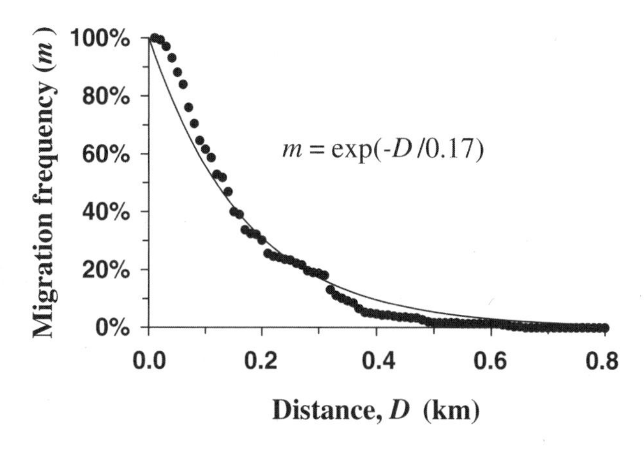

Figure 15.2 Inverse cumulative distribution (*dots*) of lifetime migration distances made by *Lopinga achine* females according to a mark-and-recapture study (Bergman 2000). The solid line represents the best fit of a negative exponential model (nonlinear regression analysis: $n = 305$; $R^2 = 97.4\%$).

resulted in a maximum emigration rate equal to 0.44 and an average emigration equal to 0.12, which is exactly the same value as that observed empirically (Bergman and Landin 2001).

By calculating cross-correlation coefficients (Spearman rank correlation: $n_{\text{correlations}} = 15$; $n_{\text{populations}} = 6$, $n_{\text{years}} = 6$) between the observed trajectories of population sizes, we found a high degree of synchrony among local populations. The average correlation coefficient (±SD) was 0.42 ± 0.28. No effect of interpatch distance on the correlation was detected. Consequently, we used the same value for all pairs of patches in the correlation matrix. This was done by setting the correlation distance function parameter *a* to 0.42, *b* to some arbitrary large number, and *c* to 0.

Results

Starting from the initial conditions observed in 1998, we simulated 1,000 replicates of population trajectories of *Lopinga achine* for each of three alternative landscape scenarios. In all scenarios, the number of local populations was predicted to increase on average from 72 to more than 93 within the next two decades (Figure 15.3). Thereafter, the scenarios start to deviate from each other more clearly. In the static scenario, the number of occupied patches will usually fluctuate around 47 to 116, sometimes reaching a maximum of 133.

In the two scenarios with changing landscape, the number of occupied patches decreases rapidly within the next 60 years as habitat quality is assumed to decrease. While no populations are expected to persist 100 years without grazing, on average 5.6 (8.0 local populations will persist if the patches that are grazed today ($n = 34$) are maintained.

The extinction risk within 100 years according to the static scenario was estimated to be 2.1 ± 1.4% (Figure 15.4). With landscape changes, the extinction risk increases significantly. If currently grazed patches remain grazed during the next 100 years, the

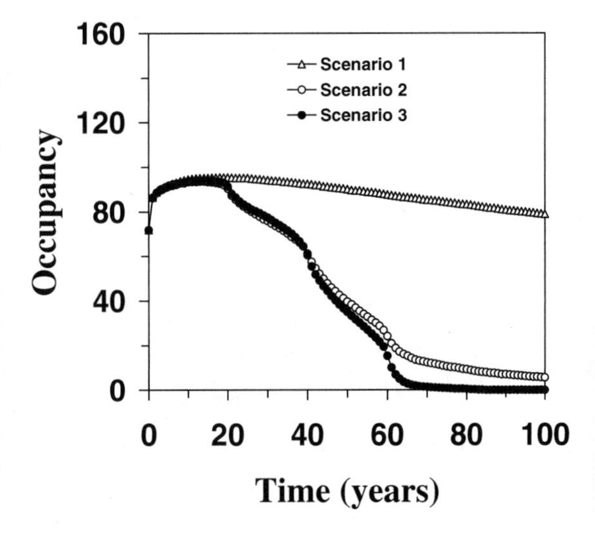

Figure 15.3 Average number of local populations of *Lopinga achine* expected within the forthcoming 100 years according to simulations of three alternative landscape scenarios: (1) a static landscape in which all patches maintain the same quality as they had in 1998; (2) a landscape in which ungrazed patches gradually become unsuitable, but all patches that are currently grazed ($n = 34$) will be maintained; (3) a changing landscape in which no patches are maintained by grazing.

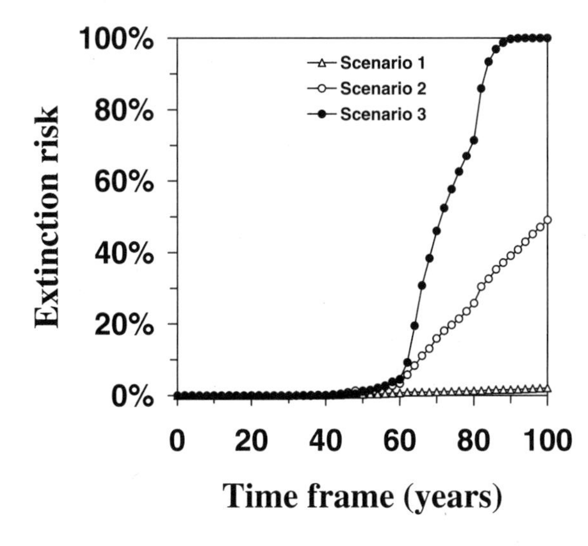

Figure 15.4 Relationship between extinction risk of *Lopinga achine* and length of the time frame according to three simulated scenarios (see Figure 15.3).

predicted risk is 49.1 ± 1.4%. If no patches are grazed, no local populations are expected to survive 100 years. In all scenarios, the extinction risk will be practically 0 within the first 50 years (Figure 15.4).

Discussion

It is apparent from the comparison of the three investigated landscape scenarios that habitat succession is a crucial factor affecting future prospects of the Östergötland population of *Lopinga achine* in Sweden. If the quality of available habitat patches is maintained by various management measures, including sustained grazing by cattle or sheep, it seems possible to effectively reduce the extinction risk. However, according to the present RAMAS model, it will not be enough to maintain current extent of grazing within the distribution area—that is, approximately 20% of the patches. More patches must be included in the conservation program to enable long-term persistence. Another important conclusion from the simulations is that we do not expect the occupancy to start decreasing until 15–20 years have passed, as the habitat changes becomes apparent, even during the current insufficient management regime. The host plant *Carex montana* is known to start decreasing as the canopy closes, and it has colonization rates (Bergman 2001). The simulations indicate that it is probably important and cost-effective to restore habitat quality in a sufficient number of unmanaged patches before it is too late.

By performing additional simulations with the model it would be possible to investigate how many, and which, patches should be managed in order to achieve an acceptable status of *L. achine*. However, this work is beyond the scope of this chapter.

Can we trust the predictions from the current model? Obviously, as the effect of landscape changes was so apparent and the empirical support for the ongoing deterioration of the habitat quality is substantial (Bergman 2001), we definitively believe that the extinction risk will be high if grazing is not applied to more patches than is the case today. However, as the empirical data used for parameterization is limited both temporarily and spatially, one cannot expect that the long-term predictions of either occupied patches or extinction risks will be accurate quantitatively. For example, the theoretical investigation made by Fieberg and Ellner (2000) suggests that quantitative predictions of extinction risks are unreliable for longer time frames than 20% of the length of period during which demographic data was sampled. In our case, demographic data were sampled during approximately 10 years, suggesting that predictions should only be made for 2 years ahead (but see Brook et al. 2000). Obviously, such a short time frame is uninteresting when trying to understand the prospects of *L. achine* as the important process of landscape change only has long-term effects on population dynamics.

References

Akçakaya, H. R., and Raphael, M. G. 1998. Assessing human impact despite uncertainty: viability of northern spotted owl metapopulation in northwestern USA. *Biodiversity and Conservation* 7: 875–894.

Bergman, K-O. 1999. Habitat utilization by *Lopinga achine* (Nymphalidae: Satyrinae) larvae and ovipositing females: implications for conservation. *Biological Conservation* 88: 69–74.

Bergman, K-O. 2000a. Ecology and conservation of the butterfly *Lopinga achine.* Ph.D. diss., No. 621, Department of Biology, Linköping University, Linköping, Sweden.

Bergman, K-O. 2000b. Oviposition, host plant choice and survival of a grass feeding butterfly, the Woodland Brown *Lopinga achine* (Nymphalidae: Satyrinae). *Journal of Research on Lepidoptera* 35: 9–21.

Bergman, K-O. 2001. Population dynamics and the importance of habitat management for conservation of the butterfly *Lopinga achine. Journal of Applied Ecology* 38: 1303–1313.

Bergman, K-O., and Landin, J. 2001. Distribution of occupied and vacant sites and migration of *Lopinga achine* (Nymphalidae: Satyrinae) in a fragmented landscape. *Biological Conservation* 102: 183–190.

Brook, B. W., O'Grady, J. J., Chapman, A. P., Burgman, M. A., Akçakaya, H. R., and Frankham R. 2000. Predictive accuracy of population viability analysis in conservation biology. *Nature* 404: 385–387.

Carlsson, A., and Kindvall, O. 2001. Spatial dynamics in a metapopulation network: recovery of a rare grasshopper *Stauroderus scalaris* from population refuges. *Ecography* 24: 452–460.

Caughley, G. 1994. Directions in conservation biology. *Journal of Animal Ecology* 63: 215–244.

Fieberg, J., and Ellner, S. P. 2000. When is it meaningful to estimate an extinction probability? *Ecology* 81: 2040–2047.

Gärdenfors, U. 2000. *The 2000 Red List of Swedish species.* ArtDatabanken, Swedish University of Agricultural Sciences, Uppsala.

Gyllenberg, M., and Hanski, I. 1997. Habitat deterioration, habitat destruction, and metapopulation persistence in a heterogeneous landscape. *Theoretical Population Biology* 52: 198–215.

Rasband, W. 1993. *NIH Image 1.52.* National Institutes of Health, Washington, D.C.

Thomas, C. D. 1991. Rare species conservation: case studies of European butterflies. Pages 149–197 in I. F. Spellerberg, F. B. Goldsmith, and M. G. Morris (eds.), *The scientific management of temperate communities for conservation.* Blackwell Scientific Publications, Oxford.

Wahlberg, N., Klemetti, T., and Hanski, I. 2002. Dynamic populations in a dynamic landscape: the metapopulation structure of the marsh fritillary butterfly. *Ecography* 25: 224–232.

16

Tapeworm *Echinococcus multilocularis* in Kazakhstan

Transmission Dynamics in a Patchy Environment

E. J. MILNER-GULLAND
PAUL TORGERSON
BLOK SHAIKENOV
ERIC MORGAN

There has been an increasing realization of the importance of incorporating spatial dynamics into models of the transmission dynamics of parasites and infectious diseases, particularly when hosts are free-living in the wild (Mollison and Levin 1995, Hess 1996, Hudson et al. 2001, Kennedy 2001). Spatial variation in parasite prevalences can reflect complex interactions between climatic suitability for free-living stages and host ecology (including host movement patterns, population dynamics, and the extent to which primary and secondary host distributions overlap).

Echinococcus multilocularis is a tapeworm that has wild carnivores as its definitive hosts and small mammals as its intermediate hosts (Eckert et al. 2001) (Figure 16.1). It is widespread throughout the Northern Hemisphere and causes fatal disease in humans if eggs are accidentally ingested. In Kazakhstan, *E. multilocularis* has been found in 18 rodent species and 4 carnivore species (mostly in foxes). Its prevalence depends strongly on the biome (Figure 16.2). This is chiefly because egg survival is strongly related to humidity and temperature (Eckert et al. 2001); under optimal conditions (4°C and high humidity) eggs can remain viable for up to 16 months. Temperatures in desert areas of Kazakhstan can reach 40°C in the summer, with very low humidity; under these conditions eggs are not likely to survive for more than a few hours. Hence extensive field surveys in arid areas of Kazakhstan found only infected rodents in patches with a humid microclimate, such as depressions or river valleys (Shaikenov and Torgerson 2002). In mountains or river valleys, up to 8% of rodents and around 30% of carnivores examined were infected. By contrast, in desert areas, <1% prevalence was found in both rodents and foxes.

In Kazakhstan's desert areas, *E. multilocularis* is found in only one rodent species *Rhombomys opimus,* the great gerbil. This species lives in colonies of 30 to 40 animals;

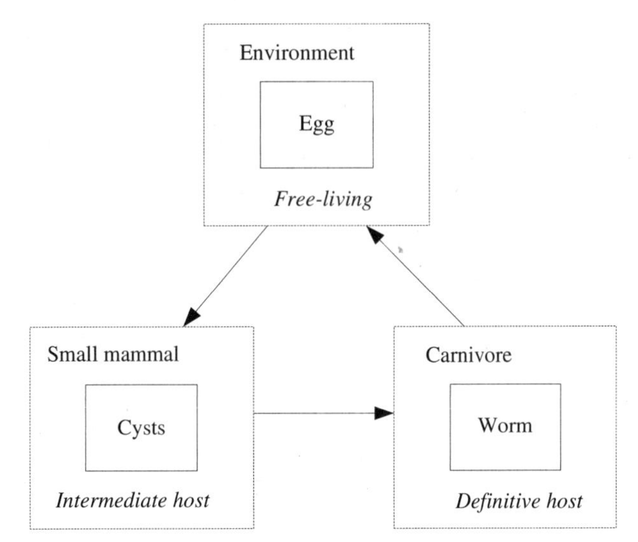

Figure 16.1 Simplified schematic representation of the life cycle of *E. multilocularis*.

although overall densities average 7/ha, in favorable locations there can be 2 to 3 colonies per hectare, leading to substantially higher densities. Each colony forages in an area of radius approximately 40 to 50 m around their burrow. Although *R. opimus* individuals can migrate considerable distances, they tend to be sedentary, particularly during the spring and summer. Foxes, too, are relatively sedentary in March through August while they raise their cubs, and in the desert regions they favor the damper microhabitats where *R. opimus* is found. Foxes migrate long distances in the autumn and live around human habitations and in wooded and riverine areas in the winter.

Hence in Kazakhstan's deserts, transmission is most likely in the spring, when both hosts are present and active and climatic conditions are suitable for *E. multilocularis* egg survival. Foxes eat adult rodents infected the year before, then there is a pre-patent period of around a month, after which eggs are excreted to be picked up by a new rodent host. There is only a 2– to 3–month period in which the climate is equable enough for transmission. There is a lag of several weeks between rodent infection and the beginning of cyst growth, but once cysts start to grow rapidly, they can damage the host relatively quickly (Devouge and Ali-Khan 1983). Rodents infected late in the season can carry cysts through the winter because low external temperatures retard cyst growth (Novak 1983). Adult worms do not live for long, so overwintering in foxes is not significant.

Approaches to Modeling *E. multilocularis* Dynamics

Traditional approaches to epidemiological modeling often use compartmental models, with the host as the focus of analysis and with differential equations governing the movement of individuals between susceptible, latent, infected, and recovered compartments (Anderson and May 1991). Models of macroparasites usually also include

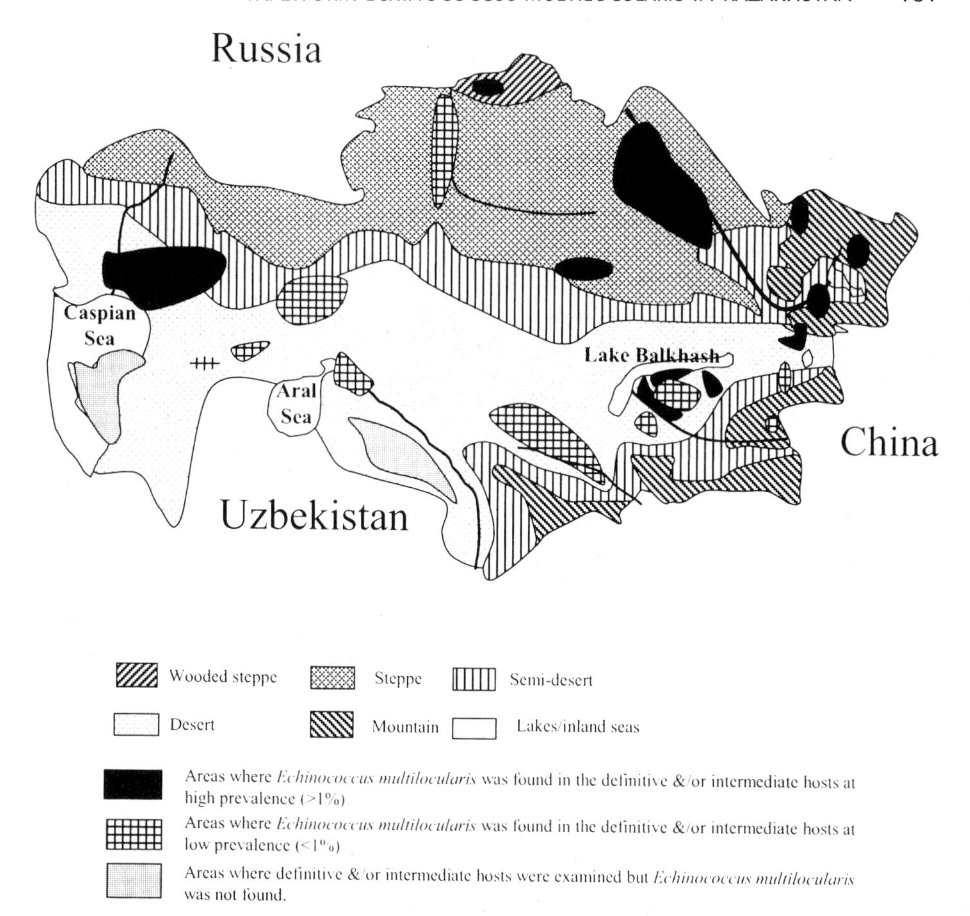

Figure 16.2 Main ecological zones of Kazakhstan and prevalence of *E. multilocularis* as found in large-scale field studies carried out throughout the country (Shaikenov and Torgerson 2001). The areas discussed in this chapter are those in which a low prevalence of *E. multilocularis* was found (*hatched areas*) in the desert region (*pale dotted areas*).

parasite numbers, either explicitly or implicitly. Roberts (1994) presents a simple compartment model for *E. multilocularis* that includes a parameter proportional to the number of eggs on the pasture and the number of worms in the definitive host. This parameter is used as the explanatory variable in functions describing the rate of acquisition and loss of immunity by hosts. Models of the dynamics of *E. granulosus,* a closely related species that cycles between dogs and domestic livestock, have fitted differential equations to data to represent prevalence at age in intermediate hosts. This allows inferences to be made about infection pressure and whether hosts acquire immunity after prolonged exposure to the parasite (Torgerson et al. 1998).

By approaching the problem of modeling *E. multilocularis* dynamics from the RAMAS GIS perspective, we take a very different view of the important features of the parasite's population dynamics. Because RAMAS only models single species, we focus on the parasite population itself rather than taking the usual host-focused approach. Because we are interested in metapopulation dynamics within a landscape, we focus on the transitional zone between the wet steppe areas, where *E. multilocularis* is found at relatively high prevalence, and the desert areas, where it is not able to persist. Hence the habitats of interest are the semiarid areas where it is found only in rodents inhabiting damper areas such as depressions. By using RAMAS, we could

- predict conditions for persistence of *E. multilocularis* in marginal habitat (i.e., in areas of desert/semidesert in Kazakhstan)
- examine the effects of climatic variability on persistence and assess the potential for persistence in desert areas that may be affected by global climate change
- predict the host communities and vegetation structures where the parasite might be able to persist, and hence help in targeting control and avoidance measures to assist in reducing human infections.

The parasite population is visualized with cysts as "individuals" within the "habitat" of a sedentary rodent population. The density of the rodent population and the vegetation type determine the carrying capacity and the habitat suitability of a patch. Dispersal between patches occurs through rodents being eaten by foxes. The parasite then develops into an adult worm and is transported to other patches by the fox. Fecundity rates are the rate of release of eggs into a patch by the adult worm. Hence, the model has three stages—egg, cyst, and worm—but only the worms can disperse and reproduce.

Using RAMAS GIS to Model *E. multilocularis*

The data that are currently available are not adequate for model parameterization. Extensive field surveys are required before the model can be parameterized with confidence and then used to address the modeling aims discussed above (Table 16.1). Here we present a preliminary illustrative model that gives an example of the dynamics of the system (Table 16.2). The output from this model is a low and variable prevalence of cysts, with regular metapopulation extinction (Figure 16.3). This is in line with our field observations.

Habitat Suitability

The landscape capability of RAMAS GIS means that we can use maps of rodent densities and vegetation type to determine the number of patches of suitable habitat for the parasite, the suitability of each patch, and thereby the number of parasite populations that exist in a given area. In our example, we used typical input maps of rodent density and vegetation cover to create a metapopulation with nine patches.

The microclimate in a shaded vegetated area differs substantially from that in surrounding areas—hence the longevity of the free-living egg is influenced by local vegetation type, as well as by overall climatic conditions. Because the rodents are effectively the habitat for the parasite, habitat suitability is a function not just of climatic conditions but also of rodent population density—hence the vegetation type (representing the length of time for which an egg is viable and so available for ingestion by a rodent) and the number of rodents in an area are combined to give a measure of habitat suitability. We model the carrying capacity of a patch for cysts as a nonlinear function of habitat

Table 16.1 Data needs for model parameterization

Component	Data Required	Use in Model
Vegetation	Map of vegetation types, including rodent densities and cumulative annual egg viability under normal conditions and in catastrophe years.	Habitat suitability map
Fox and rodent biology	Rodent densities: mean and variability, how they vary with vegetation type, and how they are affected by catastrophes	Habitat suitability function, carrying capacity
	Fox densities - mean and variability, and how they are affected by catastrophes.	Dispersal probability
	Fox movement distances: mean, shape of function and how they are affected by catastrophes	Dispersal distance function
Parasite biology	Transition probabilities between the three stages—egg, cyst, worm—with associated variability	Survival probabilities in stage matrix
	Number of eggs produced per adult worm, and effect of population-level density d ependence on this	Fecundity rates, density dependence
	Effect of catastrophes on egg survival, hence on effective number of eggs produced per adult worm	Fecundity rates

suitability. We also make intrinsic growth rate dependent on vegetation type, thus more directly capturing the relationship between microclimate and egg survival.

Density Dependence

Because parasites occupy individual hosts, density dependence happens at the intrahost level. Parasites in general tend to be overdispersed in their distributions, so that a few individuals contain large numbers of parasites and most contain very few. Density dependence takes a number of forms. In definitive hosts of tapeworms in general, the more tapeworms there are in the gut, the fewer eggs each produces (Keymer 1982). In the intermediate host, *E. multilocularis* is unusual in carrying out asexual reproduction, so that even a single ingested egg can, over time, produce a large number of infective cysts. Complex density-dependent processes are likely because an animal with a large number of cysts is both more likely to succumb to parasite-related mortality (reducing transmission rates) and to be preferentially predated (producing inverse density dependence and increasing transmission rates). The potential effects of host immunity further complicate the picture.

We cannot model these complexities explicitly. In particular, there is no mechanism for modeling overdispersion because we cannot model individual hosts. A simple population-level representation of density dependence in foxes could be included by assuming that fecundity rates vary with the total number of adult worms, which would be adequate if overdispersion was not severe. However, the literature on the related species *E. granulosus* suggests that density-dependent effects (caused by crowding of worms in the gut or

Table 16.2 Major assumptions made in the example model. The assumptions are presented in the format of the RAMAS summary file, with the inclusion of information about the patch model

Patch model

Input maps: Vegetation cover, Rodent density
Habitat suitability = thr([veg],0.2)*[rodents]
Habitat suitability threshold = 0.2
Carrying capacity = 10*(1-exp(-0.25*ahs))
Rmax = 1+[veg]

Stage-specific multipliers

Eggs	Cysts	Worms
0.5	1.0	1.0

Stages

Stage	Relative dispersal	Basis for density dependence	Breeding
Eggs	0	False	0
Cysts	0	True	0
Worms	1	False	1

Dispersal

All migration/dispersal rates are symmetric (same in both directions).
Migration/dispersal rates (Mij) are based on the following function of the distance between populations (Dij):
Mij = 1.00 * exp(-Dij ^ 2.00 / 6.00), if Dij <= 100.00
Mij = 0.0, if Dij > 100.00
Dispersal depends on target pop. K

Stage matrix

	Eggs	Cysts	Worms
Eggs	0.0	0.0	600.0
Cysts	0.005	0.0	0.0
Worms	0.0	1.0	0.1

Density dependence
Density dependence is Scramble, based on cyst abundance, and affects survival rates

Catastrophe
Name: Dzhut (harsh winter)
Extent: Regional
Probability = 0.1
Affects abundances

by an immune response) only occur with very high worm numbers and repeated exposure (Gemmell et al. 1986), while Roberts et al. (1995) state that there is no evidence for acquired immunity or parasite-induced mortality in the definitive host for *E. multilocularis.* Thus, we do not model density dependence within the definitive host.

In the intermediate host, however, density dependence is important. Cysts continue to grow and multiply in the rodent, causing severe damage and raising host mortality rates (Devouge and Ali-Khan 1983; Schantz et al. 1982). We model density dependence implicitly through the nonlinear relationship between habitat suitability and carrying capacity. We also include it explicitly through scramble-type density dependence, with cyst survival dependent on cyst abundance. This provides a simple population-level approximation to host acquired immunity and parasite-induced host mortality.

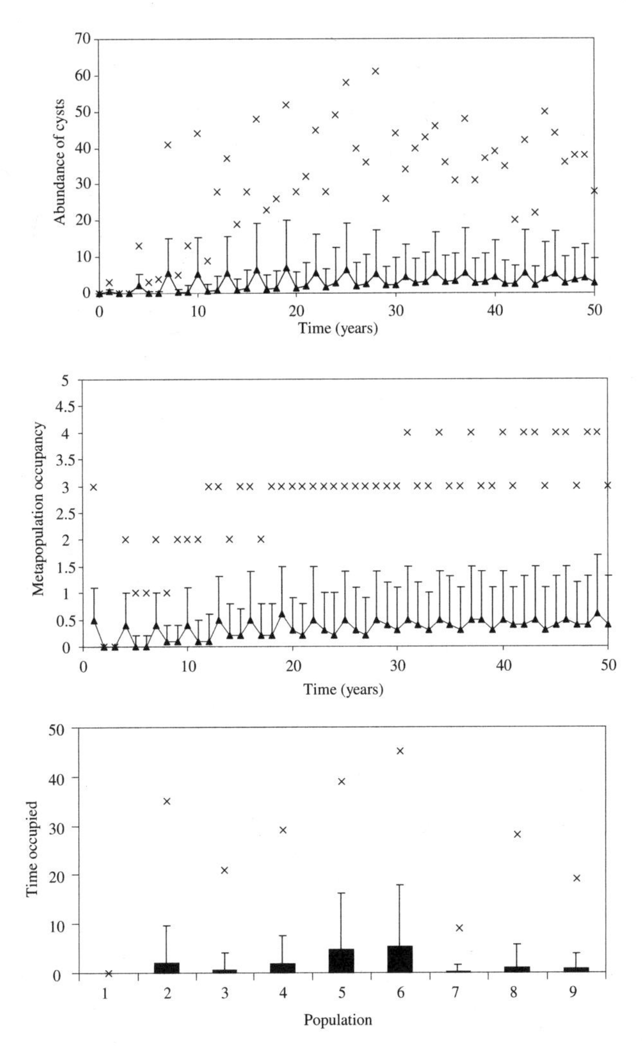

Figure 16.3 Output of the example model. In each case the mean is given as a solid line or bar, the error bars represent 1 SD, and the maximum value is given as a cross. The population is counted only as the number of cysts (excluding the eggs and adult worms). Results are from 50 iterations of 50 years duration: **(top)** abundance of cysts in the metapopulation as a whole; **(middle)** number of occupied patches (excluding year 1, in which all nine patches are occupied); **(bottom)** length of occupancy for each of the nine patches.

Variability

The semiarid areas of Kazakhstan have a highly seasonal and unpredictable climate. There are frequent drought years and occasional years of very high snowfall. Harsh winters (called *dzhuts*) may cause host mortality and may affect transmission through changes in host behavior. Egg survival may also be affected by the weather during the spring. We incorporate stochasticity into the model through variability in carrying capacity and through occasional catastrophes that cause greater egg mortality.

Dispersal

The part of the life cycle that takes place within the fox is modeled through parasite dispersal rate and fecundity. We assume dispersal in rodents is negligible. In order to capture the dependence of fox foraging behavior on rodent availability, we assume that dispersal is dependent, through carrying capacity, on the number of rodents in the receiving population. Other than this, dispersal is assumed to be related to distance.

In the spring transmission period, a fox may forage in several neighboring patches within its home range. During this period the cysts ingested in one patch can develop into adult worms within the fox and release their eggs in another patch. However, in our model we use an annual time step, so the subtleties of dispersal at two scales (within and between years) cannot be captured.

Management Actions

Because we are dealing with a parasite rather than an endangered species, management focuses on sending the metapopulation to extinction rather than preserving it. The main approach to controlling *E. multilocularis* is using bait laced with the anthelminthic praziquantel. A fox-baiting program in Europe aimed at *E. multilocularis* used 15 baits per km^2 and succeeded at reducing prevalence of the parasite in foxes from 24% to 4% (Schantz et al. 1995). We model this using the "harvest" option, assuming that if bait is left in a patch, a given proportion of the worms in that patch is killed. The increase in extinction risk from baiting in any combination of patches and at different frequencies could then be assessed and compared to cost data for a full evaluation of a strategy's cost-effectiveness.

Model Applicability

We make a number of assumptions in this study, which may not be generalizable to many other parasites (Table 16.3). For example, we assume that hosts act as habitat for the parasite and as a medium for dispersal, but that the parasite itself is not a significant driver of host dynamics. We model host dynamics implicitly to some extent, by making parasite carrying capacity and vital rates stochastic and prone to catastrophes (hence mimicking climatic effects on host populations). We model fox foraging behavior by making dispersal rates dependent on parasite density in the destination patch. But this uses parasite density as a proxy for rodent density, to which it is not linearly related. We also assume that the predator-prey relationship is not intense, so that the population dynamics of each participant is not significantly affected by the other.

These assumptions may possibly be valid in the case of *E. multilocularis* in the semi-arid regions of Kazakhstan, but not for most multiple-host parasite species. They may also be realistic for directly transmitted helminths of the gastrointestinal tract. *E. granulosus* also exists in an endemic steady state, with no significant acquired host immunity, low reproductive ratios (number of new parasites generated by each adult parasite), and low prevalence (Roberts 1994). Other parasites—for example, *Oestrus ovis*—may be maintained at a low level due to consistently low biotic potential or poor transmission success. Parasites—for example, *Skrjabinema* spp.—can also exist at high prevalence without host

Table 16.3 Assumptions made in the modeling

Assumptions	Explanation	Validity
The parasite is in an endemic steady state	Parasite population size constant, insignificant effects of density-dependence, low reproductive ratio	Probably yes: low prevalences, long-term parasite persistence
No acquired host immunity	Acquired immunity is the main density-dependent constraint, manifested as a nonlinear host age-prevalence relationship	Probably yes, but no direct evidence; rodents short-lived
Rodent population sizes within patches relatively stable	If there is a strong predator-prey relationship between foxes and rodents, then the relationships between dispersal, fecundity, and carrying capacity will be too complex for the modeling framework	Probably yes, but inadequate data
Fox population sizes stable with foraging behavior related simply to rodent density in patches and interpatch distance	See above. Foxes are partially reliant on *R. opimus* at only one particular time of year; they are sedentary at this time, with relatively stable population sizes	Probably yes, but inadequate data
Minimal parasite-related mortality or morbidity in either host	There is no feedback assumed between population size and habitat availability or transmission rate	Dubious for rodents: parasite can be found at high prevalence in foxes, but very low prevalence in rodents, which may suggest increased predation susceptibility of infected animals.
No significant loss of generality from annual rather than seasonal timestep	There are complex seasonal interactions between egg viability, host presence, and foraging behavior that we have not considered explicitly (they could be considered in RAMAS with better data availability)	Transmission is probably only significant at key points in the year (late spring)
Each host is a single species only	If multiple host species involved, complexity may increase enormously, particularly if they are biologically and behaviorally very different	Field data suggest one intermediate and one definitive host in this particular area

Note: Validity is the likelihood that the assumption holds for the particular system studied here.

immunity. In cases such as these, the modeling framework we have developed may be applicable.

One of the most interesting features of *E. multilocularis* in semiarid areas is its degree of dependence on environmental factors. The potential for parasite transmission is highly constrained by the strongly seasonal climate and is affected by climatic unpredictability. Transmission can only occur at times when the rodents are active, predators are present, and eggs are viable; this narrows the timing down to spring. The presence of patches with a favorable microclimate allows parasites to persist in the desert region. However, these patches of lusher vegetation may also have an important role in host dynamics, which could further complicate the modeling.

We have discussed the potential applicability of RAMAS GIS to the system under study, using an illustrative model. Our approach gives a very different perspective to that normally encountered in the parasite modeling literature. RAMAS allows us to emphasize the underresearched spatial aspects of parasite population dynamics, which makes it a valuable additional tool for parasite modelers. The method used in the program of scoring patches on habitat suitability could be a useful simplification where several factors important to transmission covary with habitat type. Hence in cases such as *E. multilocularis,* where these aspects are key components of system dynamics, using RAMAS GIS may produce meaningful conclusions that could be used to drive management interventions.

Acknowledgments We gratefully acknowledge the financial support of INTAS Project 97-40311 for this work.

References

Alkarmi, T. O., and Ali-Khan, Z. 1984. Chronic alveolar hydatidosis and secondary amyloidosis: pathological aspects of the disease in four strains of mice. *British Journal of Experimental Pathology* 65: 405–417.

Anderson, R. M., and May, R. M. 1991. *Infectious diseases of humans: dynamics and control.* Oxford University Press, Oxford.

Devouge, M., and Ali-Khan, Z. 1983. Intraperitoneal murine alveolar hydatidosis: relationship between the size of the larval cyst mass, immigrant inflammatory cells, splenomegaly and thymus involution. *Tropenmedizin und Parasitologie* 34: 15–20.

Eckert, J., Rausch, M. A., Gemmell, M. A., Giraudoux, P., Kamiya, M., Liu, F. J., Schantz, P. M., and Ronnig, T. 2001. Epidemiology of *Echinococcus multilocularis, Echinococcus vogeli* and *Echinococcus oligarthrus.* Pages 164–182 in *WHO/OIE Manual on Echinococcosis in Humans and Animals.* World Health Organization, Geneva.

Gemmell, M. A., Lawson, J. R., and Roberts, M. G. 1986. Population dynamics in echinococcosis and cysticercosis: biological parameters of *Echincoccus granulosus* in dogs and sheep. *Parasitology* 92: 599–620.

Hess, G. 1996. Disease in metapopulations: implications for conservation. *Ecology* 77: 1617–1632.

Hudson, P., Rizzoli, A., Grenfell, B., and Heesterbeek, H. (eds.). 2002. *Ecology of wildlife diseases.* Oxford University Press, Oxford.

Kennedy, C. R. 2001. Metapopulation and community dynamics of helminth parasites of eels, *Anguilla anguilla,* in the River Exex system. *Parasitology* 122: 689–698.

Keymer, A. 1982. Tapeworm infections. Pages 109–138 in R. M. Anderson (ed.), *Population dynamics of infectious diseases: theory and applications.* Chapman and Hall, London.

Mollison, D., and Levin, S. A. 1995. Spatial dynamics of parasitism. Pages 384–398 in B. T. Grenfell and A. P. Dobson (eds.), *Ecology of infectious diseases in natural populations.* Cambridge University Press, Cambridge.

Novak, M. 1983. Growth of *Echinococcus multilocularis* in gerbils exposed to different environmental temperature. *Experientia* 39: 414.

Roberts, M. G. 1994. Modelling of parasitic populations: cestodes. *Veterinary Parasitology* 54: 145–160.

Roberts, M. G., Smith, G., and Grenfell, B. T. 1995. Mathematical models for macroparasites of wildlife. Pages 177–208 in B. T. Grenfell and A. P. Dobson (eds.), *Ecology of infectious diseases in natural populations.* Cambridge University Press, Cambridge.

Schantz, P. M., Chai, J., Craig, P. S., Eckert, J., Jenkins, D. J., MacPherson, C. N. L., and Thakur, A. 1995. Epidemiology and control of hydatid disease. Pages 233–332 in R. C. A. Thompson and A. J. Lymbery (eds.), *Echinococcus and hydatid disease.* CAB International, Oxon.

Shaikenov, B. Sh., and Torgerson, P. R. 2002. Distribution of *Echinococcus multilocularis* in Kazakhstan. Pages 299–307 in P. S. Craig and Z. Pawlowski (eds.), C*estode zoonoses: Echinococcosis and Cysticercosis, an emergent global problem.* NATO Science Series. IOS Press, Amsterdam.

Torgerson, P. R., Williams, D. H., and Abo-Shehada, M. N. 1998. Modelling the prevalence of *Echinococcus* and *Taenia* species in small ruminants of different ages in North Jordan. *Veterinary Parasitology* 79: 35–51.

PART III

FISHES

17

Modeling Viability of Fish Populations

An Overview

CHRIS C. WOOD

Species and Life History Diversity

A fish is a poikilothermic, aquatic chordate with appendages (when present) developed as fins, whose chief respiratory organs are gills and whose body is usually covered in scales (Berra 2001). Allowing for a few exceptions, fish are classified in three phyla comprising about 25,000 species (Nelson 1994). Although not as rich in species as plants or invertebrates, fish account for about half of all vertebrate species (the subject of Parts III to VI of this book).

Fish lineages are over 400 million years old (Carrol 1988), and fish occur in virtually all aquatic habitats on Earth. Marine species are found from tropical reefs to sunless abyssal plains; freshwater species may occur in frozen arctic lakes or as resting stages in ephemeral desert ponds; others are diadromous, negotiating both marine and freshwater habitats at different life history stages. Slightly less than half (41%) of fish are freshwater species, but these live in only 0.01% of the Earth's visible water (Horn 1972).

From a modeling perspective, fish exhibit life histories that span most of those encountered in vertebrates, enough to tax even the flexibility of RAMAS. Longevity ranges from a few months in some cyprinodonts to more than a century in some rockfishes (*Sebastes*). Most fish lay eggs, but ovovivipary and vivipary occur in some taxa so that fecundity (in the sense of litter size) spans at least seven orders of magnitude; for example, the basking shark and the manta ray produce only one to two "pups" per year, whereas the ocean sunfish and greasy grouper may produce 300 million eggs. Once the fish are mature, spawning may occur almost daily in many tropical reef species, annually in most temperate species, or only once every few years in sturgeons and some sharks; other species, like Pacific salmon, die after spawning. Variability in recruitment probably scales with fecundity in most cases, but some fecund, long-lived species of *Sebastes* appear to persist only because of infrequent but very successful spawning events (Love et al. 2002), similar to mast years in some plants. Reproduction is typically gonochoristic (separate sexes) with chromosomal sex determination, but temperature-dependent sex determination, sex reversal, functional hermaphroditism, and parthenogenesis also occur in some species (Redding and Patino 1993).

Such diversity in life history makes any generalization difficult. Most marine fish are capable of widespread dispersal at some life history stage because their aquatic habitat lacks physical boundaries. Hart and Cadrin (Chapter 21 in this volume) consider the consequences of ignoring dispersal among stocks when modeling the population dynamics and management of yellowtail flounder. Fish also disperse widely within some river systems, but the size and interconnectedness of rivers and lakes ultimately determine opportunities for dispersal in freshwater.

Alternatively, fish populations can be highly structured genetically or demographically in the absence of physical barriers to dispersal because of a common tendency to home to their natal spawning habitat (philopatry). Philopatry is characteristic of many diadromous species such as the chinook salmon modeled by Ruckelshaus et al. (Chapter 19 in this volume). It may also be common in marine species, as reported for Atlantic cod (Ruzzante et al. 1998) and herring (Smedbol and Stephenson, Chapter 22 in this volume; Fu et al., Chapter 23 in this volume).

Fish differ from mammals and birds in that their growth is typically indeterminate and plastic (Weatherley 1990). Not only can fish grow faster or slower as a result of food supply or temperature, they can actually lose and recover body mass in response to rations and can continue to grow throughout their life. The greater indeterminacy and plasticity of growth in fish probably provides greater scope for density-dependent compensation of population growth rate. Even so, the potential for compensation varies widely across taxa, being lowest in the slow-growing, low-fecundity elasmobranchs (Musick 1999).

Faster growth typically results in larger body size, and hence higher fecundity, but it also can lead to earlier maturity and a greater lifetime contribution to recruitment within the population. This capacity for compensatory growth and reproduction in fish is reflected in plots of population growth rate or "recruits per spawner" versus spawning abundance that are almost invariably concave up, as assumed in the Ricker, Beverton-Holt, or ceiling stock-recruitment functions. When this is true, maximum sustainable yield (MSY) will always be provided by a spawning abundance less than half that in the population at unharvested equilibrium, assuming that the relationship itself is stationary, of course. In contrast, the same relationship might be expected to be convex up in many bird and mammal populations for reasons summarized by Sutherland and Gill (2001), implying that MSY is expected from a breeding abundance more than half that in the unharvested population.

The strong compensatory responses that seem to characterize most fish populations at intermediate to low densities do not necessarily occur at very low density. As in other taxa, there are many reasons to expect depensation at low density (Allee effects), but good empirical data are lacking (Petersen and Levitan 2001). Myers et al. (1995), and subsequently Liermann and Hilborn (1997), reviewed 128 fishery data sets yet found little evidence that depensation occurs in commercially exploited fish species. Reflecting this uncertainty, Fu et al. (Chapter 23 in this volume) evaluate conservation limit reference points for Pacific herring under different levels of possible Allee effect.

Conservation and Management Issues

Although fish are mostly hidden from view and tend to lack charisma, they can be culturally and economically important. They represent the only major food source still harvested from wild populations (Ryman et al. 1995). Failures in fisheries management have been spectacular and costly, both for freshwater fish (e.g., the introduction of Nile perch to Lake Victoria, which resulted in the extinction of 300 species of cichlids; Ogutu-Ohwayo 1990) and marine fish (e.g., the collapse of Atlantic cod; Hutchings and Myers 1994). The public is becoming increasingly aware of the importance and precarious state of aquatic biodiversity and the fundamental problems with fisheries (Harvey 2001). Concurrently, the Convention on Biological Diversity and national legislation to protect endangered species—like the Endangered Species Act in the United States, the Environment Protection and Biodiversity Conservation Act in Australia, and the Species at Risk Act in Canada—recognize the intrinsic value of wildlife species including fish and extend protection to them.

Some 30% of all fish species recently assessed by the World Conservation Union (IUCN) were designated in threatened categories: critically endangered (CR), endangered (EN), and vulnerable (VU) (http://www.redlist.org/info/tables.html). Most (84%) of the threatened fish are freshwater species. Approximately 4% of North American freshwater species are expected to be lost each decade, and this is nearly five times the rate for terrestrial species (Ricciardi and Rasmussen 1999). Freshwater fish are particularly vulnerable because their habitat is bounded by landscape features that limit their ability to avoid anthropogenic or natural disturbances. In Chapter 19 (this volume), Ruckelshaus et al. examine how the viability of a metapopulation of chinook salmon is affected by its number of component populations subject to catastrophic losses of freshwater habitat due to volcanic eruptions, landslides, or major human impacts under various levels of dispersal and spatial correlation among the populations. Suitable freshwater is itself a limited commodity often abstracted from waterways for domestic and agricultural consumption, and pollution is exacerbated in confined areas. In Chapter 18 (this volume), Wanzenböck examines the viability of one of the last remaining populations of the European mudminnow in the face of declining groundwater levels. In Chapter 20 (this volume), Nicol and Todd investigate alternative options for restoring habitat to improve the viability of one of two remaining populations of trout cod.

IUCN designations suggest that marine fish (excluding anadromous species) are less threatened than freshwater fish. As yet, there are no documented cases of a marine fish going extinct as a result of human activities (Roberts and Hawkins 1999). However, overfishing has caused local extirpation (Brander 1981, Casey and Myers 1998, Dulvy et al. 2000, Powles et al. 2000, Stevens et al. 2000) and some marine fish populations have declined by up to 99% within three generations, thereby triggering IUCN threatened designations (Reynolds et al. 2001). Total catch biomass from marine fisheries has remained relatively stable in recent decades, but the composition of the catch has shifted from large piscivorous fish (high trophic levels) to smaller planktivorous fish (lower trophic levels) in a trend known as "fishing down the food chain" (Pauly et al. 1998). Thus, the management of lower trophic level species like herring is of special concern (see Chapters 22 and 23 in this volume)

Fisheries Science and Sustainability

Fish are hard to study or census directly, so the theory and statistical tools developed for fisheries management are generally more sophisticated than in other sectors of natural resource management. This sophistication has not prevented the collapse of fisheries, however (Ludwig 2001). Harvesting generally accelerates the process of extinction, and an adequate conceptual and theoretical framework including stochasticity and uncertainty, and detailed studies of ecology are needed to improve the management of exploited species (Lande et al. 2001). More important, fisheries management needs to include uncertainty in the decision-making process (Wade 2001).

One technical problem with traditional fisheries assessments is that spatial structure and biological interactions have often been ignored or oversimplified; for example, there have been few attempts to estimate MSY by taking account of the spatial dynamics of the population (Punt and Smith 2001). Another problem is that implementation error (Francis and Shotton 1997) resulting in part from errors in assessing stock size has seldom been taken into account when developing harvest policy. It is worth noting therefore that Fu et al. (Chapter 23 in this volume) have incorporated both spatial dynamics and implementation (sampling) error into their RAMAS-based framework for evaluating alternative harvesting strategies.

Advantages and Limitations of PVA

Quantitative modeling approaches like population viability analysis (PVA) have several important advantages over other methods for developing advice on fisheries harvesting and conservation policy. To summarize from Shaffer et al. (2002) and Akçakaya (Chapter 1 in this volume), a PVA can consolidate and organize what is known and not known; it can force managers and policy makers to face the fact that size does matter; and it can make decision-making more transparent. A model whose sensitivity to assumptions can be explored is easier to contest than the judgment of experts.

RAMAS GIS provides a convenient, flexible platform to explore outcomes and evaluate relative risk of extinction under alternative management scenarios. The option to export a standardized "model summary and assumptions" file greatly improves transparency. Of particular interest for evaluating fisheries harvest policies, there are options in RAMAS to simulate "proportional threshold harvesting" (Lande et al. 2001), implementation error resulting from errors in assessing harvestable abundance, and dispersal within a metapopulation structure. Process error including demographic and environmental stochasticity and catastrophes are all easily simulated.

In practice, authors of the fish case studies had difficulty estimating temporal variation (environmental stochasticity) from historical data without including demographic stochasticity or measurement error. Failure to do this means that extinction risk will be overestimated (Akçakaya 2002, Saether and Engen 2002). Authors of the fish case studies also had difficulty choosing the density dependence parameters, known to be an influential component of fish population dynamics. Most assumed ceiling density dependence or simulated without density dependence. Both of these assumptions are conservative in that productivity is likely underestimated at low density, and, hence, the risk of extinction will be overestimated (Burgman et al. 1993).

RAMAS, and PVA in general, do have limitations, of course. To cite Beissinger (2002), the four dominant causes of uncertainty in predictions from PVA are (1) poor data, (2) difficulties in parameter estimation, (3) weak ability to validate models, and (4) effects of alternative model structures. Goodman (2002) points out that this uncertainty does not invalidate PVA; rather, it confirms that PVA is needed to inform decisions. To make this claim, however, one needs to correctly represent the uncertainty in parameter and model estimation. This is not trivial to do in RAMAS. A distribution of trajectories obtained under environmental and demographic stochasticity, assuming point estimates for the vital rate and density-dependent parameters, must still be considered a point estimate (Wade 2002). The effect of parameter uncertainty on outcomes can be bounded by simulating with parameters chosen from the lower and upper range of plausibility, but outcomes may then be so uncertain as to preclude comparison among alternative scenarios (see discussion in Chapter 1 in this volume). In numerical experiments, Taylor et al. (2002) found that outcomes ranked under a short-cut precautionary-RAMAS approach, in which vital rates were reduced by 1 SE from the best point estimate, compared reasonably well with those ranked under a full Bayesian analysis. It may also be feasible in some cases to approximate a Bayesian solution by resimulating with parameters drawn randomly from prior distributions.

Provided conclusions from PVA are properly qualified, the idea of exploring alternatives and trying to find conservation policies that are robust under the uncertainties is still as compelling as ever (Ludwig and Walters 2002). Despite obvious weaknesses in data and other uncertainties, it seems that authors for each of the fish case studies in the following chapters were able to gain insight and make useful recommendations for policy based on their simulation results.

Acknowledgments I am grateful to the following individuals for serving as referees for the chapters in Part III: R. Akçakaya, M. Bradford, D. Hart, B. Harvey, T. Johnston, R. Mohn, S. Nicol, R. Routledge, M. Ruckelshaus, A. Sinclair, J. Schweigert, K. Smedbok, M. Trudel, J. Wanzenböck, and D. Ware.

References

Akçakaya, H. R. 2002. Estimating the variance of survival rates and fecundities. *Animal Conservation* 5: 333–336.

Beissinger, S. R. 2002. Population viability analysis: past, present, future. Pages 5–17 in S. R. Beissinger and D. R. McCullough (eds.), *Population viability analysis.* University of Chicago Press, Chicago.

Berra, T. M. 2001. *Freshwater fish distribution.* Academic Press, San Diego.

Brander, K. 1981. Disappearance of common skate *Raja batis* from Irish Sea. *Nature* 290: 48–49.

Burgman, M. A., Ferson, S., and Akçakaya, H. R. 1993. *Risk assessment in conservation biology.* Chapman and Hall, London.

Carrol, R. L. 1988. *Vertebrate paleontology and evolution.* W. H. Freeman, New York.

Casey, J. M., and Myers, R. A. 1998. Near extinction of a large widely distributed fish. *Science* 281: 690–692.

Dulvy, N. K., Metcalfe, J. D., Glanville, J., Pawson, M. K., and Reynolds, J. D. 2000. Local extinctions and shifts in community structure of skates masked by fishery stability. *Conservation Biology* 14: 283–293.

Francis, R.I.C.C., and Shotton, R. 1997. "Risk" in fisheries management: a review. *Canadian Journal of Fisheries and Aquatic Sciences* 54: 1699–1715.

Goodman, D. 2002. Predictive Bayesian population viability analysis: a logic for listing criteria, delisting criteria, and recovery plans. Pages 447–469 in S. R. Beissinger and D. R. McCullough (eds.), *Population viability analysis.* University of Chicago Press, Chicago.

Harvey, B. 2001. Biodiversity and fisheries: a primer for planners. Pages 1–64 in *Blue millennium: Proceedings of an International Workshop on Managing Global Fsheries for Biodiversity, Victoria, June 2001.* World Fisheries Trust. Available at http://www.worldfish.org/proj_globe_3.htm.

Horn, M. H. 1972. The amount of space available for marine and freshwater fishes. *Fishery Bulletin* 70: 1295–1297.

Hutchings, J. A., and Myers, R. A. 1994. What can be learned from the collapse of a renewable resource? Atlantic cod, *Gadus morhua,* of Newfoundland and Labrador. *Canadian Journal of Fisheries and Aquatic Sciences* 51: 2126–2146.

Lande, R., Saether, B. E., and Engen, S. 2001. Sustainable exploitation of fluctuating populations. Pages 67–86 in J. D. Reynolds, G. M. Mace, K. H. Redford, and J. G. Robinson (eds.), *Conservation of exploited species.* Cambridge University Press, Cambridge.

Liermann, M., and Hilborn, R. 1997. Depensation in fish stocks: a hierarchic Bayesian meta-analysis. *Canadian Journal of Fisheries and Aquatic Sciences* 54: 1976–1984.

Love, M. S., Yoklavich, M., and Thorsteinson, L. 2002. *The rockfishes of the northeast Pacific.* University of California Press, Berkeley.

Ludwig, D. 2001. Can we exploit sustainably? Pages 16–38 in J. D. Reynolds, G. M. Mace, K. H. Redford, and J. G. Robinson (eds.), *Conservation of exploited species.* Cambridge University Press, Cambridge.

Ludwig, D., and Walters, C. J. 2002. Fitting population viability analysis into adaptive management. Pages 511–520 in S. R. Beissinger and D. R . McCullough (eds.), *Population viability analysis.* University of Chicago Press, Chicago.

Musick, J. A.(ed.). 1999. *Life in the slow lane: ecology and conservation of long-lived marine animals.* American Fisheries Society Symposium 23. American Fisheries Society, Bethesda, Md.

Myers, R. A., Barrowman, N. J., Hutchings, J. A., and Rosenberg, A. A. 1995. Population dynamics of exploited fish stocks at low population levels. *Science* 269: 1106–1108.

Nelson, J. S. 1994. *Fishes of the world*, 3rd ed. Wiley, New York.

Ogutu-Ohwayo, R. 1990. The decline of the native fishes of lakes Victoria and Kyoga (East Africa) and the impact of introduced species, especially the Nile perch, *Lates niloticus,* and the Nile tilapia, *Oreochromis niloticus. Environmental Biology of Fishes* 27: 81–96.

Pauly, D., Christensen, V., Dalsgaard, J., Froese, R., and Torres Jr., F. 1998. Fishing down marine food webs. *Science* 279: 860–863.

Petersen, C. W., and Levitan, D. R. 2001. Allee effects in exploited species. Pages 281–300 in J. D. Reynolds, G. M. Mace, K. H. Redford, and J. G. Robinson (eds.), *Conservation of exploited species.* Cambridge University Press, Cambridge.

Powles, H., Bradford, M. J., Bradford, R. G., Doubleday, W. G., Innes, S., and Levings, C. D. 2000. Assessing and protecting endangered marine species. *ICES Journal of Marine Science* 57: 669–676.

Punt, A. E., and Smith, A. D. M. 2001. The gospel of maximum sustainable yield in fisheries management: birth, crucifixion and reincarnation. Pages 41–66 in J. D. Reynolds, G. M. Mace, K. H. Redford, and J. G. Robinson (eds.), *Conservation of exploited species.* Cambridge University Press, Cambridge.

Redding, J. M., and Patino, R. 1993. Reproductive physiology. Pages 503–534 in D.H. Evans (ed.), *The physiology of fishes.* CRC Press, Boca Raton, Fla.

Reynolds, J. D., Jennings, S., and Dulvy, N. K. 2001. Life histories of fishes and population responses to exploitation. Pages 147–168 in J. D. Reynolds, G. M. Mace, K. H. Redford, and J. G. Robinson (eds.), *Conservation of exploited species.* Cambridge University Press, Cambridge.

Ricciardi, A., and Rasmussen, J. B. 1999. Extinction rates of North American freshwater fauna. *Conservation Biology* 15(5): 1220–1222.

Roberts, C. M., and Hawkins, J. P. 1999. Extinction risk in the sea. *Trends in Ecology and Evolution* 14: 241–246.

Ruzzante, D. E., Taggart, C. T., and Cook, D. 1998. A nuclear DNA basis for shelf- and bank-scale population structure in northwest Atlantic cod (*Gadus morhua*): Labrador to Georges Bank. *Molecular Ecology* 7: 1663–1680.

Ryman, N., Utter, F., and Laikre, L.1995. Protection of intraspecific biodiversity of exploited fishes. *Reviews in Fish Biology and Fisheries* 5: 417–446.

Saether, B. E., and Engen, S. 2002. Including uncertainties in population viability analysis using population prediction intervals. Pages 191–212 in S. R. Beissinger and D. R. McCullough (eds.), *Population viability analysis.* University of Chicago Press, Chicago.

Shaffer, M., Watchman, L. H., Snape III, W. J., and Latchis, I. K. 2002. Population viability analysis and conservation policy. Pages 123–142 in S. R. Beissinger and D. R. McCullough (eds.), *Population viability analysis.* University of Chicago Press, Chicago.

Stevens, J. D., Bonfil, R., Dulvy, N. K., and Walker, P. A. 2000. The effects of fishing on sharks, rays and chimeras (chrondrichthyans), and the implications for marine ecosystems. *ICES Journal of Marine Science* 57: 476–494.

Sutherland, W. J., and Gill, J. A. 2001. The role of behaviour in studying sustainable exploitation. Pages 259–280 in J. D. Reynolds, G. M. Mace, K. H. Redford, and J. G. Robinson (eds.), *Conservation of exploited species.* Cambridge University Press, Cambridge.

Taylor, B. L., Wade, P. R., Ramakrishnan, U., Gilpin, M., and Akçakaya. H. R. 2002. Incorporating uncertainty in population viability analyses for the purpose of classifying species by risk. Pages 239–252 in S. R. Beissinger and D. R. McCullough (eds.), *Population viability analysis.* University of Chicago Press, Chicago.

Wade, P. R. 2001. The conservation of exploited species in an uncertain world: novel methods and the failure of traditional techniques. Pages 110–143 in J. D. Reynolds, G. M. Mace, K. H. Redford, and J. G. Robinson (eds.), *Conservation of exploited species.* Cambridge University Press, Cambridge.

Weatherley, A. H. 1990. Approaches to understanding fish growth. *Transactions of the American Fisheries Society* 119: 662–672.

18

European Mudminnow (*Umbra krameri*) in the Austrian Floodplain of the River Danube

Conservation of an Indicator Species for Endangered Wetland Ecosystems in Europe

JOSEF WANZENBÖCK

The European mudminnow (*Umbra krameri* Walbaum, 1792) has a relatively small area of natural occurrence represented by the lowlands along the middle and lower River Danube including its (former vast) areas of inundation, as well as along the middle and lower River Dniestr (Figure 18.1). The mudminnow is predominantly found in swampy areas, oxbow lakes, and floodplain pools that are covered densely with aquatic vegetation (see papers in Mikschi and Wanzenböck 1995). Land reclamation and draining that began during the Middle Ages has changed the landscape dramatically, and areas suitable for mudminnows have shrunk to small spots and the narrow remnant floodplains of rivers. Consequently, the abundance and distribution of the mudminnow have diminished greatly, as described even 80 years ago (Hanko 1923).

Changes on large spatial scales are mirrored on smaller spatial scales. This case history describes the plight of the mudminnow near the western border of its distribution where it was first described (see Wanzenböck 1995) and was found regularly during the eighteenth, nineteenth, and early twentieth centuries. Sightings became less frequent in the latter half of the twentieth century, and the last specimen was reported in 1975. The species was considered extinct in Austria (Red List of Austrian Animals; Herzig-Straschil 1991) for almost 20 years until its rediscovery in 1992 (Wanzenböck and Spindler 1995).

The mudminnow was rediscovered within the relatively natural floodplain of the River Danube between Vienna and the Slovak border. Upstream, the remaining stretch of the Danube in Austria has been dammed and impounded for hydropower plants. Clearcutting of floodplain forests in the 50-km long stretch downstream of Vienna began in 1984 in preparation for construction of another power plant, but this was discontinued after a major public protest to conserve one of the last nearly natural river landscapes of Europe.

Figure 18.1 Distribution of the European mudminnow indicated by the dashed line.

Plans were made to protect this area as a national park (designated in 1996) and to improve the general ecological situation with hydroengineering measures (see http://www.donauauen.at/html/english/index.html for details).

After its rediscovery, a conservation project was started in 1993 to investigate the population dynamics of the mudminnow and the reasons for its decline, and to develop an action plan for its conservation. The continuing loss of habitat poses the most serious threat to mudminnow populations in Austria. Water regulation in the Danube (started at the end of the nineteenth century) has increased flow velocity and caused the river to cut a deeper channel, thereby lowering the groundwater level in the surrounding floodplain. As a result, the original side channel of the Danube where the mudminnow was rediscovered has been transformed into a chain of disconnected, groundwater-fed ponds. These ponds will dry out completely unless further reduction in groundwater level is prevented. A simple metapopulation model based on empirical measurements of population growth rate is used to demonstrate that this declining trend in habitat capacity is more critical to the mudminnow's persistence than are other factors that affect population growth rate or dispersal among pools.

A Metapopulation Model

This model is found in the file mudminnow.mp in the CD accompanying this volume.

Spatial Structure and Dispersal

Data on population dynamics were acquired between 1993 and 1997 by periodic inspection of ponds of various sizes within the original side channel of the River Danube (Figure 18.2). Mudminnows were found at only six of the sampling stations (2, 3, 6, 7, 9, and 10), and these have been considered together as a metapopulation in this study. All sites inhabited by mudminnows are isolated from the Danube floodplain by the main flood protection dam. Only during very wet years does the Danube flood sufficiently to reconnect some of the pools as one elongated water body. Although artificial ditches were dug to reconnect stretches of the side channel isolated by the dam (e.g., between sampling stations 8 and 9 in Figure 18.2), these were not deep enough to be effective, and they always remained dry. Therefore, dispersal of mudminnows sometimes occurs between sampling stations 2, 3, 6, and 7 but never between stations 2–7 (as a group), 9, and 10. This situation was modeled by representing the area of the side channel with 100 × 100 m raster cells. Coordinates of each population were assigned by counting the rows and columns from the origin in the upper left corner of the map. Parameters for the migration-distance function were chosen to allow a maximum dispersal distance of 1 km, which is sufficient to ensure a low rate of exchange of individuals between stations 2, 3, 6, and 7 (reflecting the rare interconnections), while stations 9 and 10 always remain isolated (see population map in the model mudminnow.mp).

Age Structure and Vital Rates

Mudminnows reach sexual maturity after their first winter so that only mature individuals were observed in spring surveys. Fecundity changes with female size, but it was considered imprudent to sacrifice individuals of this highly endangered species to obtain age- or size-specific data on fecundity and sex ratio. Instead, a simplistic model without stage or sex structure was developed, based on empirical measurements of population growth rate.

Changes in population sizes were monitored over 4 years with two surveys each year: one in spring shortly before spawning and one in autumn. Abundance was estimated using a modified DeLury depletion method (DeLury 1947). Ponds were electrofished

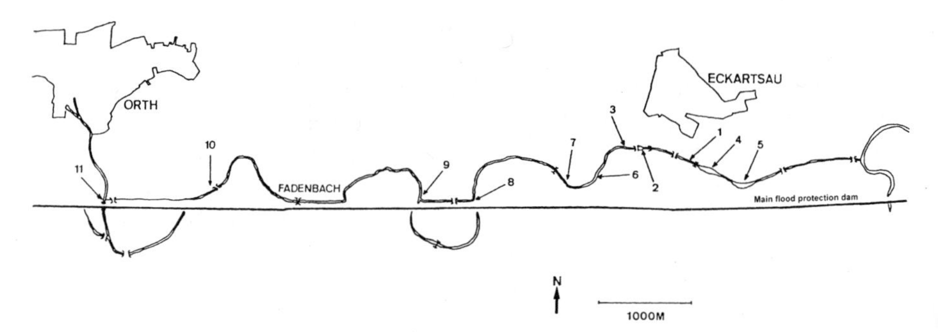

Figure 18.2 Map of the original side channel of the Danube between the villages of Orth and Eckartsau, showing locations of sampling stations and barriers to mudminnow dispersal created by the main flood protection dam (*horizontal line*). Numbers refer to sampling stations.

with two or three passes per survey. Individuals removed during a pass were kept in holding tanks until the survey was finished, and the decline in catch rate was used to estimate initial population size. The original DeLury method was modified by using the average decline in catch rate per pass over all surveys within a pond, under the assumption that the catchability coefficient was constant over successive surveys within the same pond. Initial abundance for each population in the model was set at its average population size over all eight surveys. Maximum population growth rate (R_{max}) was set at 3.2 for all six populations because the maximum increase in population size from spring to autumn observed in at least one population was more than threefold.

Density Dependence and Carrying Capacity

In the first year of investigation, 10 individuals were transferred from the largest population to a garden pond (8 m^2, 1.2 m deep) to observe their maximum density in subsequent years without competition and predation from other fish species. Such a situation may arise in natural habitats when after a winter fish kill mudminnows remain the only species in a pool (see the following section on catastrophes). Maximum density was about twice as high in the garden ponds as in natural habitats. Accordingly, the carrying capacity in the model was set to a value twice as high as the maximum abundance observed in each natural population. Population growth rate was assumed to be density dependent, falling to 1.0 when abundance reached carrying capacity. Both scramble and contest density dependence functions were used during simulations.

Temporal Variation and Catastrophes

Abundance fluctuated within each population but was generally higher in autumn surveys than in spring surveys, reflecting population increase through recruitment over the summer and population decrease because of mortality over the winter. The population increase observed each summer was correlated with hydrological conditions that year. During wet years, groundwater levels were higher and the floodplain pools were larger (implying increased habitat capacity), which resulted in faster growth and higher autumn abundance than in dry years. This situation was captured in the model as temporal variation in carrying capacity with the standard deviation of carrying capacity set to the observed standard deviation in abundance. This estimate may be conservative because mean carrying capacity was set at twice the maximum abundance observed, and variance usually scales with the mean. However, the empirical estimate of variance is sure to include measurement error that should be subtracted for this application.

Measurements by the federal water authority indicate that the groundwater level has dropped by 1 m during the last 10 years. This implies that the pools will disappear within about 50 years for the larger, deeper pools (stations 2 and 9), 45 years for the pool at station 10, 20 years for the pool at station 6, and 15 years for the pool at station 3. These predictions were modeled by specifying a linear temporal trend for each population that reduces its carrying capacity to 0 by the end of the prediction interval.

Population abundance typically declined between the autumn and spring surveys, but more so during cold winters when thick ice covered the pools for many weeks. Mortality was lower in mild winters with less ice or a shorter duration of ice cover, especially if water levels stayed high. During this investigation, cold winters caused catastrophic population declines about every other year. This was modeled by defining regional catastrophic effects that reduced autumn abundance by 80% (hence a multiplier of 0.2) with probability 0.5.

Other random effects were simulated with the demographic and environmental stochasticity options of RAMAS. A normal distribution for environmental variation was assumed initially, but this resulted in excessive truncation, so a lognormal distribution was assumed in subsequent simulations. By choosing appropriate parameters for the correlation-distance function, the spatial correlation of environmental fluctuations was specified as high between proximate populations and minimal between distant populations.

Results

Preliminary Simulations

All simulations were run with 1,000 replications and for 50 time steps (years). Assuming scramble- or contest-type density dependence had no effect on results in preliminary simulations, beyond a difference in the warning message that "Density dependence may be inconsistent with the stage matrix" (diff. = 117% and 85 % for scramble type or contest type, respectively). To improve consistency by this criterion, contest-type density dependence was used in further simulations. However, the message was considered irrelevant because the model was not stage-structured.

Warning messages in the preliminary simulations also indicated that a lognormal distribution for environmental stochasticity would be more appropriate to minimize truncations. Because there were no biological data to justify the continued use of a normal distribution, a lognormal distribution was assumed in all subsequent simulations. This increased the median time to extinction from about 18 to 24 years.

Sensitivity Analysis and Implications for Conservation Measures

Sensitivity analyses were performed using the subroutines "Sensitivity Analysis" and "Comparison of Results." Regional catastrophe probability was changed by 5%, 10%, 20%, 50%, and 100%. Slight changes in this probability (5%–20%) had little effect on time to extinction, but larger changes (50%–100%, implying that cold winters with extensive ice cover would occur in most years) had more pronounced effects, decreasing median time to extinction from 23 years to 17 and 10 years, respectively (Table 18.1). Weather conditions cannot be influenced by management, so these results suggest that it could be important to remediate the adverse effects of oxygen deficit under heavy ice cover. Theoretically this might be achieved if some water flow could be reestablished in the original side channel. Flowing water would decrease the risk of ice cover and increase the opportunity for gas exchange and oxygen saturation should ice cover occur.

Changing initial abundance by 5%, 10%, 20%, 50%, or 100% had no effect on the simulation results. This suggests that augmenting populations with captive-bred individuals would not be worthwhile. Abundance can increase quickly because of relatively

Table 18.1 Effects of changing selected parameters from values used in the basic model on the cumulative probability of extinction of mudminnows, defined as population size = zero

	Years When Cumulative Probability of Extinction Becomes					
Parameter	0.1	0.25	0.5	0.75	0.9	1
Basic model	14	18	24	30	35	46
Regional catastrophe probability +5%	13	17	23	29	33	46
Regional catastrophe probability +10%	13	17	23	29	34	46
Regional catastrophe probability +20%	12	16	22	27	32	43
Regional catastrophe probability +50%	9	12	17	22	27	39
Regional catastrophe probability +50%	7	8	10	11	14	21
Carrying capacity +5%	15	19	26	32	37	>50
Carrying capacity +10%	17	22	28	34	39	>50
Carrying capacity +20%	19	26	33	39	44	>50
Carrying capacity +50%	30	39	46	>50	>50	>50
Carrying capacity +100%	48	>50	>50	>50	>50	>50
No temporal trend in *K*	23	37	>50	>50	>50	>50

high growth rates if conditions are favorable, as was evident from empirical observations in natural habitats (R_{max} = 3.2) and in artificial habitats (10–fold population increase from one spring to the next in garden ponds).

Growth rate had relatively little influence on time to extinction. Doubling growth rate increased median time to extinction by only 2 years (from 24 to 26 years), suggesting that capacity for population increase was not limiting the viability of the mudminnow metapopulation. In contrast, changing carrying capacity by 5%, 10%, 20%, and 50% produced a corresponding increase in median time to extinction from 24 to 26, 28, 33, and 46 years, respectively (Table 18.1). A 100% increase in carrying capacity (*K*) (which may be viewed as a doubling of pond sizes) reduced extinction risk for the metapopulation to 0.12 within 50 years. Corresponding increments in the standard deviation of *K* decreased the median time to extinction from 24 to 22, 21, 19, 14, and 9 years, respectively, indicating a strong effect of variation in habitat availability on extinction risk for the metapopulation.

The temporal trend in *K* could not be varied automatically in the sensitivity module, so a sensitivity analysis was performed manually by rerunning the original model with different temporal trends. Removing the temporal trend in *K* for all populations significantly increased time to extinction for the metapopulation and reduced its extinction risk to 0.367 within 50 years (Table 18.1). This was comparable to the effect of increasing carrying capacity because with no temporal trend in *K*, pool sizes remained relatively stable (within limits defined by standard deviations).

Changing other parameters in the model ("all dispersal rates" and "all correlations") by up to 100% did not change the extinction risk. This seemed counterintuitive because predicted trends in "local occupancy" revealed that populations with connections to other populations tended to persist longer than those without connections. However, additional simulations in which maximum dispersal distance was increased by 500% (i.e., from 1 km to 5 km) confirmed that the dispersal function had no pronounced influences on extinction risk.

Conclusions and Recommendations

The central conclusion emerging from these studies is that the mudminnow metapopulation is unlikely to persist unless groundwater level is stabilized and, if possible, raised to increase habitat availability. A further recommendation is that mudminnow pools should be reconnected by creating more effective ditches along the main flood control dam. However, connectivity should not be enhanced to the point where the groundwater-fed pools become a side channel of the Danube, as this might change the fish community completely. Mudminnows are typically found in isolated pools on the margin of floodplains or in swampy areas where oxygen depletion is common during summer and under ice during winter. Only a few fish species with special adaptation to low oxygen concentrations (e.g., weatherfish, *Misgurnus fossilis;* crucian carp, *Carassius carassius;* and giebel carp, *Carassius auratus gibelio*) are found in these habitats. Mudminnows can swallow air from the surface and use oxygen from their swimbladders, which are configured as accessory respiratory organs. If pools were enlarged to provide stable habitat, other fish species would likely colonize them and mudminnows might face heavy competition or predation.

These recommendations are in accordance with plans for improving general hydrological conditions in the proposed national park. Not only mudminnows but also vegetation and many other animal species that inhabit the floodplain have suffered from falling groundwater levels. To arrest this trend, some of the longitudinal dams bordering the main channel are being opened and some backwaters and side channels in the active floodplain (inside the main flood control dam) are being reconnected to the river. This should facilitate the diversion of water to a larger area during floods and thus reduce channel cutting within the Danube. These conservation efforts began in the late 1990s and are continuing today, and their impact on the mudminnow is being monitored closely. However, success in achieving the ultimate goal of stabilizing groundwater level in the largest nearly natural floodplain of central Europe (and thus, its mudminnow metapopulation) remains to be demonstrated.

Acknowledgments The study was financed by the Austrian Federal Ministry of Environment, Youth and Family (Bundesministerium für Umwelt, Jugend und Familie) and the state government of Lower Austria (Niederösterreichische Landesregierung). Many thanks to Thomas Spindler for many years of excellent cooperation in the mudminnow project. Special thanks to M. Mann, M. Spindler, S. Wanzenböck, H. Wintersberger, E. Nemeschkal-Bauer, and I. Löffler for their help in the field.

References

DeLury, D. B. 1947. On the estimation of biological populations. *Biometrics* 3: 145–167.

Hanko, B. 1923. Über den Hundsfisch *Umbra lacustris* Grossinger (*U. krameri* Fitz.). *Zoologischer Anzeiger* 57: 88–95.

Herzig-Straschil, B. 1991. Rare and endangered fishes of Austria. *Verhandlungen der Inter-*

nationalen Vereinigung für theoretische und angewandte Limnologie (*Proceedings of the International Association of Theoretical and Applied Limnology*) 24: 2501–2504.

Mikschi, E., and Wanzenböck, J. (eds.). 1995. Proceedings of the First International Workshop on *Umbra krameri* Walbaum, 1792. *Annalen des Naturhistorischen Museums Wien (Annals of the Natural History Museum of Vienna)* 97B: 437–508.

Wanzenböck, J. 1995. Current knowledge on the European mudminnow *Umbra krameri* Walbaum, 1792. *Annalen des Naturhistorischen Museums Wien* (*Annals of the Natural History Museum of Vienna*) 97B: 439–449.

Wanzenböck, J., and Spindler, T. 1995. Rediscovery of *Umbra krameri* Walbaum, 1792 in Austria and subsequent investigations. *Annalen des Naturhistorischen Museums Wien* (*Annals of the Natural History Museum of Vienna*) 97B: 450–457.

19

Chinook Salmon (*Oncorhynchus tshawytscha*) in Puget Sound

Effects of Spatially Correlated Catastrophes on Persistence

MARY RUCKELSHAUS
PAUL McELHANY
MICHELLE McCLURE
SELINA HEPPELL

Recovery efforts for endangered species typically are focused at the level of the species. However, many species of conservation concern are composed of multiple populations, and these populations often exhibit a metapopulation-like structure that has a large effect on species viability (reviewed by Hanski and Gilpin 1997). Contributing to a species' metapopulation structure is the geographic distribution of catastrophic events; such catastrophes can have a large effect on population and, therefore, species, persistence. In practice, the effects of catastrophic population loss on species viability are difficult to estimate, in part because of the dearth of information on rates and magnitudes of catastrophic events (Hanski and Gilpin 1997). Nevertheless, it is important to consider the combined effects of the spatial distribution and connectivity among populations and their likely rates of catastrophic loss and recolonization (i.e., population turnover) in estimating species persistence. Incorporating such potential effects on species viability can lead to very different strategies for recovering species (e.g., Ralls et al. 1996). In this chapter, we develop a simple model to explore the effects of both metapopulation structure and the spatial distribution of catastrophic risks on the probability of persistence of a wide-ranging species.

Our focal species for the spatial model we developed is chinook salmon (*Oncorhynchus tshawytscha*), an anadromous fish whose geographic range extends throughout the north Pacific from Japan to southern California (Healey 1991, Myers et al. 1998). Salmon in general are characterized by a strong metapopulation-like structure because of the spatial distribution of streams they inhabit during their freshwater phase and their unique life history attributes. One of the most important of salmon life history features in their effect on population structure is their strong homing tendencies (Groot and Margolis 1991). The philopatric behavior of salmon is well documented, but it is not perfect—

salmon also "stray" at a small, but measurable, rate, returning to spawning areas that are not their natal streams (Quinn 1993). Straying by adults is an important interpopulation dispersal mechanism that allows salmon to recolonize spawning areas that can be greatly damaged or destroyed through catastrophic events. The freshwater spawning and rearing habitat of Pacific salmon is susceptible to such catastrophes as volcanoes, flooding, landslides, and other human-induced habitat destruction (Bisson et al. 1988, 1997; Leider 1989). This combination of a spatially structured species whose spawning habitats fall victim to catastrophic disturbances is the model for the population dynamics we explore here.

Because of severe declines in abundance, reduced spatial distribution of populations relative to their historical distribution, and significant loss of life history diversity, the National Marine Fisheries Service (NMFS) has listed nine separate Evolutionarily Significant Units (ESUs) of chinook under the U.S. Endangered Species Act (Nehlsen et al. 1991; National Marine Fisheries Service 1992, 1994, 1995, 1999a, 1999b; Myers et al. 1998).

As required under the Endangered Species Act, recovery planning for a number of the ESUs is under way (Ruckelshaus et al. 2002a). NMFS has defined an ESU as a reproductively isolated group of fish that represents an important part of the evolutionary legacy of the species (Waples 1991, 1995). Thus an ESU is considered to be a "distinct population segment" under the Endangered Species Act, and legally such listed units are treated as individual species that must be recovered. ESUs typically are composed of multiple populations, which are defined on the basis of their estimated degree of demographic independence (McElhany et al. 2000). For listed chinook ESUs that have been examined, 3 to 30 populations per ESU have been identified (Ford et al. 2001, Puget Sound Technical Recovery Team 2001, Myers et al. 2002).

The first steps in recovery planning for listed ESUs include defining the biological characteristics that the ESU must exhibit for delisting to occur, including the necessary distribution and number of viable populations. In the most simplistic view of a recovered ESU, a single population with a high probability of persistence could be considered to be sufficient for ESU viability. Such a conclusion ignores risks of catastrophic population loss; however, incorporating such risks increases the number of populations needed for metapopulation persistence (e.g., Ralls et al. 1996, Hanski and Gilpin 1997). One way to explore the question of how many populations are required to avoid ESU extinction is to make the simplifying assumption that populations experience independent and identical catastrophic extinction risks and then calculate the probability of ESU persistence (Ruckelshaus et al. 2002b). We know that such a simplifying assumption is not true for salmon populations, so as a next step, we choose to use simulations to explore the effects of environmentally correlated risks of population extinction due to catastrophes on ESU viability.

In this chapter, we use a simple metapopulation model in RAMAS to ask how many viable populations are needed for persistence of the Puget Sound chinook ESU, which is listed as threatened under the U.S. Endangered Species Act. To estimate the overall risk of ESU extinction, we incorporate the effects of correlation in catastrophic risks due to spatial structure into population-specific risks of catastrophe. The primary question we address is "How many viable populations are required in order for the entire ESU to persist in the face of spatially correlated catastrophes?" We used the basic metapopulation model in RAMAS 4.0 to simulate the persistence of the Puget Sound chinook ESU. Because catastrophic risks can be difficult or misleading to estimate (Coulson et al. 2001), we bracket a plausible range of the likelihood of both individual and spatially correlated catastrophic risks facing populations within the ESU.

Methods

Metapopulation Model

The Puget Sound chinook ESU historically had an estimated 21 quasi-independent populations (Puget Sound Technical Recovery Team 2001), so that is the number of populations we included in the metapopulation model. For simplicity, we did not incorporate any age or stage structure into individual population models. We define a viable population as one that is not declining and is large enough to have at most a 5% risk of extinction from normal levels of environmental variability over a 100-year period. We used 4,600 fish per population as a plausible viable population size in the simulations, based on preliminary analysis of simple population viability analyses (PVAs) (Dennis et al. 1991, Holmes 2001, McElhany and Payne 2003). Preliminary analyses of the variance in population growth rate for chinook populations in Puget Sound suggested that 0.2 is an appropriate estimate of the environmental variance in growth rate for the model (Puget Sound Technical Recovery Team, unpublished data). We chose to simulate populations with a non-negative growth rate so that the only source of population quasi extinction in the model was due to catastrophes. The average population growth rate across all 21 populations that produces a median population lambda (λ). equal to 1 is $\lambda = 1.0202$ (assuming a 0.2 estimate of the variance in population growth rate). With a median $\lambda = 1$, half of the potential population trajectories are expected to increase, and half are expected to decline. Parameter values used in the metapopulation model are shown in Table 19.1.

The Puget Sound chinook populations are connected by a relatively low level of dispersal that likely is a function of physical distance and the branching patterns of the stream network (Puget Sound Technical Recovery Team 2001; Figure 19.1). For this exercise, we estimated pairwise dispersal between populations from tag data indicating stray rates of hatchery fish (Puget Sound Technical Recovery Team 2001) and the geo-

Table 19.1 Population parameters for chinook salmon used in the metapopulation model in RAMAS 4.0

Parameter	Value
Number of stages	None
Fecundity (growth rate)	1.0202
Environmental standard deviation	0.2
Distribution of environmental variation	Lognormal
Density dependence type	Ceiling
Initial population size	4,600
Capacity without catastrophe	4,600
Capacity after catastrophe	100

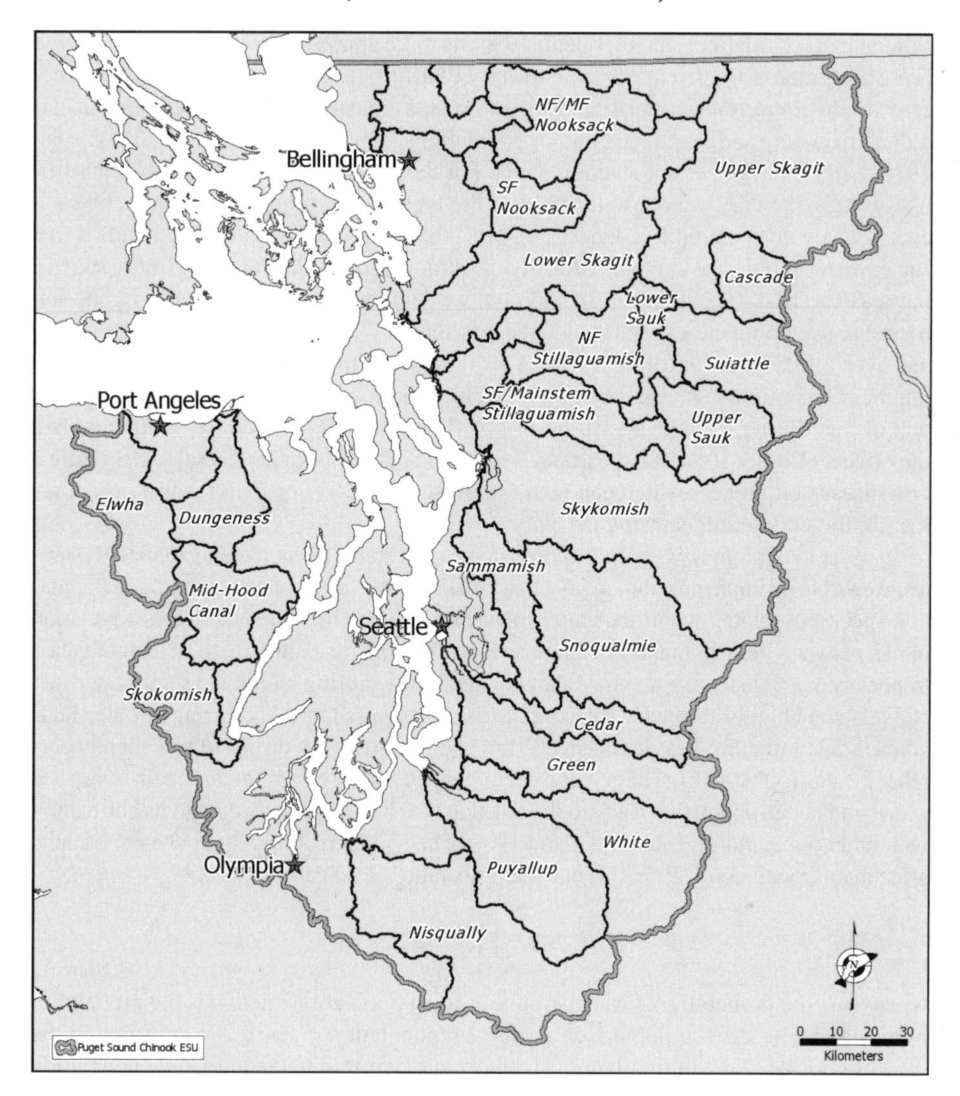

Figure 19.1 Geographic distribution of major streams in Puget Sound, Washington, and the 21 quasi-independent populations identified in the Puget Sound chinook salmon Evolutionarily Significant Units (ESU) (from Puget Sound Technical Recovery Team 2001).

graphic locations of populations. A dispersal matrix indicating the proportion of individuals exchanged between populations was input into RAMAS 4.0 for use in the metapopulation module (PugetSound.MP), available on the accompanying CD.

Incorporating Catastrophes

RAMAS 4.0 allows users to examine how catastrophic events can affect the performance of a metapopulation. Using the standard RAMAS functions, a single catastrophic event can affect either a single population or the entire metapopulation. However, for Pacific salmon metapopulations, an intermediate case is more appropriate—that is, there is some probability that a single catastrophic event will affect several populations but not necessarily the entire metapopulation. For example, a volcanic eruption could drastically reduce spawning and rearing habitat for populations in multiple watersheds (Bisson et al. 1988, Leider 1989), but not an entire ESU. For the purposes of characterizing catastrophes in this exercise, we chose to simulate the type of catastrophe represented by such events as a volcanic eruption, landslides, or a massive pollutant spill—any incident that can severely reduce the carrying capacity of salmon populations for one or more spawning seasons (Bisson et al. 1997). In addition, we simulated the types of catastrophes that do not cause complete extirpation of the population—because of adult returns spread out over 2 to 5 years and low levels of straying, it is likely in some cases that small numbers of individuals (e.g., $N = 100$; Table 19.1) may survive a catastrophic event that occurs during a single spawning season. For simplicity, we chose not to simulate the effects of other sorts of catastrophes—events such as droughts, lethal temperatures, or a disease outbreak—which can result in drastic mortality of individuals without affecting the population carrying capacity.

In order to explore how spatially correlated catastrophes affect metapopulation dynamics, we wrote a simple program in java that simulated the effect of catastrophes on population-specific capacities within the Puget Sound ESU, given a hypothesized spatial correlation in risk between these populations. Because local information on the spatial extent of catastrophes is unavailable, we assumed that populations occurring within the same watershed had relatively highly correlated catastrophic risks, and populations occurring in watersheds whose stream mouths were less than 200 km apart (by network distance) had slightly correlated catastrophic risks. The java program applied catastrophes stochastically over 100 years, with a per-population catastrophe risk that was increased according to its correlation with other populations (Tables 19.2 and 19.3). The annual risk of at least one catastrophe affecting a population j, P_j, is 1 minus the probability of no catastrophes,

$$P_j = 1 - \prod_{i=1}^{21} \left(1 - \phi_i \gamma_{ij}\right)$$

where ϕ_i is the probability of a catastrophe at focal population i and γ_{ij} is the probability of a catastrophe at focal population i affecting population j. The ϕ_i are specified in the scenarios we used in the simulation, and the γ_{ij} are depicted in the individual population rows and columns of Table 19.3. The lower portion of Table 19.3 shows the P_j at $\phi_i = 0.001$ for all populations and at $\phi_i = 0.003$ for all populations.

The output of the program is a capacity text file for each population (Capacity*n*.KCH) that lists the capacity of the population at each time step. Unless a catastrophe strikes, capacities for each population during simulations were identical. The process was made more efficient by having the capacity program simply replace the capacity files in the RAMAS folder between each simulation. We used these capacities in the RAMAS metapopulation model and incorporated ceiling-type density dependence in predictions of metapopulation (ESU) quasi extinction.

Simulation Scenarios

We explored ESU persistence to a specified quasi-extinction threshold over 100 years and varied the number of the ESU's 21 populations that were initially viable. We ex-

Table 19.2 Input parameters for the simulation program (SimCatastrophes) that generates a risk of catastrophe for each population at each time step

Parameter	Value	Description
Number of populations	21	Total number of possible viable populations in the metapopulation
Number of populations with zero capacity	Variable	Number of populations that are removed from the metapopulation in an effort to determine metapopulation persistence as a function of the number of populations; the program randomly selects populations to remove from the metapopulation
Base capacity	4,600	Capacity of a viable population that has not experienced a catastrophe
Catastrophe capacity	100	Capacity of a population after a catastrophe: once a population experiences a catastrophe, the capacity of the population never recovers in the simulation (i.e., catastrophes are permanent)
Number of years for simulation	100	Time horizon for the assessment of ESU persistence
Output file	Variable	The output file is the base output file name; the program outputs one capacity file for each population, which is named by concatenating a population number onto the base file name (i.e., CAPACITY*n*)
Probability of population catastrophe file	Variable	Text file that describes the per-population catastrophic risk. For the Puget Sound simulations we set every population with the same individual risk of catastrophe.
Probability of multiple catastrophe file	Variable	Text file that is a matrix describing the probability that neighboring populations also suffer a catastrophic event if a focal population experienced a catastrophic event. The probability of multiple catastrophes is determined by the spatial correlation in catastrophic risks facing populations (e.g., Table 19.3).

Note: The output files (CAPACITY*n*.kch) from SimCatastrophes are input files for the RAMAS 4.0 metapopulation model.

plored a total of four scenarios in which we varied the per-population risk of catastrophe and the quasi-extinction threshold. We examined the annual probability of catastrophe for each population (ϕ_i) at both the 0.001 and 0.003 levels. Those annual risks correspond to a mean time of 1,000 and 333 years between catastrophic events, respectively. For each of these probabilities of catastrophe, we also examined two quasi-extinction thresholds: 10,000 fish (2% of estimated historical ESU abundance) and 20,000 fish (4% of estimated historical ESU abundance) (Myers et al. 1998). In each scenario, we assessed the probability that the ESU would drop below the specified abundance threshold in 100 years if it included a given number (from 3 to 21) of populations that were initially viable. The populations that were initially viable for each simulation were selected randomly from the set of 21 historical populations. We ran the metapopulation model 20 times for each combination of initially viable populations, catastrophe level, and ESU quasi-extinction threshold, and then we tallied the number of times the total number of fish in the ESU reached the specified abundance threshold.

The probability of ESU quasi extinction was obtained from the interval extinction risk results table in RAMAS. Because catastrophes in the simulation were permanent, the ESU would always go extinct if the simulation was run for a long enough period

Table 19.3 Matrix depicting probabilities of multiple catastrophes among populations

Population	1	2	3	4	5	6	7	8	9	10	11	12	13	14	15	16	17	18	19	20	21
1	1	0.25	0	0	0	0	0	0	0	0	0	0	0	0	0	0	0	0	0	0	0
2	0.25	1	0	0	0	0	0	0	0	0	0	0	0	0	0	0	0	0	0	0	0
3	0	0	1	0.25	0.25	0.25	0.25	0.25	0.1	0.1	0.1	0.1	0	0	0	0	0	0	0	0	0
4	0	0	0.25	1	0.25	0.25	0.25	0.25	0.1	0.1	0.1	0.1	0	0	0	0	0	0	0	0	0
5	0	0	0.25	0.25	1	0.25	0.25	0.25	0.1	0.1	0.1	0.1	0	0	0	0	0	0	0	0	0
6	0	0	0.25	0.25	0.25	1	0.25	0.25	0.1	0.1	0.1	0.1	0	0	0	0	0	0	0	0	0
7	0	0	0.25	0.25	0.25	0.25	1	0.25	0.1	0.1	0.1	0.1	0	0	0	0	0	0	0	0	0
8	0	0	0.25	0.25	0.25	0.25	0.25	1	0.1	0.1	0.1	0.1	0	0	0	0	0	0	0	0	0
9	0	0	0.1	0.1	0.1	0.1	0.1	0	1	0.25	0.1	0.1	0	0	0	0	0	0	0	0	0
10	0	0	0.1	0.1	0.1	0.1	0.1	0.1	0.25	1	0.1	0.1	0	0	0	0	0	0	0	0	0
11	0	0	0.1	0.1	0.1	0.1	0.1	0.1	0.1	0.1	1	0.25	0	0	0	0	0	0	0	0	0
12	0	0	0.1	0.1	0.1	0.1	0.1	0.1	0.1	0.1	0.25	1	0	0	0	0	0	0	0	0	0
13	0	0	0	0	0	0	0	0	0	0	0	0	1	0.25	0.1	0	0	0	0	0	0
14	0	0	0	0	0	0	0	0	0	0	0	0	0.25	1	0.1	0	0	0	0	0	0
15	0	0	0	0	0	0	0	0	0	0	0	0	0.1	0.1	1	0.1	0.1	0.1	0	0	0
16	0	0	0	0	0	0	0	0	0	0	0	0	0	0	0.1	1	0.25	0.1	0	0	0
17	0	0	0	0	0	0	0	0	0	0	0	0	0	0	0.1	0.25	1	0.1	0	0	0
18	0	0	0	0	0	0	0	0	0	0	0	0	0	0	0.1	0.1	0.1	1	0	0	0
19	0	0	0	0	0	0	0	0	0	0	0	0	0	0	0	0	0	0	1	0	0
20	0	0	0	0	0	0	0	0	0	0	0	0	0	0	0	0	0	0	0	1	0.1
21	0	0	0	0	0	0	0	0	0	0	0	0	0	0	0	0	0	0	0	0.1	1
Total @ $\phi = 0.0001$ ($\times 10^3$)	1.25	1.25	2.65	2.65	2.65	2.65	2.65	2.65	2.05	2.05	2.05	2.05	1.35	1.35	1.5	1.45	1.45	1.3	1	1.1	1.1
Total @ $\phi = 0.0003$ ($\times 10^3$)	3.75	3.75	7.92	7.92	7.92	7.92	7.92	7.92	6.14	6.14	6.14	6.14	4.05	4.05	4.49	4.35	4.35	3.9	3	3.3	3.3

Note: If a focal population (row) experiences a catastrophe, this table indicated the probability that neighboring populations (columns) will also experience a catastrophe in the same time step. The probability matrix is based on the spatial distribution of Puget Sound chinook populations and the Puget Sound stream network (Fig.19.1). This matrix is an input file used in the catastrophe simulation described in Table 19.2. The values at the bottom of the table describe the total annual catastrophe risk experienced by each population for two different focal population extinction risk probabilities. Extinction probabilities in the last 2 rows are expressed in units of 10^{-3}.

of time; thus, our explorations were relatively pessimistic. Our rationale in posing the question in this way was to explore to what extent more viable populations increases the chance that the ESU will be viable. Furthermore, we hoped that framing the question as estimating "How many viable populations are necessary for ESU persistence at the beginning of a recovery period?" would increase the likelihood that there is a reasonable chance of rebuilding the ESU. We reasoned that if we recommend starting with enough populations in a threatened ESU, there would be time for some conservation planning course corrections, even if catastrophes were a severe problem.

We examined the effect of correlation in catastrophic risk on quasi extinction by simulating the four scenarios using either the correlated probability matrix shown in the body of Table 19.3 or by treating catastrophes as independent events with the per-population annual risk shown at the bottom of Table 19.3.

Results

The probability of ESU quasi extinction varied as a function of the initial number of viable populations for correlated catastrophes. Results from simulations where catastrophes were treated as independent events (not shown) were similar to the results where catastrophes were correlated (Figure 19.2). Generally, the risk of ESU quasi extinction declines with increasing numbers of viable populations within the ESU, as expected. If an ESU has a less frequent catastrophic loss of populations (dashed lines), fewer viable populations are required to achieve a specified level of ESU persistence than in an ESU experiencing a higher risk of catastrophic population loss (solid lines). The threshold ESU abundance level chosen to define ESU quasi extinction also has a predictable effect on the number of populations needed for ESU persistence: when 20,000 chinook are required for the ESU to persist, more viable populations are required than when 10,000 chinook are sufficient for ESU viability.

Determining the number of viable populations required for ESU persistence depends greatly on the acceptable level of risk of ESU quasi extinction, in addition to the abundance level considered to be quasi extinction and the estimated risk of catastrophic population losses. For example, if policy dictates that ESU persistence be estimated with 95% assurance, the simple simulations we performed in this example indicate that from 10 to 21 viable populations are required for ESU viability (Figure 19.2). Lower ESU quasi-extinction thresholds, age- or stage-structured population models, and less frequent catastrophic risks would reduce the numbers of viable populations needed for ESU persistence at a given level of acceptable risk of ESU quasi extinction.

The variability in simulation results depicting ESU persistence is greater under the condition of more frequent catastrophic risks (Figure 19.2). The increase in variance may be due, in part, to the increase in the variance of cumulative catastrophic risk shown at the bottom of Table 19.3, although the coefficient of variation is the same for $\phi_i = 0.001$ and $\phi_i = 0.003$. We speculate that as catastrophic risk for each individual population increases, the connectivity of those populations becomes more important and affects the variance of metapopulation extinction risk; further simulations are needed to explore this relationship more thoroughly.

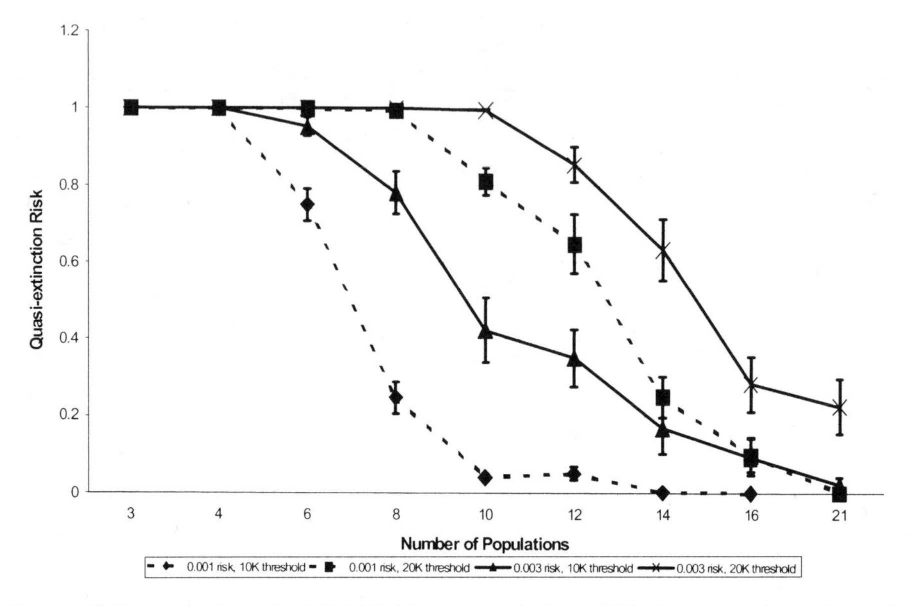

Figure 19.2 Results from the RAMAS 4.0 metapopulation model. The curves depict the probability of chinook salmon Evolutionarily Significant Unit (ESU) quasi extinction as a function of the number of viable populations within the ESU. The effects of ESU quasi-extinction thresholds at 10,000 (*diamonds and triangles*) and 20,000 (*squares and crosses*) fish and frequency of catastrophic risks with probability 0.001 (*broken lines*) and 0.003 (*solid lines*) per population per year on ESU persistence were explored. The bars on the figure represent standard errors of the mean of the 20 simulations.

Conclusions

The spatial distribution of populations within a species (or in our case, an ESU) has been shown to be important for viability (e.g., Hanski and Gilpin 1997). One of our primary aims for this exercise was to explore the importance of correlated catastrophes to predictions of chinook salmon ESU viability. Although these simulations do not indicate a large effect of correlations in catastrophes on quasi-extinction risk, we suspect the result is a product of randomly selecting initially viable populations. In reality, the probability of ESU persistence is likely to improve if populations that are uncorrelated in their catastrophic risks are managed for viability. Future research should focus on the consequences of strategically targeting specific populations for achieving a viable ESU.

This exercise shows how simple metapopulation models can be useful in generating ballpark ranges on the numbers of salmon populations necessary for ESU delisting. Even our simple formulation illustrates how interpopulation demographic processes such as dispersal and catastrophic population loss can affect the number of populations required for metapopulation persistence.

The model we developed was purposefully simple—we did not incorporate model uncertainty or different degrees of spatial correlation in the catastrophic risks, to name a few additional details. Furthermore, because we simplified the within-population dynamics and did not include age or stage structure in the model, our ESU persistence results are likely to be somewhat pessimistic. Even this very simple metapopulation construct to describe chinook ESU dynamics was very difficult to parameterize and therefore is based on educated guesses from almost no data.

Because of the great uncertainties in parameterizing even the simplest metapopulation model, in practice, our approach to deciding which combination of population characteristics constitutes a viable salmonid ESU is twofold. First, we are developing other, less quantitative criteria—such as historical levels of diversity and geographic regions over which catastrophic risks to populations are expected to be correlated—to choose sets of populations critical to ESU viability. These criteria for diversity and ESU spatial structure are likely to require more populations for ESU persistence than a simple metapopulation model can prescribe (McElhany et al. 2000, Ruckelshaus et al. 2002b). Second, we can use the metapopulation modeling approach such as we have illustrated here to give us greater confidence in the numbers of populations we estimate are needed to achieve ESU viability goals. Results presented here suggest that in addition to the commonly discussed threats of increased environmental variability and reduced marine survival on ESU persistence (e.g., Lawson 1993, National Research Council 1996, Mantua et al. 1997, Lichatowich 1999), reductions in the number of viable populations in an ESU can result in a significant increase in the risk of ESU extinction.

Acknowledgments We thank Resit Akçakaya for inviting us to contribute this essay. Ideas we explore in this example have benefited from discussions with Mike Ford, Eli Holmes, Peter Kareiva, John Payne, and Robin Waples. Comments from Chris Wood, Richard Routledge, Tom Good, Jeff Hard, Rich Zabel, and an anonymous reviewer greatly improved the manuscript.

References

Bisson, P. A., Nielsen, J. L., and Ward, J. W. 1988. Summer production of coho salmon stocked in Mount St. Helens streams 3–6 years after the eruption. *Transactions of the American Fisheries Society* 117: 322–355.

Bisson, P. A., Reeves, G. H., Bilby, R. E., and Naiman, R. J. 1997. Watershed management and Pacific salmon: Desired future conditions. Pages 447–474 in D. J. Stouder, P. A. Bisson, and R. J. Naiman (eds.), *Pacific salmon and their ecosystems.* Chapman and Hall, New York.

Coulson, T. G., Mace, M., Hudson, E., and Possingham, H. 2001. The use and abuse of population viability analysis. *TREE* 16: 219–221.

Dennis, B., Munholland, P. L., and Scott, J. M. 1991. Estimation of growth and extinction parameters for endangered species. *Ecological Monographs* 61: 115–143.

Ford, M. J., Budy, P., Busack, C., Chapman, D., Cooney, T., Fisher, T., et al. 2001. *Upper Columbia River steelhead and spring chinook salmon population structure and biological requirements.* A report prepared by the Upper Columbia River Steelhead and Spring Chinook Salmon Biological Requirements Committee. Available at http://www.nwfsc.noaa.gov/

Groot, C., and Margolis, L. 1991. *Pacific salmon life histories.* University of British Columbia Press, Vancouver.

Hanski, I., and Gilpin, M. E. 1997 *Metapopulation biology: ecology, genetics and evolution.* Academic Press, San Diego.

Healey, M. C. 1991. The life history of chinook salmon (*Oncorhynchus tshawytscha*). Pages 313–393 in C. Groot and L. Margolis (eds.), *Life history of Pacific salmon.* University of British Columbia Press, Vancouver.

Holmes, E. 2001. Estimating risks in declining populations with poor data. *Proceedings of the National Academy of Sciences* 98: 5072–5077.

Lawson, P. W. 1993. Cycles in ocean productivity, trends in habitat quality, and the restoration of salmon runs in Oregon. *Fisheries* 18: 6–10.

Leider, S. A. 1989. Increased straying by adult steelhead trout, *Salmo gairdneri*, following the 1980 eruption of Mt. St. Helens. *Environmental Biology of Fishes* 24: 219–229.

Lichatowich, J. 1999. *Salmon without rivers.* Island Press, Washington, D.C.

Mantua, N. J., Hare, S. R., Zhang, Y., Wallace, J. M., and Francis, R. C. 1997. A Pacific interdecadal climate oscillation with impacts on salmon production. *American Meteorology Society Bulletin* 78: 1069–1079

McElhany, P., and Payne, J. 2003. Defining recovery: a Pacific salmon example. Unpublished ms.

McElhany, P., Ruckelshaus, M., Ford, M. J., Wainwright, T., and Bjorkstedt, E. 2000. *Viable salmonid populations and the recovery of Evolutionarily Significant Units.* NOAA Technical Memorandum NMFS-NWFSC-42. U.S. Department of Commerce, Washington, D.C.

Myers, J. M., and 10 others. 1998. *Status review of chinook salmon from Washington, Idaho, Oregon and California.* NOAA Technical Memorandum NMFS-NWFSC-35. U.S. Department of Commerce, Washington, D.C.

Myers, J., Busack, C., and Rawding, D. 2002. Identifying historical populations of chinook and chum salmon and steelhead with the Lower Columbia River and Upper Willamette River Evolutionarily Significant Units. Technical Review Draft. Willamette/Lower Columbia River Technical Recovery Team, Seattle.

National Research Council (NRC). 1996. *Upstream: salmon and society in the Pacific Northwest.* National Academy Press, Washington, D.C.

Nehlsen, W., Williams, J. E., and Lichatowich, J. A. 1991. Pacific salmon at the crossroads: stocks at risk from California, Oregon, Idaho and Washington. *Fisheries* 16: 4–21.

National Marine Fisheries Service (NMFS). 1992 Final Rule: Endangered and Threatened Species; Threatened Status for Snake River Spring/Summer Chinook Salmon, Threatened Status for Snake River Fall Chinook Salmon. *Federal Register* 57: 14653.

National Marine Fisheries Service (NMFS). 1994 Final Rule: Endangered and Threatened Species; Status of Sacramento River Winter-run Chinook Salmon. *Federal Register* 59: 440.

National Marine Fisheries Service (NMFS). 1995 Final Rule: Endangered and Threatened Species; Status of Snake River Spring/Summer Chinook Salmon and Snake River Fall Chinook Salmon. *Federal Register* 60: 19342.

National Marine Fisheries Service (NMFS). 1999a. Final Rule: Endangered and Threatened Species; Threatened Status for Two Chinook Salmon Evolutionarily Significant Units (ESUs) in California. *Federal Register* 64: 50393.

National Marine Fisheries Service (NMFS). 1999b. Final Rule: Endangered and Threatened Species; Threatened Status for Three Chinook Salmon Evolutionarily Significant Units (ESUs) in Washington and Oregon, and Endangered Status for One Chinook Salmon ESU in Washington. *Federal Register* 64: 14308.

Puget Sound Technical Recovery Team (PSTRT). 2001. *Independent populations of chinook salmon in Puget Sound.* Public review draft. April, 2001. Available at http://www.nwfsc.noaa.gov/cbd/trt/trt_puget.htm

Quinn, T. P. 1993. A review of homing and straying of wild and hatchery-produced salmon. *Fisheries Research* 18: 29–44.

Ralls, K., Demaster, D. P., and Estes, J. A. 1996. Developing a criterion for delisting the southern sea otter under the U.S. Endangered Species Act. *Conservation Biology* 10: 1528–1537.

Ruckelshaus, M., Levin, P. Johnson, J. and Kareiva, P. 2002a. The Pacific salmon wars: what science brings to the challenge of recovering species. *Annual Review of Ecology and Systematics* 33: 665–706.

Ruckelshaus, M., McElhany, P., and Ford, M. J. 2002b. Recovering species of conservation concern: are populations expendable? Pages 305–329 in P. Kareiva and S. Levin (eds.), *The importance of species: perspectives on expendability and triage.* Princeton University Press, Princeton, N.J.

Waples, R. S. 1991. Pacific salmon, *Oncorhynchus* spp., and the definition of "species" under the Endangered Species Act. *Marine Fisheries Review* 53: 11–22.

Waples, R. S. 1995. Evolutionarily significant units and the conservation of biological diversity under the Endangered Species Act. *American Fisheries Society Symposium* 17: 8–27.

20

Trout Cod (*Mucculochella macquariensis*) in the Murray River, Southeast Australia

Prioritizing Rehabilitation Efforts

SIMON NICOL
CHARLES TODD

Trout cod (*Mucculochella macquariensis* Cuvier) were formerly distributed over a substantial proportion of the Murray-Darling Basin in eastern Australia (Douglas et al. 1994). In the last 50 years they have undergone a dramatic contraction in range and are now classified as critically endangered (Australian Society for Fish Biology 2001). At present only two breeding populations are known: a naturally occurring population in a 200-km reach of the Murray River below Lake Mulwala, and a translocated population in an 8-km reach of Seven Creeks below Polly McQuinns Weir (Richardson and Ingram 1989, Ingram et al. 1990).

Alteration or destruction of habitat is widely recognized as a major cause of decline in fish native to Australia (Cadwallader 1978; Koehn and O'Connor 1990a, 1990b; Lintermans 1991). In combination with angling pressure, it is thought to be the primary threat to remaining populations (Douglas et al. 1994). Fish habitat has been degraded in the Murray-Darling Basin by the construction of instream barriers to fish passage, alteration of flow regimes and water extraction, reduction in water quality, loss of floodplain habitat, and the removal of instream physical habitat such as woody debris and pools (Lawrence 1991). Large woody debris from trees that have fallen into a river is an important habitat attribute for many Australian freshwater fish species (Treadwell et al. 1999), including trout cod (Koehn and Nicol 1998). The reach of the Murray River where trout cod still occur is large, deep, and slow flowing; has a sand, silt, and clay substrate; and contains abundant large woody debris (Australian Capitol Territory Government 1999). There has been little removal of large woody debris in this reach (Department of Land and Water Resources, unpublished records) in contrast to areas where trout cod no longer occur (Treadwell et al. 1999). Removal of large woody debris from much of the former range is thought to be a significant factor in the declining range of the species (Scientific Advisory Committee 1991).

To assist the recovery of trout cod, reintroduction of large woody debris is being considered as a way to rehabilitate degraded reaches in the Murray River (Murray-Darling Basin Commission 2002). Potential sites for rehabilitation are all downstream as a weir wall immediately upstream creates a barrier to upstream dispersal and colonization. In this chapter we describe the use of RAMAS Metapop to examine whether conservation benefits for trout cod are maximized by directing conservation resources to increase habitat and carrying capacity at the existing site or to create additional new habitat downstream of the current population.

Conceptually there are potential advantages for the long-term viability of a species when spatially independent populations are connected as a metapopulation. The premise behind these advantages is that migration between local populations stabilizes local population dynamics (Hanski and Simberloff 1997). In the case of trout cod, only one population remains in the Murray River, and its loss would mean extinction from the basin. Creating a metapopulation by rehabilitating one or more populations at sites downstream might increase the viability of trout cod in the Murray River if the additional populations were not affected by the same adverse circumstances and could provide a source of individuals for recolonization.

Population Biology of Trout Cod

Movement and Dispersal

Adults are sedentary once they have established a home range (Koehn and Nicol 1998). Larvae are known to drift downstream (Brown and Nicol 1998), and juveniles are known to move distances of up to 60 km (Faragher et al. 1993). Dispersal and recolonization occur mainly during the juvenile stages.

Reproduction

Sexual maturity is reached between the ages of 3 and 5 years (0.75–1.5 kg). Males are younger and smaller than females at first maturity (Douglas et al. 1994). Spawning occurs in spring in response to increasing water temperatures and day length (Cadwallader 1977). A minimum water temperature of 18°C is required for spawning (Koehn and O'Connor 1990b), and eggs are deposited on a hard substrate. Although the fecundity of trout cod in the wild is unknown, in hatcheries 1,188 to 11,388 eggs have been stripped from individual broodfish (Ingram and Rimmer 1992). Growth rates and survival in the wild are unknown.

Age Structure

Although the maximum reported age is 12 years (Geoff Gooley, unpublished data, Snobs Creek Hatchery, Department of Natural Resources and Environment), the species is believed to live to 20 years (Department of Natural Resources and Environment 1997). Age-specific survivorships are uncertain, but Todd (2002) estimated annual survival using 5-year census data (Simon Nicol, unpublished data) from the Murray River where the coefficient of variation was estimated to decrease from 25% to 13.6% with increasing age (see Table 20.1).

Carrying Capacity

Large woody debris can influence the carrying capacity of river systems by increasing access to food resources and by minimizing predation (Fausch 1993, Crook and Robertson 1999). In the Murray River, trout cod presence depends on large woody debris (Koehn and Nicol 1998). Trout cod are thought to be ambush predators that shelter behind large woody debris in areas of still water to minimize the energy costs of swimming and to provide visual isolation for ambushing prey or avoiding predators. The habitat immediately upstream and downstream of the Murray River population has been substantially degraded, largely through the removal of large woody debris, resulting in extremely low carrying capacity in these areas. This is thought to limit the ability of this population to expand its geographic range and abundance (Brown et al. 1998).

Methods

Stage Structure

Only females are represented in the models because there is no evidence to suggest that dynamics differ between males and females or that the sex ratio is unequal. Populations include seven stages to describe the population dynamics. Stages 1 to 4 represent the juvenile age classes prior to sexual maturity at age 5. Trout cod mature between 3 and 5 years of age with females maturing older than males (Douglas et al. 1994). Consequently, we modeled all females maturing at 5 years of age. Hatchery data suggest that

Table 20.1 Stage (in boldface) and standard deviation (in italics) matrices for a seven-stage trout cod model

	J1	J2	J3	J4	A5	A6	A7+
J1	0	0	0	0	**0.97** *0.6*	**0.97** *0.6*	**2.55** *1.0*
J2	**0.51** *0.13*	0	0	0	0	0	0
J3	0	**0.660** *0.13*	0	0	0	0	
J4	0	0	**0.73** *0.15*	0	0	0	0
A5	0	0	0	**0.78** *0.16*	0	0	0
A6	0	0	0	0	**0.81** *0.12*	0	0
A7+	0	0	0	0	0	**0.83** *0.12*	**0.88** *0.11*

females become more fecund as they increase in size and age (Ingram and Rimmer 1992). To allow consideration of age-related fecundity, particularly for young adults, the model includes three adult stages, representing age 5, age 6, and a final stage comprising ages 7 and older. Fecundity rates specified account for a female population only. Three early life history stages (eggs, larvae, and early juvenile fish) have been aggregated (or collapsed) into one time step (see Figure 20.1). The vital rates for each stage class are reported in Table 20.1. Elements of the first row represent the contribution of individual female fish of specified age to the female population in which they live.

Stochasticity

Both demographic and environmental stochasticity are included in the model. Demographic stochasticity occurs because populations comprise individuals and because chance variations among individuals alter the risk of extinction of a population (Akçakaya 1991, Burgman et al. 1993). Environmental stochasticity is incorporated to model a variable environment that may be interpreted as sequences of good and bad years and is sampled from a lognormal distribution restricted to the unit interval. Standard deviations, calculated from the census data (Simon Nicol, unpublished data), for the lognormal distributions are included in Table 20.1. These estimates are likely to include measurement error, which might result in overestimated standard deviations and thus precautionary risk results. Survivorship, fecundity, and carrying capacity within populations were assumed to be uncorrelated as there is no evidence to indicate any correlation between survivorship and fecundity. In some scenarios, a catastrophic failure in recruitment (no stage 1 fish) was simulated with a probability of 20% at each time step in each population.

Density Dependence and Carrying Capacity

Contest competition was specified in the model and limited to survival rates as there is no evidence for density dependence in fecundity for trout cod. Contest competition

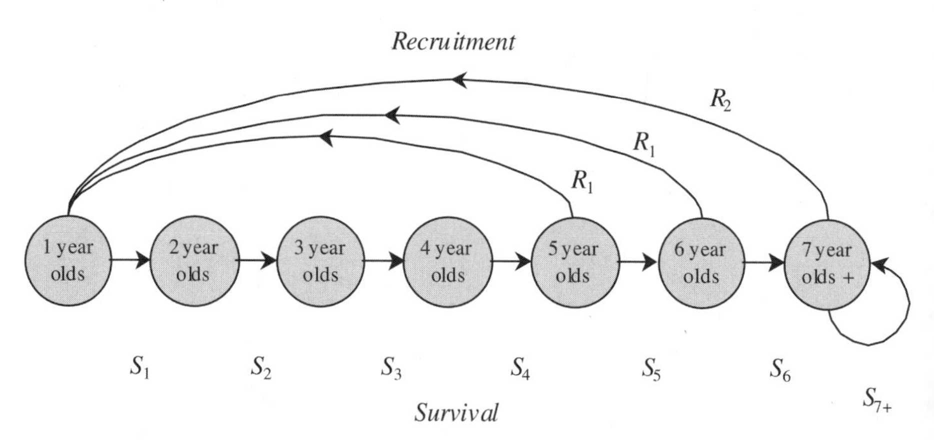

Figure 20.1 Age/stage network diagram of the structure adopted for a single population trout cod model, where recruitment is a combination of egg, larval, and early juvenile fish stages.

describes the division of resources unequally with some individuals out competing others for those resources (Burgman et al. 1993). Generally, density dependence in fish has been modeled by limiting recruitment (Saila et al. 1991, Hilborn and Walters 1992, Hilborn et al. 1994), but it is not clear that this approach is appropriate for trout cod. Field observations indicate that the critical limiting factor is the establishment of a territory and this occurs at age 4. The negative feedback of density dependence should limit recruitment indirectly by affecting the survival of fish aged 4 and older. We also observe pulse recruitment of fish aged 1 to 3 in the Murray River population, and excluding young fish from the density-dependent mechanism allows the model to exhibit this behavior. Consequently, contest competition was assumed to occur only among fish aged 4 and older. The maximum growth rate was set as 1.21, chosen to reflect our belief that it is possible that the population could double in a 4-year period if conditions were suitable. Carrying capacity for fish aged 4 and older in the Murray River population was estimated at 6,000 with an initial abundance of 5,000.

Dispersal

Our model describes the scenario where fish aged 1 to 3 disperse both upstream and downstream, and dispersal rate decreases with distance (see Table 20.2). Although trout cod larvae are known to drift downstream, larval drift was ignored as the contribution to downstream populations from larval drift could not be quantified.

Variation in the dispersal rate was modeled with a coefficient of variation of 20%. Dispersal was density-dependent, with emigration increasing as the number of individuals in a population increased. Immigration was also density-dependent with immigration rate decreasing linearly once the target population density exceeded two-third of its carrying capacity (*K*). These estimates were derived from a limited mark-recapture dataset.

Metapopulation Structure

The model metapopulation was a linear system of populations except under scenario 11 (Figure 20.2). The correlation between populations was assumed to be high as any change that may affect the source (existing population) is likely to affect any additional sites

Table 20.2 Dispersal (in boldface) and correlation (in italics) between populations (except in scenario 11)

	Pop 1	Pop 2	Pop 3	Pop 4
Pop 1	0	**0.053** *0.972*	**0.014** *0.923*	**0.005** *0.863*
Pop 2	**0.053** *0.972*	0	**0.053** *0.972*	**0.014** *0.923*
Pop 3	**0.014** *0.923*	**0.053** *0.972*	0	**0.053** *0.972*
Pop 4	**0.005** *0.863*	**0.014** *0.923*	**0.005** *0.972*	0

downstream. Rivers are linear systems, and environmental variability may be limited within a river's length. In the Murray River, environmental conditions are known to take several hundred kilometers to change (Koehn and Nicol 1998). For this reason, the correlation between populations was modeled as between 0.86 and 0.97. We considered a number of metapopulation scenarios (Table 20.3). For example, in scenario 11 we tested the influence of correlation structure by notionally placing one of the populations in a tributary to the Murray River such that its correlation with other populations could be reduced to 0.1 (Table 20.4). Initial abundance was set to 0 for each of the proposed new populations (populations 2–4 in Figure 20.2).

Simulation Scenarios

To represent alternative rehabilitation scenarios, the model included the present Murray River population and up to three new populations farther downstream. Viability of the metapopulation was assessed under the alternative scenarios in which carrying capacity (for ages 4–7+) was manipulated for some or all populations (see Table 20.4). Scenarios were initially simulated without catastrophes (except scenario 11) and then simulated with a 20% chance of catastrophic recruitment failure.

Results

The simulation period was 50 years with 10,000 replications for each scenario. The performance of alternative restoration scenarios was assessed by the reduction in interval quasi-extinction risk. Interval quasi-extinction risk (IQR) refers to the number of trajectories (replications) that fall to or below a specified threshold at least once within the simulation period. We chose a threshold of 1,000 adult females (IQR 1,000) as the appropriate level for management concern.

Increasing carrying capacity (*K*) for trout cod had an appreciable conservation benefit by decreasing the probability of falling to the IQR 1,000 threshold (light grey blocks in Figure 20.3). However, the same benefit was realized whether carrying capacity was added at the existing site or at other sites downstream. The greatest conservation gain occurred for the initial increment in carrying capacity (IQR 2,000). Additional increments provided smaller and smaller reductions in quasi-extinction risk. The addition of one or more populations downstream did not improve resilience in the presence of local

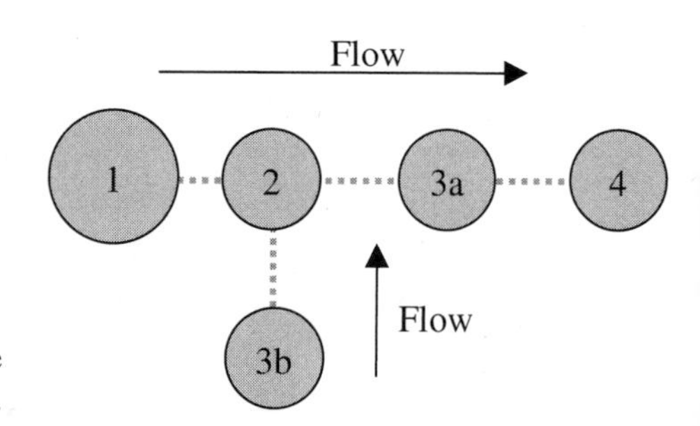

Figure 20.2 Spatial structure of the metapopulation.

Table 20.3 Large woody debris rehabilitation scenarios modeled

	Carrying Capacity (Age 4+)			
Scenario	Pop 1	Pop 2	Pop 3	Pop 4
1 (status quo)	6,000			
2	8,000			
3	9,000			
4	12,000			
5	6,000	2,000		
6	6,000	3,000		
7	6,000	6,000		
8	6,000	1,000	1,000	
9	6,000	1,500	1,500	
10	6,000	3,000	3,000	
11[a]	6,000	3,000	3,000	
12	6,000	667	667	667
13	6,000	1,000	1,000	1,000
14	6,000	2,000	2,000	2,000

[a]Scenario 11 is modeled with a population added to a tributary of the Murray River, site 3b in Figure 20.2.

catastrophes (dark grey blocks in Figure 20.3). However, this conclusion is sensitive to assumptions about the spatial correlation structure for environmental fluctuations. Quasi-extinction risk decreased dramatically in scenario 11 where we reduced the correlation between population 3 and other populations (Figure 20.4).

Discussion

Our simulation results indicate that, in principle, the probability of trout cod extirpation can be reduced by increasing carrying capacity at the existing population site. However, this may not be feasible. The quantity of large woody debris at the existing popu-

Table 20.4 Correlation (in boldface) and dispersal (in italics) between populations for scenario 11

	Pop 1	Pop 2	Pop 3
Pop 1	0	**0.97** *0.053*	**0.1** *0.014*
Pop 2	**0.97** *0.053*	0	**0.1** *0.053*
Pop 3	**0.1** *0.014*	**0.1** *0.053*	0

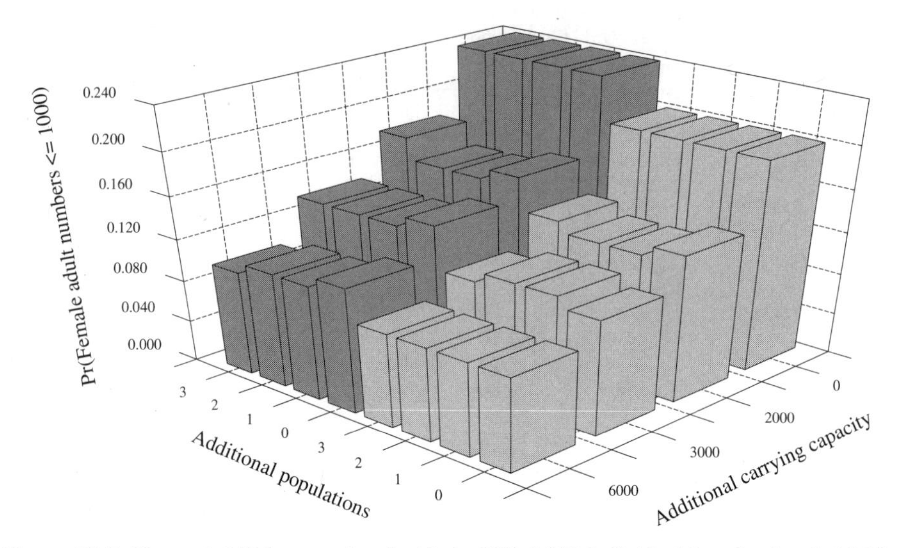

Figure 20.3 The probabilities associated with the IQR 1,000 individual females threshold. The light gray boxes are without catastrophes, and the dark gray boxes include catastrophes. The current population with no management is indicated by the 0 label on the additional carrying capacity axis. Under the no-management scenario, the carrying capacity of the additional populations is zero, and all quasi-extinction probabilities are identical.

lation site is already at its naturally occurring maximum, so adding more large woody debris may not increase carrying capacity. It may be much more effective to increase carrying capacity by adding large woody debris to sites downstream where this habitat structure has been removed in the past. Our simulation results indicate that any comparable increase in total carrying capacity through rehabilitation of sites downstream should deliver similar conservation benefits.

Creating populations by adding new habitat downstream might also provide the additional conservation benefits of metapopulation structure in that environmental fluctuations or catastrophic events might affect some populations but not all. However, by assuming high spatial correlations in environmental fluctuations between populations, we effectively removed any possible benefit of metapopulation structure. We assumed these high correlations because environmental impacts on trout cod most commonly occur from variation in flow and water quality. We considered it unlikely that such detrimental conditions would dissipate at the spatial scale that we have considered for a river system that is essentially linear in structure. Even so, the advantage of metapopulation structure became evident in scenario 11, which included local catastrophes and one rehabilitation site in a tributary with reduced correlation with other sites. This demonstrates the advantage of developing a strategy where at least one of the additional habitat sites is connected to, but minimally correlated with other sites.

In a highly correlated environment, where environmental stochasticity acts largely in unison across all populations, then demographic stochasticity may play an important role in the persistence of a metapopulation. Planning and management decisions must be sensitive to this situation and involve contingency plans to mitigate against conditions that could drive the metapopulation to low numbers. For example, we believe that

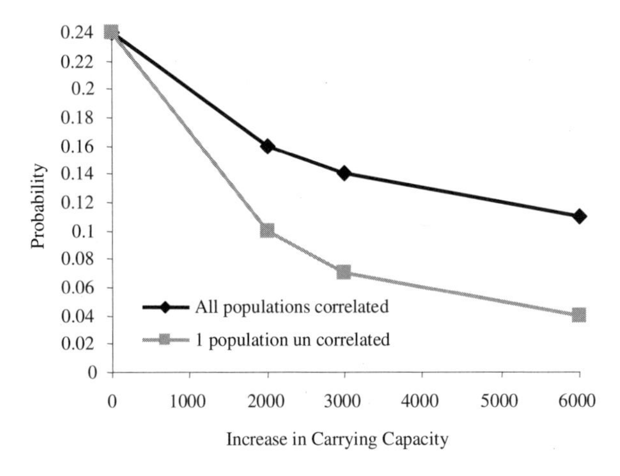

Figure 20.4 The differences in the IQR 1,000 individual females threshold probabilities when the high correlation assumptions are relaxed by one additional population in a tributary river, see scenario 11 and Table 20.4.

poor water quality conditions could affect the metapopulation as a whole and would require immediate action.

Creating extra habitat to increase carrying capacity decreased the risk of quasi extinction regardless of metapopulation structure. However, the rate at which quasi-extinction risk decreased was not directly proportional to the increase in carrying capacity. The diminishing returns from increased carrying capacity imply an optimal balance between costs and benefit. To decrease quasi-extinction risk further, management actions may have to address other processes that limit vital rates in addition to efforts to increase carrying capacity.

Our results may be pessimistic because of several simplifying assumptions. First, we assumed that only fish from age 1 to age 3 would move among populations. In reality we believe that upstream populations would contribute larvae to downstream populations through larval drift. We ignored larval dispersal because we could not specify the stage at which it would be most likely to occur. Second, we aggregated a number of life stages to specify the average number of age 1 female fish produced by an adult female. This has the effect of perfectly correlating the stages egg, larvae, and early juvenile survival, which may not be the case. This construction leads to a poor recruitment event (no female fish aged 1) being overly pessimistic. Third, the dispersal mechanism may be unrealistic for the colonisation of new habitat. We do not believe that fish dispersing to new or vacant habitat would continue past this habitat in the manner in which the dispersal matrix requires. We would expect that the metapopulation model would predict a lower quasi-extinction risk were we able to model this behavior more explicitly.

Todd (2002) performed a sensitivity analysis for a stage-structured single population model and found adult survivorship to be the most sensitive parameter for both the deterministic model and stochastic simulations. Before the model explored here is used to inform the planning process for habitat rehabilitation, further analysis is required to determine the sensitivity of risk predictions to assumptions about spatial correlations, dispersal rates, and the number of populations. Those making decisions about conservation and restoration activities need to be cognizant of these assumptions. The results from Todd (2002) identify the need for more work on the role of the annual environmental factors that affect survival at the basin level, such as an intensive mark-recapture study.

We found the modeling exercise useful to develop a set of expectations, based on current knowledge, on how the viability of trout cod in the Murray River might be increased by augmenting its habitat. The convenient metapopulation framework of RAMAS has allowed us to explore the benefits of undertaking rehabilitation work at sites downstream of the current single population. Otherwise, given our current beliefs about the likely correlation in environmental fluctuations between upstream and downstream sites, we may not have considered a metapopulation approach. After exploring the implications of our assumptions, we now recommend including a minimally correlated but connected population as an important part of the rehabilitation process. Furthermore, the metapopulation approach has also allowed us to identify the important role of environmental variation given our current beliefs about spatial correlation among populations in the Murray River. It also underscores the importance of careful monitoring of environmental quality in a highly correlated aquatic metapopulation.

References

Akçakaya, H. R. 1991. A method for simulating demographic stochasticity. *Ecological Modelling* 54: 133–136.

Australian Capitol Territory Government. 1999. *Trout cod* (Muccullochella macquariensis): *an endangered species*. Action Plan No.12. Environment ACT, Canberra.

Australian Society for Fish Biology, Inc. (ASFB). 2001. Conservation status of Australian fishes, 2001. *ASFB Newsletter* 31(2): 37–41

Brown, A., and Nicol, S. 1998. *Trout cod recovery plan final report to Environment Australia.* Department of Natural Resources and Environment, Melbourne.

Brown, A., Nicol, S., and Koehn, J. 1998. *Recovery plan for the trout cod,* Maccullochella macquariensis. Department of Natural Resources and Environment, Melbourne.

Burgman, M. A., Ferson, S., and Akçakaya, H. R. 1993. *Risk assessment in conservation biology*. Chapman and Hall, London.

Cadwallader, P. L. 1977. *J. O. Langtry's 1949–50 Murray River investigations*. Fisheries and Wildlife Paper, Victoria No. 13. Ministry for Conservation, Melbourne.

Cadwallader, P. L. 1978. Some causes of the decline in range and abundance of native fish in the Murray-Darling River System. *Proceedings of the Royal Society of Victoria* 90: 211–224.

Crook, D. A., and Robertson, A. I. 1999. Relationships between riverine fish and woody debris: implications for lowland rivers. *Marine and Freshwater Research* 50: 941–953

Department of Natural Resources and Environment. 1997. *Trout cod workshop*. Proceedings of a workshop held on the recovery of trout cod in the Murray-Darling Basin. Arthur Rylah Institute for Environmental Research, May 1997. Department of Natural Resources and Environment, Melbourne.

Douglas, J. W., Gooley, G. J., and Ingram, B. A. 1994. *Trout cod,* Maccullochella macquariensis *(Cuvier) (Pisces: Percichthyidae) resource handbook and research and recovery plan.* Department of Conservation and Natural Resources, Melbourne.

Faragher, R. A., Brown, P., and Harris, J. H. 1993. *Population surveys of the endangered fish species trout cod* (Maccullochella macquariensis) *and eastern cod (M. ikei)*. New South Wales Fisheries, Fisheries Research Institute, Cronulla.

Fausch, K. D. 1993. Experimental analysis of microhabitat selection by juvenile steelhead (*Oncorhyncus mykiss*) and coho salmon (*O. kisutch*) in a British Columbia stream. *Canadian Journal of Fisheries and Aquatic Sciences* 50: 1198–1207.

Hanski, I., and Simberloff, D. 1997. The metapopulation approach, its history, conceptual domain, and application to conservation. Pages 5–26 in I. A. Hanski and M. E. Gilpin (eds.), *Metapopulation biology: ecology, genetics and evolution*. Academic Press, San Diego.

Hilborn, R., and Walters, C. J. 1992. *Quantitative fisheries stock assessment: choice, dynamics and uncertainty*. Chapman and Hall, New York.

Hilborn, R., Pikitch, E. K., and McAllister, M. K. 1994. A bayesian estimation and decision analysis for an age-structured model using biomass survey data. *Fisheries Research* 19: 17–30.

Ingram, B. A., and Rimmer, M. A. 1992. Induced breeding and larval rearing of the endangered Australian freshwater fish trout cod, *Maccullochella macquariensis* (Cuvier) (Percichthyidae). *Aquaculture and Fisheries Management* 24: 7–17.

Ingram, B. A., Barlow, C. G., Burchmore, J. J., Gooley, G. J., Rowland, S. J., and Sanger A. C. 1990. Threatened native freshwater fish in Australia: some case histories. *Journal of Fish Biology* 37(Suppl. A): 175–182

Koehn, J., and Nicol, S. 1998. Habitat and movement requirements of fish. Pages 1–6 in R. J. Banens and R. Lehane (eds.), *Proceedings of the riverine environment research forum, Brisbane, October 1996*. Murray-Darling Basin Commission, Canberra.

Koehn, J. D., and O'Connor, W. G. 1990a. Threats to Victorian native freshwater fish. *Victorian Naturalist* 107: 5–12.

Koehn, J. D., and O'Connor, W. G. 1990b. *Biological information for management of native freshwater fish in Victoria*. Government Printer, Melbourne.

Lawrence, B. W. 1991. *Fish management plan*. Murray-Darling Basin Commission, Canberra.

Lintermans, M. 1991. The decline of native fish in the Canberra region: the effects of habitat modification. *Bogong* 12(3): 4–7.

Murray-Darling Basin Commission (MDBC). 2002. Native fish strategy for the Murray-Darling Basin. Murray-Darling Basin Commission, Canberra.

Richardson, B. A., and Ingram, B. A. 1989. *Trout cod*. New South Wales Agriculture and Fisheries Agfact F3.2.6. New South Wales Fisheries, Sydney.

Scientific Advisory Committee (SAC). 1991. *Final recommendation on a nomination for listing:* Muccollochella macquariensis (*Nomination No.4*). Department of Conservation and Environment, Victoria.

Saila, S., Martin, B., Ferson, S., and Ginzburg, L. 1991. *Demographic modeling of selected fish species, with results from RAMAS 3*. Electric Power Research Institute, Palo Alto, Calif.

Todd, C. R. 2002. Tools for the conservation management of wildlife under uncertainty. Ph.D. diss., University of Melbourne, Melbourne, Victoria.

Treadwell, S., Koehn, J., and Bunn, S. 1999. Large woody debris and other aquatic habitat. Pages 79–97 in S. Lovett and P. Price (eds.), *Riparian land management technical guidelines*. Vol. 1, Part A. *Principles of sound management*. Land and Water Resources Research and Development Corporation, Canberra.

21

Yellowtail Flounder (*Limanda ferruginea*) off the Northeastern United States

Implications of Movement among Stocks

DEBORAH R. HART
STEVEN X. CADRIN

Models and assessments of fishery resources typically assume that fish stocks are closed populations (i.e., that there is neither immigration nor emigration). In reality, some exchange among stocks often occurs, yet the implications of such exchanges have rarely been investigated. Yellowtail flounder, *Limanda ferruginea,* is an example of a fish species for which such movement has been documented (Royce et al. 1959, Lux 1963). The purpose of this chapter is to demonstrate the use of the RAMAS Metapop model (v. 4.0) as an exploratory tool to investigate the consequences of exchanges among stocks for the population dynamics of yellowtail flounder.

Yellowtail flounder is one of the most important commercially exploited groundfish off the northeastern United States. It inhabits relatively shallow waters (20–100 m) in the northwest Atlantic from Labrador to the Chesapeake Bay. Spawning occurs in spring and early summer. The pelagic egg and larval stages last about 2 months, during which time they can be transported considerable distances by currents. Postlarval juveniles and adults are benthic and feed primarily on small arthropods and polychaete worms (Bigelow and Schroeder 1953, Collette and Klein-MacPhee 2002).

The exploitation history of yellowtail flounder resources varies geographically with several episodes of stock abundance, collapse, and rebuilding. A fishery for yellowtail flounder developed off southern New England in the 1930s, following a decline in winter flounder abundance, and expanded with the demand for food products during World War II (Royce et al. 1959). Landings increased to over 40,000 metric tons (mt) in 1942 as fishing for yellowtail extended on to Georges Bank and off Cape Cod. After a brief decline in landings during the early 1950s, landings from southern New England and Georges Bank increased to over 50,000 mt in the middle 1960s (Overholtz and Cadrin 1998). A fishery for yellowtail also developed on the Grand Bank off Newfoundland during the 1960s (Walsh et al. 1999). However, yellowtail abundances generally declined throughout its range from the 1970s through the mid-1990s due to excessive fishing mortality. In 1994, the Grand Bank yellowtail fishery was closed, and large areas on Georges Bank and off southern New England were closed to fishing because of low abundance of groundfish (Walsh et al. 1999, Murawski et al. 2000). Yellowtail on the Grand Bank and Georges Bank have responded favorably to reductions in fishing mortality with substantial increases in abundance (Walsh et al. 1999, Murawski et al. 2000, Stone et al. 2001). However, yellowtail biomass in southern New England has remained low despite reduced fishing mortality (Cadrin 2000).

Over the past 25 years, the fisheries for yellowtail flounder have been managed under several jurisdictions. From 1971 to 1976 the International Commission for the Northwest Atlantic Fisheries (ICNAF) allocated national quotas. Since 1976, the United States and Canada managed yellowtail resources within national waters separately, despite transboundary movement of adult and larval yellowtail on Georges Bank. Regulations in U.S. and Canadian fisheries have become increasingly restrictive in order to limit fishing mortality. These measures include minimum size limits (28 cm in 1982, 30 cm in 1986, and 33 cm in 1989), minimum net mesh sizes (130 mm in 1982, 140 mm in 1983, 152 mm in 1994, and 165 mm in 1999), limitations on the number of days fished, and closed areas.

Scientific assessments of the U.S. yellowtail resources have been conducted for the last four decades using various models for estimating abundance and mortality from fishery and survey data. Status of the three major U.S. yellowtail stocks are currently based on virtual population analysis (VPA) (see, e.g., Hilborn and Walters 1992) and biomass dynamics models (Prager 1994). For example, the 2001 assessments indicate that the Georges Bank stock rapidly increased in the late 1990s due to low fishing mortality rates, the southern New England stock is slowly rebuilding at moderate fishing mortality rates, and the Cape Cod stock is increasing despite high fishing mortality (Northeast Fisheries Science Center 2001).

Information on geographic distribution, geographic variation, and movements among areas has been synthesized to identify management units of yellowtail flounder off the northeast United States. These stock definitions are used for assessing and managing the yellowtail flounder resources and fisheries. The principal stocks of yellowtail off the northeast United States are in southern New England waters, on Georges Bank, and off Cape Cod (Royce et al. 1959, Lux 1963) (Figure 21.1).

Traditionally, each yellowtail flounder stock has been modeled in isolation, assuming that they are closed populations. However, tagging experiments have indicated that there is some exchange of adults among the stocks (e.g., Royce et al. 1959, Lux 1963). Moreover, hydrodynamic models suggest the possibility that some larvae from the Cape Cod stock may be transported into the southern New England stock, and also that some larval fish from Georges Bank may be transported into the Cape Cod and southern New England stock areas (Royce et al. 1959, Evseenko and Nevinskiy 1980, Yevseynko and Nevinskiy 1981). The main objective of this study is to investigate the implications of larval and adult interchanges for the population dynamics of the major U.S. yellowtail flounder stocks.

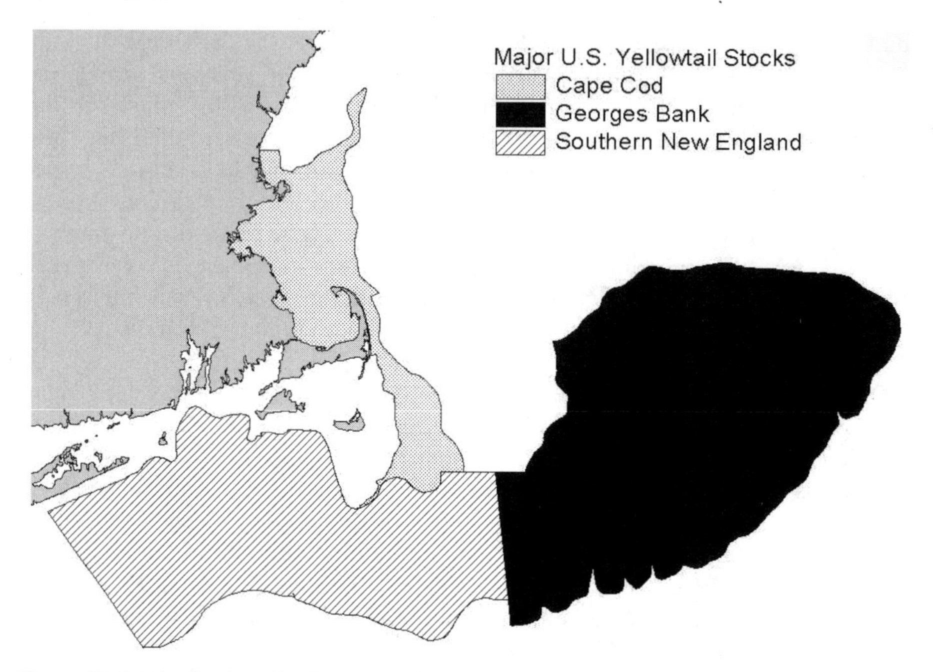

Figure 21.1 Distribution of major stocks of yellowtail flounder off the northeastern United States.

Methods

Metapopulation Structure

Our metapopulation model consists of three populations: Georges Bank, Cape Cod, and southern New England yellowtail flounder, the three major stocks in U.S. waters. Other stocks occur in Canadian waters, in the Gulf of Maine, and in the Mid-Atlantic Bight. Other than the portion of the Georges Bank stock in the Canadian exclusive economic zone (which is included in our model as part of the Georges Bank stock), there is likely to be little exchange between U.S. and Canadian yellowtail; thus Canadian stocks are not modeled here. The Gulf of Maine and Mid-Atlantic stocks are relatively minor, and no analytic age-structured assessment of them has been performed. For lack of information, these stocks will not be included in our model; given their relatively small size, neglecting to include these stocks would be unlikely to substantially affect the population dynamics of the major stocks modeled here.

Stage Matrix

An age-structured approach was used for modeling each of the stocks, using seven age groups (age 0 to age 6+). The number of fish and fishing mortality at each age (except age 0) has been estimated using VPA since 1973 for the southern New England and Georges Bank stocks, and since 1985 for the Cape Cod stock (Northeast

Fisheries Science Center 2001). Natural mortality for individuals of age 1 and greater has been estimated at $M = 0.2$ (Lux 1969, Brown and Hennemuth 1971). Natural mortality of fish at age 0 is unknown; for convenience only, it is also assumed to be 0.2. This assumption does not affect any of our results, however, since we report population sizes of ages 1 or greater, and fecundity was derived from observed abundance at age 1 (see later in this chapter). Population dynamics at low density (i.e., no density dependence), in the absence of fishing, is thus governed by a Leslie matrix with first row (fecundity) elements $0, f_1, \ldots, f_{6+}$, and subdiagonal and right-hand corner elements $\exp(-M)$, representing survival to the next age group and yearly survival of the plus group, respectively.

The fecundity coefficients f_i were calculated to be proportional to female fecundity at age estimates e_i (Howell and Kessler 1977). However, survivorship of eggs to age 1 is very low and somewhat uncertain. To estimate these quantities, we used Beverton-Holt stock-recruitment relationships that have been developed for each stock (Beverton and Holt 1957; parameter estimates from Overholtz et al. 1999). Spawning stock biomass (SSB) of each age class is defined to be the product of mean weight at age i with the proportion of the fish at that age that are sexually mature. The predicted number of age-1 recruits R is given by

$$R = \frac{aS}{b + S} \tag{1}$$

where S is the total SSB in the stock. At low densities, the number of 1-year-old recruits produced by S is thus $R \approx aS/b$. The fecundity coefficients f_i were set by the relationship $f_i = ce_i$, with c chosen so that the number of 1-year-old recruits as calculated from the Leslie matrices match that from the stock-recruitment relationships (eq. (1)) at low density and the equilibrium age structure.

Stochasticity

The variability in fecundity was estimated by calculating the mean squared deviation of observed recruitment from that predicted by the stock-recruitment curves. These estimates are likely to include measurement error, which might result in overestimated standard deviations.

Density Dependence and Carrying Capacity

The maximal growth rates at low density (r_{max}) for each stock was taken as the dominant eigenvalue of the corresponding Leslie matrix. Carrying capacity was calculated, in units of millions of eggs, as the product of mean historical recruitment and the expected number of eggs per recruit in the absence of fishing mortality, in a manner analogous to that used to calculate yellowtail flounder reference points in Northeast Fisheries Science Center (2002). While density dependence may modestly affect growth and mortality, the main effect would be to reduce fecundity. Thus, in our model, density dependence was assumed to affect fecundity only. Estimates of fecundity coefficients, maximal growth rates, and carrying capacity for each stock are given in Table 21.1.

Table 21.1 RAMAS input parameters: fecundity, growth, and carrying capacity parameters

	f_1	f_2	f_3	f_4	f_5	f_6	r_{max}	K
Mean fecundities at low densities								
GB	0.0	1.03	5.19	9.21	11.06	12.73		
SNE	0.0	0.65	9.82	18.24	21.16	22.73		
CC	0.0	0.26	9.46	24.28	37.67	51.83		
Standard deviation of fecundities								
GB	0.0	0.68	3.44	6.10	7.32	8.43		
SNE	0.0	0.30	4.62	8.57	9.95	10.68		
CC	0.0	0.05	1.99	5.1	7.91	10.89		
Growth and carrying capacities								
GB							1.71	2887830
SNE							1.68	2086900
CC							2.04	1060700

Harvest Rates

The exploitation rates used in the simulations are given in Table 21.2. Fishing mortality rates for years since 1973 for southern New England and Georges Bank yellowtail flounder, and since 1985 for the Cape Cod yellowtail stock (Northeast Fisheries Science Center 2001), have been estimated using VPA. For simplicity, we divided these years into periods with similar fishing mortality characteristics. For the 1973–2000 simulations, there were three periods: 1973–1980, when there was high fishing mortalities on all yellowtail of ages 2 or more; 1981–1994, when age-2 mortality was reduced due to increases in mesh size, but mortality of ages 3+ remained high; and 1995–2000, when fishing mortalities decreased due to effort restrictions and the closure of large areas on Georges Bank and southern New England to fishing for groundfish (Murawski et al. 2000). The 1985–2000 retrospective simulations were divided into two periods: 1985–1994 and 1995–2000. The forward projections (1997–2002) were performed assuming a constant fishing mortality rate of $F_{0.1}$, a conventional fishery management target that

Table 21.2 RAMAS input parameters: exploitation rates

Period	Ages	GB	SNE	CC
1973–1981	2+	0.51	0.59	NA
1981–1994	2	0.34	0.47	NA
1981–1994	3+	0.60	0.75	NA
1985–1994	2	0.34	0.47	0.34
1985–1994	3+	0.58	0.74	0.64
1995–2000	2	0.04	0.08	0.11
1995–2000	3+	0.22	0.40	0.51
$F_{0.1}$	2	0.04	0.05	0.04
$F_{0.1}$	3+	0.2	0.215	0.17

is expected to achieve nearly maximum yield per recruit and to maintain a relatively abundant stock.

Dispersal

Movement of yellowtail among stock areas was quantified by several field studies. Royce et al. (1959) tagged and released 2,597 yellowtail flounder on U.S. fishing grounds from 1942 to 1949 and recovered 377 tags over nearly 6 years; they concluded that groups of yellowtail are relatively localized (e.g., most tagged fish were recovered within 80 km of the release site) with short seasonal migrations and that little mixing occurs among fishing grounds. Lux (1963) tagged and released 1,800 yellowtail flounder on the three major U.S. fishing grounds from 1955 to 1957 and recovered 431 tags over 4 years. With subsequent recoveries through 1962 and an additional 3,160 fish tagged in 1959, a total of 4,960 releases and 1,020 recoveries were reported by Lux and Porter (1963). Lux (1963) concluded that groups of yellowtail move seasonally within fishing grounds with a small amount of seasonal mixing among groups. In 1963, Lux (unpublished data) tagged 411 yellowtail flounder off Cape Ann and recorded location of 45 recaptures through 1965. All recaptures were near the release site, except one that moved northward 50 km to the Isles of Shoals. Tagging studies from Canadian waters confirm that yellowtail flounder are relatively sedentary: the longest observed movement from an unpublished tagging study on the northeast Scotian Shelf was less than 50 km (Neilson et al. 1986), and fish recaptured up to 8 years after their release on the Grand Bank traveled an average of 60 km (Morgan and Walsh 1999). A summary of all documented yellowtail movements off the northeastern U.S. is given in Cadrin et al. (1999). These data were adjusted by time at large to obtain annual estimates of dispersal that were used in the model runs where adult dispersal was assumed (Table 21.3).

In addition to movements of juvenile and adult yellowtail, dispersal of eggs and larvae among areas may also be significant. Yellowtail flounder eggs are buoyant with approximately a 5-day incubation period at 10°C (Bigelow and Schroeder 1953, Collette and Klein-MacPhee 2002), and larvae are pelagic for approximately 2 months (Miller et al. 1991). However, no quantitative estimates of larval dispersal for yellowtail flounder are available.

Table 21.3 RAMAS input parameters: dispersal matrices (columns are the donor populations)

	CC	GB	SNE
Adult dispersal			
CC	0.86	0.0	0.07
GB	0.08	0.93	0.06
SNE	0.06	0.07	0.87
Larval dispersal			
CC	0.9	0.1	0.0
GB	0.0	0.8	0.0
SNE	0.1	0.1	1

Evseenko and Nevinsky (1980) and Yevseyenko and Nevinsky (1981) compared geographic distribution of yellowtail flounder eggs and larvae with prevailing currents to infer drift of eggs and larvae. They concluded that yellowtail flounder generally spawn in sufficiently closed water masses with gyronic flows that retain larvae in favorable habitats, such as Georges Bank and the southern New England shelf. However, some leakage may occur from Georges Bank to Cape Cod and southern New England. Butman et al. (1987) reported results from drogued drifters that were released on Georges Bank: of the nine drifters released during yellowtail spawning season, five circled the bank in approximately 2 months, three crossed the Great South Channel into southern New England, and one drifted into the central Gulf of Maine. Sinclair and Iles (1986) and Sinclair (1988) reviewed information on distribution of spawning yellowtail flounder, ichthyoplankton distribution, larval behavior, and oceanographic patterns to conclude that larval retention areas form discrete stocks off southern New England/mid-Atlantic, on Georges Bank, off Cape Cod, and on Browns Bank.

For the model runs that assumed larval transport among the stocks, we assumed that 10% of the larvae from Georges Bank settled in the Cape Cod stock area, and 10% settled in the southern New England stock area. Additionally, 10% of the larvae produced by the Cape Cod stock were assumed to settle in the southern New England stock area (Table 21.3). The assumption of larval transport creates some inconsistencies for our parameterizations, since the stock-recruitment relationships (eq. (1)) were constructed under the assumption that each stock is a closed population. For example, recruitment from the Georges Bank "source" population is defined in terms of 1-year-old fish observed in Georges Bank. Thus, if a percentage of the larvae are being transported to other stocks, the stock-recruitment curve will have underestimated the fecundity of the Georges Bank stock by that percentage. To account for this, in the model runs where larval transport was assumed, the fecundity of the Georges Bank stock was adjusted so as to produce the same number of 1-year-old recruits in Georges Bank as do the simulations where no larval exchange is assumed. Because larval transport among the stocks would affect both the stock-recruitment relationships and the VPAs, our simulations that assume larval exchange must be considered as initial explorations to the sensitivity of population dynamics to such exchange, rather than giving conclusive findings. Accordingly, results are presented to compare general trends only, and projected abundance and rebuilding rates should be considered to be merely approximations.

Simulation Scenarios

Three sets of model runs (each consisting of 10,000 stochastic simulations) were performed. The first set is a simulation of the Georges Bank and southern New England yellowtail stocks for the 27-year period 1973–2000, assuming no dispersal between these stocks. The second consists of simulations of all three stocks during 1985–2000, both with and without dispersal. These retrospective projections are useful for comparison with the VPA estimates of abundance during these periods and for estimating the effects of dispersal during periods of relatively low abundance and high exploitation rates. The last set of simulations are forward projections from 1997 to 2012, with and without dispersal, assuming fishing at $F_{0.1}$. These simulations are used to indicate the effects of dispersal on rebuilding stocks at much lower mortality rates than the retrospective simulations.

Results

Retrospective Simulations

Except for the first few years of the simulations, the standard deviations of the model runs were on the order of 50% to 100% of the mean, due to the high variability in recruitment. Because of the high variability of the model runs, it should not be expected that the mean of these runs would always closely fit the observed abundances. However, model performance can be evaluated by considering whether the observed data remains within the range of values predicted by the model, and whether the residuals are reasonably balanced between positive and negative values.

Model simulations of the Georges Bank and southern New England stocks match observations reasonably well, especially considering that the parameters were estimated external to the model and that model fishing mortalities were held constant for long periods (Figures 21.2 and 21.3). The model means are in good agreement with the observed abundances on Georges Bank except that the model means underestimate the recent dramatic increases in abundance. This may be due to stochastic variations in recruitment, as the observed abundance in 2000 is less than 1 SD above the mean of the model runs. The mean model predictions for the southern New England stock underestimate observed abundances in the 1980s, when there were two large year classes, and then overestimates abundances in the late 1990s, when recruitment was poor. These discrepancies are most likely primarily due to the high variability of recruitment as the observed abundances remain well within the minimum and maximum model runs. However, larval or adult dispersal may also play a role in the lack of good model fit.

By contrast to the other two stocks, the means of the model runs consistently underestimate the observed biomass of Cape Cod yellowtail (Figure 21.3c). This might be due to inaccuracy of parameters, such as an underestimation of r_{max} or K or an overestimation of fishing mortality. Another possibility is that the discrepancy is caused in whole or in part by dispersal of adults or larvae into the Cape Cod stock.

To explore the latter possibilities, the 1985–2000 simulations were re-run assuming dispersal of either adults or larvae, or both, as discussed in the methods section (Figure 21.3). Model runs that include adult or larval dispersal predict higher abundances for the Cape Cod stock. Unlike the other two stocks, model projections of Cape Cod yellowtail are quite sensitive to assumptions about dispersal. This is especially striking in the case of larval dispersal because it would normally be expected that the "sink" population (southern New England yellowtail) would be most enhanced by dispersal. However, the gain of larvae to Cape Cod from Georges Bank considerably outweighs the loss to southern New England, because the Cape Cod stock is much smaller than the other two stocks. Southern New England also gains larvae from Georges Bank, but it is much less affected by this transport than is Cape Cod because it is much larger than the Cape Cod stock.

Rebuilding Simulations

While the Georges Bank yellowtail stock has rebuilt considerably from low abundance, the biomasses of both the Cape Cod and southern New England stocks are well below

(a)

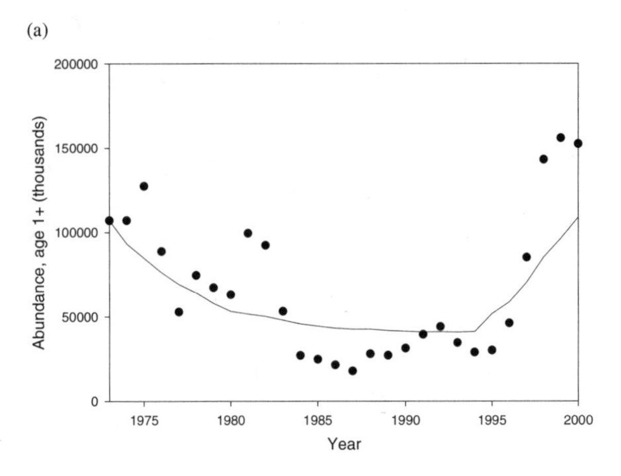

(b)

Figure 21.2 Retrospective projections for **(a)** Georges Bank (GB) and **(b)** southern New England (SNE) (1973–2000), without dispersal, compared to abundances estimated by virtual population analyses. The lines give the mean of the model projections; the solid circles indicate estimated abundances.

their rebuilding targets (Northeast Fisheries Science Center 2001). Here we will use the RAMAS model to examine a possible harvesting strategy for the years 1997–2012 (1997 is the most recent year in which abundance at age estimates are available for all three stocks), with fishing mortality set at $F_{0.1}$. Because fishing mortalities in each of the stocks were different than $F_{0.1}$ in the years 1997–2002, these simulations cannot be regarded as forecasts of future abundances. Rather, our goal is to explore the effects of larval and adult exchanges on the extent and speed of rebuilding and to investigate the implications of such exchanges at relatively low exploitation rates and large population sizes.

Mean model projections, under the assumptions of no dispersal, adult dispersal, or larval dispersal, and combined adult and larval dispersal, are given in Figure 21.4. Model projections of the Georges Bank stock underestimate the increases in abundance that were observed between 1997 and 2000. This is in part due to recent fishing mortality rates being less than $F_{0.1}$. Asymptotic projected abundances were near the observed 1999–2000 population sizes, suggesting that this stock is approaching the expected equilibrium abundance associated with $F_{0.1}$. Model projections for the Georges Bank stock appear insensitive to assumptions regarding dispersal.

(a)

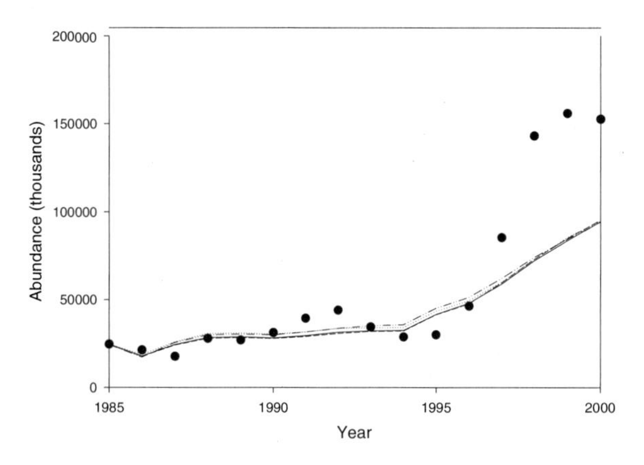

(b)

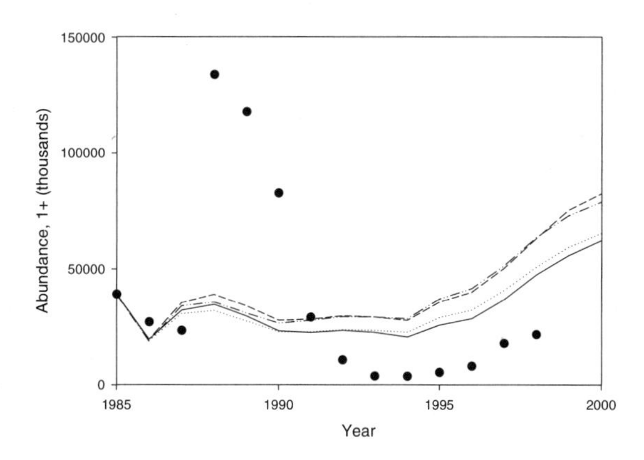

(c)

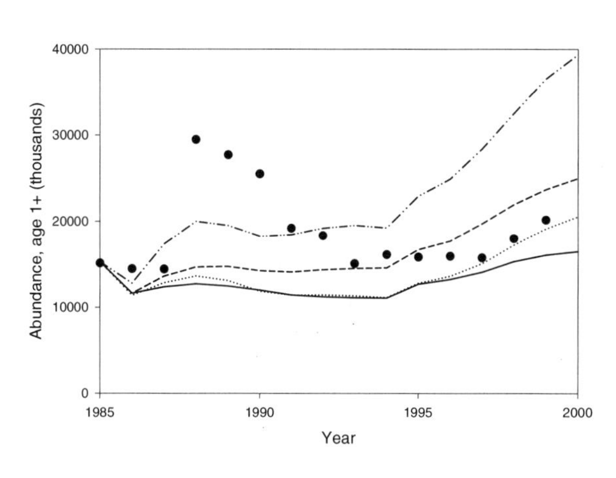

Figure 21.3 Model runs for the years 1985–2000 for (a) Georges Bank, (b) southern New England, and (c) Cape Cod yellowtail flounder. The lines are model means under the assumption of no dispersal (*solid lines*), age 1+ movement (*dotted lines*), larval dispersal (*dashed lines*), and both larval and age 1+ dispersal (*dashed-dotted lines*). Estimated abundances from virtual population analyses are indicated by the solid circles.

(a)

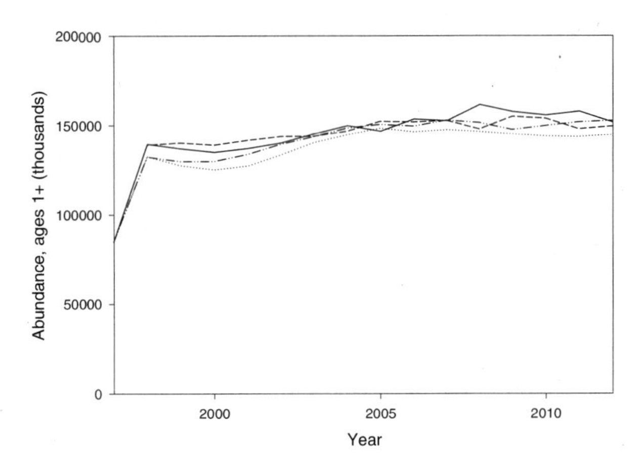

(b)

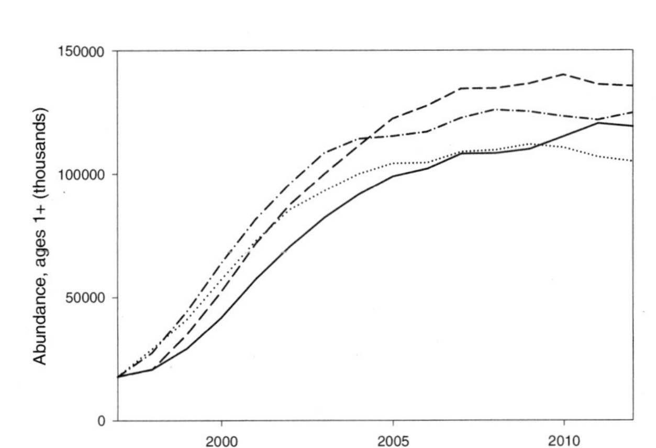

(c)

Figure 21.4 Projections of stock abundances from 1997 to 2012 of (a) Georges Bank, (b) southern New England, and (c) Cape Cod yellowtail flounder, assuming that each stock is fished at $F_{0.1}$. The solid line was calculated under the assumption of no dispersal, the dotted line assumes adult dispersal, the dashed line assumes larval dispersal, and the dashed-dotted line assumes both larval and adult dispersal.

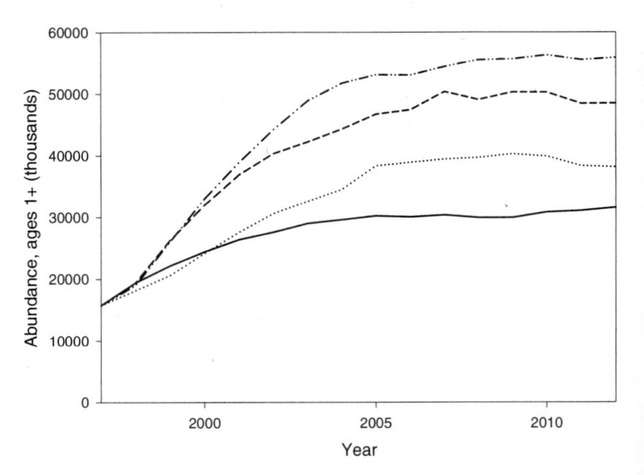

Model projections of the southern New England stock are only slightly more sensitive to assumptions about dispersal. Note that in the early years of the simulations that assume adult dispersal, there is a net flux of adults into southern New England, as this stock's population is considerably smaller than that of Georges Bank. However, after the southern New England's population rebuilds, and becomes comparable in size to the Georges Bank stock, there will be a net loss through emigration away from southern New England.

Model projections of the Cape Cod stock are very sensitive to dispersal assumptions. The mean of the model runs assuming no dispersal is considerably smaller than those that assume either adult or larval dispersal. The sensitivity to adult dispersal tends to increase over time, because immigration from the southern New England stock will increase as it rebuilds.

Sensitivity Analysis

Table 21.4 gives the percentage of gain or loss for each stock due to dispersal over that with no dispersal for the 1985–2000 and 1997–2012 model runs. The Georges Bank stock appears to be insensitive to dispersal, and the slight changes given in the table are likely due to the stochastic variability of model runs. The effect of dispersal on southern New England yellowtail was more positive in the retrospective simulations than in the $F_{0.1}$ runs. This is due to the relatively low abundance of this stock at the end of the 1985–2000 simulations, whereas it is nearly the size of the Georges Bank stock at the end of the 1997–2012 simulations. Thus, the relative subsidy received by the southern New England stock due to dispersal from other stocks is higher in the retrospective simulations. For the same reason, adult-only dispersal induces a gain in the southern New England stock in the 1985–2000 simulations but a loss in the 1997–2012 simulations. Note also the nonlinear effect of combining adult and larval dispersal in this stock in the retrospective simulations. The increased abundance induced by larval dispersal into southern New England causes a greater loss of adults from this stock. Thus, combining adult with larval dispersal in this stock results in a smaller gain than larval dispersal alone, even though larval and adult dispersal individually induce increases in this stock's

Table 21.4 Sensitivity of stocks to dispersal assumptions 1985–2000 retrospective simulations.

	GB	SNE	CC
1985–2000 retrospective simulations			
Larval dispersal	1	32	52
Adult dispersal	0	5	25
Larval and adult dispersal	1	27	141
1997–2012 rebuilding simulations			
Larval dispersal	–1	14	54
Adult dispersal	–4	–12	21
Larval and adult dispersal	1	5	77

Note: Percentage change in mean abundance at the end of the simulation run to dispersal compared to the corresponding simulation with no dispersal. GB = Georges Bank; SNE = Southern New England; CC = Cape Cod.

abundance. As discussed in this chapter, the Cape Cod yellowtail stock shows the largest effects to both larval and adult dispersal.

Discussion

There are a number of difficulties that complicate the modeling of exploited marine populations, including highly variable recruitment and the uncertainty found in life history parameters and exploitation rates. Given these inherent problems, and the fact that no within-model parameter fitting was performed, the retrospective projections of the Georges Bank stock fit the observed biomass levels reasonably well. The southern New England stock biomass was characterized by two strong year classes in the 1980s, followed by poor recruitment in the 1990s. Because the mean of a number of model runs cannot be expected to conform to a specific highly variable recruitment pattern, we can conclude that the model also gave an acceptable fit to the southern New England stock data. The model was less successful in modeling the Cape Cod yellowtail stock because model projections consistently underestimated the stock abundances.

Both the retrospective and $F_{0.1}$ projections of the Georges Bank stock were relatively insensitive to assumptions regarding adult or larval dispersal. The southern New England stock projections were only modestly more sensitive to dispersal assumptions; these assumptions are only important when the Georges Bank stock is considerably larger than southern New England (as is currently the case). The extent of the recovery of the southern New England stock in the next few years may therefore give clues as to the extent to which these dispersal processes are actually occurring.

Our most important finding was that model projections of the Cape Cod yellowtail stock were highly sensitive to assumptions regarding adult or larval dispersal. This is due to the fact that the Cape Cod stock is much smaller than the other two stocks. Thus, even a small percentage of adults or larvae that immigrate from the other two stocks would constitute a much larger percentage of the Cape Cod stock and hence cause sensitivity to dispersal assumptions. Dispersal between stocks may explain why single-stock analysis of Cape Cod yellowtail has been problematic (Cadrin et al. 1999).

One major assumption in our analyses is that the published mark-recapture observations can be used to derive dispersion rates among areas. However, the referenced tagging experiments were not explicitly designed for such estimates, and therefore dispersion estimates may be biased. As mentioned, another reason that our results should be taken cautiously as initial explorations only is that the model parameters and the VPA estimates of abundances were all derived under the assumption that each stock is a closed population. If in fact dispersal among the stocks is occurring to the extent considered here, then these parameters and estimates will need to be reevaluated. In particular, adult dispersal into the Cape Cod stock may cause the VPA estimates of this stock to be inflated. This may be another source of the discrepancy between our model and the VPA estimates of abundance.

Traditionally, stock assessment scientists have had the choice of modeling stocks as closed individual populations or modeling all of the stocks together as a single "well-mixed" population. Neither of these alternatives may be satisfactory for yellowtail flounder. Our analysis has demonstrated that the population dynamics of some stocks can be strongly sensitive to even modest levels of dispersal among stocks, which may thus cause difficulties for models that assume no such dispersal.

The alternative of modeling the stocks as a single population is no better. Exploitation patterns can differ considerably among stocks. For example, the fishing mortality rate on Georges Bank yellowtail is currently much less than that of either the Cape Cod or southern New England stocks; this fact gives at least a partial explanation as to the more rapid recovery of the Georges Bank population. Such a differential recovery rate would not be seen in a model where all three stocks were contained in a single well-mixed compartment. Furthermore, life history attributes vary substantially among areas (e.g., Cape Cod yellowtail grow slower, attain greater maximum size, and mature slower). Therefore, critical management parameters such as $F_{0.1}$ vary among stocks, and the aggregated average of this quantity would vary over time according to the relative abundance of each component.

A metapopulation approach, such as that explored here, may thus be the best framework for modeling yellowtail flounder or similar species that may have modest levels of movement among individual stocks. Such an approach can model dispersal while still being able to take into account variations in exploitation rates and life history parameters among individual stock components.

Acknowledgments We thank the many researchers who have been involved with yellowtail flounder stock assessments through the NEFSC Demersal Working Groups and the Transboundary Assessment Working Group. We also thank Steve Murawski, Fred Serchuk, Chris Legault, Paul Rago, Resit Akçakaya, Chris Wood, and two anonymous reviewers for their comments on drafts of this manuscript.

References

Beverton, R. J. H., and Holt, S. J. 1957. *On the dynamics of exploited fish populations.* Chapman and Hall, London.

Bigelow, H. B., and Schroeder, W. C. 1953. Fishes of the Gulf of Maine. *Fisheries Bulletin* 53 (special issue).

Brown, B. E., and Hennemuth, R. C. 1971. *Assessment of the yellowtail flounder fishery in subarea 5.* Res. Doc. 71/14. International Commission for the Northwest Atlantic Fisheries, Dartmouth, Nova Scotia, Canada.

Butman, B., Loder, J. W., and Beardsley, R. C. 1987. The seasonal mean circulation: observation and theory. Pages 125–138 in R. H. Backus (ed.), *Georges Bank.* MIT Press, Cambridge, Mass.

Cadrin, S. X. 2000. Southern New England yellowtail flounder. Pages 65–82 in *Assessment of 11 Northeast Groundfish Stocks through 1999.* NEFSC Ref. Doc. 00–05. Northeast Fisheries Science Center, Woods Hole, Mass.

Cadrin, S. X., King, J., and Suslowicz, L. E. 1999. *Status of the Cape Cod yellowtail flounder stock for 1998.* NEFSC Ref. Doc. 99–04. Northeast Fisheries Science Center, Woods Hole, Mass.

Collette, B. B., and Klein-MacPhee, G. 2002. *Bigelow and Schroeder's Fishes of the Gulf of Maine.* Smithsonian Press, Washington, D.C.

Evseenko, S. A., and Nevinskiy, M. M. 1980. *Drift of eggs and larvae of yellowtail flounder* (Limanda ferruginea (*Storer*)) *in the Northwest Atlantic.* NAFO SCR Doc. 80/IX/118.Northwest Atlantic Fisheries Organization, Dartmouth, Nova Scotia, Canada.

Hilborn, R., and Walters, C. J. 1992. *Quantitative fisheries stock assessment: choice, dynamics, and uncertainty.* Chapman and Hall, New York.

Howell, W. H., and Kessler, D. H. 1977. Fecundity of the southern New England stock of yellowtail flounder, *Limanda ferruginea. Fisheries Bulletin* 75: 877–880.

Lux, F. E. 1963. Identification of New England yellowtail flounder groups. *Fisheries Bulletin* 63: 1–10.

Lux, F. E. 1969. Landings per unit effort, age composition, and total mortality of yellowtail flounder, *Limanda ferruginea* (Storer) off New England. *ICNAF Research Bulletin* 6: 47–69.

Lux, F. E., and Porter Jr., L. R. 1963. *Tagging and tag recovery data for yellowtail flounder tagged in 1955, 1957, and 1959.* Woods Hole Lab. Rep. 63–1. Woods Hole Laboratory, Mass.

Miller, J. M., Burke, J. S., and Fitzhugh, G. R. 1991. Early life history patterns of Atlantic North American flatfish: likely (and unlikely) factors controlling recruitment. *Netherlands Journal of Sea Research* 27: 261–275.

Morgan, M. J., and Walsh, S. J. 1999. An update of results of tagging experiments with juvenile yellowtail flounder in NAFO divisions 3LNO. NAFO SCR Doc. 99/23. Northwest Atlantic Fisheries Organization, Dartmouth, Nova Scotia, Canada.

Murawski, S. A., Brown, R., Lin, H. L., Rago, P. J., and Hendrickson, L.. 2000. Large-scale closed areas as a fishery-management tool in temperate marine systems: the Georges Bank experience. *Bulletin of Marine Science* 66: 775–798.

Neilson, J. D., Hurley, P., and Perry, R. I. 1986. *Stock structure of yellowtail flounder in the Gulf of Maine area: implications for management.* CAFSAC Res. Doc. 86/64. Canadian Department of Fisheries and Oceans, Ottawa, Ontario, Canada.

Northeast Fisheries Science Center (NEFSC). 2001. *Assessment of 19 Northeast Groundfish Stocks through 2000.* NEFSC Ref. Doc. 01–20. Northeast Fisheries Science Center, Woods Hole, Mass.

Northeast Fisheries Science Center (NEFSC). 2002. Working group on re-evaluation of biological reference points for New England groundfish. NEFSC Ref. Doc. 02-04. Northeast Fisheries Science Center, Woods Hole, Mass.

Overholtz, W., and Cadrin, S. 1998. Yellowtail flounder. Pages 70–74 in S. H. Clark (ed.), *Status of the fishery resources off the northeastern United States for 1998.* NOAA Tech. Mem. NMFS-NE-115. National Marine Fisheries Service, Woods Hole, Mass.

Overholtz, W. J., Murawski, S.A., Rago, P. J., Gabriel, W. L., Terceiro, M., and Brodziak, J. K. T. 1999. *Ten-year projections of landings, spawning stock biomass, and recruitment for five New England groundfish stocks.* NEFSC Ref. Doc. 99–05. Northeast Fisheries Science Center, Woods Hole, Mass.

Prager, M. H. 1994. A suite of extensions to a nonequilibrium surplus-production model. *Fisheries Bulletin* 92: 374–389.

Royce, W. F., Buller, R. F., and Premetz, E. D. 1959. Decline of the yellowtail flounder (*Limanda ferruginea*) off New England. *Fisheries Bulletin* 146: 1–267.

Sinclair, M. 1988. *Marine populations: an essay on population regulation and speciation*. University of Washington Press, Seattle.

Sinclair, M., and Iles, T. D. 1986. *Population richness of marine fish species.* ICES C.M. 1986/M:22. International Council for the Exploration of the Sea, Copenhagen, Denmark.

Stone, H., Legault, C., Cadrin, S., Gavaris, S., Neilson, J., and Perley, P. 2001. *Stock assessment of Georges Bank (5Zjmnh) yellowtail flounder for 2001.* Canada DFO Res. Doc. 2001/068. Canadian Department of Fisheries and Oceans, Ottawa, Ontario, Canada.

Yevseyenko, S. A., and Nevinskiy, M. M. 1981. On the development of the eggs and larvae of the yellow-tailed dab *Limanda ferruginea* Storer and their passive transport in the Northwest Atlantic. *Journal of Icthyology* 21(5): 65–74.

Walsh, S. J., Cadrin, S. X., Rivard, D., Cook, R. M., Casey, J., and Brodie. W. B. 1999. *Parameter estimates of fishing mortality and biological reference points using research survey data: yellowtail flounder in NAFO Div 3LNO, a preliminary analysis.* NAFO SCR Doc. 99/5. Northwest Atlantic Fisheries Organization, Dartmouth, Nova Scotia, Canada.

22

Atlantic Herring (*Clupea harengus*) in the Northwest Atlantic Ocean

Dynamics of Nested Population Components under Several Harvest Regimes

R. KENT SMEDBOL
ROBERT L. STEPHENSON

The Atlantic herring, *Clupea harengus* L., is an abundant marine fish species with a distribution that encompasses coastal and continental shelf regions on both sides of the North Atlantic Ocean. Herring are pelagic, and individuals usually aggregate in schools and can form mammoth shoals during the spawning and overwintering seasons. Herring are considered "population rich" (Sinclair 1988) in that they form many populations or population "complexes" within their transoceanic distribution, including a number of separate populations in Canadian jurisdictional waters (Iles and Sinclair 1982, Stephenson et al. 2001). Iles and Sinclair (1982) have suggested that geographically stable oceanographic features serve as larval retention areas and determine the distribution of herring populations. These features may serve additionally as incomplete barriers to larval exchange among major population complexes. Herring populations may mix during overwintering and feeding seasons, but then separate during the spawning period. Adults return to discrete spawning areas, some of which are used every year and others only occasionally, and deposit eggs directly upon the seabed. Spawners are considered to exhibit a fair degree of interannual fidelity to specific spawning grounds and, thus, are presumed to "home" to their natal spawning areas (e.g., Iles and Sinclair 1982, Sinclair 1988). McQuinn (1997) proposed that this benthic spawning distribution and population dynamics of the Atlantic herring fit within the metapopulation concept. Ware and Schweigert (2001) used a metapopulation approach to model the structure and dynamics of Pacific herring (*Clupea pallasi*), a closely related species.

Stephenson et al. (2001) and Smedbol and Stephenson (2001) have proposed conceptual models to describe the "nested" spatial structure of herring populations. Efforts are under way to determine the level of reproductive isolation among spawning groups (McPherson et al. 2003 and in press), but the basis for this is not clear. Accordingly, we have avoided the usual "population-subpopulation" terminology in this chapter and instead refer to the four spawning groups modeled in this study as "spawning components."

Failure to incorporate nested spatial structure in fishery management plans can lead to overexploitation even if the overall harvest level is conservative. During the 1980s, fishing was concentrated on individual spawning areas within the Nova Scotia/Bay of Fundy management unit (Northwest Atlantic Fisheries Organization Subdivision 4X), and this resulted in the serial collapse of individual spawning components (Stephenson et al. 2001). An example is the collapse of the spawning component associated with Trinity Ledge, a spawning ground that was the target of intense fishing during the 1980s (Figure 22.1). Herring caught on Trinity Ledge accounted for nearly 40% of total summer catches from the management unit in 1985. By 1989, the Trinity Ledge spawning component had collapsed in abundance (Chang et al. 1995), even though the overall level of exploitation in the management unit was considered appropriate. Closure of this area to fishing has led to some recovery during the past 10 years, but abundance is still well below that observed in the mid-1980s (Stephenson et al. 2000). Such progressive erosion of spawning components is recognized to have occurred in other stocks as well (e.g., North Sea herring; Stephenson 2001).

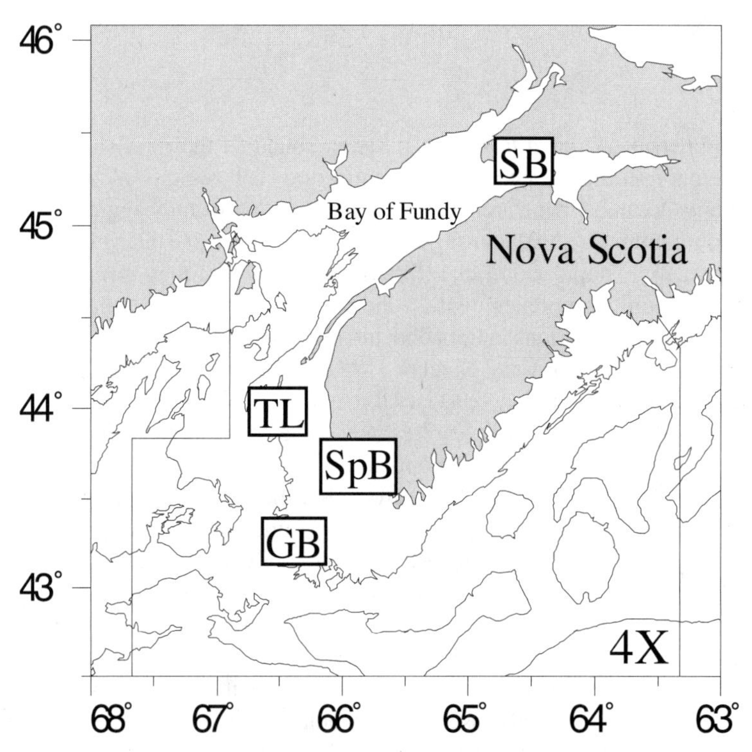

Figure 22.1 Bathymetric chart of the waters around southwest Nova Scotia and the Bay of Fundy. Solid line = boundary of Northwest Atlantic Fisheries Organization (NAFO) Subdivsion 4X; long-dashed line = 100–m isobath; short-dashed line = 200–m isobath. Locations of spawning areas included in simulation: Scots Bay (SB), Trinity Ledge (TL), German Bank (GB), and Spectacle Buoy (SpB).

Recognition of the nested structure of spawning components in Southwest Nova Scotia/Bay of Fundy and the problems associated with spatial concentration of fishing effort has led to fishery management at smaller spatial scales. Assessment of population status now includes consideration of viability of components associated with individual spawning grounds within each major complex (see Stephenson et al. 1999). Current management aims to maintain reproductive capacity of the population complex, which requires the persistence of all spawning components in the management unit. To protect each spawning component, the total allowable catch (TAC) for the entire management unit is spread among spawning grounds according to the relative size of each spawning component. Protection of individual spawning components was enhanced by "in-season" management whereby decisions concerning the timing and distribution of the fishery are based on the most current information (Stephenson et al. 1999). Industry vessels undertake these surveys, and the information collected is used to calculate the portion of the TAC that can be harvested at a particular time and location. Under this protocol, no more than 20% of any spawning aggregation can be removed.

In the case study reported here, we used RAMAS Metapop Version 4.0 to simulate the dynamics of spawning components in the Southwest Nova Scotia and the Bay of Fundy (SWNS/BF) population complex within Northwest Atlantic Fisheries Organization (NAFO) Subdivision 4WX (Figure 22.1). This complex contains four relatively large spawning components, associated with Scots Bay in the Bay of Fundy, German Bank, Trinity Ledge, and Spectacle Buoy off Southwest Nova Scotia. The bulk of landings from the fishery within NAFO subdivision 4WX are caught over these spawning grounds. These spawning components also constitute the majority of herring resident in the management unit (Stephenson et al. 2000). Our purpose was to investigate the population dynamics of the four spawning components under harvest and explore the adequacy of the current management strategy under a reversed burden of proof. We simulated population dynamics under various levels of fishing mortality, natural mortality, and exchange of individuals among spawning components to determine under what conditions it becomes prudent to spread the harvest among components on a proportional basis.

Methods

Stage Matrix

Information used in the simulation is derived from several sources. Commercial catch data were obtained from fishery landings during the period of 1963–1999, and sampling of the commercial catch in 1999. Estimates of population abundance were derived from sequential population analysis (SPA) for the years 1965–2000 and from acoustic surveys presented in recent stock status evaluations including the years 1997–2000. All estimates of abundance include both males and females. No sex structure was assumed because the acoustic estimates of biomass do not differentiate between sexes and both sexes have the same dynamics. Ten stages were used in the simulation, ages 1 to 10+. Fish in their first year of life (< age 1) were not represented explicitly in the simulations because their abundance has not been assessed. Fish aged < 1 year exhibit a different spatial distribution from that of older fish, and they are underrepresented in catches and sampling programs. All ages 10 and up were combined into a single stage (age 10+), due to their low abundance.

Within the study region, approximately 28% of individuals are mature at age 3, and all individuals attain maturity by age 4 (Power and Iles 2001). Annual survivorship values of 0.8, 0.7, and 0.6 were used in population simulations. Fisheries scientists assume an instantaneous natural mortality of 0.2 (e.g., Stephenson et al. 2000), implying an annual survivorship of approximately 0.8. Natural mortality has not been studied directly for herring populations in the area, but herring is an important forage species for many fish, mammal, and bird species, and thus natural mortality may be higher than is regularly assumed. Early assessments of herring stocks in the region (e.g., Gulf of Maine) included alternative analyses in which instantaneous natural mortality was set at 0.2 or 0.3 (International Commission of Northwest Atlantic Fisheries 1972). The instantaneous mortality rate estimated from longevity data (method of Hoenig 1983) is 0.29, which is equivalent to a constant annual survivorship of approximately 0.7. Thus, our minimum trial value for survivorship (0.6) corresponds roughly to a doubling of current estimates of natural mortality.

Stochasticity and Error

Demographic stochasticity was not modeled since the quasi-extinction threshold for each spawning component was set at 10,000 individuals. This threshold may seem relatively high, but it reflects the difficulty of detecting small herring schools given the large geographic range of this study.

Recruitment (reproductive success) is related largely to environmental influences and exhibits high variance under typical conditions. However, the modeled relationship between recruitment and spawner biomass indicates that, on average, recruitment does not increase, and may even decrease, when spawning biomass increases beyond 200,000 tons (Figure 22.2). Accordingly, variability in reproductive rates (first-row elements of the Stage Matrix, including survival to age 1) was estimated after taking spawning biomass into account using a Ricker stock-recruit function of the form

$$R = aSe^{-bS} \tag{1}$$

where R is the number of recruits, S is the biomass of the spawning stock, and a and b are shape parameters. Total recruitment was attributed to each age class, using the number of individuals at age and the mean weight at age. Annual variation in age-specific reproductive rate was estimated as the sum of squared deviations between the observed and predicted recruitment due to each age class, at each year in the spawner-recruit time series.

One catastrophe was included in the simulations to simulate the effects of environmentally induced recruitment failure. Each spawning component was exposed each year to a 20% probability of recruitment failure that decreased age-1 abundance to 5% of the value expected if there had been no catastrophe.

Initial Abundances

Initial abundances at age for each of the four spawning components were derived from acoustic biomass estimates and catch at age used in annual assessments of SWNS/BF herring (Stephenson et al. 2000). The abundance estimates represent minimum levels of biomass (Department of Fisheries and Oceans 2001) because the survey does not cover

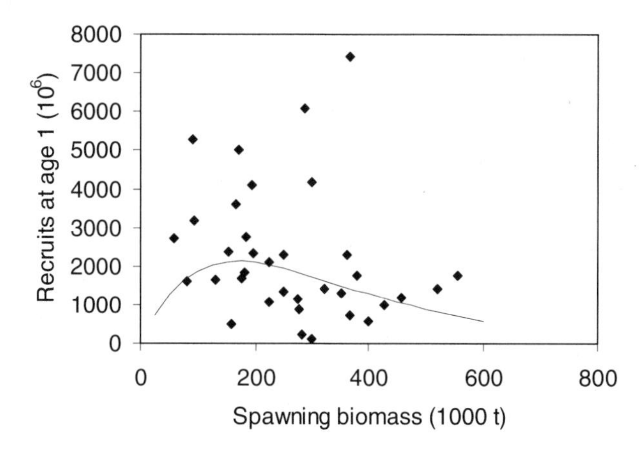

Figure 22.2 Stock–recruitment relationship in southwest Nova Scotia and the Bay of Fundy. The solid line represents the fitted Ricker stock-recruit function.

the total area of the spawning grounds or the entire duration of the known spawning period. See Melvin et al. (2001) for details of the acoustic methods.

Acoustic estimates of herring biomass were averaged over the survey years (1997–1998 for Spectacle Buoy and 1997–2000 for others; Department of Fisheries and Oceans 2001). The averaged biomass estimates were converted to numbers at age for each spawning component in three steps. First, biomass associated with each spawning component was expressed as a proportion of the summed total biomass of the four spawning components in the SWNS/BF spawning complex. Second, total numbers at age for the complex were calculated through SPA of commercial catch data from 1999 (the most recent SPA; Stephenson et al. 2000). Third and finally, total numbers at age from the SPA were multiplied by the biomass proportions to generate estimates of the numbers at age for each of the four spawning areas (Table 22.1). Weights at age and the proportion of individuals at each age were assumed to be identical for all four spawning groups. This assumption was reasonable because year class strength was similar for all spawning areas, and catches from all spawning areas associated with the SWNS/BF complex were aggregated in the SPA. We used acoustic estimates from 2000 instead of 1999, as the sampling coverage was superior. Age structure is unlikely to have changed substantially in the course of one year, and any mismatch in year of acoustic estimate and SPA will not be misleading as the numbers at age derived from the SPA are only approximate.

Dispersal

Three of the four spawning components—German Bank, Trinity Ledge, and Spectacle Buoy—are in relatively close proximity, whereas the Scots Bay spawning ground is more distant (Figure 22.1). Genetic variation has been detected among these four spawning components (McPherson et al. in press), suggesting some degree of reproductive isolation. Mark-recapture studies involving SWNS/BF herring are under way, but as of yet the number of recaptures is too low to be of use in dispersal studies. Studies of straying rates in Atlantic herring found along the east coast of Newfoundland provided estimates of 10% exchange among spawning areas (Wheeler and Winters 1984). Investigations of straying in Pacific herring off British Columbia reported exchanges rates of 4% to 25% among local populations (Ware et al. 2000). Given this range of straying rates in

Table 22.1 Mean weights-at-age and initial abundances by spawning component

		Abundance			
Age (years)	Mean Weight (g)	Scots Bay	Trinity Ledge	German Bank	Spectacle Buoy
1	20	183.08	16.53	784.69	15.71
2	42	18.31	1.65	78.47	1.57
3	75	427.30	38.58	1831.45	36.67
4	120	457.87	41.34	1962.50	39.29
5	172	183.08	16.53	784.69	15.71
6	220	159.82	14.43	685.03	13.71
7	263	84.95	7.67	364.09	7.29
8	304	18.86	1.70	80.82	1.62
9	344	2.01	0.18	8.63	0.17
10+	378	0.18	0.02	0.78	0.02
Total		1,535.45	138.63	6,581.15	131.76

Note: Mean weights at age are equal for all four spawning components. Numbers expressed in millions of individuals (10^6).

Pacific herring, and the preliminary results of population level genetic studies involving SWNS/BF herring, simulation scenarios were designed to investigate the effects of annual dispersal (exchange) rates of 10%, 20%, or 30% of the individuals in the spawning component of origin.

Density Dependence

Ceiling-type density dependence was assumed to affect all vital rates in each spawning component. Carrying capacity (*K*) was defined as double the highest annual biomass observed since the first estimates of abundance in 1965. It is reasonable to assume that *K* is greater than the highest estimates of biomass because SWNS/BF herring have been harvested for centuries (e.g., Iles 1993, Stephenson et al. 1993), and exploitation has been relatively heavy since 1965 (Stephenson et al. 2000). Most simulation scenarios included annual harvest that prevented abundance from reaching the ceiling defined by *K*.

Management Actions

Atlantic herring are harvested intensively by commercial fisheries. In 2000, approximately 85,250 tons of herring were harvested from the SWNS/BF complex (Department of Fisheries and Oceans 2001). Purse seiners targeting spawning, summer feeding, and overwintering aggregations take most of the harvest. A small gillnet fishery captures mainly adult herring near the spawning grounds. A relatively small portion of the total catch comprises juvenile herring taken in the coastal weir fishery. Overall, these fisheries result in relatively low exploitation of age 1 and age 2 herring and higher exploitation of ages 3+.

Harvest in the simulation was structured to reflect age selectivity by the fishery. Age 1 fish were not harvested in the model, age 2 herring were always exposed to an annual harvest rate of 10%, and fully recruited fish (age 3+) were harvested at target rates of 20%, 30%, or 40%, depending on the scenario. The 20% harvest rate scenario is of particular interest because it approximates current fishing limits for the 4WX complex

(Stephenson et al. 2000). Two management scenarios were examined: mixed-stock and individual-stock management. In the first, all four spawning components were harvested as a single unit of pooled abundance; in the second, each spawning component was harvested separately. Harvest occurred prior to dispersal, provided that at least one individual remained in the spawning component.

Simulation Projections

Simulations were projected for 50 years, with a time step of 1 year. Each simulation condition was replicated 1,000 times. Thresholds for metapopulation quasi extinction and explosion were set at 10^4 and 10^5, respectively.

Results

In general, the probability of extinction was less affected by dispersal rate than by survivorship or harvest rate, regardless of management scenario (mixed versus separate fisheries on spawning components) (Table 22.2). With dispersal, extinction occurred only under combinations of elevated harvest levels (0.3–0.4) and reduced survivorship (0.6–0.7). The duration of these local extinctions tended to be relatively short but increased under more pessimistic situations of relatively high exploitation and low natural survivorship. Local population (spawning component) occupancy and the duration of local population extinction were similar across the four spawning components (Figure 22.3).

Without dispersal, all combinations of harvest and survivorship resulted in extinction in at least some of the trials. In simulations without harvest, all spawning components rebuilt to carrying capacity and no extinction events were recorded.

Discussion

The current management approach for the herring fishery in this region distributes fishing effort among spawning areas according to their relative abundance, with the view that such an approach should prevent overfishing of any one particular spawning component (Stephenson et al. 1999). Nevertheless, simulations based on our present understanding of herring population dynamics (as specified in this model) highlight several issues under this management approach. Given current total biomass estimates for herring in the management unit, a 20% harvest rate is consistent with the $F_{0.1}$ harvest strategy (Stephenson et al. 2000). (See Deriso 1987 for a discussion of the $F_{0.1}$ harvest reference point.) Model runs undertaken with a 20% harvest rate did not result in any extinction events over the range of dispersal and survivorship values used, whether fishing effort is pooled or operated separately on spawning components. However, this harvest rate did not allow for a substantial increase in abundance in any spawning component, regardless of the other model parameter values. All spawning components did exhibit relatively rapid recovery in the absence of exploitation (e.g., fishing moratoria). Typically, for extinction to occur, survivorship had to be relatively low and exploitation had to be greater than what is allowed under the current management plan (Table 22.2).

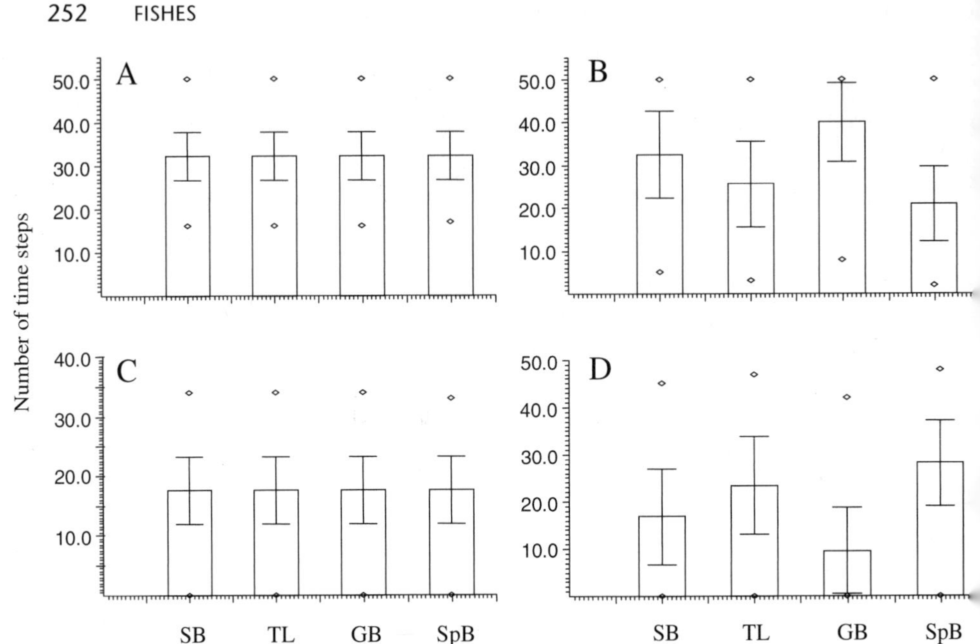

Figure 22.3 Example of the effect of dispersal on local occupancy and duration of extinction of the four spawning components of southwest Nova Scotia and the Bay of Fundy: Scots Bay (SB), Trinity Ledge (TL), German Bank (GB), and Spectacle Buoy (SpB). (A) Local occupancy, dispersal = 0.3. (B) Local occupancy, dispersal = 0. (C) Duration of extinction, dispersal = 0.3. (D) Duration of extinction, dispersal = 0. Harvest was undertaken on the pooled (mixed) abundance of the spawning components and set at 0.4; survivorship was set to 0.6. Mean value = hollow bar; error bars = ± 1 SD; upper diamond = maximum value; lower diamond = minimum value.

The similarity among outcomes from the mixed versus individual stock management scenarios may have been influenced by how dispersal and harvest were structured in the model. Each spawning component exchanged an equivalent proportion of individuals with other components so that the largest spawning components contributed more individuals to other components than they received. Under this process, component abundances tended to equilibrate. Harvest removed a constant proportion of a spawning component, regardless of local abundance. Losses of actual numbers of individuals to harvest were relatively high initially and subsequently decreased as abundance declined. Thus proportional loss functioned as a decay model, and resulting population trajectories may have exhibited a tendency to converge at lower abundance levels. These two characteristics may have contributed to the similarity in responses of the spawning components.

Another reason for convergence is that all four of the spawning components were assumed to exhibit identical population dynamics. No data exist concerning the vital rates of individual spawning components. Past population assessments have pooled the four components in calculations of numbers, length, and weight at age. However, relatively large differences in vital rates among spawning components are not expected.

Table 22.2 Model scenarios that resulted in the extinction of individual spawning components

	Input Parameter Suites						
Proportional harvest on pooled spawning components (mixed fishery)							
Dispersal	0.1	0.1	0.2	0.2	0.3	0.3	0.3
Survivorship	0.6	0.6	0.6	0.6	0.7	0.6	0.6
Harvest	0.3	0.4	0.3	0.4	0.4	0.3	0.4
Proportional harvest on individual spawning components (separate fishery)							
Dispersal	0.1	0.1	0.2	0.2	0.3	0.3	0.3
Survivorship	0.6	0.6	0.6	0.6	0.7	0.6	0.6
Harvest	0.3	0.4	0.3	0.4	0.4	0.3	0.4

Note: Equivalent parameter suites for the two model scenarios are aligned vertically.

While some differences in environmental conditions at spawning sites may exist, Atlantic herring are highly mobile, and SWNS/BF herring are known to travel widely throughout their range during the year.

This modeling study did not address a previously major problem with the exploitation of this population complex: the serial targeting of individual spawning components (Stephenson et al. 2001). This particular issue was the main reason for initiating a management approach that spread fishing effort among existing spawning components. Model runs were undertaken wherein 50% of total fishing effort (as proportional harvest) was directed at individual spawning areas for short periods (e.g., 5 years). The results were predictable, and a rapid collapse in abundance was common and expected.

A note of caution: this study is exploratory in nature, and the results should be interpreted with caution. Population growth projections depend on both fecundity and survivorship at age. Natural mortality (survivorship) is unknown for this population complex, and a range of plausible values was used. Numbers of individuals at age were based on current minimum biomass estimates and previous numbers at age calculated from a sequential population analysis that was poorly resolved. Recruitment in Atlantic herring is poorly understood and known to be highly variable (Figure 22.2). Two consecutive years of strong (or weak) recruitment can have a large impact on the demography and subsequent dynamics of a herring population (see Hjort 1914 for the classic example). However, responsible management cannot expect or rely on such strong year classes when striving for the recovery of an overexploited fish population. Conservative estimation of population abundance and dynamics is prudent because strong recruitment is the exception rather than the rule, when adult population sizes are low (Myers et al. 1995). Recovery has been slow following a serious decline in abundance in collapsed Atlantic herring populations (e.g., Stephenson 1997 and references therein; Overholtz and Friedland 2002).

References

Chang, B. D., Stephenson, R. L., Wildish, D. J., and Watson-Wright, W. M. 1995. Protecting regionally significant marine habitats in the Gulf of Maine: a Canadian perspective. Pages 121–146 in *Improving interactions between coastal science and policy: proceedings of the Gulf of Maine Symposium, Kennebunkport, ME, 1–3 November 1994.* National Academy Press, Washington, D.C.

Department of Fisheries and Oceans. 2001. *4VWX Herring*. Science Stock Status Report B3-03 (2001). DFO, Ottawa.

Deriso, R. B. 1987. Optimal $F_{0.1}$ criteria and their relationship to maximum sustainable yield. *Canadian Journal of Fisheries and Aquatic Sciences* 44(Suppl. 2): 339–348.

Hjort, J. 1914. Fluctuations in the great fisheries of northern Europe viewed in the light of biological research. *Rapports et Procés-Verbaux des Réunions, Conseil International pour l'Exploration de la Mer* 20: 1–228.

Hoenig, J. M. 1983. Empirical use of longevity data to estimate mortality rates. *Fisheries Bulletin of the United States* 81: 898–903.

International Commission of Northwest Atlantic Fisheries (ICNAF). 1972. Report of the Standing Committee on Research and Statistics. Appendix II. Report of the herring working group. Pages 43–66 in *International Commission Northwest Atlantic Fisheries (ICNAF) Redbook 1972*. ICNAF, Dartmouth, Ottawa.

Iles, T. D. 1993. The management of the Canadian Atlantic herring. Pages 123–150 in L. S. Parsons and W. H. Lear (eds.), *Perspectives on Canadian marine fisheries management.* Canadian Bulletin in Fisheries and Aquatic Sciences 226. NRC Press, Ottawa.

Iles, T. D., and Sinclair, M. 1982. Atlantic herring: stock discreteness and abundance. *Science* 215: 627–633.

McPherson, A. A., O'Reilly, P. T., and Taggart, C. T. In press. Genetic differentiation, temporal stability and the absence of isolation by distance among Atlantic herring populations. *Transactions of the American Fisheries Society.*

McPherson, A. A., Stephenson, R. L., and Taggart, C. T. 2003. A genetic basis for Atlantic herring *Clupea harengus* spawning waves. *Marine Ecology Progress Series* 247: 303–309.

McQuinn, I. H. 1997. Metapopulations and the Atlantic herring. *Reviews in Fish Biology and Fisheries* 7: 297–329.

Melvin, G. D., Stephenson, R. L., Power, M. J., Fife, F. J., and Clark, K. J. 2001. Industry acoustic surveys as a basis for in-season decisions in a co-management regime. Pages 675–688 in F. Funk, J. Blackburn, D. Hay, A. J. Paul, R. Stephenson, R. Toresen, and D. Witherell (eds.), *Herring: expectations for a new millennium.* AK-SG-01–04. University of Alaska Sea Grant Program, Fairbanks.

Myers, R. A., Barrowman, N. J., Hutchings, J. A., and Rosenberg, A. A. 1995. Population dynamics of exploited fish stocks at low population levels. *Science* 269: 1106–1108.

Overholtz, W. J., and Friedland, K. D. 2002. Recovery of the Gulf of Maine–Georges Bank Atlantic herring (*Clupea harengus*) complex: perspectives based on bottom trawl survey data. *Fishery Bulletin* 100: 593–608.

Power, M. J., and Iles, T. D. 2001. Biological characteristics of 4WX herring. Pages 135–152 in F. Funk, J. Blackburn, D. Hay, A. J. Paul, R. Stephenson, R. Toresen, and D. Witherell (eds.), *Herring: expectations for a new millennium.* AK-SG-01–04. University of Alaska Sea Grant Program, Fairbanks.

Sinclair, M. 1988. *Marine populations: an essay on population regulation and speciation.* University of Washington Press, Seattle.

Smedbol, R. K., and Stephenson, R. L. 2001. The importance of managing within-species diversity in cod and herring fisheries of the north-western Atlantic. *Journal of Fish Biology* 59(Suppl. A): 109–128.

Stephenson, R. L. 1997. Successes and failures in the management of Atlantic herring fisheries: do we know why some have collapsed and others survived? Pages 49–54 in D. A. Hancock, D. C. Smith, A. Grant, and J. P. Beumer (eds.), *Developing and sustaining world fisheries resources: the state of science and management. Proceedings of the 2nd World Fisheries Congress.* CSIRO Publishing, Melbourne.

Stephenson, R. L. 2002. Stock structure and management structure: an ongoing challenge for ICES. *International Commission for the Exploration of the Seas (ICES) Marine Science Symposium* 215: 305–314.

Stephenson, R. L., Lane, D. E., Aldous, D. G., and Nowak, R. 1993. Management of the 4WX Atlantic herring (*Clupea harengus*) fishery: an evaluation of recent events. *Canadian Journal of Fisheries and Aquatic Sciences* 50: 2742–2757.

Stephenson, R. L., Rodman, K., Aldous, D. G., and Lane, D. E. 1999. An in-season approach to management under uncertainty: the case of the SW Nova Scotia herring fishery. *International Commission for the Exploration of the Seas (ICES) Journal of Marine Science* 56: 1005–1013.

Stephenson, R. L., Power, M. J., Clark, K. J., Melvin, G. D., Fife, F. J., Scheidl, T., Waters, C. L., and Arsenault, S. 2000. *2000 evaluation of 4VWX herring*. Research Document 2000/65. Canadian Stock Assessment Secretariat, Ottawa.

Stephenson, R. L., Clark, K. J., Power, M. J., Fife, F. J., and Melvin, G. D. 2001. Herring stock structure, stock discreteness and biodiversity. Pages 559–571 in F. Funk, J. Blackburn, D. Hay, A. J. Paul, R. Stephenson, R. Toresen, and D. Witherell (eds.), *Herring: expectations for a new millennium*. AK-SG-01–04. University of Alaska Sea Grant Program, Fairbanks.

Ware, D., Tovey, C., Hay, D., and McCarter, B. 2000. *Straying rates and stock structure of British Columbia herring*. Research Document 2000/006. Canadian Stock Assessment Secretariat, Ottawa.

Ware, D., and Schweigert, J. 2001. *Metapopulation structure and dynamics of British Columbia herring*. Research Document 2001/127. Canadian Science Advisory Secretariat, Ottawa.

Wheeler, J. P., and Winters, G. H. 1984. Homing of Atlantic herring (*Clupea harengus harengus*) in Newfoundland waters as indicated by tagging data. *Canadian Journal of Fisheries and Aquatic Sciences* 41: 108–117.

23

Pacific Herring (*Clupea pallasi*) in Canada

Generic Framework for Evaluating Conservation Limits and Harvest Strategies

CAIHONG FU
CHRIS C. WOOD
JAKE SCHWEIGERT

Fishery managers often ask how low the abundance of a population can go without causing irreversible harm. The concept of a minimum viable population size is problematic, particularly for pelagic fishes like herring that exhibit high variation in productivity and abundance. However, it should be feasible to determine a conservation limit for population size (N^L) that would satisfy stated conservation criteria. For instance, we could choose N^L high enough to ensure acceptably low probabilities of triggering World Conservation Union (IUCN) listing criteria (Mace and Stuart 1994) or of falling below a threshold or quasi-extinction level of abundance in the long term. A population reduced to N^L should also have an acceptably high probability of recovery to a sustainable target level within a shorter time period. For example, the U.S. Magnuson-Stevens Fishery Conservation and Management Act specifies a legal requirement for recovery within 10 years or as soon as possible (Restrepo et al. 1998) and Johnston et al. (2000) propose a limit to allow recovery within a single generation. Similarly, it should be feasible to determine a corresponding conservation limit for harvest rate (μ^L). N^L and μ^L values that satisfy these requirements can be determined in Monte Carlo simulations by evaluating outcomes for a wide range of population sizes and harvest rates.

In practice, fishery managers are interested in harvest strategies that not only reduce conservation concerns but also achieve maximal benefits, perhaps by increasing total catch over the long term or reducing the probability of low catch in a particular season. Optimal harvest strategies often involve "proportional threshold harvesting" (Lande et al. 2001) in which no harvest occurs when abundance is below a management threshold (N^{Cutoff}). When the management threshold is set to satisfy conservation objectives, N^{Cutoff} can be considered as the conservation limit. Harvest strategies with cutoff thresholds are more robust to adverse environmental regimes in which average survival rate falls below the long-term average for extended periods (Fu et al. 2000), although they may entail a high variance in annual catch (Lande et al. 1997). Cutoff thresholds are often set at 20% of unfished equilibrium abundance (Francis 1993, Thompson 1993). Somewhat arbitrarily, in 1985 a cutoff threshold of 25% of an early estimate of unfished equilibrium biomass was introduced for herring fisheries in British Columbia to restrict harvest during periods of reduced abundance (Schweigert and Fort 1999).

Risk analyses are necessary to evaluate the likely consequences and trade-offs associated with alternative decisions about harvest rates and cutoff thresholds. Ideally Monte Carlo simulations to evaluate these trade-offs should be performed with models designed to capture the unique features of a particular fishery and its target species (e.g., Hollowed and Megrey 1993, Zheng et al. 1993, Fu et al. 2000). However, the time required to design species-specific programs may impede progress where conservation limits and harvest strategies are required for a large number of species. Moreover, individual species-specific programs often fail to consistently address more complicated problems that involve environmental variability, spatial structure, and changes in productivity due to habitat modification.

In the case study reported here, we show that RAMAS GIS software (Akçakaya 2002) is adequate to evaluate conservation limits and proportional harvesting controls for Pacific herring populations, and it has sufficient power to address ecological problems, including demographic and environmental stochasticity, spatial structure, and habitat issues. We developed a generic simulation framework to derive conservation limits N^L and μ^L, and the optimal harvest strategy (combination), using RAMAS Metapop to evaluate performance measures over a broad spectrum of values for initial population size and the control variables.

Methods

Stage Structure

After hatching in February and March, Pacific herring spend 6 months inshore as age 0 juveniles. During August and September, many age 0 herring migrate offshore and join schools of immature age 1 and 2 herring. As they mature, most at age 3, they join schools of older mature fish on their fall/winter inshore migration to the spawning grounds (Ware 1996). Age 1 abundance has not been estimated in stock assessments to date, so we modeled nine stages (age 2, 3, . . . , 10+) for each of five herring stocks. We used the subscripts i, t, and a to represent stock, year, and age, respectively. Fecundity varies with weight such that an average of 200,000 eggs per kilogram of mature herring are produced annually ($\bar{P}$) in each stock (Hay 1985). Mean weight at age was averaged over all stocks ($W_{i,a}$; Table 23.1), because empirical measurements showed little variation among stocks; in contrast, maturity parameters ($O_{i,a}$, the proportion of individuals maturing at age a, Table 23.1) did vary among stocks (Schweigert 2001). Accordingly, egg production for stock i in year t was calculated as

$$E_{i,t} = \sum_{a} N_{i,t,a} W_a O_{i,a} \bar{P}$$

where $N_{i,t,a}$ are estimates of abundance from an age-structured stock assessment (Schweigert 2001).

Table 23.1 Average weight at age (kg) across all five stocks (W_a), maturity parameters (the proportion that mature, $O_{i,a}$), age composition ($\sigma_{i,a}$), and recruitment to age 2 ($R_{i,a}$) at age a for each stock i

	Age (years)								
	2	3	4	5	6⁺	7	8	9	10
W_a	0.05	0.08	0.11	0.13	0.14	0.16	0.17	0.18	0.18
Maturity proportion at age a ($O_{i,a}$)									
QCI	0.1	0.4	0.64	0.90	1	1	1	1	1
PRD	0.10	0.48	0.71	0.90	1	1	1	1	1
CC	0.15	0.56	0.78	0.97	1	1	1	1	1
GS	0.12	0.71	0.94	0.97	1	1	1	1	1
WCVI	0.09	0.69	0.90	0.99	1	1	1	1	1
Age composition ($\sigma_{i,a}$)									
QCI	0.0	0.27	0.25	0.21	0.27				
PRD	0.0	0.21	0.23	0.21	0.35				
CC	0.0	0.17	0.2	0.2	0.42				
GS	0.0	0.29	0.26	0.2	0.24				
WCVI	0.0	0.25	0.24	0.21	0.3				
Recruitment to age 2 per individual at age a ($R_{i,a}$)									
QCI	0.06	0.37	0.78	1.32	1.65	1.80	1.93	2.03	2.09
PRD	0.04	0.30	0.58	0.89	1.11	1.21	1.30	1.37	1.41
CC	0.05	0.27	0.49	0.73	0.85	0.93	0.99	1.05	1.08
GS	0.07	0.62	1.08	1.35	1.56	1.71	1.83	1.93	1.98
WCVI	0.03	0.38	0.64	0.86	0.98	1.07	1.15	1.21	1.24

Note: QCI = Queen Charlotte Islands; PRD = Prince Rupert District; CC = Central Coast; GS = Strait of Georgia; WCVI = West Coast of Vancouver Island.

Survival rate from egg to age 2 recruits was calculated as

$$S_{i,t,0} = \frac{N_{i,t+2,2}}{E_{i,t}}$$

The first row elements of the transition matrix (recruitment to age 2) for stock i at age a were calculated as

$$R_{i,a} = W_a O_{i,a} \bar{P}\bar{S}_{i,o}$$

where $\bar{S}_{i,o}$ is the average of $S_{i,t,o}$ across the time period from 1977 to 2001 (Table 23.1). Diagonal elements of the transition matrix (survival rate) were assumed not to change with age after age 2 but were estimated independently for each stock by Schweigert (2001).

Spatial Structure and Dispersal

The herring metapopulation in British Columbia (B.C.) comprises five major "stocks" denoted by area: the Queen Charlotte Islands (QCI), the Prince Rupert District (PRD), the Central Coast (CC), the Strait of Georgia (GS), and the west coast of Vancouver Island (WCVI) (Beacham et al. 2001, Ware and Schweigert 2001). Ware and Schweigert (2001) concluded that despite variations in spawner biomass, the dispersal pattern for the B.C. herring metapopulation matched an isolation by distance model, with most herring straying to nearby stocks and a few to the most distant ones. Accordingly, we assumed that dispersal rates among the five stocks of B.C. herring depended only on the distance $D_{i,j}$ between stocks so that

$$m_{i,j} = \exp\left(\frac{-D_{i,j}}{b}\right)$$

(Table 23.2). To be conservative about the prospects for rescue from adjacent stocks, we set $b = 100$, implying a maximum rate of dispersal among stocks of 14%; this corresponds to the lower bound for dispersal estimated by Ware and Schweigert (2001).

Density Dependence

No density dependence was evident for any of the stocks in regressions of $\sqrt{N_{t+1}/N_{t-1}}$ versus N_t (Akçakaya 2002) based on estimates of abundance from the stock assessment model. Except for the PRD stock, the relationship between age 2 abundance (i.e., recruits) and their spawning biomass for B.C. herring resembled the "ceiling" model (now usually referred to as the "hockey-stick" model in fisheries research; e.g., Barrowman and Myers 2000), with recruits increasing linearly with spawning biomass until the latter reached a threshold biomass (Ware and Schweigert 2001). Therefore, the ceiling density-dependent (d-d) model was used for all stocks and all life stages. The Allee parameter (A) for each stock was set at a very low level, 0.02% of its carrying capacity (K_i), because there is no empirical evidence for this effect in herring. The K_i level was set at 110% of the historical maximum abundance for each stock (HistMax_i; Figure 23.1) but results were relatively insensitive to values for K_i within ±30% of HistMax_i.

Stochasticity

Both demographic stochasticity and lognormal environmental stochasticity were included in the model. Coefficients of variation in recruitment (first-row elements) were set at $CV_i = 0.5$ based on temporal variation in estimates of early survival ($S_{i,t,0}$) averaged across all stocks, excluding QCI where suspected errors in estimates of abundance would have inflated the estimates of process error. Variability in survival (diagonal elements of transition matrix) is unknown and was assumed to be the same for each stock,

Table 23.2 Matrix of distance (km)/dispersal rates among five stocks

	QCI	PRD	CC	GS	WCVI
QCI	0 / 0	197 / 0.140	202 / 0.133	622 / 0.002	596 / 0.003
PRD	197 / 0.140	0 / 0	303 / 0.048	776 / 0.0004	713 / 0.0008
CC	202 / 0.133	303 / 0.048	0 / 0	434 / 0.013	447 / 0.011
GS	622 / 0.002	776 / 0.0004	434 / 0.013	0 / 0	274 / 0.065
WCVI	596 / 0.003	713 / 0.0008	447 / 0.011	274 / 0.065	0 / 0

Note: QCI = Queen Charlotte Islands; PRD = Prince Rupert District; CC = Central Coast; GS = Strait of Georgia; WCVI = West Coast of Vancouver Island.

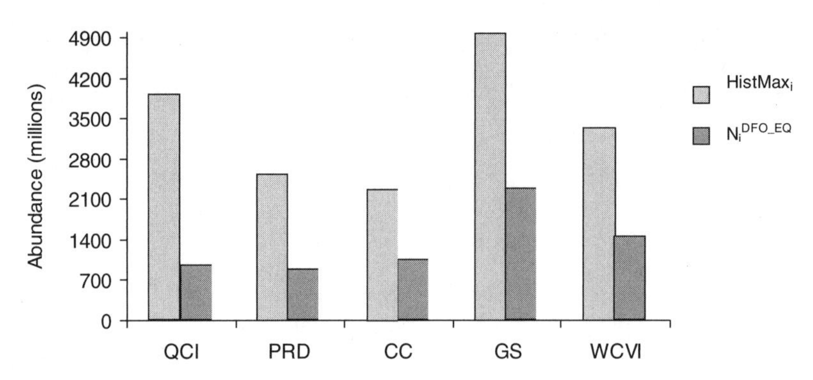

Figure 23.1 Plots of historical maximum abundance ($HistMax_i$) and equilibrium abundance ($N_i^{DFO\_EQ}$) for five stocks: Queen Charlotte Islands (QCI), Prince Rupert District (PRD), Central Coast (CC), Strait of Georgia (GS), and West Coast of Vancouver Island (WCVI).

with $CV_i = 0.2$, a level that is commonly used to model marine fish population dynamics. Parameters $R_{i,a}$, $\bar{S}_{i,a}$ and carrying capacity K_i were assumed uncorrrelated, and the CV_i for K_i was set at 0.1. We assumed $CV_i = 0.2$ for dispersal and $CV_i = 0.3$ for errors in measuring abundance. No catastrophes were simulated.

Population Management

RAMAS Metapop Version 4.0 offers the capability to investigate dual population management strategies in that harvest rate (μ_i) can vary with stock-specific abundance. In the study reported here, we used the following harvest control rule:

$$\text{harvest rate} = \begin{cases} (1) \quad 0 \text{ if } N_i \leq N_i^{\text{Cutoff}} \\ (2) \quad \mu_i \left(N_i - N_i^{\text{Cutoff}}\right) / \left(1.05 \cdot N_i^{\text{Cutoff}} - N_i^{\text{Cutoff}}\right) \text{ if } N_i^{\text{Cutoff}} < N_i \leq 1.05 \cdot N_i^{\text{Cutoff}} \\ (3) \quad \mu_i \text{ if } N_i > 1.05 \cdot N_i^{\text{Cutoff}} \end{cases}$$

No harvest occurs below a cutoff threshold (N_i^{Cutoff}) predefined for each stock. When abundance exceeds the cutoff threshold, harvest rate is proportional to abundance up to a predefined value. We used a restricted transition zone (N_i^{Cutoff} to $1.05 \cdot N_i^{\text{Cutoff}}$) to reflect current policy and to ensure clear contrast between scenarios with the different levels of N_i^{Cutoff} and μ_i used to evaluate the performances of different harvest strategies.

To account for partial recruitment to the fishery, age-specific harvest rate was calculated as $\mu_{i,a} = \mu_i \delta_{i,a}$, where $\delta_{i,a}$ is the gear selectivity parameter for age a in stock i as estimated from the stock assessment model and μ_i is the fully selected harvest rate for stock i. Each age-specific cutoff threshold ($N_{i,a}^{\text{Cutoff}}$) was calculated as $N_{i,a}^{\text{Cutoff}} = \delta_{i,a}\sigma_{i,a}N_i^{\text{Cutoff}}$, where $\sigma_{i,a}$ is the relative abundance of the five age groups from 2 to 6^+. Current Fisheries and Oceans Canada (DFO) policy sets the cutoff threshold at 25% of estimated "unfished equilibrium biomass." We now know (after attempting to parameterize Metapop) that these early estimates must underestimate the true unfished equilibrium biomass. However, to simplify comparisons with current DFO policy, we still express the cutoff threshold as a percentage of the equilibrium abundance (denoted $N_i^{\text{DFO\_EQ}}$) that corresponds to the early estimates of unfished equilibrium biomass for each stock. Therefore, $N_{i,a}^{\text{Cutoff}} = \delta_{i,a}\sigma_{i,a} N_i^{\text{DFO\_EQ}}\pi_i$, where π_i is the percentage.

Simulation Scenarios

Simulation Frame

Simulated scenarios have a duration of 100 years (20 generations), long enough to approximate the total catch over this period as the maximum sustainable yield (MSY), which is often employed to evaluate harvest strategies. Generation time was calculated according to

$$\sum_{a=1}^{a_A} a r_a N_a \Big/ \sum_{a=1}^{a_A} r_a N_a$$

(Restrepo et al. 1998), where $a_A = 10$ (the oldest age in a pristine condition), resulting in an average value of 4.98 (approximated at 5) for the five herring stocks. To reduce computation time given the large number of conditions evaluated, each scenario was replicated 200 times rather than 1,000 (as recommended). However, our general conclusions about conservation limits and harvest strategies should not be seriously affected by this compromise.

To simulate different levels for initial population size, harvest rate, and cutoff threshold, our program loops over trial values with a step value of 5%, ranging from 5% · $N_i^{\text{DFO\_EQ}}$ to 100% · $N_i^{\text{DFO\_EQ}}$ for $N_{i,o}$, 0.0 to 0.95 for μ_i, and 0% to 65% · $N_i^{\text{DFO\_EQ}}$ for N_i^{Cutoff}.

Performance Measures

Five types of performance measure were chosen to measure how effectively alternative conservation limits and harvest strategies achieved sustainable benefits and the conservation objective of maintaining populations and species within bounds of natural variability, as proposed by DFO's National Policy Committee. Sustainability of benefits from the fishery was measured by the first two performance measures: the average annual catch from all stocks over 100 years (C_{AvgAnn}), and the probability that total catch in any year would fall below 0.5% of $N_i^{\text{DFO\_EQ}}$ (P_{LowCatch}).

For each stock in the B.C. herring metapopulation, we also wanted to know the probability that abundance would decline by 50% within three generations, which is the threshold for triggering an IUCN endangered (EN) designation for a closed population under criterion A. This third performance measure, denoted $P_{\text{TriggerEN},i}$, was estimated as the probability of 50% terminal decline in simulations of 15 years (three generations) duration.

Our fourth performance measure, denoted $P_{\text{LowN},i}$, is the probability that abundance in a particular stock would fall below an unacceptable threshold within 50 years (10 generations). The unacceptable threshold was defined as the current DFO cutoff threshold (25% of $N_i^{\text{DFO\_EQ}}$). Within RAMAS, $P_{\text{LowN},i}$ is computed as the probability of quasi extinction, although the threshold is much higher than what is normally used for quasi extinction.

Our fifth performance measure, denoted $P_{\text{recovery},i}$, is the probability of recovery within 10 years (two generations) from a specified initial population size to an acceptable level, arbitrarily defined as 70% of $N_i^{\text{DFO\_EQ}}$. This is computed as the probability of "quasi explosion" within RAMAS.

Sixth and finally, an objective function is formed to evaluate overall performance by combining the five individual performance measures:

$$f_{i,j} = \alpha \frac{C_{\text{AvgAnn},j}}{C_{\text{AvgAnn,max}}} - \gamma P_{\text{LowCatch},j} - \xi P_{\text{TriggerEN},i,j} - \beta P_{\text{LowN},i,j} + \chi P_{\text{Recovery},i,j},$$

where the subscript j denotes a specific combination of $N_{i,0}$, μ_i, and N_i^{Cutoff}. $C_{\text{AvgAnn,max}}$ is the maximum value of $C_{\text{AvgAnn},j}$. In this heuristic example, the weighting parameters α, γ, ξ, β, and χ were arbitrarily set at 1.0 to illustrate the procedure and possible use of objective values in decision-making.

Results

The conservation benefit of the current cutoff threshold is illustrated in Figure 23.2 by comparing average trajectories of abundance with and without the cutoff control rule under a 20% harvest rate (current policy). Trajectories for the weakest (WCVI) stock were simulated for 100 years, starting at both high ($N_i^{\text{DFO\_EQ}}$) and low abundance (25% of $N_i^{\text{DFO\_EQ}}$). Without the cutoff threshold, abundance declined, on average, even from a high initial abundance. With a cutoff threshold at 25% of $N_i^{\text{DFO\_EQ}}$, average abundance was maintained or recovered from low initial abundance.

To summarize the effects of initial population size and harvest rate, we looped over a range of $N_{i,0}$ and μ_i, setting N_i^{Cutoff} at the level currently specified by DFO (25% of $N_i^{\text{DFO\_EQ}}$). Average annual catch over the simulation period, C_{AvgAnn}, increased steadily

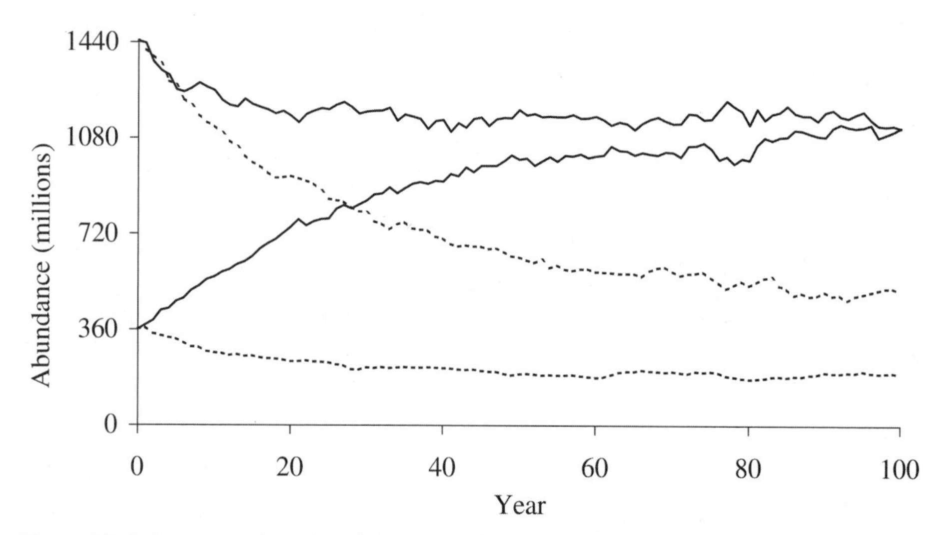

Figure 23.2 Average trajectories of abundance for the WCVI stock from two initial population sizes (100% and 25% of $N_i^{\text{DFO\_EQ}}$) under 20% harvest rate with (*solid lines*) and without (*dashed lines*) a cutoff threshold control rule.

as μ_i increased from 0 to about 0.2, then declined quickly as μ_i reached 0.25 (Figure 23.3). $P_{LowCatch}$ was negligible for $\mu_i \leq 0.2$ and $N_{i,0} \geq 50\%$ of $N_i^{DFO\_EQ}$, but it increased drastically as $N_{i,0}$ decreased. $P_{TriggerEN,i}$ was highest for the highest $N_{i,0}$ and μ_i. To ensure $P_{LowN,i} \leq 5\%$, it was necessary to set the conservation limit N_i^L at 35% of $N_i^{DFO\_EQ}$ for the metapopulation and at 50% of $N_i^{DFO\_EQ}$ for the WCVI stock. However, to ensure $P_{Recovery,i} \geq 95\%$, the conservation limit N_i^L must be 50% of $N_i^{DFO\_EQ}$ for the metapopulation. The example objective function values indicated an optimal range for μ_i that became narrower as $N_{i,0}$ decreased.

When the cutoff was set at 0, high C_{AvgAnn} was achieved only at high $N_{i,0}$, and $P_{LowCatch}$ was negligible only when $\mu_i \leq 0.2$ (Figure 23.4). $P_{TriggerEN,i}$ for the metapopulation was likely to be below 5% only when $\mu_i \leq 0.2$, and harvest rate had to be reduced by half for the WCVI stock in order to achieve the same desirably low $P_{TriggerEN,i}$. At high $N_{i,0}$ (e.g., 90% of $N_i^{DFO\_EQ}$), the conservation limit for harvest rate μ_i^L could be set at 0.2 for the metapopulation, but it had to be reduced by half for the WCVI stock. Highest objective values could be achieved at high $N_{i,0}$ and $\mu_i \leq 0.2$.

An optimal harvest strategy must facilitate recovery when a population falls below some target level. To address this issue, we started the simulations at $N_{i,0}$ = 50%

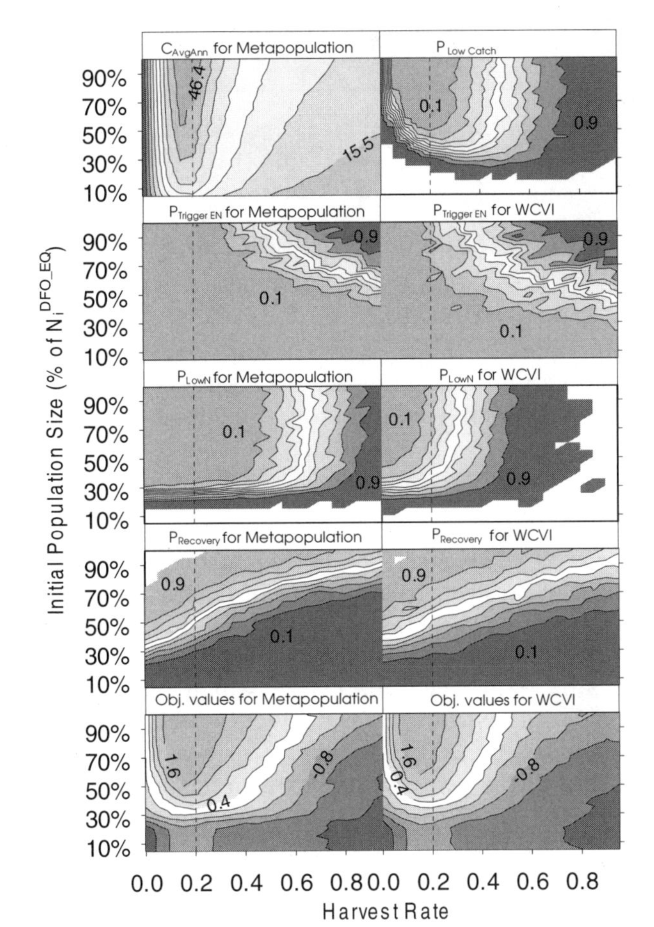

Figure 23.3 Plots of C_{AvgAnn}, $P_{LowCatch}$, $P_{TriggerEN,i}$, $P_{LowN,i}$, and $P_{Recovery,i}$, and objective function values for the metapopulation and the WCVI stock under various combinations of harvest rate and initial population size and the current cutoff threshold (25% of $N_i^{DFO\_EQ}$). Dashed line represents 20% harvest rate as specified under current management policy.

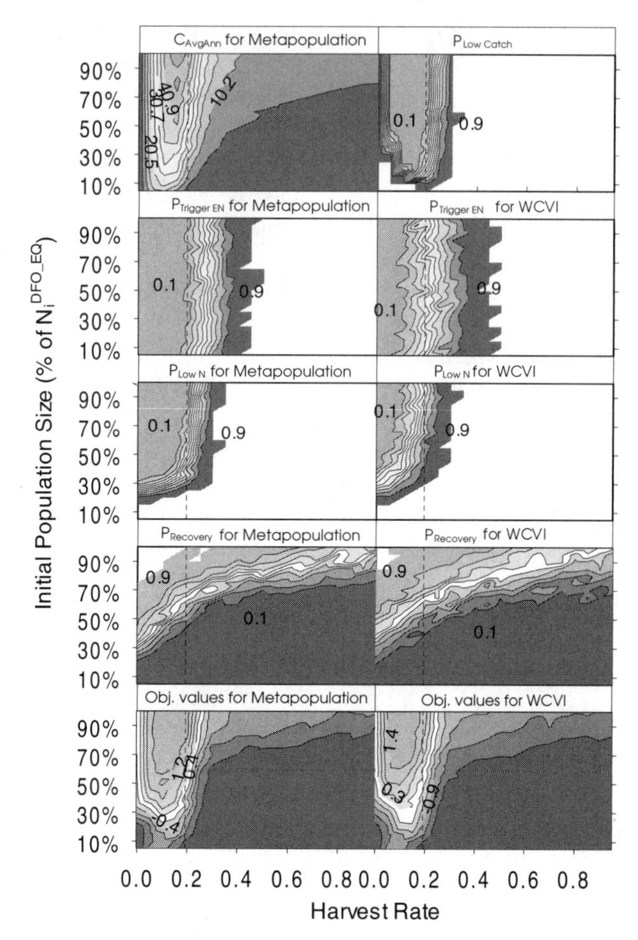

Figure 23.4 Plots of C_{AvgAnn}, P_{LowCatch}, $P_{\text{TriggerEN},i}$, $P_{\text{LowN},i}$, and $P_{\text{Recovery},i}$, and objective function values for the metapopulation and the WCVI stock under various combinations of harvest rate and initial population size and no cutoff threshold. Dashed line represents 20% harvest rate as specified under current management policy.

of $N_i^{\text{DFO\_EQ}}$ and looped over μ_i and N_i^{Cutoff} (Figure 23.5). In the long run (100 years), setting N_i^{Cutoff} above 30% of $N_i^{\text{DFO\_EQ}}$ was beneficial in terms of C_{AvgAnn}, which increased steadily as μ_i increased from 0 to about 0.2, declining thereafter, but more slowly for high N_i^{Cutoff} than low N_i^{Cutoff}. However, P_{LowCatch} was only negligible at low N_i^{Cutoff} (<25%) and at $0.05 \leq \mu_i \leq 0.2$. Without a cutoff ($N_i^{\text{Cutoff}} = 0$), μ_i must remain below 0.15 for the metapopulation, and even lower for the WCVI stock, to ensure $P_{\text{LowN},i} \leq 5\%$. As N_i^{Cutoff} increased above 30% of $N_i^{\text{DFO\_EQ}}$, the metapopulation (but not the WCVI stock) was secure against overharvest regardless of μ_i. In contrast $P_{\text{Recovery},i}$ was enhanced by reducing μ_i and increasing N_i^{Cutoff}. The objective function values were maximal at intermediate levels of μ_i that were positively correlated with the choice of N_i^{Cutoff}.

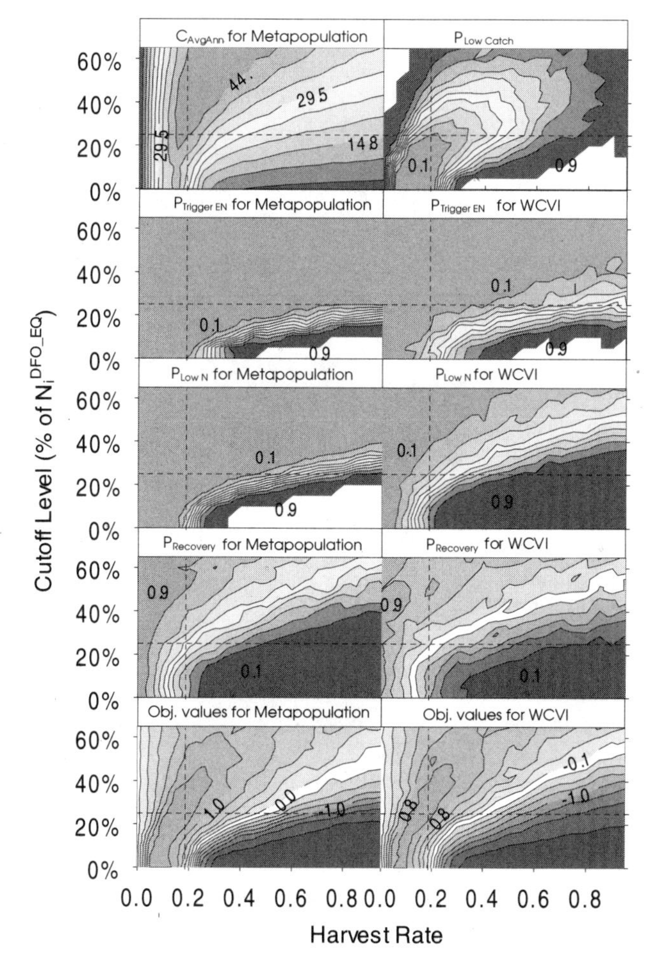

Figure 23.5 Plots of C_{AvgAnn}, $P_{LowCatch}$, $P_{TriggerEN,i}$, $P_{LowN,i}$, and $P_{Recovery,i}$, and objective function values for the metapopulation and the WCVI stock under various combinations of harvest rate and cutoff threshold. Dashed lines represent the harvest rate and cutoff threshold under current management policy.

Finally we set harvest rate at the level specified by current management policy (μ_i = 0.2), and evaluated outcomes by looping over a range of $N_{i,0}$ and N_i^{Cutoff} (Figure 23.6). C_{AvgAnn} increased steadily with $N_{i,0}$ and N_i^{Cutoff} for N_i^{Cutoff} < 50% of $N_i^{DFO\_EQ}$. Without a cutoff, N_i^L must exceed 90% of $N_i^{DFO\_EQ}$ to ensure $P_{LowN,i} \leq 5\%$ for the metapopulation and to maximize C_{AvgAnn}. Without imposing a cutoff, no conservation limit could be found to satisfy the conservation criteria of $P_{LowN,i} \leq 5\%$ for the WCVI stock within the range of $N_{i,0}$ examined. When N_i^{Cutoff} exceeded 20% of $N_i^{DFO\_EQ}$, $P_{LowN,i}$ depended only on $N_{i,0}$. In contrast, $P_{Recovery,i}$ varied with both N_i^{Cutoff} and $N_{i,0}$. Objective function values were maximized by intermediate levels of N_i^{Cutoff} that were negatively correlated with $N_{i,0}$ if N_i^{Cutoff} < 50% of $N_i^{DFO\_EQ}$. Objective function values increased with $N_{i,0}$ regardless of N_i^{Cutoff}.

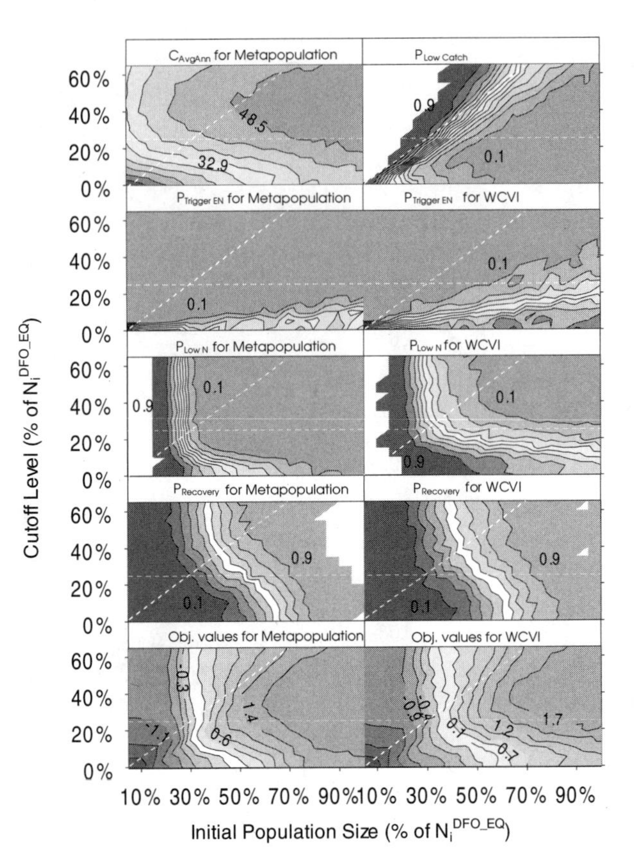

Figure 23.6 Plots of C_{AvgAnn}, P_{LowCatch}, $P_{\text{TriggerEN},i}$, $P_{\text{LowN},i}$, and $P_{\text{Recovery},i}$, and objective function values for the metapopulation and the WCVI stock under various combinations of initial population size and cutoff threshold. Dashed white lines represent the cutoff threshold under current management policy and the isopleth where initial population size is the same as the cutoff threshold.

Discussion

The simulations with an initial population size $N_{i,0}$ = 50% of $N_i^{\text{DFO\_EQ}}$ indicated that the current policy achieved the highest possible C_{AvgAnn} (44.3) at low P_{LowCatch} and $P_{\text{TriggerEN},i}$ ($\leq$ 0.1), low $P_{\text{LowN},i}$ for the metapopulation ($\leq$ 0.1), and the highest objective function values both for the metapopulation (1.2) and the weakest (WCVI) stock (1.0) (Figure 23.5). However, $P_{\text{LowN},i}$ for WCVI stock was $\geq$ 0.2, and $P_{\text{Recovery},i}$ for both the metapopulation and WCVI stock was only around 0.5. Therefore, although the current policy should be adequate to conserve the metapopulation ($P_{\text{TriggerEn},i}$ and $P_{\text{LowN},i} \leq 0.1$), it may not be adequate to conserve the weaker WCVI stock, and it may hinder recovery to higher population levels such as the target abundances explored as trial values in this study. It seems advisable therefore to raise the cutoff thresholds for weaker stocks (e.g., to 50% of $N_i^{\text{DFO\_EQ}}$ for the WCVI stock). Options will be explored in more detailed assessments using this framework.

Overall, RAMAS Metapop provides a flexible and convenient way to tackle problems concerning conservation limits and optimal harvest strategies. We expect that the framework could be readily applied to a variety of fish species, regardless of the degree of exploitation or extent of data available. Of course, for species with only limited data, parameter values must be based on professional judgments. In such cases, it will be important to consider the effect of model uncertainty (e.g., Goodman 2002), perhaps by reducing best estimates for vital rates by 1 SE (estimated) (as in P-RAMAS scenarios examined by Taylor et al. 2002) or by resimulating with initial values drawn from prior distributions for vital rates.

The flexibility and convenience of this generic approach lie in mainly three areas. First, RAMAS Metapop provides options to allow for age (or stage) and spatial structure in a metapopulation, to include demographic and environmental stochasticities and measurement errors, and to incorporate temporal variation in factors such as vital rates and carrying capacity. Our simulations illustrate that spatial structure analysis is especially informative regarding weak stocks. Ignoring spatial structure can produce an overly optimistic picture of the population dynamics and thus can result in decline or even extinction of weak stocks. Second, RAMAS Metapop provides various ways to diagnose risk. Performance measures such as C_{AvgAnn}, P_{LowCatch}, $P_{\mathrm{TriggerEN},i}$, $P_{\mathrm{LowN},i}$, and $P_{\mathrm{Recovery},i}$ are particularly useful in fisheries management. Third, RAMAS Metapop can be used in tandem with other computer programs to examine the performance of conservation limits and optimal harvest strategies over a wide range of scenarios involving combinations of state variables (e.g., $N_{i,0}$, μ_i, and N_i^{Cutoff}), thus facilitating a higher level of synthesis.

Nevertheless, the interpretations of conservation limits and optimal harvest strategies are open to many uncertainties, including those in the estimates of vital rates, carrying capacity and density dependence, Allee effect, dispersal rates among the stocks, measurement errors, and fishery implementation errors. All these uncertainties may have affected our conclusions on conservation limits and optimal harvest strategies. Sensitivity analysis is needed to prioritize research to reduce these uncertainties. The interpretation of conservation limits also depends on the choice of "quasi-extinction" and "quasi-explosion" thresholds—that is, the tolerable level of $P_{\mathrm{LowN,i}}$ and the desirable level of $P_{\mathrm{Recovery},i}$. In addition, the interpretation of optimal harvest strategies necessarily involves a value judgment about the risks and benefits of harvest. This could be explored explicitly in the simulation framework by adjusting the weighting of individual indicators within the objective function.

Acknowledgments The chapter is the result of research sponsored by the Objective Based Fisheries Management initiative of Fisheries and Oceans Canada. We are grateful to Tom Johnston, Paul McElhany, and Mary Ruckelshaus for reviewing an earlier draft of the manuscript.

References

Akçakaya, H. R. 2002. *RAMAS GIS: linking landscape data with population viability analysis (version 4.0)*. Applied Biomathematics, Setauket, N.Y.

Barrowman, N. J., and Myers, R. A. 2000. Still more spawner-recruit curves: the hockey stick and its generalizations. *Canadian Journal of Fisheries and Aquatic Sciences* 57: 665–676.

Beacham, T. D., Schweigert, J. F., MacCjonnachie, C., Le, K. D., Labaree, K., and Miller, K. M. 2001. *Population structure of herring* (Clupea pallasi) *in British Columbia: an analysis using microsatellite loci*. Canadian Stock Assessment Secretariat Research Document 2001/128. Available at http://www.dfo-mpo.gc.ca/csas/English/Publications/Research_Doc_e. htm.

Francis, R. I. C. C. 1993. Monte Carlo evaluation of risks for biological reference points used in

New Zealand fishery assessments. Pages 221–230 in S. J. Smith, J. J. Hunt, and D. Rivard (eds.), *Risk evaluation and biological reference points for fisheries management*. Canadian Special Publication of Fisheries and Aquatic Sciences, 120. Fisheries and Oceans Canada, Ottawa.

Fu, C., Quinn II, T. J., and Kruse. G. H. 2000. Simulation analyses of harvest strategies for pandalid shrimp populations. *Journal of Northwest Atlantic Fishery Science* 27: 247–260.

Goodman, D. 2002. Predictive Bayesian population viability analysis: a logic of listing criteria, delisting criteria, and recovery plans. Pages 447–469 in S. R. Beissinger and D. R. McCullough (eds.), *Population viability analysis*. University of Chicago Press, Chicago.

Hay, D. E. 1985. Reproductive biology of Pacific herring (*Clupea harengus pallasi*). *Canadian Journal of Fisheries and Aquatic Sciences* 42: 111–126.

Hollowed, A. B., and Megrey, B. A. 1993. Harvest strategies for Gulf of Alaska walleye pollock. Pages 291–320 in G. H. Kruse, D. M. Eggers, R. J. Marasco, C. Pautzke, and T. J. Quinn II (eds.), *Proceedings of the International Symposium on Management Strategies for Explored Fish Populations*. University of Alaska Sea Grant College Program Report 93–02, University of Alaska, Fairbanks.

Johnston, N. T., Parkinson, E. A., Tautz, A. F., and Ward, B. R. 2000. *Biological reference points for the conservation and management of steelhead,* Oncorhynchus mykiss. Canadian Stock Assessment Secretariat Research Document 2000/126. Available at http://www.dfo-mpo.gc.ca/csas/English/Publications/Research_Doc_e.htm.

Lande, R., Saether, B.-E., and Engen, S.. 1997. Threshold harvesting for sustainability of fluctuating resources. *Ecology* 78: 1341–1350.

Lande, R., B.-E. Saether, and S.Engen. 2001. Sustainable exploitation of fluctuating populations. Pages 67–86 in J. D. Reynolds, G. M. Mace, K. H. Redford, and J. G. Robinson (eds.), *Conservation of exploited species*. Conservation Biology 6. Cambridge University Press, Cambridge.

Mace, G. M., and Stuart, S. N. 1994. Draft IUCN Red List Categories, Version 2.2. Species 21–22: 13–24. IUCN, Gland, Switzerland.

Restrepo, V. R., Thompson, G. G., Mace, P. M., Gabriel, W. L., MacCall, A. D., Methot, R. D., Powers, J. E., Taylor, B. L., Wade, P. R., and Witzig, J. F. 1998. *Technical guidance on the use of precautionary approaches to implementing national standard 1 of the Magnuson-Stevens fishery conservation and management act*. NOAA Technical Memorandum NMFS-F/SPO-31. National Marine Fisheries Service, U.S. Department of Commerce, Washington, D.C.

Schweigert, J. 2001. Stock assessment for British Columbia herring in 2001 and forecasts of the potential catch in 2002. Canadian Stock Assessment Secretariat Research Document 99/178. Available at http://www.dfo-mpo.gc.ca/csas/English/Publications/Research_Doc_e.htm.

Schweigert, J., and Fort, C. 1999. Stock assessment for British Columbia herring in 1999 and forecasts of the potential catch in 2000. Canadian Stock Assessment Secretariat Research Document 99/178. Available at http://www.dfo-mpo.gc.ca/csas/English/Publications/Research_Doc_e.htm

Taylor, B. L, Wade, P. R., Ramakrishnan, U., Gilpin, M., and Akçakaya, H. R. 2002. Incorporating uncertainty in population viability analyses for the purpose of classifying species by risk. Pages 239–252 in S. R. Beissinger and D. R. McCullough (eds.), *Population viability analysis*. University of Chicago Press, Chicago.

Thompson, G. G. 1993. A proposal for a threshold stock size and maximum fishing mortality rate. Pages 303–320 in S. J. Smith, J. J. Hunt, and D. Rivard (eds.), *Risk evaluation and biological reference points for fisheries management*. Canadian Special Publication of Fisheries and Aquatic Sciences, 120. Fisheries and Oceans Canada, Ottawa.

Ware, D. M. 1996. Herring carrying capacity and sustainable harvest rates in different climate regimes. Working Paper H96–3. Pacific Scientific Advice Review Committee, Nanaimo, Canada.

Ware, D. M., and Schweigert, J. 2001. Metapopulation structure and dynamics of British Columbia herring. Canadian Stock Assessment Secretariat Research Document 2001/127. Available at http://www.dfo-mpo.gc.ca/csas/English/Publications/Research_Doc_e.htm.

Zheng, J., Funk, F. C., Kruse, G. H., and Fagen, R. 1993. Threshold management strategies for Pacific herring in Alaska. Pages 141–165 in G. H. Kruse, D. M. Eggers, R. J. Marasco, C. Pautzke, and T. J. Quinn II (eds.), *Proceedings of the International Symposium on Management Strategies for Explored Fish Populations*. University of Alaska Sea Grant College Program Report 93–02. University of Alaska, Fairbanks.

PART IV

AMPHIBIANS AND REPTILES

24

Modeling Amphibians and Reptiles

An Overview

PER SJÖGREN-GULVE

Population viability analyses of amphibians and reptiles are few but will probably increase greatly in the future due to improved computer software and more case studies that highlight useful modeling approaches for a variety of species. In this section, six case studies provide examples of how population viability analyses (PVAs) of amphibians and reptiles can be implemented with different data sources, varying uncertainty in the data, and various management questions and conservation problems. They illustrate how the PVA process reveals uncertainties and needs of complementary data collection, as well as suggest management actions and further investigations (Akçakaya and Sjögren-Gulve 2000, Burgman and Possingham 2000).

Reptiles and amphibians have a wide variety of life histories (e.g., Spellerberg 1982, Stebbins and Cohen 1995, Pough et al. 2001), of which this overview and the section's case studies cover just a small sample. Two of these case studies are on amphibians (Griffiths, Chapter 25 on great crested newt; Hatfield et al., Chapter 26 on Houston toad), and four are on reptiles (Breininger et al., Chapter 27 on eastern indigo snake; Brook and Griffiths, Chapter 28 on frillneck lizard; Berglind, Chapter 29 on sand lizard; and Chaloupka, Chapter 30 on green sea turtle). However, the modeling approaches of these studies may be relevant for other taxa in similar situations. This overview provides a brief, but not exhaustive, review of published PVAs of amphibians and reptiles.

PVA models can be broadly categorized into occupancy models (based on data on population presence or absence), structured or unstructured (scalar) demographic models, and individual-based demographic models (Akçakaya and Sjögren-Gulve 2000). My overview here focuses largely on PVAs with structured demographic models (e.g., Akçakaya 2000). Studies using occupancy models (which use population presence/absence data from repeated inventories of entire regions; Sjögren-Gulve and Hanski 2000) have been conducted for two amphibian species in the temperate zone: the pool frog (*Rana lessonae*) in Sweden (Sjögren-Gulve and Ray 1996), and the tree frog (*Hyla arborea*) in the Netherlands (Vos et al. 2000). Moreover, the predictive precision of the incidence-function occupancy model was compared to that of a structured demographic (RAMAS) model for the Florida scrub lizard by Hokit et al. (2001), who found that the demographic model predicted which habitat patches were occupied and unoccupied, respectively, slightly better (80% correct) than the incidence function model (77% correct) (see also Kindvall 2000). Individual-based models have been used in studies of, for example, the Australian arboreal gecko *Oedura reticulata* (e.g., Weigand et al. 2002), the European wall lizard (*Podarcis muralis;* Bender et al. 1996) and for two amphibian and seven reptile taxa in population and habitat viability analyses (PHVAs) published by the Conservation Breeding Specialist Group (e.g., the Wyoming toad, *Bufo baxteri;* Conservation Breeding Specialist Group 2001). In general, few of these studies have been able to use the full power of the individual-based models (e.g., examining effects of inbreeding and/or certain behaviors on population viability), usually due to inadequate data.

PVAs using Structured or Scalar Demographic Models

In these PVAs of amphibians and reptiles, many of the published studies have used deterministic matrix models and calculated asymptotic population growth rate (λ or R, finite rate of population increase) and used elasticity analysis (e.g., Mills and Lindberg 2002) to explore which vital rates affect λ the most. Demographic and environmental stochasticities are generally included in simulations of the matrix models to assess extinction risk with a specific time frame or to examine whole-model sensitivity (Mills and Lindberg 2002) to variation in single vital rates. Some models also consider density dependence, but few have included effects of inbreeding depression.

Gibbs (1993) used a spatially explicit scalar model (RAMAS Space) to simulate the effect of hypothetical loss of small wetlands on the persistence of metapopulations of five wetland-dependent species in Maine, U.S.A: snapping turtle (*Chelydra serpentina*), bullfrog (*Rana catesbeiana*), green frog (*Rana clamitans*), spotted salamander (*Ambystoma maculatum*), and red-spotted newt (*Notophthalmus viridescens*). Population growth rate (R) was estimated from simple annual population censuses. Gibbs (1993) simulated population viability of "frogs," "salamanders," and snapping turtles in a wetland mosaic of five large and eight small wetlands in a hypothetical landscape covering almost 20 km^2 by using published data on densities of females/ha of wetland and observed dispersal distances of these species. The models included environmental stochasticity (standard deviation of R) and logistic density dependence, and they assumed that 10% of each population dispersed each year. The simulations were then repeated without the small wetlands ("post-loss" scenario), and quasi-extinction probabilities (Akçakaya 2000) of the metapopulation under these two scenarios were compared. Under these exploratory premises, Gibbs (1993) concluded that small wetlands seem more important for the persistence of certain species than wetland area alone might imply, and he suggested that this should be further explored using more realistic models.

Since the late 1980s, worldwide declines and losses of amphibian populations have been highlighted and discussed in the scientific literature (e.g., Barinaga 1990, Alford and Richards 2000, Houlahan et al. 2000). Results from repeated inventories, population monitoring, and experimental studies of egg and larval survival have primarily fueled this discussion. Biek et al. (2002) stressed that very few demographic studies have addressed amphibian population declines. They performed elasticity analysis and life-stage simulation analysis (LSA) (Mills and Lindberg 2002) of population matrix models of the western toad (*Bufo boreas*), the red-legged frog (*Rana aurora*), and the common frog (*R. temporaria*) to identify vital rates that contribute fundamentally to population growth rate (λ).These three anuran species have declined in all or parts of their geographic ranges. Biek et al. (2002) reported that variation in postmetamorphosis vital rates (survival rates of metamorphs, juveniles, and adults; adult fecundity) affected λ more than did premetamorphosis survival rates (survival of embryos and larvae). This applied to all three species with only one exception: the ample variation in larval survival in the common frog had a major effect on λ according to the LSA, but this did not show in the elasticity value. Since most amphibian-decline studies have addressed mortality among embryos and larvae, and only few the survival of metamorphs, juveniles, or adults, Biek et al. (2002) cautioned that critical factors affecting postmetamorphosis survival may have been overlooked and urgently need to be investigated further. Most likely, management that increases the yearly survival among juvenile and adult amphibians may reduce or even reverse population declines.

One such amphibian that has declined rapidly in many parts of its geographic range is the great crested newt (*Triturus cristatus*) in Europe (e.g., Corbett 1989). Griffiths and Williams (2000) used RAMAS Metapop ver. 3.0 to explore the effects of pond isolation, drought, habitat fragmentation, and dispersal on newt population persistence in southeastern England. They only considered ponds that were suitable for great crested newts, (i.e., without predatory fish). In simulations, the risk of metapopulation extinction increased as the frequency of droughts increased, as rates of dispersal declined between the local populations, and as populations became smaller and fewer. The lowest probability of regional extinction during 50 years was scored with the highest number (10) of local populations and with 10% dispersal between them. Griffiths (Chapter 25 in this volume) presents a further-developed PVA model that includes distance-dependent dispersal between populations, separate sexes, mating structure, and decreasing carrying capacity of breeding ponds with time due to plant succession. All of these analyses suggest that a metapopulation approach is valid and important for great crested newt conservation, and they highlight the importance of further examining juvenile survival and interpopulation dispersal.

The importance of juvenile survival for population persistence in amphibians is further exemplified by Hels and Nachman's (2002) study of a spadefoot toad metapopulation (*Pelobates fuscus*) in Denmark using an age-based model with separate sexes. Using whole-model sensitivity analysis of vital rates and dispersal rates, they reported that variation in juvenile survival had a stronger effect on population persistence than did adult survival rate and fecundity. In the study system of five ponds, one pond had higher yearly survival of juvenile spadefoot toads than the others, and the yearly dispersal rates of adults was 1.09% (Hels 2002). Simulating the system, these local populations seemed to function as a "source-sink" (Pulliam 1988) system, sustained primarily by the pond with higher juvenile survival. This pattern was corroborated by the fact that the spadefoot toads at one pond isolated by a trafficked road survived much worse in simulations when the barrier effect of the road increased.

Using a stage-based RAMAS model, Sutherland et al. (2000) found that population growth rate (λ) of the tailed frog (*Ascaphus truei*) in British Columbia also was most sensitive to the probability of survival with growth in larval and prereproductive stages. Tailed-frog larvae may take 1 to 4 years to reach metamorphosis, depending on geographic location and features of their natal stream. Sutherland et al. (2000) emphasized that harvest rotation in the forestry is not enough to save such populations from extinction and that the integrity of riparian habitats needs to be maintained for tailed-frog conservation.

Among reptiles, Ferrière et al. (1996) used an age-structured matrix model for analysis of a French population of Orsini's viper (*Vipera ursinii ursinii*), one of the most endangered reptiles in Europe (Corbett 1989). Variation in age-specific survival rates affected λ (estimated at 0.89) more than did both age-specific fecundity and probability of breeding. Survival at ages 1 to 4 had the highest elasticities, with 4 being the age at first reproduction. The extinction probability was estimated at 90% during 1985–2050 using a simulation of the matrix model that included demographic and environmental stochasticity. Ferrière et al. (1996) thus suggested that management should focus on measures to enhance population growth rate (primarily improve juvenile and subadult survival), supply adult females with high reproductive value to increase population size, and if possible, reduce temporal variation in population growth. They also stressed the importance of close monitoring of population size and of the targeted vital rates.

For tortoises and turtles, adult survival may be more important for population persistence than in the other examples given here. Doak et al. (1994) modeled population viability of the desert tortoise *Gopherus agassizii* in the western Mojave desert (United States) using a size-structured demographic model. Their results were in agreement with the population decline observed in the field. Elasticity analysis of the population model indicated that population growth rate was most sensitive to variation in the survival of large adult females. Doak et al. (1994) argued that improving the annual survival of such females to reputably "pristine" rates could reverse population declines, whereas large improvements in any other vital rate alone would not. Thus, they recommended that reduction of adult mortality caused by shooting, off-road vehicles, and upper-respiratory tract disease, should be the primary focus of management actions to prevent further decline. They also recommended further investigations of the epidemiology of the respiratory disease and the effects of livestock grazing on tortoise survival and reproduction, along with collection of data to estimate adult mortality and its temporal variation with greater accuracy.

For sea turtles, Chaloupka and Musick (1997) provide an extensive review of population dynamics modeling and further modeling aspects, and useful references are also found in Chaloupka (Chapter 30 in this volume). In another turtle study, Heppell et al. (1996) used deterministic matrix models to evaluate "headstarting" (i.e., the captive rearing of hatchlings from eggs collected in the field) as a management tool for the endangered Kemp's ridley sea turtle (*Lepidochelys kempi*) and the nonthreatened yellow mud turtle (*Kinosternon flavescens*). Elasticity analysis indicated that annual survival rates of subadult and adult turtles affected population growth rate more than did other vital rates, suggesting that headstarting alone would not be the most efficient measure to increase population size. Efforts are also needed to increase and safeguard the survival of subadult and adult turtles. A similar pattern was found in eight other turtle species analyzed using a similar approach (Heppell 1998).

Headstarting was also one of the management alternatives for preserving Swedish sand lizard populations (*Lacerta agilis*) evaluated by Berglind (2000). Using an age-structured matrix simulation model, he compared the effect of five management scenarios, which included combinations of habitat management, headstarting during 5 or 10 years, "captive raising" (i.e., headstarting plus captive hibernation), and "captive breeding" during 5 or 10 years with hatchlings released into the wild each year. Elasticity analysis of the matrix model indicated that juvenile survival had a major effect on population growth rate, which was estimated at 0.94 and 1.03 for two study populations. Using stochastic simulations of a single population without density dependence, but including catastrophes, Berglind (2000) reported that with no action taken, the population extinction risk was 40% within 20 years. The captive breeding and captive raising management measures reduced the risk of population decline to critically low abundances most efficiently. Berglind (2000) thus recommended that management be initiated and combined with habitat restoration and continued monitoring. Berglind (Chapter 29 in this volume) has now expanded the sand lizard single-population model into a metapopulation simulation model that includes density dependence, as well as examining the importance of patch geometry and number, dispersal, and reintroduction strategy for restocking of Swedish sand lizard metapopulations.

Discussion

While many of the above studies have used elasticity analysis to examine how much small proportional changes in individual vital rates (i.e., matrix transition elements, p_{ij}) affect population growth rate, whole-model sensitivity analysis of the effect of variation or uncertainty in individual model parameters has now become more common (for review, see Mills and Lindberg 2002). In Part IV, the amphibians and reptiles section, conducting a PVA in the face of uncertainty in model parameters is illustrated primarily by Breininger et al.'s study of the viability of the eastern indigo snake (*Drymarchon corias couperi*) in Florida (Chapter 27 in this volume). Similar approaches can be found for individual parameters in Hatfield et al.'s PVA of the Houston toad (*Bufo houstonensis*) in Texas (Chapter 26 in this volume), Chaloupka's analysis of effects of indigenous local harvesting of green sea turtles (*Chelonia mydas*) (Chapter 30 in this volume), Berglind's comparison of density dependence effects (Chapter 29 in this volume), and Brook and Griffiths' analysis of the effects of various forms of fire management on the viability of a frillneck lizard population (*Chlamydosaurus kingii*) in northern Australia (Chapter 28 in this volume). Despite the uncertainty in the estimates of some model parameters, many PVA studies can robustly rank alternative management strategies and determine which would be highly likely to benefit or threaten the populations of concern or to identify data needed to improve model predictions (Akçakaya and Sjögren-Gulve 2000; Burgman and Possingham 2000; Akçakaya, Chapter 1 in this volume). For example, in the PVA of the Houston toad (Hatfield et al., Chapter 26 in this volume), the survival of offspring from egg or embryo to 1 year of age ("juveniles") and the fraction of females breeding as 2-year-olds were estimated and subjected to sensitivity analysis. These analyses showed that model predictions of population extinction were highly sensitive to variation in juvenile survival rate, and also to the fraction of 2-year-old females breeding in simulations where juvenile survival was low and probability of catastrophes was high. Hence, collecting data to estimate these two parameters with acceptable certainty are among the primary recommendations of Hatfield et al. (Chapter 26 in this volume) for continuing the Houston toad PVA process.

Future Directions

Amphibians and reptiles are taxonomic groups with large interspecific variation in life history traits and which live in a wide variety of environments (Spellerberg 1982, Stebbins and Cohen 1995, Pough et al. 2001). Because many of the species are highly fecund, and do not show extensive individual variation in behavior, age- or stage-structured demographic models using projection matrices often seem more appropriate than individual-based models for population modeling. The structured models are also highly flexible (Akçakaya 2000; Chapter 1 in this volume). However, modeling that requires greater detail, such as inclusion of inbreeding depression and important behavioral components for species reproduction demands individual-based models (see, e.g., Lacy 2000, Beissinger and McCullough 2002).

The broad variety of life histories among amphibians and reptiles is poorly represented, both taxonomically and geographically, among the PVAs carried out thus far. PVAs have virtually been restricted to taxa occurring in developed countries, probably because necessary data are difficult or too expensive to collect. One challenge of PVAs of amphibians and reptiles is the various modes of reproduction, which by use of appropriate assumptions is quite flexible to model with demographically structured models and RAMAS Metapop (Akçakaya and Root 2002). A small group of taxa (e.g., certain species of the lizard genera *Cnemidophorus* and *Hemidactylus*) are parthenogenic. All crocodilians, some lizards (geckos and lacertids), and many turtle species have temperature-dependent sex determination, and some taxa such as the poison-dart frogs (family Dendrobatidae) have relatively low fecundity and behaviorally complex reproduction, while others are highly fecund explosive breeders (e.g., some *Bufo* and *Rana* spp.) (see Spellerberg 1982, Stebbins and Cohen 1995, Pough et al. 2001). Many variants of reproduction and life-stage transitions can be handled with structured demographic models (using the Lefkovitch matrix) and with RAMAS Metapop (e.g., Burgman, Chapter 2 in this volume). The catastrophes option allows modeling of various effects on reproduction and survival in single or multiple populations (e.g., Brook and Griffiths, Chapter 28 in this volume) that can simulate both stochastic and deterministic events that affect population growth and composition (effects of predation, unfavorable weather, etc.). Thus, many assumptions made in other (often simpler analytical) models can be relaxed, but it is important to explain how the modeling was conducted and under what assumptions (e.g., Akçakaya and Sjögren-Gulve 2000, Burgman and Possingham 2000); the case studies in this volume and literature review provide several examples.

One issue in theoretical and practical conservation is whether or not there are classical metapopulations and to what extent metapopulation theory applies to natural populations (e.g., Harrison 1991, Sjögren-Gulve and Hanski 2000, Marsh 2001, Trenham et al. 2001). With the flexibility of current demographic and occupancy PVA models, many of the unrealistic assumptions of the seminal metapopulation model (Levins 1969) can be relaxed. A system can thus be modeled as a more or less partitioned population (e.g., Breininger et al., Chapter 27 in this volume), with more or less connected and demographically correlated subpopulations ("patchy population" or some form of metapopulation, according to Harrison 1991), as seems appropriate to mimic patch characteristics and habitat heterogeneity.

Acknowledgments I thank Reşit Akçakaya, Sven-Åke Berglind, Milani Chaloupka, Angela Fuller, Richard Griffiths, and Mats Höggren for constructive input to and comments on this chapter. As subject editor of the amphibians and reptiles section, I also want to thank the authors and peer reviewers of the essays for their stimulating contributions.

References

Akçakaya, H. R. 2000. Population viability analyses with demographically and spatially structured models. *Ecological Bulletins* 48: 23–38.

Akçakaya, H. R., and Root, W. 2002. *RAMAS Metapop: viability analyses for stage-structured metapopulations (version 4.0)*. Applied Biomathematics, Setauket, N.Y.

Akçakaya, H. R., and Sjögren-Gulve, P. 2000. Population viability analyses in conservation planning: an overview. *Ecological Bulletins* 48: 9–21.

Alford, R. A., and Richards, S. J. 2000. Global amphibian declines: a problem in applied ecology. *Annual Review of Ecology and Systematics* 31: 133–165.

Barinaga, M. 1990. Where have all the froggies gone? *Science* 247: 1033–1034.

Beissinger, S. R., and McCullough, D. R. (eds.). 2002. *Population viability analysis*. University of Chicago Press, Chicago.

Bender, C., Hildenbrandt, H., Schmidt-Loske, K., Grimm, V., Wissel, C., and Henle. K. 1996. Consolidation of vineyards, mitigations, and survival of the common wall lizard (*Podarcis muralis*) in isolated habitat fragments. Pages 248–261 in J. Settele, C. R. Margules, P. Poschlod, and K. Henle (eds.), *Species survival in fragmented landscapes*. Kluwer Academic, Amsterdam.

Berglind, S.-Å. 2000. Demography and management of relict sand lizard *Lacerta agilis* populations on the edge of extinction. *Ecological Bulletins* 48: 123–142.

Biek, R., Funk, W. C., Maxell, B. A., and Mills, L. S. 2002. What is missing in amphibian decline research: insights from ecological sensitivity analysis. *Conservation Biology* 16: 728–734.

Burgman, M. A., and Possingham, H. 2000. Population viability analysis for conservation: the good, the bad and the undescribed. Pages 97–112 in A. G. Young and G. M. Clarke (eds.), Genetics, demography and viability of fragmented populations. Cambridge University Press, Cambridge.

Chaloupka, M. Y., and Musick, J. A. 1997. Age, growth and population dynamics. Pages 233–276 in P. J. Lutz and J. A. Musick (eds.), *The biology of sea turtles*. CRC Marine Science Series. CRC Press, Boca Raton, Fla.

Conservation Breeding Specialist Group (CBSG). 2001. Wyoming toad (*Bufo baxteri*): population and habitat viability assessment. Conservation Breeding Specialist Group, Apple Valley, Minn. Available at http://www.cbsg.org/reports/report.php?type=phva.

Corbett, K. F. 1989. *Conservation of European reptiles and amphibians*. Christopher Helm, London.

Doak, D., Kareiva, P., and Klepetka, B. 1994. Modeling population viability for the desert tortoise in the western Mojave desert. *Ecological Applications* 4: 446–460.

Ferrière, R., Sarrazin, F., Legendre, S., and Baron, J.-P. 1996. Matrix population models applied to viability analysis and conservation: theory and practice using the ULM software. *Acta Oecologica* 17: 629–656.

Gibbs, J. P. 1993. Importance of small wetlands for the persistence of local populations of wetland-associated animals. *Wetlands* 13: 25–31.

Griffiths, R. A., and Williams, C. 2000. Modelling population dynamics of great crested newts (*Triturus cristatus*): a population viability analysis. *Herpetological Journal* 10: 157–163.

Harrison, S. 1991. Local extinction in a metapopulation context: an empirical evaluation. *Biological Journal of the Linnean Society* 42: 73–88.

Hels, T. 2002. Population dynamics in a Danish metapopulation of spadefoot toads *Pelobates fuscus*. *Ecography* 25: 303–313.

Hels, T., and Nachman, G. 2002. Simulating viability of a spadefoot toad (*Pelobates fuscus*) metapopulation in a landscape fragmented by a road. *Ecography* 25: 730–744.

Heppell, S. S. 1998. Application of life-history theory and population model analysis to turtle conservation. *Copeia* 1998: 367–375.

Heppell, S. S., Crowder, L. B., and Crouse, D. T. 1996. Models to evaluate headstarting as a management tool for long-lived turtles. *Ecological Applications* 6: 556–565.

Hokit, D. G., Stith, B. M., and Branch, L. C. 2001. Comparison of two types of metapopulation models in real and artificial landscapes. *Conservation Biology* 15: 1102–1113.

Houlahan, J. E., Findlay, C. S., Schmidt, B. R., Meyer, A. H., and Kuzmin, S. I. 2000. Quantitative evidence for global amphibian population declines. *Nature* 404: 752–755.

Kindvall, O. 2000. Comparative precision of three spatially realistic simulation models of metapopulation dynamics. *Ecological Bulletins* 48: 101–110.

Lacy, R. C. 2000. Considering threats to the viability of small populations using individual-based models. *Ecological Bulletins* 48: 39–51.

Levins, R. 1969. Some demographic and genetic consequences of environmental heterogeneity for biological control. *Bulletin of the Entomological Society of America* 15: 237–240.

Marsh, D. M. 2001. Fluctuations in amphibian populations: a meta-analysis. *Biological Conservation* 101: 327–335.

Mills, L. S., and Lindberg, M. S. 2002. Sensitivity analysis to evaluate the consequences of conservation actions. Pages 338–366 in S. R. Beissinger and D. R. McCullough (eds.), *Population viability analysis*. University of Chicago Press, Chicago.

Pough, F. H., Andrews, R. M., Cadle, J. E., Crump, M. L., Savitzky, A. H., and Wells, K. D. 2001. *Herpetology*. Prentice Hall, Upper Saddle River, N.J.

Pulliam, H. R. 1988. Sources, sinks and population regulation. *American Naturalist* 132: 652–661.

Sjögren-Gulve, P., and Hanski, I. 2000. Metapopulation viability analysis using occupancy models. *Ecological Bulletins* 48: 53–71.

Sjögren-Gulve, P., and Ray, C. 1996. Using logistic regression to model metapopulation dynamics: large-scale forestry extirpates the pool frog. Pages 111–137 in D. R. McCullough (ed.), *Metapopulations and wildlife conservation*. Island Press, Washington D.C.

Spellerberg, I. F. 1982. *Biology of reptiles: an ecological approach*. Blackie and Son, London.

Stebbins, R. C., and Cohen, N. W. 1995. *A natural history of amphibians*. Princeton University Press, Princeton, N.J.

Sutherland, G. D., Richardson, J. S., and Bunnell, F. L. 2000. Uncertainties linking tailed frog habitat and population dynamics with riparian management. Pages 477–484 in L. M. Darling (ed.), *Proceedings of a conference on the biology and management of species and habitats at risk, Kamloops, B.C., 15–19 Feb. 1999*. Vol. 2. B. C. Ministry of Environment, Lands and Parks, Victoria, B.C. Available at http://142.103.180.19/richardson/abstracts/RE%2002%20Sutherland.pdf

Trenham, P. C., Koenig, W. D., and Shaffer, H. B. 2001. Spatially autocorrelated demography and interpond dispersal in the salamander *Ambystoma californiense*. *Ecology* 82: 3519–3530.

Vos, C. C., ter Braak, C. J. F., and Nieuwenhuizen, W. 2000. Incidence function modelling and conservation of the tree frog *Hyla arborea* in the Netherlands. *Ecological Bulletins* 48: 165–180.

Wiegand, K., Henle, K., and Sarre, S. D. 2002. Extinction and spatial structure in simulation models. *Conservation Biology* 16: 117–128.

25

Great Crested Newts (*Triturus cristatus*) in Europe

Effects of Metapopulation Structure and Juvenile Dispersal on Population Persistence

RICHARD A. GRIFFITHS

Despite a resurgence of interest in amphibian declines that has resulted in a proliferation of population case studies in recent years (e.g., Richards et al. 1993, Kuzmin 1994, Drost and Fellers 1996, Green 1997, Lips 1998), population viability analysis has not been widely used as a management tool for these species. Although a diverse group, which encompasses species that are entirely aquatic as well as entirely terrestrial, amphibians are frequently small, cryptic animals that are inconspicuous for a large part of their life cycle. This makes many amphibian population studies problematic, and the determination of reliable demographic parameters is difficult. The vast majority of amphibian population studies have therefore consisted of assessments of adults at breeding foci, as this phase of the life cycle is the one in which animals are most conspicuous. Such breeding foci—usually at ponds, streams, or lakes—are spatially distinct units that are convenient for testing hypotheses about subdivided populations. As Marsh and Trenham (2000) point out, however, adopting a "ponds-as-patches" approach to amphibian spatial dynamics may be oversimplistic, as apparent population turnovers may occur for reasons other than stochastic processes. Indeed, such turnovers may be ultimately related to wider landscape-level factors. Population viability analysis can be used as a tool to explore the interaction between deterministic (e.g., pond desiccation, habitat change, fish introductions) and stochastic processes in amphibian population dynamics. Equally, it can also be used to cautiously explore some of the lacunae that exist in amphibian life history data, through sensitivity analyses of some of the more elusive demographic and environmental parameters.

The great crested newt (*Triturus cristatus*) is one of the best-studied amphibian species in Europe. It is a species that spends most of its life on land, but adults return to ponds to breed for a few months in the spring and early summer. Eggs are individually wrapped in the leaves of aquatic plants, and the resulting larvae pass through an aquatic phase lasting 3 to 4 months, although some may remain aquatic for longer and may overwinter in water. The species has been the subject of two recent symposia proceedings (Cummins and Griffiths 2000, Krone 2001), as well as a general natural history monograph (Thiesmeier and Kupfer 2000). This high level of interest stems largely from the protection afforded to it within Europe under the Convention on the Conservation of European Wildlife and Natural Habitats (known as the Bern Convention) and the Council Directive on the Conservation of Natural Habitats and of Wild Flora and Fauna (known as the Habitats Directive). Both of these legal instruments place certain obligations on member states to implement measures to protect the species. Consequently, the great crested newt is one of the very few amphibian species that has been the subject of previous modeling studies. Using a stochastic difference equation approach, Halley et al. (1996) predicted that great crested newt populations would persist if they supported more than 40 females, although viable populations smaller than this were possible if they were close to a source of immigrants. Griffiths and Williams (2000, 2001) used RAMAS Metapop 3.0 in conjunction with published demographic data to simulate the effects of sequential fragmentation on great crested newt populations. They found that even relatively large (i.e., 100–200 newts) populations had a relatively high risk of extinction if they remained isolated for 50 years or more, but that increased dispersal decreased the extinction risk of a hypothetical metapopulation.

In this chapter, the models described by Griffiths and Williams (2000, 2001) are developed further using RAMAS Metapop 4.0 to incorporate the effects of differences in subpopulation size and subpopulation number within a metapopulation system. These new models incorporate juvenile dispersal as a function of distance between patches and then compare the viability of metapopulations of different sizes. In addition, a stable initial age distribution is used, and sex structure is incorporated by including both males and females in the models.

Materials and Methods

Metapopulation Model

Stage Structure

Great crested newts do not reach sexual maturity until they are 2 or 3 years old. Whereas about half of all 2-year-old newts return to ponds to breed, all 3-year-old newts are capable of breeding. However, some subadult animals (i.e., 1-year-old and nonbreeding 2-year-old newts) may return to the water each year, along with breeding adults. The models therefore used three age classes: 1-year-olds, 2-year-olds, and >3-year-olds. All 1-year-old newts were juveniles that did not breed. Half of the 2-year-old newts, along with all newts aged 3 or more, were classified as "breeding."

Survival

Annual survival of great crested newts has been calculated using a number of methods, including age structure analysis (Williams 1999), mark-recapture analysis (Hagström

1979, Arntzen and Teunis 1993, Baker 1999, Williams 1999) and direct tracking of individuals (Cummins and Swan 2000). By combining data from the study by Arntzen and Teunis (1993) with his own information from Leicestershire, Oldham (1994) constructed a survivorship curve for great crested newts, which assumed that great crested newts breed for the first time at 2 or 3 years and then breed every year until death. From this survivorship curve, juvenile survival was estimated at 0.2, and annual adult survival was estimated at about 0.68. These calculations are based on return rates of juveniles and adults, respectively, and do not take emigration into account. As juveniles appear to be the main dispersers (see the discussion later in this chapter), in isolated populations the value of 0.2 is likely to be a reliable estimator of survival if all dispersers return to the natal pond. However, if some juveniles disperse but die without returning to the pond, then 0.2 survival may be an overestimate. Acknowledging this caveat, the Leslie matrix of survival-fecundity schedules in the RAMAS models used a survival rate of 0.2 for 1- to 2-year-olds, and 0.68 for annual survival thereafter, for both males and females. As published estimates of adult annual survival vary from about 0.31 to 1.00, a standard deviation of 0.3 was attached to adult survival (coefficient of variation = 44%). Juvenile survival is more variable than adult survival, and a standard deviation of 0.176 was attached to the estimate of juvenile survival (coefficient of variation = 88%).

Fecundity

As the models were based on postmetamorphic age classes only, "fecundity" was modeled as the number of new individuals (= 1-year-old newts) produced per individual within each age class (i.e., "recruitment"; see, e.g., Burgman et al. 1993). Published estimates of fecundity range from 189 to 220 eggs per female (see citations by Arntzen and Teunis 1993). Assuming a 1:1 sex ratio, and taking account of the 50% egg abortion (see Horner and MacGregor 1985), 95% egg/larval mortality—based on data provided by Arntzen and Teunis (1993) and Oldham (1994)—this translates to about five new recruits per clutch of eggs (= 2.5 recruits of each sex per female), with an estimated standard deviation of 0.2.

Sex Structure

Males and females were included as separate matrices, but they had identical age structures and survival schedules. As male newts may mate with several females during the course of a breeding season (Hedlund 1990), a polygynous mating system was incorporated with each male mating with up to four females.

Stochasticity

Environmental stochasticity was modeled by drawing values randomly from lognormal distributions described by the fecundity and survival values and their associated standard deviations as described here. The effects of stochasticity on fecundity, survival, and carrying capacity were assumed to be correlated within a population, and the extinction threshold for each population was set to one remaining newt.

At three ponds supporting a metapopulation in Kent, the correlations in population size over a 6-year period were low (unpublished data), suggesting that great crested newt populations fluctuate independently of each other. The variation in yearly fecundity and survival in the subpopulations in the models was therefore modeled as being uncorrelated between populations.

Catastrophes

In dry summers, pond desiccation may occur and severely reduce larval survival. Even if the pond does not dry completely, desiccation will lead to a reduction or loss of recruits as a result of increased predation or intraspecific competition within a crowded pool. Observations in Kent suggest that ponds suffer episodes of desiccation about once every 3 years. Shallower ponds are likely to suffer from desiccation more than deeper ponds, and the effects are likely to be synchronous on a regional basis, with all shallow ponds undergoing some degree of desiccation within the same year. In the models, half of the ponds were assumed to have a desiccation risk of 0.3. This was modeled as a regional catastrophe, affecting all the vulnerable subpopulations at the same time. In these "desiccation" years the recruitment of 1-year-old newts was reduced by 80% in vulnerable subpopulations.

Initial Abundances

Published population estimates of great crested newts vary considerably. For example, Arntzen and Teunis (1993) cite studies that give population sizes that vary from $N < 10$ to $N = 1{,}000$. However, populations chosen for research purposes may be biased toward small population sizes as these are more amenable for ecological investigation. Compounding the problem is the fact that (1) most studies census the population of breeding adults only, and (2) the effective number of breeding adults and the effective population size are an order of magnitude lower than the census population size (Jehle et al. 2001). The starting population sizes used in the models were 50, 100, and 200 newts, with the stable age distribution. These may lie toward the lower end of the range of known censused adult populations, but they may be adequate representations of great crested newt populations in a fragmented situation.

Spatial Structure and Dispersal

The models simulated groups of 2, 4, 8, or 16 isolated populations and compared these to populations with the same spatial structure but linked by dispersal. Although adults occasionally move between ponds (Arntzen and Teunis 1993, Williams 1999, Kupfer and Kneitz 2000), the main dispersal phase of the great crested newt is the juvenile (i.e., 1-year-old to 2-year-old) phase. It is not known what proportion of metamorphs disperse to new ponds or what proportion return to the natal pond. It was therefore assumed that 50% of 1-year-old newts and 25% of 2-year-old newts disperse to new subpopulations. Likewise, the problem of permanently marking large numbers of small metamorphs means that there are few data on the dispersal distances of juveniles. Out of 176 juvenile newts that were captured leaving a pond in Germany, 35 were recaptured at distances varying from 10 to 860 m (Kupfer and Kneitz 2000). On the basis of these data, the average dispersal distance is 254 m. Kupfer and Kneitz's (2000) maximum dispersal distance of 860 m is consistent with other studies that have suggested that great crested newts can disperse up to 1 km from the breeding pond (e.g., see Oldham and Swan 1997). Using these values, a distance-dependent juvenile dispersal function (see Chapter 1) was used with $a = c = 1$, $b = 254$, and $D_{max} = 1$ km.

Density Dependence

Density-dependent effects are poorly understood in postmetamorphic stage newts. A simple ceiling model was therefore used that affected all vital rates and that was based on the abundance of all stages. The carrying capacity (K) was initially set at 50% higher than the initial population size (i.e., $K = 1.5\ N_0$), but declined linearly by 33% over the 100-year period (e.g., a carrying capacity of 300 would decline to 200 after 100 years). This was to simulate a gradual deterioration in habitat quality as a result of pond succession and loss of terrestrial habitat.

Simulation Scenarios

Models were constructed to compare the risk of extinction in metapopulations with different numbers of subpopulations (i.e., 2, 4, 8, and 16 subpopulations) and with different subpopulation sizes (i.e., $N_0 = 50$, 100, or 200). The effects of dispersal were investigated by comparing metapopulations with juvenile dispersal between subpopulations with those in which there was no dispersal and the component populations were therefore isolated. All models used a stochastic simulation that ran for 100 years (that is, 15–20 generations of great crested newts), and calculated 1,000 estimates of projected population size for each year over this period. As the population age structures included both terrestrial and aquatic stages, the term *population* or *subpopulation* is used here to refer to a unit that encompasses a breeding pond and its immediate terrestrial hinterland.

The model mapped the subpopulations into hypothetical "landscapes" as follows: 2 subpopulations, one row of two subpopulations; 4 subpopulations, two rows of two subpopulations; 8 subpopulations, two rows of four subpopulations; and 16 subpopulations, four rows of four subpopulations. Within these landscapes, the perpendicular distance between a subpopulation and its nearest neighbor(s) was set at 290 m, based on the average distance between ponds occupied by great crested newts at a study site in Kent (Williams 1999).

Results

Populations with No Dispersal

Groups of isolated ponds had a relatively high extinction risk (Table 25.1). Without effects of density dependence and dispersal, the asymptotic growth rate (λ) of a local population was 1.12. In simulations, the extinction risk declined as the number of populations within the group increased. This was probably due to different populations going extinct at different times; as the threshold for extinction was the last remaining newt in the last remaining population, the more populations considered, the longer was the persistence time. Groups of small populations ($N_0 = 50$) had low viabilities and extinction

Table 25.1 Predicted extinction risks and median time to extinction of groups of isolated great crested newt populations

Population Parameters				
No. of Populations in Group	N_o	N_{100}	Extinction Risk (*P*)	Time to Extinction (years)
2	50	0.9±6.36	0.969	30.7
4	50	1.3±7.35	0.961	46.0
8	50	3.4±12.62	0.896	58.3
16	50	6.6±17.05	0.799	73.0
2	100	8.4±25.14	0.862	48.5
4	100	16.0±34.8	0.754	68.3
8	100	34.8±52.11	0.549	92.7
16	100	64.8±69.67	0.333	>100
2	200	34.2±65.26	0.689	70.0
4	200	81.2±99.72	0.426	>100
8	200	150.7±135.94	0.201	>100
16	200	309.1±204.07	0.048	>100

Note: N_o, starting population size of *each* local population; N_{100}, mean (±SD) overall regional population size (n = 1,000 simulations) after 100 years.

risks close to $p = 1.00$ in all cases. Standard deviations were high, but large populations ($N_0 = 200$) also declined in size, and after 100 years the mean population size was less than 10% of that at the start of the simulations. Although the persistence time increased as population size increased, at least 16 populations of 100, or 4 populations of 200, were needed to achieve persistence times of longer than 100 years.

Populations with Dispersal

Connecting local populations by allowing the dispersal of juveniles reduced extinction risks and increased persistence times (Table 25.2). However, groups comprising two subpopulations with 100 or fewer newts in each had low viabilities and persisted for less than 70 years. Extinction risks declined—and persistence times increased—with increasing size of the subpopulations and increasing number of subpopulations. Standard deviations remained high, and with a steady decline in carrying capacity, even large metapopulations declined by at least 35% over 100 years. If an extinction risk of <10% over 100 years is regarded as acceptably low for management purposes, then based on the level of juvenile dispersal modeled here, viable metapopulations must comprise at least 16 subpopulations of 50 newts, or 4 to 8 subpopulations of at least 100 newts.

Discussion

Population Parameters

PVA exercises fall broadly into two categories. First are models that attempt to simulate the dynamics of real populations: comparisons between real and simulated systems

Table 25.2 Predicted extinction risks and median time to extinction of great crested newt populations connected by juvenile dispersal

Population Parameters				
No. of Subpopulations	N_o	N_{100}	Extinction Rrisk (P)	Time to Extinction (years)
2	50	2.3±11.66	0.937	37.9
4	50	32.0±51.97	0.608	80.3
8	50	117.9±101.50	0.237	>100
16	50	272.1±159.2	0.079	>100
2	100	28.2±53.30	0.694	66.5
4	100	176.8±120.93	0.165	>100
8	100	451.0±146.8	0.010	>100
16	100	928.6±199.4	<0.001	>100
2	200	116.1±131.17	0.394	>100
4	200	467.3±193.87	0.033	>100
8	200	1016.0±24.5	<0.001	>100
16	200	2035.8±366.81	<0.001	>100

Note: N_o, starting size of *each* subpopulation; N_{100}, mean (±SD) overall metapopulation size (n = 1,000 simulations) after 100 years. Juvenile dispersal was modeled using a dispersal-distance function (see text).

can serve as validations of the modeling exercise, or they can help identify values of unknown population parameters through sensitivity analysis (e.g., Brook et al. 1997, 2000; McCarthy and Broom 2000). If the simulation uses population parameters obtained from the population to which it is being compared, then the validation may be flawed by a degree of circularity (Coulson et al. 2001, McCarthy et al. 2001). Second are models that collate population data from a variety of case studies and use it to make predictions about new, hypothetical scenarios (e.g., Lindenmayer and Lacy 1995, Halley et al. 1996). Although such exercises may suffer from a lack of realism, they may be more useful in informing management decisions that concern the future of a population. The models described here fall into this second category, partly because complete demographic and environmental data are not yet available for any single great crested newt regional population, necessitating the collation of data from several studies. In terms of the population parameters used, the models were based on survival and fecundity schedules collated from existing studies, which appear to provide reliable estimates of these variables. Studies of within-population variation in these variables are scarce, however, so the standard deviations that were used to generate lognormal distributions of vital rates and environmental stochasticity were based more on estimates of between-population variation. Further long-term work is needed to determine whether within-population variation in vital rates is similar to between-population variation of the same parameters. What is much more speculative for great crested newt populations is the role of density dependence and carrying capacity in regulating populations. There are no reliable estimators of these parameters in the literature, beyond the general observation that large ponds often contain larger populations than small ponds. What is clear is that egg and larval mortality is erratic and that this leads to wide year-to-year fluctuations in population size (e.g., Arntzen and Teunis 1993, Miaud et al. 1993, Cooke 1995, Baker 1999, Williams 1999). This was reflected by the population trajectories produced by the models, which contained high standard deviations. In the field, variation in recruitment is related to environmental variability and occasional catastrophes (e.g., fish introduction, pond desiccation) and may mean that adult population sizes are maintained below potential carrying capacities. Further exploration of density-related effects on adult newts is therefore needed.

Metapopulations or Patchy Populations?

It is now widely accepted that few—if any—subdivided populations strictly conform to Levins's (1970) classical model of metapopulation dynamics (e.g., Hanski 1991a, 1991b; Harrison 1991, 1994; Sjögren-Gulve and Hanski 2000). In nature, patches vary in size and quality, and the dynamics of extinction and colonization are more likely to be governed by deterministic than stochastic factors. Moreover, dispersal between local populations may be too low to balance extinctions, or, at the other extreme, high dispersal rates or correlations (or both) between local populations in their dynamics may make the system indistinguishable from a single large population. In reality, then, there are several different types of subdivided population that may intergrade into each other (Harrison 1991, Thomas and Kunin 1999).

Despite their intrinsic appeal as models for metapopulation studies, then, identifying whether a spatially structured amphibian system conforms to a metapopulation—as opposed to a patchy population (Harrison 1991)—remains problematical. As far as the great crested newt is concerned, there is some indirect evidence that metapopulation processes may be in play. At the landscape level, for example, regional populations of great crested newts are more prevalent in areas where there are high densities of ponds, and they often display a clustered distribution (e.g., Swan and Oldham 1993). Moreover, patch occupancy seems to be determined by a combination of habitat and isolation variables: newts are unlikely to be found in isolated ponds, even if the aquatic and terrestrial habitat is otherwise suitable (Williams 1999). Because local populations of newts can fluctuate considerably—and asynchronously—in size, it seems that the maintenance of connectivity between adjacent patches to enable dispersal and conserve terrestrial habitat is important to maintain viable regional populations (Joly et al. 2001). In contrast, deterministic processes such as habitat change or disturbance, or colonization by predators, are much more likely to be responsible for local extinctions than are stochastic events. As Thomas (1994) points out, in such cases it is unlikely that an empty patch will become available for colonization again quickly, if at all. The impact of such deterministic processes was probably underestimated in the models presented here, as it was assumed that all populations could potentially persist for 100 years, although a general decline in habitat quality over this period was modeled by a linear decline in carrying capacity. Such a scenario may be overoptimistic: if the current rate of pond loss in Britain continues unabated, it is likely that at least 30% of ponds will be lost altogether within the next century (e.g., see Oldham and Swan 1997).

Population Management Implications

Although it is illegal to disturb great crested newts or their habitat in Britain, the species is still relatively widespread across the country. This means that it is frequently the subject of conflict between development and conservation. When such conflict arises, some form of mitigation activity is often proposed, involving limiting the development, enhancing remaining habitat, or—in some cases—translocation of animals to a new site. Although translocation of newts should be a last resort, a recent review by Oldham and Humphries (2000) revealed that some 178 translocation exercises had been carried out up until 1996. Although these were all carried out legally, there is no legal obligation to ensure that these exercises result in viable populations or, for that matter, to even carry out any sort of viability assessment. In fact, some 31% of the translocated populations were not even monitored, and in over half the cases there was insufficient evidence for judging "success" (Oldham and Humphries 2000). In an earlier study by May (1996), the number of newts translocated was fewer than 50 animals in 51 out of 90 exercises. Clearly, population viability analysis has a potentially valuable role to play here in guiding better-informed translocation exercises. According to the models described here, much larger numbers of newts need to be used to establish viable populations than have been used in the past, and greater consideration needs to be given to landscape-level population management and ensuring juvenile dispersal between subpopulations. Although PVA has been criticized for having little useful predictive value when the reliability of population data is questionable (e.g., Fieberg and Ellner 2000, Coulson et al. 2001, Ellner et al. 2002), as other authors have pointed out, alternative approaches to estimating viability may be less objective, even vaguer than PVA and unable to use all existing information (Akçakaya and Sjögren-Gulve 2000, Brook et al. 2002). It is therefore hoped that population viability analysis will eventually become an integral part of the mitigation planning process for great crested newts, as this is kept under review by government agencies (e.g., see English Nature 2001).

Acknowledgments Thanks go to Clair Williams for gathering demographic and environmental data on newt populations in Kent and for carrying out some preliminary PVA analyses. The work was licensed by English Nature and funded by the Natural Environment Research Council (NERC) and the University of Kent. Reşit Akçakaya, Per Sjögren-Gulve, and two anonymous referees provided insightful comments that improved both the models and the manuscript.

References

Akçakaya, H. R., and Sjögren-Gulve, P. 2000. Population viability analysis in conservation planning: an overview. *Ecological Bulletins* 48: 9–21.

Arntzen, J. W., and Teunis, S. F. M. 1993. A six year study on the population dynamics of the crested newt (Triturus cristatus) following the colonization of a newly created pond. *Herpetological Journal* 3: 99–110.

Baker, J. M. R. 1999. Abundance and survival rates of great crested newts (*Triturus cristatus*) at a pond in central England: monitoring individuals. *Herpetological Journal* 9: 1–8.

Brook, B. W., Lim, L., Harden, R., and Frankham, R.1997. Does population viability analysis software predict the behaviour of real populations? A retrospective study on the Lord Howe Island Woodhen *Tricholimnas sylvestris* (Sclater). *Biological Conservation* 82: 119–128.

Brook, B. W., O'Grady, J. J., Chapman, A. P., Burgman, M. A., Akçakaya, H. R., and Frankham, R. 2000. Predictive accuracy of population viability analysis in conservation biology. *Nature* 404: 385–387.

Brook, B. W., Burgman, M. A., Akçakaya, H. R., O'Grady, J. J., and Frankham, R. 2002. Critiques of PVA ask the wrong questions: throwing the heuristic baby out with the numerical bath water. *Conservation Biology* 16: 262–263.

Burgman, M. A., Ferson, S., and Akçakaya, H. R. 1993. *Risk assessment in conservation biology*. Chapman and Hall, London.

Cooke, A. S. 1995. A comparison of survey methods for crested newts (*Triturus cristatus*) and night counts at a secure site, 1983–1993. *Herpetological Journal* 5: 221–228.

Coulson, T., Mace, G. M., Hudson, E., and Possingham, H. 2001. The use and abuse of population viability analysis. *Trends in Ecology and Evolution* 16: 219–221.

Cummins, C. P., and Griffiths, R. A. (Eds.). 2000. Editorial. Scientific studies of the great crested newt: its ecology and management. *Herpetological Journal* 10(4) (Special issue).

Cummins, C. P., and Swan, M. J. S. 2000. Long-term survival and growth of free-living great crested newts (*Triturus cristatus*) PIT-tagged at metamorphosis. *Herpetological Journal* 10: 177–182.

Drost, C. A., and Fellers, G. M. 1996. Collapse of a regional frog fauna in the Yosemite area of the California Sierra Nevada, USA. *Conservation Biology* 10: 414–425.

Ellner, S. P., Fieberg, J., Ludwig, D., and Wilcox, C. 2002. Precision of population viability analysis. *Conservation Biology* 16: 258–261.

English Nature. 2001. *Great crested newt mitigation guidelines*. English Nature, Peterborough.

Fieberg, J., and Ellner, S. P. 2000. When is it meaningful to estimate an extinction probability? *Ecology* 81: 2040–2047.

Green, D. M. (Ed.). 1997. *Amphibians in decline: Canadian studies of a global problem*. Herpetological Conservation no. 1. St Louis: Society for the Study of Amphibians and Reptiles.

Griffiths, R. A., and Williams, C. 2000. Modelling population dynamics of great crested newts (*Triturus cristatus*): a population viability analysis. *Herpetological Journal* 10: 157–163.

Griffiths, R. A., and Williams, C. 2001. Population modelling of great crested newts (*Triturus cristatus*). *Rana* 4: 239–247.

Hagström, T. 1979. Population ecology of *Triturus cristatus* and *T. vulgaris* (Urodela) in SW Sweden. *Holarctic Ecology* 2: 108–114.

Halley, J. M., Oldham, R. S., and Arntzen, J. W. 1996. Predicting the persistence of amphibian populations with the help of a spatial model. *Journal of Applied Ecology* 33: 455–470.

Hanski, I. 1991a. Metapopulation dynamics: brief history and conceptual domain. *Biological Journal of the Linnean Society* 42: 3–16.

Hanski, I. 1991b. Single-species metapopulation dynamics: concepts, models and observations. *Biological Journal of the Linnean Society* 42: 17–38.

Harrison, S. 1991. Local extinction in a metapopulation context: an empirical evaluation. *Biological Journal of the Linnean Society* 42: 73–88.

Harrison, S. 1994. Metapopulations and conservation. Pages 111–128 in P. J. Edwards, R. M. May, and N. R. Webb (eds.), *Large-scale ecology and conservation biology*. Oxford, Blackwell Science.

Hedlund, L. 1990. Courtship display in a natural population of crested newts, *Triturus cristatus*. *Ethology* 85: 279–288.

Horner, H. A., and MacGregor, H. C. 1985. Normal development in newts (*Triturus*) and its arrest as a consequence of an unusual chromosome situation. *Journal of Herpetology* 19: 261–270.

Jehle, R., Arntzen, J. W., Burke, T., Krupa, A. P., and Hödl, W. 2001. The annual number of breeding adults and the effective population size of syntopic newts (*Triturus cristatus, T. marmoratus*). *Molecular Ecology* 10: 839–850.

Joly, P., Miaud, C., Lehmann, A., and Grolet, O. 2001. Habitat matrix effects on pond occupancy in newts. *Conservation Biology* 15: 239–248.

Krone, A. (Ed.). 2001. Der Kammolch (*Triturus cristatus*). Verbreitung, Biologie, Ökologie und Schutz. *Rana* 4. Rangsdorf: NABU.

Kupfer, A., and Kneitz, S. 2000. Population ecology of the great crested newt (*Triturus cristatus*) in an agricultural landscape: dynamics, pond fidelity and dispersal. *Herpetological Journal* 10: 165–171.

Kuzmin, S. L. 1994. The problem of declining amphibian populations in the Commonwealth of Independent states and adjacent territories. *Alytes* 12: 123–134.

Levins, R. 1970. Extinction. Pages 77–107 in M. Gerstenhaber (ed.), *Some mathematical questions in biology*. American Mathematical Society, Providence, R.I.

Lindenmayer, D. B., and Lacy, R. C. 1995. A simulation study of the impacts of population subdivision on the mountain brushtail possum *Trichosurus caninus* Ogilby (Phalangeridae: Marsupialia) in south-eastern Australia. I. Demographic stability and population persistence. *Biological Conservation* 73: 119–129.

Lips, K. R. 1998. Decline of a tropical montane amphibian fauna. *Conservation Biology* 102: 106–117.

Marsh, D. M., and Trenham, P. C. 2000. Metapopulation dynamics and amphibian conservation. *Conservation Biology* 15: 40–49.

May, R. 1996. The translocation of great crested newts, a protected species. M.Sc. thesis, University of Wales.

McCarthy, M. A., and Broome, L. S. 2000. A method for validating stochastic models of population viability: a case study of the mountain pygmy-possum (*Burramys parvus*). *Journal of Animal Ecology* 69: 599–607.

McCarthy, M. A., Possingham, H. P., Day, J. R., and Tyre, A. J. 2001. Testing the accuracy of population viability analysis. *Conservation Biology* 15: 1030–1038.

Miaud, C., Joly, P., and Castanet , J. 1993. Variation in age structures in a subdivided population of *Triturus cristatus. Canadian Journal of Zoology* 71: 1874–1879.

Oldham, R. S. 1994. Habitat assessment and population ecology. Pages 45–67 in T. Gent and R. Bray (eds.), *Conservation and management of great crested newts: proceedings of a symposium held on 11 January 1994 at Kew Gardens, Richmond, Surrey.* Science Reports no. 20. English Nature, Peterborough, U.K.

Oldham, R. S., and Humphries, R. N. 2000. Evaluating the success of great crested newt (*Triturus cristatus*) translocation. *Herpetological Journal* 10: 183–190.

Oldham, R. S., and Swan, M. J. S. 1997. Pond loss and amphibians: historical perspectives. Pages 3–16 in J. Boothby (ed.), *British pond landscapes: action for protection and enhancement.* PondLife Project, Lancashire, U.K.

Richards, S. J., McDonald, K. R., and Alford, R. A. 1993. Declines in populations of Australia's endemic tropical rainforest frogs. *Pacific Conservation Biology* 1: 66–77.

Sjögren-Gulve, P., and Hanski, I. 2000. Metapopulation viability analysis using occupancy models. *Ecological Bulletins* 48: 53–71.

Swan, M. J. S., and Oldham, R. S. 1993. *Herptile sites.* Vol. 1. *National amphibian survey final report.* Research Report No. 38. English Nature, Peterborough, U.K.

Thiesmeier, B., and Kupfer, A. 2000. *Der Kammolch: Ein Wasserdrache in Gefahr.* Laurenti Verlag, Bochum.

Thomas, C. D. 1994. Extinction, colonization, and metapopulations: environmental tracking by rare species. *Conservation Biology* 8: 373–378.

Thomas, C. D., and Kunin, W. E. 1999. The spatial structure of populations. *Journal of Animal Ecology* 68: 647–657.

Williams, C. 1999. Metapopulation dynamics of the crested newt (*Triturus cristatus*). Ph.D. diss., University of Kent.

26

Houston Toad (*Bufo houstonensis*) in Bastrop County, Texas

Need for Protecting Multiple Subpopulations

JEFF S. HATFIELD
ANDREW H. PRICE
DAVID D. DIAMOND
C. DIANE TRUE

The Houston toad (*Bufo houstonensis*) was listed as an endangered species in 1970 under the U.S. Endangered Species Act and is currently found in nine counties in east-central Texas (U.S. Fish and Wildlife Service 2003). This species is thought to require habitat consisting of deep sandy soils among pine (*Pinus*) or oak (*Quercus*) woodlands. The reasons for its decline are believed to be habitat loss, habitat fragmentation, competition and hybridization with other species of *Bufo* (Hillis et al. 1984), and competition or predation by introduced species (e.g., fire ants, *Solenopsis invicta*). Bastrop County, Texas, is believed to support the largest remaining population of the Houston toad. Because this area (and nearby Austin) is undergoing significant development, habitat loss and fragmentation are major concerns, especially because of the effect of roads on some amphibian populations (Hels and Buchwald 2001).

Currently, two relatively large, forested tracts of land within Bastrop County, separated by a minimum distance of 2222 m, contain breeding populations of the Houston toad (Figure 26.1). Although it is unknown whether either area will be able to support this species indefinitely, it is clear that management actions within and between these areas may make extinction less likely. For example, maintaining corridors for dispersal between these two areas might be beneficial for the long-term maintenance of this species in Bastrop County, although this option may be difficult because there is a two-lane highway separating these areas.

Methods

Population viability analysis (PVA) provides an objective tool for evaluating different management scenarios concerning the future management of a given species. RAMAS

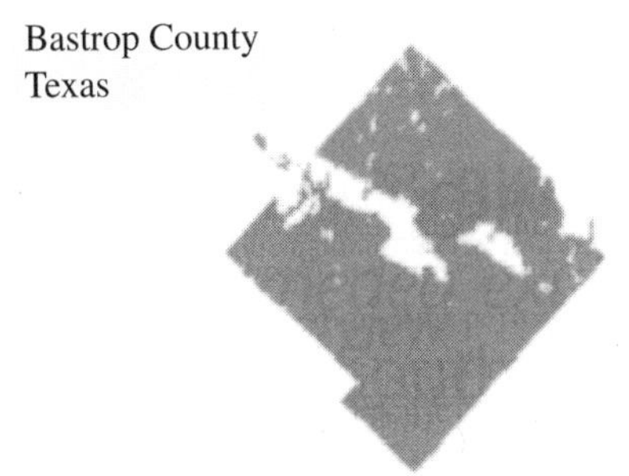

Griffith League Ranch: Total Area=1965 ha; Forest Area=1883ha

Bastrop State Park: total Area=2207 ha; forest Area=1684 ha

Figure 26.1 Forested area within Griffith League Ranch and Bastrop State Park, Texas. Both sites are known to support breeding populations of the Houston toad (*Bufo houstonensis*).

Metapop software (Akçakaya 2002) was used to conduct PVA simulations of the Houston toad because it allows for different age classes of toads to have different vital rates, such as fecundity, survival, and temporal variability (i.e., year-to-year variability in vital rates). An age-structured, prebreeding census model was chosen, with the two age classes being 1-year-olds and individuals that are 2 years old or older. Only females were simulated because information about fecundity exists for only this sex in the Houston toad, and because the ratio of males to females is much greater than 1 (A. Price, pers. comm.), implying that females are limiting. A total of 1,000 replications were run for each simulation, and each replication was projected for 100 years into the future. The probability of extinction was then estimated to be the proportion of simulations, out of 1,000 replications, that became extinct.

To estimate survival rates for the Houston toad to use in the simulations, we analyzed data from a capture-recapture study conducted at Bastrop State Park from 1990 to 2001 (A. Price, unpublished data). A variety of Cormack-Jolly-Seber models (Nichols 1992) were fit to the female capture-recapture data using the MARK program (White and Burnham 1999). Because certain parameters for the more heavily parameterized models were nonidentifiable, however, the most viable model was the one that allowed for time-specific survival and age-specific capture probability. Overdispersion was identified under this general model ($\hat{c}$ = 3.644) using a bootstrap goodness-of-fit test (800 replications). Thus, model selection was executed using quasi-likelihood Akaike information criterion (QAIC*c*), which adjusts the information criterion used in model selection and the estimated variances and covariances using the variance inflation factor (Burnham and Anderson 1998). The model that allowed for time-specific survival rates and constant capture probability (over time and among groups) was selected as the best model (i.e., the model with the lowest QAICc value, 175.752). Under this model, the mean survival rate per year was estimated to be 0.2022. To estimate the temporal variance in survival rates, we used the method of Link and Nichols (1994) as modified for survival rates by Gould and Nichols (1998), which yielded a variance estimate of 0.0194.

For an estimate of fecundity, we used data from Kennedy (1962) and Quinn and Mengden (1984) to estimate a mean clutch size of 1,772.71 eggs per female ($n = 7$ females), which implies 886.36 female eggs per female, assuming a 50:50 sex ratio. The variance of this estimate (322,817.15) was used as the estimate of the temporal variance in clutch size. This is probably a conservative estimate of temporal variance in clutch size because sampling variance is included in the estimate.

Two parameters—proportion of second-year females breeding and the survival rate of juveniles for their first year (from egg to beginning of second year)—were unknown, and therefore we ran two different scenarios for juvenile survival rate (1% and 2% for the first-year survival rate of juveniles) and three different scenarios for proportion of 1-year-old females breeding (0, 0.01, and 0.33 as the proportions of the 1-year-old females breeding). All females that were 2 years old or older were assumed to be able to breed.

We ran a variety of scenarios for the carrying capacity (K) of the number of females (1 year old and older) in the population. This allows habitat limitations to be incorporated into the simulations of the dynamics of the population. The range of carrying capacities was $K = 500, 1{,}000, \ldots, 5{,}000$ female toads, and each simulation was started with 500 females in each of these populations. The number of 1-year-old and older females in the starting population of 500 was determined by using the stable age distribution for each model.

We also ran three different scenarios of catastrophes: one scenario was no catastrophes, and the other two scenarios were a high probability of catastrophe (50% chance per year) and a low probability of catastrophe (10% chance per year). Catastrophes occur periodically when a drought or disease wipes out a major portion of the population. When a catastrophic year occurs, the result in our simulations is a 50% reduction in fecundity and survival during each catastrophic year.

Several technical assumptions, explained in detail in Akçakaya (2002), were used in our simulations. Demographic stochasticity was used, and we assumed independent, lognormal distributions of fecundity and survival as these rates varied over time.

Results

The probabilities of extinction for the various single-population models are reported in Table 26.1. Clearly, juvenile survival is important because, with a low juvenile survival rate of 1%, the single populations often have a high probability of extinction, and with a juvenile survival rate of 2%, many of the simulations have a low probability of extinction.

The proportion of 1-year-old females that breed also has a large effect, especially when juvenile survival rate is low and probability of catastrophes is high. In scenarios where the probability of extinction was high without breeding by 1-year-old females, the scenarios with breeding by 1-year-old females lowers the probability of extinction considerably.

Table 26.1 Probability of extinction in 100 years for a single population of the Houston toad (*Bufo houstonensis*), with an annual survival rate of 20%, varying the carrying capacity (K) of the total number of females that are one year old or older

	Low Juvenile Survival (1%)									High Juvenile Survival (2%)								
	0.5[a]			0.1			0			0.5			0.1			0		
K	0[b]	0.01	0.33	0	0.01	0.33	0	0.01	0.33	0	0.01	0.33	0	0.01	0.33	0	0.01	0.33
500	1.000	1.000	0.048	0.657	0.552	0.000	0.132	0.089	0.000	0.973	0.931	0.000	0.077	0.063	0.000	0.004	0.000	0.000
1000	1.000	1.000	0.028	0.531	0.448	0.000	0.074	0.044	0.000	0.928	0.881	0.000	0.041	0.027	0.000	0.000	0.000	0.000
1500	1.000	1.000	0.018	0.462	0.371	0.000	0.048	0.031	0.000	0.914	0.844	0.000	0.027	0.019	0.000	0.000	0.000	0.000
2000	1.000	1.000	0.011	0.403	0.319	0.000	0.038	0.028	0.000	0.887	0.812	0.000	0.016	0.013	0.000	0.000	0.000	0.000
2500	1.000	1.000	0.010	0.392	0.302	0.000	0.043	0.020	0.000	0.873	0.802	0.000	0.012	0.012	0.000	0.000	0.000	0.000
3000	1.000	1.000	0.006	0.374	0.302	0.000	0.037	0.023	0.000	0.854	0.755	0.000	0.013	0.011	0.000	0.000	0.000	0.000
3500	1.000	1.000	0.006	0.389	0.276	0.000	0.026	0.021	0.000	0.862	0.777	0.000	0.010	0.011	0.000	0.000	0.000	0.000
4000	1.000	1.000	0.007	0.380	0.282	0.000	0.026	0.014	0.000	0.844	0.739	0.000	0.007	0.008	0.000	0.000	0.000	0.000
4500	1.000	1.000	0.003	0.328	0.258	0.000	0.024	0.012	0.000	0.855	0.742	0.000	0.007	0.006	0.000	0.000	0.000	0.000
5000	1.000	1.000	0.005	0.325	0.237	0.000	0.019	0.010	0.000	0.816	0.748	0.000	0.013	0.009	0.000	0.000	0.000	0.000

Note: See text for detailed descriptions of these simulations.

[a]Probability per year of a catastrophe

[b]Proportion of second-year females breeding

Catastrophes have a large effect on the probabilities of extinction as well, at least in the way we simulated them, with the effect of the catastrophes particularly important for the 1% juvenile survival rate scenarios.

To investigate the effect of having multiple populations on the probability of extinction, we took one of the models and simulated it with two and three different subpopulations, each with a carrying capacity of $K = 500$ to 5,000 females. We used the model with low juvenile survival (1%), low proportion of 1-year-old females breeding (0.01), and low probability of catastrophes (0.10). This model had probabilities of extinction of 0.237 to 0.552 for a single population with $K = 500$ to 5,000 females (see Table 26.1, column 5, and Figure 26.2, upper curve). We assumed 1% of individuals dispersing between populations and 50% correlation among vital rates in these simulations. The overall probability of extinction dropped considerably, as shown in Figure 26.2, when there was a metapopulation consisting of two or three subpopulations with dispersal among these subpopulations.

Conclusions

The results and conclusions presented here depend on the assumptions we made in our simulations, as well as on the parameter estimates we used. In particular, for the metapopulation simulations, we assumed the vital rates of the additional populations would be the same—in particular, that none of the populations were "sinks" with low average vital rates. Also, dispersal among subpopulations would need to be at least as high as we assumed in this model, and since this parameter has not been estimated for this species, we do not know if our assumption is valid. Furthermore, although we assumed a relatively high level of correlation among vital rates of the subpopulations, it is conceivable that it could be even larger due to climate or disease or other factors that could affect all subpopulations simultaneously. The result of greater correlation among vital rates of the subpopulations would be higher probabilities of extinction for the metapopulation models portrayed in Figure 26.2.

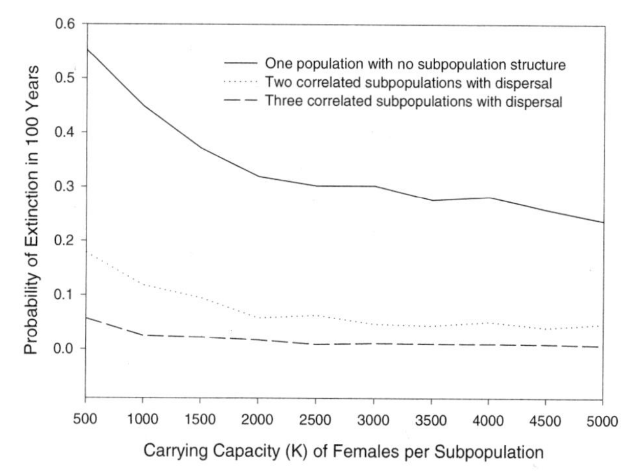

Figure 26.2 Probability of extinction in 100 years versus the carrying capacity (K) of females for a single population of the Houston toad (*Bufo houstonensis*), and metapopulatons consisting of either two correlated subpopulations with some dispersal or three correlated subpopulations with some dispersal. See the text for detailed descriptions of the simulations presented here.

However, several recommendations of model revision result from these exploratory simulations. First, estimating the juvenile survival rate of females from egg to 1 year old would help us to determine whether our simulations are closer to the low or high juvenile survival scenarios. It might be possible to estimate this survival rate by estimating the number of eggs laid at a given pond, estimating the number of toadlets that metamorphose from the pond, marking the toadlets, and using capture-recapture methods to estimate their survival rate to the beginning of their second year (see Reading 1991 for such a study on the common toad, *Bufo bufo*). Also, management activities geared toward improving juvenile survival rate, such as increasing the number of breeding ponds and removal of predators from the ponds, should lower the probability of extinction.

Second, estimating the proportion of 1-year-old females that breed is also important, and this might be possible with a more intensive field effort in the capture-recapture studies (Reading 1991). It would also be useful to estimate vital rates, dispersal, and correlation among vital rates for at least two subpopulations, rather than just the one population at Bastrop State Park, because then we could determine whether our assumptions for the metapopulation simulations are reasonable.

However, given the simulations we have performed, and the caveats stated here, it seems that a relatively low probability of extinction (< 5% per 100 years) might be achieved in Bastrop County if two or three subpopulations are protected and managed for Houston toads, each with habitat to support several thousand 1-year-old and older female toads, and allowing for dispersal of individuals among the subpopulations. The probability of dispersal between areas should be studied, but if protection of natural corridors for dispersal is not an option, then some mechanism should be put into place to allow for artificial dispersal among the subpopulations. For example, if one area was having a good year for productivity and another area was not, it would be relatively simple to translocate eggs, larvae, or toadlets from one area to another. However, transmission of disease among subpopulations may be a concern if this type of management is undertaken in the future (Seigel and Dodd 2002).

Acknowledgments Thanks go to James T. Dureka for helping to produce Figure 26.1, to William R. Gould for analyzing the capture-recapture data, to Lisa O'Donnell for suggesting we do this study, and to two anonymous reviewers for providing helpful comments.

References

Akçakaya, H. R. 2002. *RAMAS Metapop: viability analysis for stage-structured metapopulations (version 4.0)*. Applied Biomathematics, Setauket, N.Y.

Burnham, K. P., and Anderson, D. R. 1998. *Model selection and inference: a practical information-theoretic approach.* Springer, New York.

Gould, W. R., and Nichols, J. D. 1998. Estimation of temporal variability of survival in animal populations. *Ecology* 79: 2531–2538.

Hels, T., and Buchwald, E. 2001. The effect of road kills on amphibian populations. *Biological Conservation* 99: 331–340.

Hillis, D.M., Hillis, A. M., and Martin, R. E. 1984. Reproductive ecology and hybridization of the endangered Houston toad (*Bufo houstonensis*). *Journal of Herpetology* 18: 56–72.

Kennedy, J. P. 1962. Spawning season and experimental hybridization of the Houston toad, *Bufo houstonensis. Herpetologica* 17: 239–245.

Link, W. A., and Nichols, J. D. 1994. On the importance of sampling variance to investigations of temporal variation in animal population size. *Oikos* 69: 539–544.

Nichols, J. D. 1992. Capture-recapture models. *Bioscience* 42: 94–102.

Quinn, H. R., and Mengden, G. 1984. Reproduction and growth of *Bufo houstonensis* (Bufonidae). *Southwestern Naturalist* 29: 189–195.

Reading, C. J. 1991. The relationship between body length, age and sexual maturity in the common toad, *Bufo bufo*. *Holarctic Ecology* 14: 245–249.

Seigel, R. A., and Dodd Jr., C. K. 2002. Translocations of amphibians: proven management method or experimental technique? *Conservation Biology* 16: 552–554.

U.S. Fish and Wildlife Service. 2003. Houston toad recovery plan (revised). U.S. Fish and Wildlife Service, Albuquerque, New Mexico. Unpublished report.

White, G. C., and Burnham, K. P. 1999. Program MARK: survival estimation from populations of marked individuals. *Bird Study* 46(Suppl.): S120–139.

27

Eastern Indigo Snakes (*Drymarchon couperi*) in Florida

Influence of Edge Effects on Population Viability

DAVID R. BREININGER
MICHAEL L. LEGARE
REBECCA B. SMITH

The eastern indigo snake (*Drymarchon couperi*) is threatened with extinction because of direct mortality caused by humans and rapid urbanization in Florida, which comprises most of the current range (U.S. Fish and Wildlife Service 1998). The eastern indigo snake is the longest nonvenomous snake in the United States; it eats a variety of prey such as snakes, small mammals, birds, and frogs. There is little published regarding its population ecology, and there is no reliable method to survey populations because the species is a wide-ranging habitat generalist that occurs in low densities and which uses gopher tortoise (*Gopherus polyphemus*) burrows, debris piles, or dense vegetation for its cover.

The killing of snakes by humans is a worldwide concern (Dodd 1987). Rudolph et al. (1999) reported that densities of large snakes were low, within 450 m of roads, and increased steeply up to 850 m from roads without evidence of an asymptote. The population viability of wide-ranging animals that undergo conflicts with people along edges of conservation areas is not just a function of population size (Woodroffe and Ginsberg 1998). Increased mortality along boundaries and emigration of animals into human-dominated landscapes that are ecological traps can diminish population persistence (Stamps et al. 1987).

Our objectives were to stimulate a population-based dialog for conserving indigo snake populations and to prioritize data needed to quantify the effects of habitat fragmentation. Objectives were also to investigate the possibilities of using existing information to evaluate the risks of urbanization and the possibilities for establishing viable eastern indigo snake populations before all of their habitat becomes too fragmented.

Methods

Geographic Population Structure

Our study included the upper St. Johns River basin east to the Atlantic Ocean. The area was generally rural west of the St. Johns River and urban to the north and east (Figure 27.1). We generated an ARCVIEW shapefile (Environmental Systems Research Institute 1999) comprised of 120-ha grid cells, representing the average size of an adult male home range (M. Legare et al., unpublished data). We overlaid the grid cells on ARCVIEW shapefiles of landcover, roads, and potential conservation areas that we downloaded from the Florida Department of Environmental Protection web site. We classified grid cells as suitable habitat if they were dominated by natural communities or grazing land (but not open water or marshes) and if primary roads did not intersect them. During telemetry studies we found that indigo snakes living along primary roads

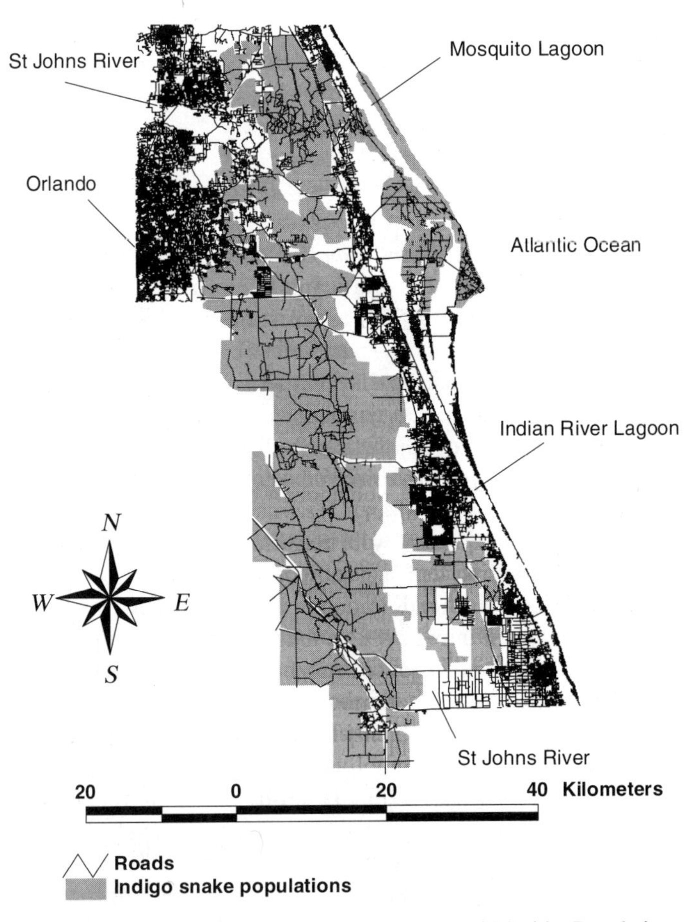

Figure 27.1 Potential indigo snake populations in east central Florida. Populations were subdivided by highways, urban areas, large marshes (St. Johns River), open water (e.g., Indian River lagoon), and proposed conservation status. The roads provided a general measure of urbanization.

soon died (Legare et al., unpublished data). We classified suitable habitat cells into separate populations by grouping contiguous suitable habitat cells into the same population, except that we considered groups of contiguous suitable habitat cells within and outside potential population reserves as separate populations to evaluate management scenarios. We generated 54 populations with >10 contiguous cells and 36 small populations with <10 contiguous cells separated from other populations by >1 grid cell of unsuitable habitat. We deleted small populations to simplify analyses by assuming that all habitat associated with these grids would soon be destroyed or would become too isolated to help sustain a regional population. Our telemetry studies suggest occupancy is temporary in such habitat fragments because mortality rates are great. In fragments, we usually find mostly females, probably because the wider ranging males already died (Legare et al., unpublished data). Dispersal in snakes is presumed low, with minimal exchanges among populations (Turner 1977, Parker and Plummer 1987). Even short distances (e.g., 1.6 km) of hostile habitat can restrict gene flow (Prior et al. 1997) and result in elevated mortality (Bonnett et al. 1999).

Edge Effects

Attributes identified for every cell included the presence or absence of suitable habitat, a unique population identifier, potential conservation land status, the identifier of any adjacent population, and edge attributes (Table 27.1). We defined edge effects as one-way movements of snakes from populations into ecological traps comprised of primary roads or human-dominated areas. We assumed that edge effects and exchanges among populations were proportional to the number of grid cells that occurred along population boundaries. The sum of all suitable habitat cells within a population i was K_i. Along the largest highways, we identified bordering cells as urban/road edge (Ce) except at bridges or culverts where we identified possible dispersal (Cur, Cpr).

Population Model and Sensitivity Analyses

Within-population dynamics were modeled using a sex- and stage-structured matrix of annual demographic rates (Table 27.2). The matrix was parameterized according to a prereproductive census involving 1-year-olds (young), 2-year-olds (subadults), 3-year-olds (possible breeding adults), and 4+-year-olds (breeders). The default stage struc-

Table 27.1 Edge attributes used to classify each grid cell

Edge Category	Characteristics
Interior (Cn)	Bordered by suitable habitat, open water, or large marshes
Urban/road edge (Ce)	Bordered a primary road or urban area
Reserve edge (Cp)	Bordered by another population that was a potential reserve
Road edge/reserve (Cpr)	Bordered by a road with a potential reserve on the other roadside
Unprotected population (Cu)	Bordered a population not proposed for conservation
Road/unprotected population (Cur)	Bordered a road with a population not proposed for conservation on the other roadside

Note: The edge category was the total number of cells for that category in every population.

Table 27.2 Stage structure matrix

	Female 1	Female 2	Female 3	Female 4+	Male 1	Male 2	Male 3	Male 4+
Female 1	0.0	0.0	0.5 S_hMP_3	0.5 S_hMP_4	0.0	0.0	0.0	0.0
Female 2	Sf_1	0.0	0.0	0.0	0.0	0.0	0.0	0.0
Female 3	0.0	Sf_2	0.0	0.0	0.0	0.0	0.0	0.0
Female 4+	0.0	0.0	Sf_3	Sf_4	0.0	0.0	0.0	0.0
Male 1	0.0	0.0	0.5 S_hMP_3	0.5 S_hMP_4	0.0	0.0	0.0	0.0
Male 2	0.0	0.0	0.0	0.0	Sm_1	0.0	0.0	0.0
Male 3	0.0	0.0	0.0	0.0	0.0	Sm_2	0.0	0.0
Male 4+	0.0	0.0	0.0	0.0	0.0	0.0	Sm_3	Sm_4

Note: Sf_x = the survival rate of female x-year-olds; Sm_x = the survival rate of male x-year-olds; S_h = survival of hatchlings to the pre-breeding census period; M = maternity (number of male and female hatchlings per female); P_x = the proportion of x-year old females that bred.

ture included least favorable, best estimates, and most favorable estimates of parameters to incorporate their uncertainties (Table 27.3). Most empirical data specific to indigo snakes included clutch sizes, the proportion of adults capable of breeding, the proportion of females with eggs, annual adult survival, habitat use, home range characteristics, and movements (Smith 1987; Moler 1985; Layne and Steiner 1996; Speake et al. 1978; Legare et al., unpublished data). Little data on indigo snakes were available for recruitment and hatchling survival, so we relied on information from other late-

Table 27.3 Categories to simulate uncertainties in model parameters

Model Assumption	Least Favorable	Best Estimate	Most Favorable
Number of females a male can mate with	2	3	6
Proportion of female 3-year-olds breeding (P_3)	0	0.25	0.50
Proportion of female 4+-year-olds breeding (P_4)	0.41	0.76	1.00
Proportion of male 3-year-olds breeding (P_3)	0.80	0.90	1.00
Proportion of male 4+-year-olds breeding (P_4)	0. 80	0.90	1.00
Hatchlings/breeder female (M)	3	4.5	6
Survival of hatchlings to first breeding census (S_h)	0.14	0.34	0.55
Young survival (S_1)	0.55	0.70	0.88
Subadult survival (S_2)	0.75	0.88 (females), 0.80 (males)	0.95
Adult female survival (S_{f3}, S_{f4+})[a]	0.75	0.88 (0.67)	0.95
Adult male survival (S_{m3}, S_{m4+})[a]	0.75	0.80 (0.53)	0.95
Edge effects (Em)	1.00	0.50	0.00
Proportion that cross roads (Dr)	0.0	0.5	1.0
Proportion that disperse (Pd)	0.0	0.5	1.0
Stages influenced by ceiling	All stages	Adults, subadults	Adults only
Stages that disperse	Adults only	Adults, subadults	All stages
Carrying capacity/cell (N_k)	2	6	10
Correlation in vital rates among populations	1.00	0.50	0.00
Annual fecundity SD	0.50	0.30	0.10
CV for survival	0.15	0.10	0.06

[a]Values outside parentheses represented annual survivals in natural landscapes and were used for best and intermediate management scenarios (see text). Values inside parentheses represented annual survivals in small habitat fragments and were used for the least favorable management scenario.

maturing colubrids (e.g., Turner 1977, Parker and Plummer 1987). We assumed fecundity ranged between the low and midpoint of late-maturing colubrids because some indigo snake eggs are infertile, some predation almost certainly occurs, and the mean indigo clutch size is near the midpoint of fecundity for late-maturing colubrids.

We used data on edges (Table 27.1) to compute population-specific survival and dispersal rates for each of the 54 populations, or only 30 populations in some management scenarios (see the following discussion in "Analyses of Uncertainty and Landscape Management Scenarios"). For every population, we simulated edge effects using relative survival (RS_i), which was calculated as 1 minus the product of the edge mortality rate (Em) (Table 27.3) and the proportion of edge home ranges E_i in population i. A population-specific E_i was calculated using data described in Table 27.2 by the equation

$$E_i = \{Ce + Cu + [(1 - Dr)*(Cpr + Cur)]\} / K_i$$

where Dr was the proportion of snakes that successfully dispersed beyond primary roads into an adjacent population and 1 – Dr was the snakes that died.

Dispersal rates between populations D_{ij} were calculated using the equation

$$D_{ij} = Pd*[(Cp + Cu) + Dr*(Cpr + Cur)]$$

where Pd was the proportion of snakes in edge cells that dispersed (Table 27.3).

There was some evidence of cannibalism and evidence that home ranges of adults of the same sex had little overlap (Layne and Steiner 1996; Legare et al., unpublished data). Therefore, density dependence was implemented using a ceiling on carrying capacity following the advice of Ginzburg et al. (1990) to not use other density dependence (i.e., scramble or contest) formulations because we lacked evidence that populations would recover from low densities by mechanisms other than those explicitly described by the stage matrix. The product of N_k and K_i provided the specific ceiling for every population. Our best estimate was that one grid cell could support a combined total of six adults and subadults. Approaches to incorporate the uncertainties in density dependence included varying which stages were influenced by a ceiling and by varying the number of snakes (N_k) (Table 27.3) that could be supported by each K_i. When investigating uncertainty in which stages were influenced by density dependence, we kept the influence of K constant for each population using the stable age distribution. We assumed that each cell could support 4.3 adults if only adults were affected by a ceiling, and we assumed that each cell could support 8.9 individuals if all stages were affected. When we simulated the influence of K, we assumed only adults and subadults were affected by varying values of K.

Analyses of Uncertainty and Landscape Management Scenarios

We distinguished among a broad range of scenarios, given great uncertainty about indigo snake population biology and the landscape possibilities that existed. The most favorable management alternative involved no further habitat fragmentation or destruction within the 54 populations. We began the simulations assuming that all snakes outside the 54 populations were dead because of urbanization. The intermediate management alternative assumed that all 30 populations in potential conservation reserves were protected from further fragmentation but that snakes in the other 24 populations were dead and all habitat was destroyed. The least favorable management alternative also considered only the 30 populations in potential conservation lands but assumed that these populations would become fragmented by roads to accommodate the rapidly expanding human population. To simulate the least favorable management alternative, we used adult survival rates equivalent to those we measured in fragmented landscapes (Table 27.3). We also assumed that subadult survival rates were the same as adult survival rates in fragmented landscapes.

We used the intermediate management scenario to compare the uncertainties of model assumptions. To investigate which uncertainties had the greatest influence, we used the best estimate for all model assumptions and varied one assumption at a time using the least favorable or most favorable estimates. We investigated differences in the three management scenarios by using least favorable, best, and most favorable estimates for edge effects (Em). We used Em as one example of how uncertainty influenced management scenarios because it had direct management applications. For the intermediate management scenario, we used a range of edge effects (Em) to investigate how population size and edge/area ratios (E_i) influenced vulnerabilities of decline. We used Spearman rank correlations (Statistical Package for the Social Sciences 1999) to determine whether the population occupancy was influenced by area (K_i) or edge/area (E_i) for the intermediate scenario using least favorable, best, and most favorable values of edge effects (Em).

Each simulation consisted of 1,000 replications. Simulations were initialized using a total abundance of all snakes that was equal to the product of K_i and N_k in each subpopulation. The stable stage distribution of the best-estimate matrix was used to determine the relative proportion of individuals in each stage to initialize individual stage abundances. The initial numbers of adults and subadults were almost always at 65% of their carrying capacity because the population ceiling was applicable only to subadults and adults for nearly all simulations, and because 35% of all individuals in the stable stage distribution were young. Initializing adults and subadults below carrying capacity allowed for population growth from the initial distribution under favorable circumstances.

Results

Few populations exceeded 120,000 ha because of extensive fragmentation by primary roads (Figure 27.2). The Spearman rank correlation coefficient between K_i and E_i was only –0.39 ($p = 0.003$). Many smaller populations had a small E_i because they bordered large marshes or open water; several middle-sized populations had high edge/area ratios. Initial population sizes ranged from 42 to 2,504 adults and subadults when all populations were considered.

Approximately half of the model assumptions influenced uncertainty, and the other half had little influence. Quasi extinction (decline below 10 adults and subadults) occurred in many populations that were potential reserves regardless of the assumptions (Table 27.4). Results of assumptions with great influence ranged from no risk of overall population decline to great risk of substantial population decline and quasi extinction of most populations.

The least and most favorable management alternatives were distinct, assuming that variation in Em captured uncertainty (Table 27.5). When Em was 0.5, the respective risks of the metapopulation declining by 40% at the end of the simulation (terminal risk

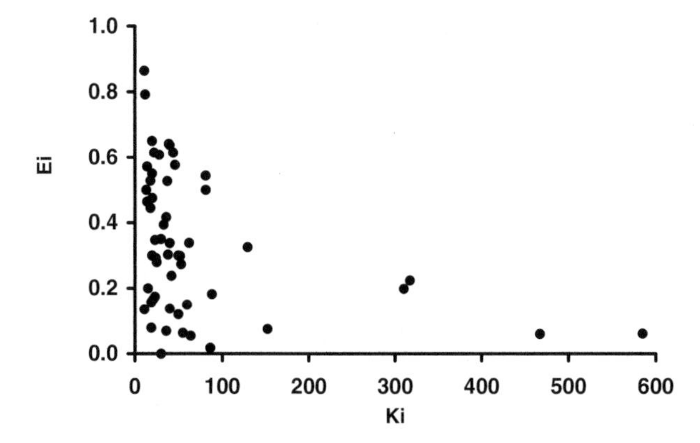

Figure 27.2 The number of grid cells (K_i) and edge/area ratios (E_i) of all 54 populations. Each grid cell was 120 ha and could provide habitat for 2 to10 eastern indigo snakes. Edges refer to urban areas and primary roads.

of decline) were 0.02 and 0.65, respectively, for the best and intermediate management scenarios. There was a reduction in the occupied range and increased population fragmentation for the intermediate management scenario when best estimates for all parameters were applied (Figure 27.3). However, indigo snake population viability was greatly influenced by uncertainty in edge effects for the intermediate management scenario. Simulations for the least favorable management scenario showed that populations had little chance of persisting if adults had the survival rates we measured for fragmented populations.

For the intermediate scenario, Spearman correlation coefficients between population values of the number of grid cells (K_i) and the mean number of time steps occupied were 0.38 ($p = 0.036$) and 0.45 ($p = 0.013$) for respective values of edge effects (Em) of 0.5 and 1.0. Spearman correlation coefficients between population values of edge/area ratios (E_i) and the mean number of time steps occupied were –0.85 ($p < 0.001$) and –0.89 ($p < 0.001$) for respective values of edge effects (Em) of 0.5 and 1.0. All populations had mean occupancy rates of 50 when Em = 0.0, so we only presented results for least favorable and best estimates of E_i (Figure 27.4).

Discussion

We know of no population viability analyses (PVA) published on snakes that we can compare this model to, although the use of PVAs for snake conservation has long been recommended (Dodd 1987). We report quantitative results but emphasize the comparison of management scenarios because there was much uncertainty in model assumptions and the results were not proven with empirical data. Even though it is impossible to restore the indigo snake population to its former range, it appeared that indigo snake populations in east central Florida had low extinction risk if there was no further fragmentation of the remaining populations. The opportunities to set aside large indigo snake populations were greater than exist for many listed species (see Elphick et al. 2001). The least favorable management scenario suggested that indigo snake populations would decline steeply and have high extinction risk if populations were restricted to the proposed conservation reserves and if these became fragmented. Rapid urbanization makes

Table 27.4 Comparative uncertainties of model assumptions for the intermediate management scenario where only proposed conservation reserves were protected

Assumption	Least Favorable Value			Most Favorable Value		
	Interval Percent Decline			Interval Percent Decline		
	40%	80%	Metapopulation Occupancy	40%	80%	Metapopulation Occupancy
Hatchling survival (S_h)	1.00	1.00	2	0.13	<0.01	21
Adult female survival	1.00	1.00	4	0.60	<0.01	19
Edge effects (Em)	1.00	1.00	5	<0.01	<0.01	30
Proportion of 4-year-old females that breed	1.00	1.00	5	0.67	<0.01	19
Hatchlings/clutch	1.00	0.89	9	0.55	<0.01	19
Probability of crossing a road (Dr)	1.00	0.07	10	0.96	0.01	17
Carrying capacity/cell (N_k)	1.00	0.06	11	0.99	0.16	17
Young survival (S_1)	1.00	0.48	12	0.81	<0.01	18
Proportion that disperse (Pd)	1.00	0.02	12	0.99	0.07	18
Subadult survival	1.00	0.26	14	0.97	0.01	17
Annual fecundity SD	1.00	0.21	15	0.89	<0.01	17
Proportion of 3-year-old females that breed	1.00	0.08	15	0.69	<0.01	19
Stages influenced by ceiling	1.00	0.08	16	0.84	0.02	16
Correlation in vital rates among populations	0.99	0.13	16	0.99	<0.01	16
Coefficient of variation for annual survival	1.00	0.11	16	0.98	<0.01	17
Number of females a male can mate with	0.99	0.05	16	0.99	0.04	16
Adult male survival	0.99	0.04	16	0.97	0.02	17
Stages that disperse	0.99	0.04	16	0.99	0.04	16
Proportion of 4-year-old males that breed	0.99	0.04	16	0.99	0.04	16
Proportion of 3-year-old males that breed	0.99	0.04	16	0.99	0.04	16

Note: Each simulation started with 30 populations and ran for 50 years where each assumption was varied by the least and most favorable estimates while using the best estimate for all other assumptions. "Interval percent decline" referred to the probability that adults and subadults decline by 40% and 80% at least once during 50 years. "Metapopulation occupancy" referred to the number of populations with >10 adults and subadults at the end of the simulation. The probabilities were 0.99 and 0.04, respectively, for 40% and 80% interval percent declines; metapopulation occupancy was 16 when best estimates of all model assumptions were applied.

Table 27.5 Comparing management scenarios using three levels of uncertainty regarding edge effects (Em)

	All Populations Protected (n = 54)[a]			All Proposed Reserves Protected (n = 30)[b]			All Proposed Reserves Fragmented (n = 30)[b, c]		
	Em = 0.0	Em = 0.5	Em = 1.0	Em = 0.0	Em = 0.5	Em = 1.0	Em = 0.0	Em = 0.5	Em = 1.0
Starting population[d]	15608	15608	15608	6794	6794	6794	6794	6794	6794
Ending population[d]	21325	13305	7820	9404	3557	460	86	0	0
Risk of 40% decline[e]	<0.01	0.29	1.00	<0.01	0.99	1.00	1.00	1.00	1.00
Risk of 80% decline[e]	<0.01	<0.01	0.01	<0.01	0.04	1.00	1.00	1.00	1.00
Risk of declining below 5,000[f]	<0.01	<0.01	0.20	<0.01	1.00	1.00	1.00	1.00	1.00
Risk of declining below 500[f]	<0.01	<0.01	<0.01	<0.01	<0.01	0.77	0.99	1.00	1.00
Number of populations occupied after 50 years[g]	54	40	18	30	16	5	2	0	0

[a]Populations represented contiguous suitable habitat >1200 ha.

[b]Proposed reserves included existing and proposed public lands designated in the Florida Department of Environmental Protection website as potential conservation lands.

[c]Fragmented populations are subject to adult survival rates equal to data from areas with low to moderate human housing densities (Legare et al., unpublished data).

[d]Adults and subadults (excludes hatchlings and young) at the start and end of 50-year simulations.

[e]Percent chance of declining to the specified percentage of the starting population at least once during the simulation (interval risk of decline).

[f]Percent chance of declining to the specified population size during the simulation (interval quasiextinction probability).

[g]Number of populations with >10 adults and subadults at the end of the simulation.

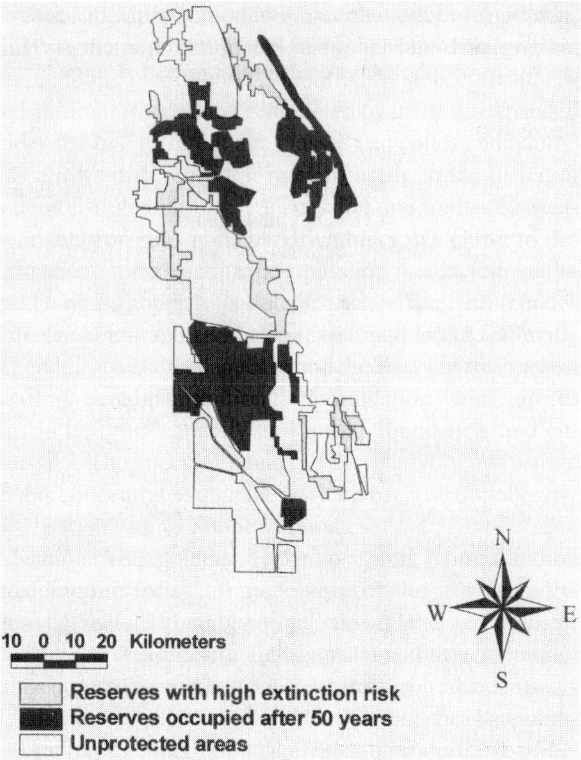

Figure 27.3 Occupied populations (with >10 adults and subadults) and unoccupied populations at the end of 50-year simulations for the intermediate landscape scenario where only proposed lands were protected. All habitat in unprotected areas was assumed to have been destroyed before the simulations started.

this scenario realistic because acquisition of these lands is incomplete. Furthermore, most lands have multiple use objectives, and permanent wildlife conservation is not the primary reason for acquisition.

Our model required the assumption that indigo snakes shared similar life histories to other late-maturing colubrids, which needs testing. The rate of habitat fragmentation and a low priority rating for funding indigo snake studies point out that conservation measures may need implementation before all data are available. There was too much uncertainty to depend only on proposed reserves to support viable indigo snake populations, particularly because the best estimates of model parameters suggested that populations remaining within reserves would undergo substantial declines. Several populations not proposed for conservation were large and had value in connecting other public lands (Figure 27.4). It would be reasonable to investigate the potential conservation value of these large unprotected areas. Many other species would benefit by increased conservation of the uplands adjacent to the protected St. Johns River marshes.

The most important functions of PVA are often to prioritize research and monitoring (Breininger et al. 2002). Parker and Plummer (1987) review that survival from fertilization until 1 year later is the most glaring data deficiency in snake population studies. Our results were consistent with this data gap because we observed that collecting annual survival data on hatchlings represented the most important parameter needing study.

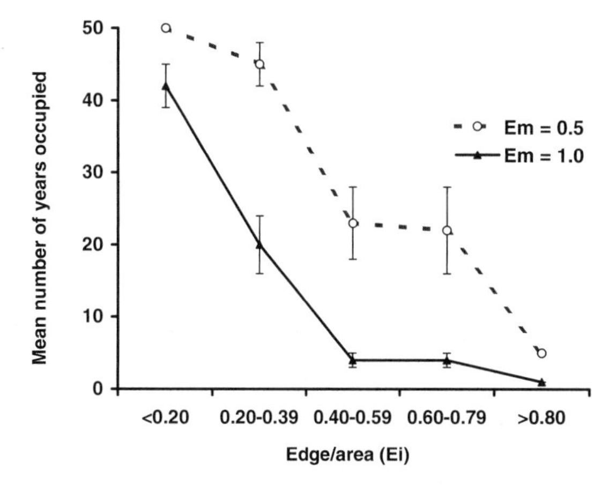

Figure 27.4 Population occupancy for 5 categories of edge/area (Ei) at two levels of edge effects (Em) for the intermediate management scenario. Occupied populations had >10 adults and subadults after 50-year simulations. Bars are standard errors.

We observed that female breeder survival, the proportion of 3+-year-old females that breed, and the number of hatchlings produced per female were also important, indicating that additional radio tracking studies of adult females are also needed. These studies should emphasize tracking young and females along the edges and interiors of potential reserves while considering a range of road densities and types. The indigo snake might be a good umbrella species to quantify fragmentation effects because they are wide-ranging and have reduced dispersal abilities in landscapes fragmented by humans. Additional sensitivity analyses studying the interaction effects among model parameters might also enhance the prioritization of research.

Simulations suggested that edge/area relationships greatly influenced indigo snake population occupancy and that edge/area effects were more important than area alone. The greater influence of edge/area effects relative to area effects alone was consistent with other models (Stamps et al. 1987) and empirical studies (Woodroffe and Ginsberg 1998). It might be possible to test such hypotheses by comparing captures of large snakes using traps developed by Rudolph et al. (1999). Preliminary studies suggest these traps are suitable for capturing indigo snakes (M. Legare, unpublished data). We could have added additional detail to include a relationship that describes the tendency of snakes to emigrate from a patch based on the characteristics of the edge (Stamps et al. 1987), but we lacked necessary data. Our preliminary studies suggested that indigo snakes readily enter urban areas and cross roads, but more intensive radio tracking studies might be able to quantify how edge characteristics influence survival and permeability. Research is increasing on techniques to discourage biota from entering highways except at safe corridors (Forman 1999). Given that mitigative measures might be expensive and have limited success, the greatest benefit might be to begin to conserve snake populations in the largest upland systems that connect other large reserves while keeping edge/area ratios low. These should be coupled with extensive research and monitoring, given that extensive habitat fragmentation and urbanization will continue and our knowledge of long-term effects is poor. Our results indicate that improving the recovery efforts is warranted because the species has great potential to remain viable but such potential is diminishing.

Acknowledgments The Bailey Wildlife Foundation, National Aeronautics and Space Administration, and United States Fish and Wildlife Service funded this study and related empirical work. We thank R. Akçakaya, D. Alessandrini, G. Bailey, M. Bailey, J. Berish, M. Burgman, K. Gorman, C. Hall, R. Hinkle, W. Knott III, L. LaClaire, P. Moler, R. Seigel, P. Sjögren-Gulve, B. Summerfield, and two anonymous reviewers. We also thank about 50 natural resource professionals who helped us capture indigo snakes for radio tracking studies.

References

Bonnett, X., Naulleau, G., and Shine, R. 1999. The dangers of leaving home: dispersal and mortality in snakes. *Biological Conservation* 89: 39–50.

Breininger, D. R., Burgman, M. A., Akçakaya, H. R., and O'Connell, M. O. 2002. Use of metapopulation models in conservation planning, Pages 405–427 in K. J. Gutzwiller (ed.), *Concepts and applications of landscape ecology in biological conservation*. Springer-Verlag, New York.

Dodd, C. K., Jr. 1987. Status, conservation, and management. Pages 478–512 in R. A. Seigel, S. S. Novak, and J. T. Collins (eds.), *Snakes: ecology and evolutionary biology*. McGraw-Hill, New York.

Elphlick, C. S., Reed, J. M., and Marcelo Bonta, J. 2001. Correlates of population recovery goals in endangered birds. *Conservation Biology* 15: 1285–1291.

Environmental Systems Research Institute (ESRI). 1999. *Using ARCVIEW GIS guide 3.2: the geographic information systems software*. Environmental Systems Research Institute, Redlands, Ca.

Forman, T. T. 1999. Spatial models as an emerging foundation of road system ecology and a handle for transportation planning and policy. Pages 119–123 in G. L. Evink, P. Garrett, and D. Ziegler (eds.), *Proceedings of the Third International Conference on Wildlife Ecology and Transportation*. FL-ER-73–99. Florida Department of Transportation, Tallahassee.

Ginzburg, L. R., Ferson, S., and Akcakaya, H. R. 1990. Reconstructibility of density dependence and the conservative assessment of extinction risks. *Conservation Biology* 4: 63–70.

Layne, J. N., and Steiner, T. M. 1996. *Eastern indigo snake* (Drymarchon corais couperi)*: summary of research conducted on Archbold Biological Station*. Report no. 43910–6–0134. U.S. Fish and Wildlife Service, Jackson, Miss.

Moler, P. E. 1985. *Home range and seasonal activity of the eastern indigo snake* (Drymarchon corais couperi), *in northern Florida*. Final performance report, Study E-1–06, III-A-5. Florida Game and Freshwater Fish Commission, Tallahassee.

Parker, W. S., and Plummer, M. V. 1987. Population ecology. Pages 253–301 in R. A. Seigel, S. S. Novak, and J. T. Collins (eds.), *Snakes: ecology and evolutionary biology*. McGraw-Hill, New York.

Prior, K. A., Gibbs, H. L., and Weatherhead, P. J. 1997. Population genetic structure in the black rat snake: implications for management. *Conservation Biology* 11: 1147–1158.

Rudolph, C., Burgdorf, S., Conner, R., and Schaefer, R. 1999. Preliminary evaluation of the impact of roads and associated vehicular traffic on snake populations in eastern Texas. Pages 129–136 in G. L. Evink, P. Garrett, and D. Ziegler (eds.), *Proceedings of the Third International Conference on Wildlife Ecology and Transportation*. FL-ER-73–99. Florida Department of Transportation, Tallahassee.

Smith, C. R. 1987. Ecology of juvenile and gravid Eastern Indigo Snakes in North Florida. Ms. thesis, Auburn University.

Speake, D. W., McGlincy, J. A., and Colvin, T. R. 1978. Ecology and management of the Eastern Indigo Snake in Georgia: a progress report. Pages 64–73 in R. R. Odum and L. Landers (eds.), *Proceedings of Rare and Endangered Wildlife Symposium*. Game and Fish Division Technical Bulletin, WL4. Georgia Department of Natural Resources, Atlanta.

Stamps, J. A., Buechner, M., and Krishnan, V. V. 1987. The effects of edge permeability and habitat geometry on emigration from patches of habitat. *American Naturalist* 129: 533–552.

Statistical Package for the Social Sciences (SPSS). 1999. *SPSS/PC+Statistics. Version 9.0*. SPSS, Chicago.

Turner, F. B. 1977. The dynamics of populations of Squamates, Crocodilians, and Rhynchocephalians. Pages 157–264 in C. Gans and D. W. Tinkle (eds.), *Biology of the reptilia*. Academic Press, New York.

U.S. Fish and Wildlife Service. 1998. Eastern indigo snake. Pages 4.579–4.581 In *Multispecies recovery plan for the threatened and endangered species of South Florida*. Vol. 1. U.S. Fish and Wildlife Service, Atlanta, Ga.

Woodroffe, R., and Ginsberg, J. R. 1998. Edge effects and the extinction of populations inside protected areas. *Science* 280: 2126–2128.

28

Frillneck Lizard (*Chlamydosaurus kingii*) in Northern Australia

Determining Optimal Fire-Management Regimes

BARRY W. BROOK
ANTHONY D. GRIFFITHS

Fire is a dominant factor determining the distribution and abundance of fauna and flora in northern Australia. It occurs regularly over extensive areas of the savanna-woodland environment. For instance, at present, over 50% of the landscape of the "Top End" of the Northern Territory is burned annually, and only 12% of the entire area of this region remains unburned for more than 6 years (Williams et al. 2001). Fire is, and has been, used extensively by Aboriginal people to manage this landscape (Bowman 1998). The exodus of Aboriginal people from many parts of the country over the last 50 to 100 years after European colonization has dramatically changed fire regimes in these areas, almost certainly altering the timing and increasing the frequency and extent of bushfires.

Fire is one of the main tools used in the management of wildlife by government conservation agencies in northern Australia (Press et al. 1995). It can be used to manipulate habitat in the vast open landscapes of northern Australia, and every year it affects at least 40% to 50% of the 20,000 km^2 area that encompasses Kakadu National Park (Russell-Smith et al. 1997). The main natural ignition source for fire in the Top End is lightning strikes, which occur predominantly in the late dry season (September to December). Yet the majority of fires are deliberately lit, mainly by land managers (i.e., National Parks staff and pastoralists) as prescribed fires, but also by residents (both Aboriginal and non-Aboriginal) and visitors. Land managers regularly use aerial ignition (from light planes or helicopters) and ground ignition (using roads and tracks for access) plus firebreaks as the most commonly used methods of ignition (Dyer et al. 2001). Most previous biological studies in the region, particularly those on reptiles, have focused on community composition of fauna in relation to timing, intensity, and frequency of fire in savanna environments. This work has demonstrated the importance of habitat structure and environmental gradients, interacting with different fire variables, in determining species distribution and abundance (see Braithwaite 1987, Trainor and Woinarski 1994, Woinarski et al. 1999). However, to be able to examine these factors at the population level requires more detailed information on life history, behavior, and habitat selection for individual species. Research geared toward collecting this type of data has been undertaken for the frillneck lizard, one of the 55 species of reptiles in Kakadu National Park. Previous studies have investigated the short- and long-term effects of fire on the frillneck (see Griffiths and Christian 1996b), but until now, this information has not been effectively developed into a management framework. This chapter outlines our attempt to remedy this deficiency.

The frillneck lizard (*Chlamydosaurus kingii* Gray 1825) is an arboreal and insectivorous species, widespread and common across northern Australia and Papua New Guinea's tropical savannas, favoring open forests and woodlands (Shine and Lambeck 1989). During seasonally dry periods (May to October), frillneck lizards reduce their metabolic rate and activity to conserve energy. This is contrasted by increased activity, growth, and reproduction during the seasonally wet period of November to April (Christian et al. 1996, Griffiths and Christian 1996a). Frillneck lizards are well adapted to cope with regular fire in this tropical environment, and they show a clear preference for recently burned habitat. They are able to increase the volume of food ingested after a fire, probably because of the increased visibility of prey after shrub and grass layers have been removed, and they show poor body condition in habitat that has been unburned for over 5 years. However, in more intense fires, up to 30% of the population can be killed directly, so both absence of fire and intense frequent fires over long periods may limit populations (Griffiths and Christian 1996b).

Here we describe a detailed population and habitat model for the frillneck lizard in a well-studied portion of Kakadu National Park known as "Kapalga" (36,500 ha of open forest, which constitutes about 2% of the total area of the park). Our model is presented as a spatially explicit population model, rather than a true metapopulation model, since patch boundaries are not truly discrete and dispersal is high. The Kapalga fire experiment "compartments" (see Andersen et al. 1998) form the basis of the spatial population structure, with 22 patches in total (Figure 28.1) (mean patch size = 16.6 km^2). A landscape-scale experiment conducted in the early 1990s subjected these patches to a range of fire frequency and intensity treatments. Frillneck lizard population dynamics are modeled with a primary focus on the "catastrophes" aspect of RAMAS Metapop. The effect of fire on the population was modeled using catastrophes, rather than part of a continuous spectrum of environmental variation, because fires occur as discrete disturbance events (albeit common), are potentially a major driver of the frillneck lizard populations in Australian tropical savannas and are directly amenable to management intervention.

The main fire-management dilemma tackled in this model of the frillneck lizard (a relatively fire-tolerant species) is the inability of populations to thrive or persist in relatively large areas that are burned frequently and homogeneously, even though fire creates preferred habitat. We investigate this apparent paradox by considering (a) what will happen to frillneck populations if current management practices prevail in the future (i.e., 50% of the landscape), and (b) whether different fire regimes, given the current amount of burning, will either promote population stability or, alternatively, lead to major depopulation and reduction in densities.

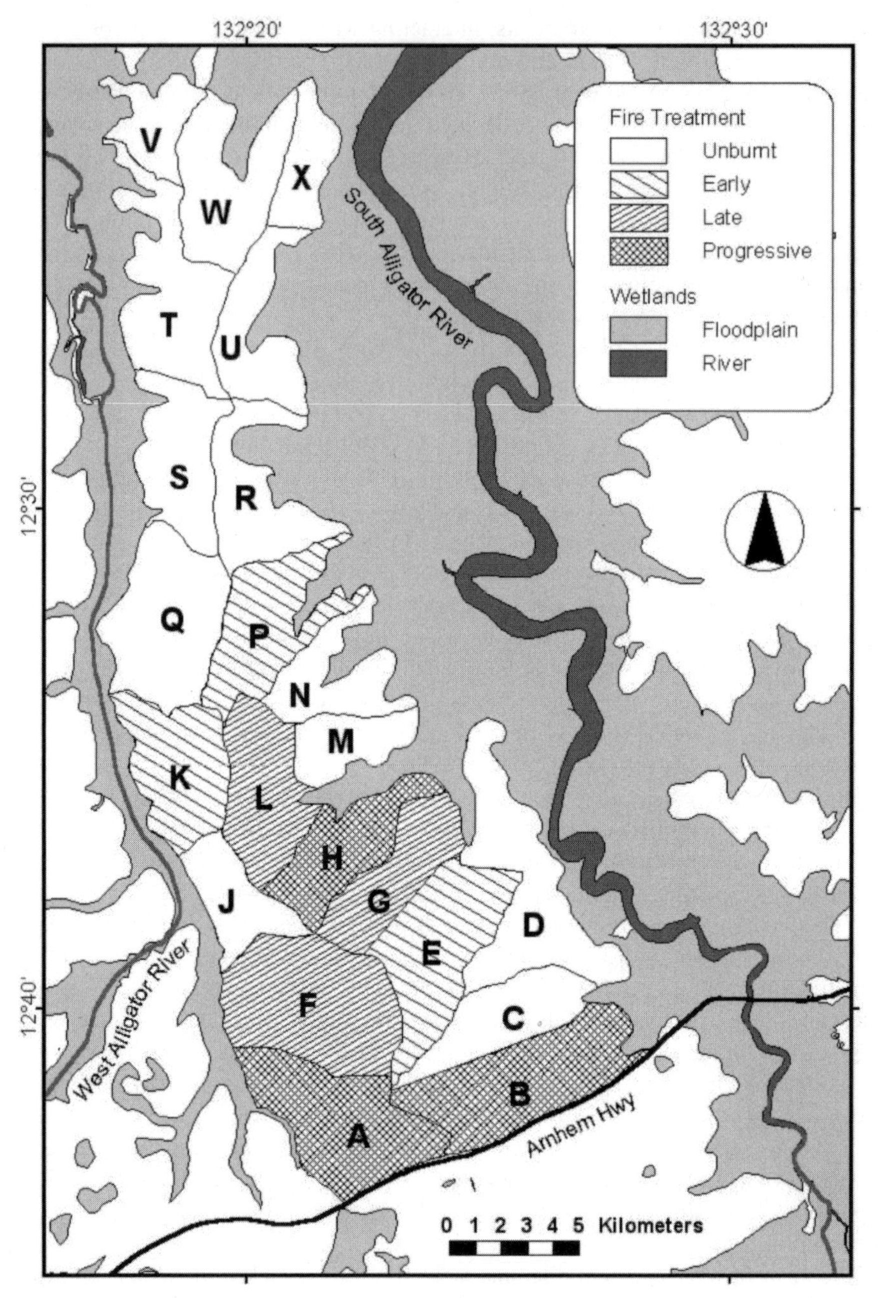

Figure 28.1 Map of the Kapalga study area, Kakadu National Park, Northern Territory, Australia, showing the experimental fire compartments (A–X) of *Eucalyptus* forest and savanna woodland, surrounded by a matrix of treeless riparian floodplain.

Methods

Model Structure and Output

The basic matrix population model included two stages (juvenile, adult), sex-structure, demographic and environmental stochasticity, two types of disturbance (cool and hot fires), ceiling-type density dependence, and a metapopulation-like spatial structure with 22 patches (the Kapalga "compartments," labeled A through to X on Figure 28.1.) connected through relatively high dispersal rates. The Kapalga metapopulation of frillneck lizards is surrounded by an inhospitable "matrix" (the floodplain) on three sides, but it is open on the southern boundary. Because we modeled this as a closed population, we probably underestimated some migration from neighboring areas into the southernmost compartments (Figure 28.1.).

The stage matrix was set up according to a prebreeding census (see Caswell 2001). Details of parameterization process are given in the succeeding sections and are summarized in Table 28.1.

Scenarios were projected for 50 years and iterated 1,000 times to minimize sampling variation. The simulation model outputs we examined were mean total population size (*N*), mean final population size for individual patches (population structure), probability of quasi extinction (i.e., risk of decline below a specified threshold population size), and time to quasi extinction (50% of initial *N*).

Demographics

Survival rates were estimated from a 2-year mark-recapture study (eight capture events spaced at 3-month intervals), undertaken in compartments Q (unburned), A/E (early burn), and F/G (late burn). A total of 183 individual lizards were marked, and 33 were recaptured at least once. Given the relatively small and unbounded areas of this study, most single captures probably represented transient individuals. These were excluded from the survival analyses because their inclusion led to entirely unrealistic survival rates, based on our prior knowledge of the frillneck's biology and behavior. Analyses were performed in Program MARK ver. 1.9 (White and Burnham 1999). A sex-specific model, with time-invariant survival and time-specific recapture rates, was best supported by the data. Female survival (87.6%) was found to be higher than male (65.4%), presumably because males expend considerable energy in establishing and defending territories and growing bigger and faster (Shine 1990).

Age of first breeding is approximately 2 years. The components of the fecundity element of the matrix were (a) proportion of females breeding = 39% (Griffiths 1999), (b) mean clutch size = 15.1 (Bedford et al. 1993), (c) survival from egg to 1 year of age (10%; A. D. Griffiths, unpublished data), and (d) sex ratio = 0.5. Because we have no age-specific data for fecundity, we used the same rate for all age classes. We assume a polygynous mating system. Although this assumption can reduce effective population size (N_e), our study does not incorporate genetic factors such as inbreeding depression. Nevertheless, demographic effects of a polygamous mating system, such as the maintenance of high fertility rates in females even when males are relatively scarce, are successfully captured by RAMAS Metapop.

Table 28.1 Summary of biological detail and key input parameters for the frillneck lizard population model in RAMAS Metapop

Life History Attribute or Model Parameter	Value or Estimate
Basic biology	
Family	Agamidae
Body size	400–870 g
Diet	Invertebrates
Territory size	1.8 ha
Sex ratio	1:1
Key RAMAS GIS input parameters	
Age of first breeding	2 years
Breeding system	Polygamous
Proportion of females breeding	39%
Mean clutch size	15
Survival age 0–1 (incl. clutch success)	10%
Coefficient of variation in fecundity	21%
Subadult/adult female survival	88%
Subadult/adult male survival	65%
Density dependence type	Ceiling
Cool/hot fire frequency (mutually exclusive)	50%
Cool/hot fire impact on carrying capacity (*K*)	Reset to max.
Hot fire impact on direct mortality (0% for cool)	30%

We have no direct information on year-to-year environmental variation (EV), though some is likely to be present (e.g., activity of animals associated with the timing of the first wet-season rains or total amount of rain). We have conservatively set EV to be a coefficient of variation (CV) of 21% on mean fecundity (because timing of rainfall events likely influence the survival of hatchlings at both ends of the wet season and probably also affects proportion of females breeding). The value of 21%—the CV for the average length of wet season (see Taylor and Tulloch 1985)—was used in the absence of any direct quantitative measurements of this relationship. Adult survival is likely to be relatively constant, as adult lizards are extremely resilient to fluctuating weather patterns (because they are an arid-adapted agamid species; Christian et al. 1996). However, we derived a rough estimate of environmental variation in survival (SD [F] = 0.11; SD [M] = 0.02) using a MARK model in which survival was set to be constant within a year, but different across the 2 years of the mark-recapture study. Correlation in EV across patches was established using the RAMAS distance-correlation function, with $a = c = 1$ and $b = 130$, such that there was 100% correlation between adjacent points on the ground and 75% correlation between the most distant patches (B–V).

Population Structure, Habitat and Movement

Initial population size and carrying capacity (*K*) for each patch was based on the observation that males defend a territory of 1.8 ha, and females have a foraging home range of 0.56 ha (Griffiths 1999). In addition, we used a "rating" to reflect the relative quality of each compartment. This multiplier on *K* was based on the proportion of a compartment's perimeter bordering on the riparian floodplain, since frillnecks tend to avoid densely vegetated areas around creeks, rivers, and the periphery of floodplains (Shine and Lambeck 1989):

- Prime = 1.0 = 0%–10% of the perimeter is floodplain edge (A, B, E, F, and L)
- Suboptimal A = 0.8 = >10%–30% of the perimeter is floodplain edge (C and G–J)
- Suboptimal B = 0.5 = >30% of the perimeter is floodplain edge (D, K, M–X)

Dispersal among compartments was modeled using the RAMAS dispersal-distance function, with the parameters $a = 0.286$, $b = 1.5$, and $c = 1$. Maximum dispersal using this function approximated 7 km. Actual dispersal out of a given compartment depended on its relative "connectedness" (i.e., the number of surrounding compartments and their distance). For instance, in the most "connected" compartment (M), a total of 25% of individuals dispersed, whereas in the least connected compartment (V), only 4.2% dispersed to other compartments in a given year. Juveniles have a greater propensity to disperse than adults because the latter have established (and are willing to defend) their home range (males and females) or territory (males). We therefore used relative dispersal weightings of 1.0 for juveniles and 0.5 for adults. All these rates represent maximum dispersal when K of source patch is at its maximum value.

Dispersal into a given locale will be greatest immediately after the area has been burned and subsequently will decline as the understory vegetation thickens (Griffiths and Christian 1996b). We therefore set Target Population Threshold $K = K_{max}$, which equals the carrying capacity of a compartment immediately after a fire. Dispersal into a given area thus declines linearly as K is reduced (as a function of time since last fire, see the following discussion on "Fire Modeling"). It is not possible in RAMAS Metapop ver. 4 to explicitly model dispersal out of a patch in relation to the K of the resident's patch, though this is implicitly captured (to a degree) by the Target Population Threshold K function just described.

Fire Modeling

Two distinct types of fire were modeled. Early dry season ("cool") fires typically occur between April and June; they are patchy, have low intensity (1,000 kW m^{-1}), and exhibit a relatively slow rate of spread (<0.5 m sec^{-1}). By contrast, late dry season ("hot") fires typically occur between July and September; they are highly intense (10,000 kW m^{-1}) and spread rapidly (>1 m sec^{-1}) (see Williams et al. 1998). The total annual frequency of burning of a given area of ground in northern Australia (whether by cool or hot fires) has been estimated by various workers (e.g., Edwards et al. 2001, Gill et al. 2000, Press 1988) to be 40% to 60%. We used the midpoint of this range as the baseline fire probability in our model, with cool and hot fires equally likely, but mutually exclusive in a given year (i.e., the "maximum negative correlation between catastrophes" option was used). The joint occurrence of both a cool and hot fire in the one season is unlikely, as both types of fire clear the ground litter and reduce fuel load. RAMAS has the options of "no," "maximum positive," and "maximum negative" correlations between different catastrophes, with the latter option being most realistic in our situation. The spread of fire from one compartment to another was not modeled. This is because the spread of catastrophes using the RAMAS vector function takes one time step (a year in our model), whereas the spread of fire across the landscape is effectively instantaneous.

Hot fires are responsible for roughly 30% direct mortality in frillneck lizards, whereas cool fires have no discernible effect on mortality (Griffiths and Christian 1996b). However, both cool and hot fires act to clear low-lying shrub, along with grass and leaf-litter layers, thereby greatly increasing the visibility of prey items for frillnecks and hence habitat quality. To model this effect on the habitat, K was reset to maximum (K_{max}) immediately following a fire event (cool or hot), thereafter declining linearly to a minimum of 0.5 K_{max} after 5 years with no fire.

Sensitivity Analysis of the Basic Model

We first undertook a simple sensitivity analysis of the baseline model, to (a) gauge the importance of our (somewhat intuitive) assumptions regarding dispersal and the spatial correlation of environmental variation and (b) gain a better understanding of how dispersal and spatial correlations influence model output relative to the more fundamental fecundity and survival parameters.

The following scenarios were run: half/double dispersal, equal juvenile and adult dispersal, 50% versus 100% correlation in environmental variation, no correlation between occurrence of cool and hot fires, ±10% in fecundity, ±5% in survival, and half/double environmental variation in vital rates (a total of 12 variations). Relative sensitivity was measured by the change in the expected minimum abundance (EMA) compared to that of the baseline model. The EMA provides a useful means of characterizing risk, as it is equivalent to the area under the quasi-extinction curve and thus captures a large amount of information in a single output metric (see McCarthy and Thompson 2001).

Management Scenarios

Five management scenarios were examined, based on different realistic combinations of the frequency, extent, and mix of fire regimes:

1. The baseline model, with a fire frequency of 50% per annum (one fire every 2 years, on average) and an equal probability of cool or hot fires, allocated randomly to each of the 22 compartments.
2. Reduced random fire frequency, to (a) 33% (one fire every 3 years) and (b) 10% (one fire every 10 years). Other parameters as per (1).
3. Change in the mix of cool and hot fires with same total random frequency as in (1), with (a) 10% probability of a cool fire and 40% probability of a hot fire and (b) the inverse.
4. A perpetuation of the Kapalga study design (see Andersen et al. 1998), with no fires in compartments C, D, J, M, N, and Q–X; cool fires in E, K and P; hot fires in F, G, and L; and progressive burning (random mix of cool and hot fires) in A, B, and H. All fire regimes occur with a 100% probability per annum in this scenario (except unburned).
5. Patch-specific stratified fire frequencies and intensities, where compartments A–F (close to the Arnhem Highway) are burned every year (i.e., nonrandom); G–L are burned every second year, on average (i.e., random); and M–X are burned every fifth year (random). Cool and hot fires occur with equal probability.

Results

Sensitivity Analysis

The frillneck lizard model was insensitive to quite substantial changes in the dispersal rate and the degree of environmental correlation between patches (the most uncertain parameters), as described in Table 28.2. Surprisingly, the model results were also relatively unaffected when cool and hot fires were no longer specified to be mutually exclusive in a given year, although this simplification did increase risk, as expected. The greatest degree of sensitivity was found in the mean and variance of the basic demographic parameters (survival and fecundity), which have been relatively well studied.

Management Scenarios

Projections of the baseline scenario (50% fire frequency, cool and hot fires with equal likelihood) show a substantial decline in densities of frillneck lizards over time, with a halving of total population size in 17 years (Table 28.3). However, the magnitude of the population decline is strongly related to the relative mix of cool and hot fires, rather than to the overall fire frequency, as illustrated in Figure 28.2. Of particular interest is the implication that efforts to decrease the overall fire frequency to very low levels (e.g., one fire every 10 years) would be a much less effective management option than maintaining regular and frequent fires while limiting the number of hot, damaging late dry season fires (by promulgating low intensity burns early in the dry season).

When comparing the compartment-specific fire regimes (scenarios 4 and 5), it is instructive to consider the relative fates of each patch. Figure 28.3 illustrates the distri-

Table 28.2 Results of the sensitivity analysis of the baseline frillneck lizard model over a 50-year projection period

Model Parameter Changed	EMA	Rank
1. Baseline model (50% fire frequency)—F50%.MP	20,630	N/A
2. Dispersal rate halved—HFL-DISP.MP	–2%	10
3. Dispersal rate doubled—DBL-DISP.MP	–1%	11
4. Juvenile and adult dispersal equally likely—EQL-DISP.MP	0%	12
5. Spatial correlation in EV halved—50%EVC.MP	+4%	8
6. Complete (100%) spatial correlation in EV—100%EVC.MP	–3%	9
7. EV in vital rates halved—HLF-EV.MP	+43%	4
8. EV in vital rates doubled—DBL-EV.MP	–49%	2
9. 10% increase in fecundity rate—+10%FEC.MP	+21%	6
10. 10% decrease in fecundity rate— –10%FEC.MP	–28%	5
11. 5% increase in survival rates—+5%SURV.MP	+46%	3
12. 5% decrease in survival rates— –5%SURV.MP	–54%	1
13. Fire events not mutually exclusive—CF+HF.MP	–7%	7

Note: The expected minimum abundance (EMA) is given for the baseline model, and thereafter the percentage change in EMA is relative to the baseline model. "Rank" compares parameter changes, where 1 = largest change from baseline model. Scenarios 2–4 examine dispersal assumptions, 5–8 consider environmental variation (EV), 9–12 examine vital rates, and 13 looks at the mutual exclusivity of early and late dry-season fires. RAMAS input file (*.MP) names are also given.

Table 28.3 Results of fire frequency sensitivity analysis for the frillneck lizards of Kapalga

Scenario	Mean N		Quasi-extinction Probability (%)		Median Time to QuE
	25 years	50 years	$0.5N_i$ (34,457)	$0.25N_i$ (17,229)	($0.5N_i$) (years)
1. Baseline model (50-% fire frequency (F50%.MP)	33,900	28,600	96	36	17
2. a. 33% fire frequency (F33%.MP)	36,400	34,800	95	20	17
b. 10% fire frequency (F10%.MP)	34,300	33,800	100	16	11
3. a. 10% cool + 40% hot fire frequency (F10%C+40%H.MP)	17,600	8,100	100	94	9
b. 40% cool + 10% hot fire frequency (F40%C+10%H.MP)	48,100	47,800	50	9	50
4. Kapalga study design perpetuated (KAPALGA.MP)	27,600	25,800	100	30	8
5. Patch-specific fire frequency (PCH-SPC.MP)	24,600	18,800	100	66	10

Note: The mean final population size (all compartments) after 25 and 50 years; the probability of declining to less than 50% and 25% of the initial total population size (N_i = 68,914) over a 50-year period; and the median time taken to decline to less than 50% cf intitial population size. Fires occur randomly across all patches in scenarios 1 to 3, are fixed for each patch in scenario 4, and are stratified random for scenario 5. RAMAS input file (*.MP) names are given after the scenario description.

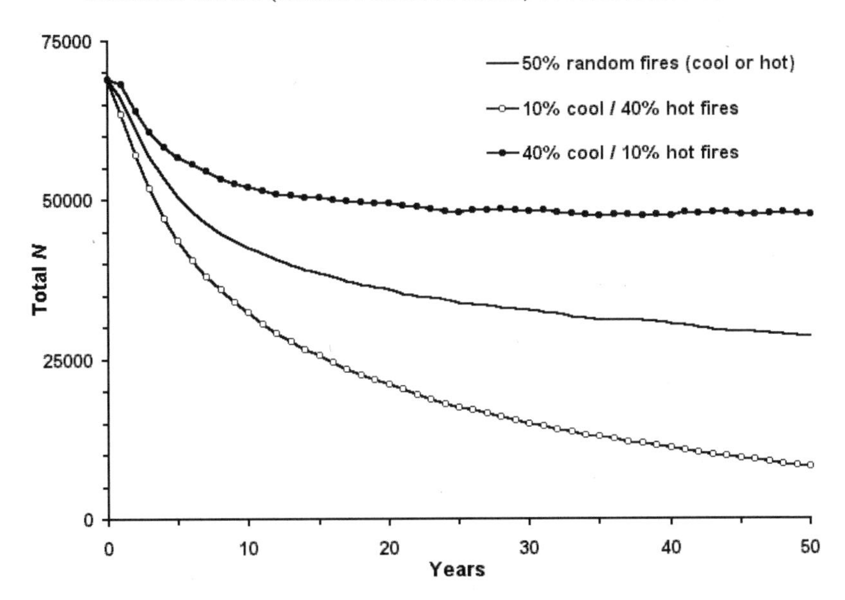

Figure 28.2 Change in total mean population size of frillneck lizards in Kapalga over a 50-year time period, under the influence of three different fire regimes.

bution of population sizes across all 22 compartments for the initial population structure and three fire-management scenarios.

A high-frequency, random mix of cool and hot fires is clearly uniformly deleterious across all compartments (scenario 1). If the pattern of burning regimes used in the Kapalga fire experiment is perpetuated for a further 50 years (scenario 4), the impact is generally less severe than a random patterning of fire, with a few compartments maintaining partially intact populations. These tend to be either those compartments that were burned early in the dry season (e.g., E, K) or patches that were not directly burned but were situated close to burned patches (e.g., D, P) and thereby maintained by dispersal from these healthy subpopulations. Populations in compartments that remained entirely unburned (e.g., V, X) also declined, as habitat quality decreased when shrub and grass layers thicken in the absence of fire. Nevertheless, the relatively high levels of dispersal tend to smear any strong compartment-specific signals. Finally, when fires were most frequent close to the main human access points (along the highway) and less frequent at the opposite end of the Kapalga peninsula, the southerly compartments (A–F) closer to the highway suffered the most substantial declines.

Discussion

Evaluating RAMAS Metapop ver. 4 for the Frillneck Lizard

The suitability of RAMAS Metapop ver. 4 for modeling the frillneck lizard is evaluated by comparing the capabilities of the program to earlier versions and to other potential approaches.

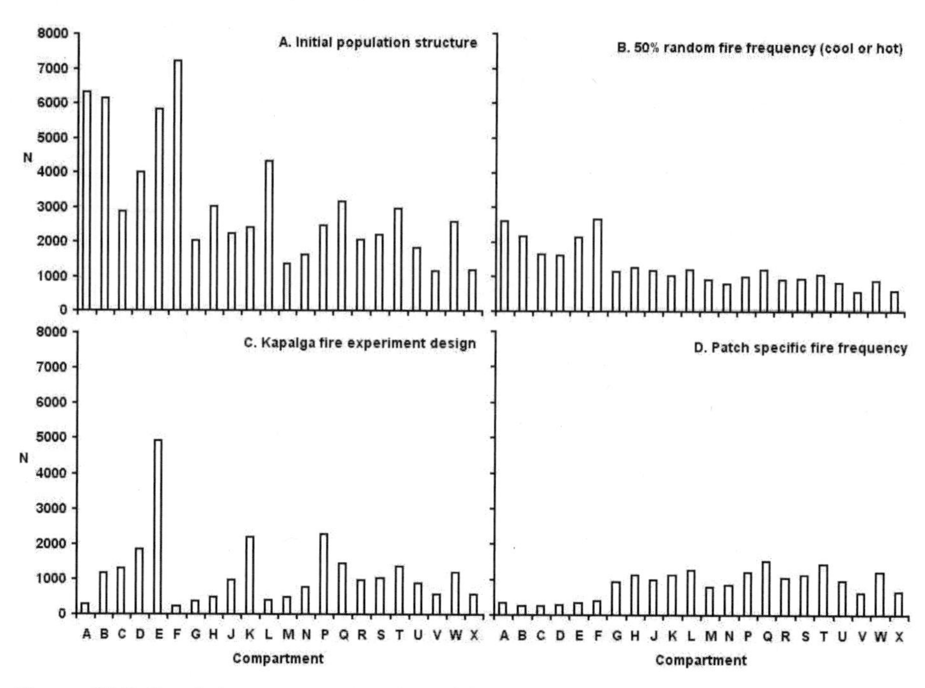

Figure 28.3 Population structure (number of frillneck lizards in each of the 22 management compartments) for **(A)** initial population structure for all scenarios; **(B)** a 50% total probability of fire in each compartment each year, with an equal but mutually exclusive likelihood of the intensity being either cool or hot; **(C)** distribution of fire regimes following the experimental regime described in Figure 28.1; and **(D)** patch-specific fire frequencies, where compartments closer to the highway (A–F) burn more regularly than those at a greater distance (G–X).

It would have been impossible, using earlier versions of RAMAS, to realistically model the interaction between fire and the frillneck lizard population dynamics. For instance, in RAMAS ver. 3, catastrophes could not be correlated, making it impossible to specify the mutual exclusivity of cool and hot fires in a single year. The "multiple effects of a catastrophe" feature meant that we could consider simultaneously the impact of fire on mortality, habitat quality, and dispersal (version 3 allowed fire effects on either mortality or habitat quality).

An alternative to RAMAS would be to use a different generic "PVA package," such as the individual-based simulation program, VORTEX ver. 8 (Miller and Lacy 1999). RAMAS and VORTEX produce highly concordant results when applied in a comparable manner to the same species (Brook et al. 1999, 2000). The basic interface of VORTEX is tightly structured and somewhat restrictive in its scope, but the facility to define complex functions for most input parameters makes it a highly flexible modeling tool (see Lacy 2000). Using parameter functions, essentially all of the features of the RAMAS frillneck model could have been incorporated into a VORTEX counterpart. In many instances, however, the required functions would have needed to be quite complex (see Miller and Lacy 1999 for details). RAMAS offers a more user-friendly and intuitive interface for designing a spatially explicit, disturbance-driven model such as ours. Further, RAMAS has the advantage of permitting much larger population sizes to be run (our initial population size of 69K would not have been possible in VORTEX).

Another option would be to develop a tailored, species-specific simulation model, from scratch. In this case it would be possible to provide exactly the level of model complexity that could be supported by available data. However, this task generally requires advanced computer programming skills and the willingness to devote considerable time and resources into software development, debugging, and testing and is therefore not feasible for most wildlife managers.

Meaning of Results for Frillneck Management

The results of this modeling exercise are somewhat alarming, but not unexpected, given the high mortality from late dry season fires. Repeated burning late in the dry season in the same location at a frequency of once every 2 years, is predicted to substantially reduce the population of frillneck lizards. An increase in fuel loads caused by the spread of introduced grasses such as gamba (*Andropogon gayanensis*) will only increase this problem. However, it seems clear that although repeated fires will suppress populations, dispersal at this scale is sufficient to allow the overall population to persist, albeit at suppressed densities.

The scenario best approximating a passively managed landscape (scenario 5) produced close to the worst result for the frillneck lizard population. Annual fires lit from the highway and carried through the southern part of the Kapalga peninsula by the prevailing southeast winds, while protecting the northern part from frequent fires, produced substantially reduced populations within 10 km of the highway. The effect of this fire regime was evident after a relatively short period of time. In contrast, the scenarios based on early dry season burns at a frequency of once every 2 (or fewer) years, and spread across the whole area in a random manner, produced the most stable populations.

Based on this exercise, guidelines for fire management of frillneck lizard populations can be summarized as follows. First, encourage early dry season fires to cover proportionally more area than late dry season fires. Second, limit late dry season fire frequency in individual patches to no more than once every 3 years. It is important to note here that these results and management recommendations are limited to a relatively simple set of scenarios based on current amounts of burning and do not represent the complete set of possible answers to the dilemma of managing wildlife in fire-prone environments. However, it is quite evident and logical that repeated high-intensity fires in the same location (especially at this scale of management) would adversely affect almost any member of the vertebrate fauna. Why did frillneck lizard numbers decline under all scenarios? The main reason is likely to be that the initial population size was set equal to the carrying capacity, yet when there is environmental variation in a system, the equilibrium population size sits below the maximum (ceiling) population size (this is because populations are free to fluctuate below the ceiling K, but not above it).

A number of important problems still remain and may be tackled effectively using RAMAS. For instance:

- Do fires occurring at larger spatial scales produce a negative result, as dispersal would decrease between defined "patches"? Areas over 1000 km^2 in remote regions can be burned in a single fire. Sensitivity analysis of the dispersal function at larger spatial scales may provide some insight.

- It is difficult to realize the full effect of spatial heterogeneity in the current model. Remotely sensed or other data may enhance the accuracy of the configuration of burned patches. We limited this model to the compartments used in the Kapalga fire experiment (which were effectively small stream catchments), and we assumed uniform burning of a compartment once ignition occurred within it. A more accurate representation would reflect the important geographical features that limit the spread of fire, along with variation in the relative effectiveness of those barriers, and hence better reflect the patchiness of cool and hot fires. This could be addressed with the RAMAS GIS Habitat Dynamics program, using fire maps produced from remotely sensed data.

It would be worth considering the usefulness of this approach for different species in the tropical savannas. Data from the Kapalga fire experiment suggest that a number of small mammals show a clear response to different fire treatments, with some mid-successional species showing a strong negative response to repeated burning (Pardon et al. 2003). Recent evidence of a decline in the small mammal populations at Kapalga in Kakadu National Park is perplexing (Woinarski et al. 2001). Similar models to the one presented here could be applied to small-mammal populations from Kapalga, with a view to examining implications of the total range of options available to conservation managers in northern Australia.

Acknowledgments This work was undertaken by the Key Centre for Tropical Wildlife Management, Northern Territory University, under funding from the Australian Research Council. Many thanks to Peter Whitehead, Per Sjögren-Gulve, Reşit Akçakaya, Alan Andersen, and two anonymous referees for their comments and suggestions.

References

Andersen, A. N., Braithwaite, R. W., Cook, G. D., Corbett, L. K., Williams, R. J., Douglas, M. M., Gill, A. M., Setterfield, S. A., and Muller, W. J. 1998. Fire research for conservation management in tropical savannas: introducing the Kapalga fire experiment. *Australian Journal of Ecology* 23: 95–110.

Bedford, G. S., Christian, K. A., and Griffiths, A. D. 1993. Preliminary investigations on the reproduction of the frillneck lizard *Chlamydosaurus kingii* in the Northern Territory. Pages 127–131 in D. Lunney and D. Ayres (eds.), *Herpetology in Australia: a diverse discipline.* Surrey Beatty and Sons, Sydney.

Bowman, D. M. J. S. 1998. Tansley Review No. 101: the impact of Aboriginal landscape burning on the Australian biota. *New Phytologist* 140: 385–410.

Braithwaite, R. W. 1987. Effects of fire regimes on lizards in the wet-dry tropics of Australia. *Journal of Tropical Ecology* 3: 265–275.

Brook, B. W., Cannon, J. R., Lacy, R. C., Mirande, C., and Frankham, R. 1999. A comparison of the population viability analysis packages GAPPS, INMAT, RAMAS and VORTEX for the Whooping crane (*Grus americana*). *Animal Conservation* 2: 23–31.

Brook, B. W., O'Grady, J. J., Chapman, A. P. , Burgman, M. A., Akçakaya, H. R., and Frankham, R. 2000. Predictive accuracy of population viability analysis in conservation biology. *Nature* 404: 385–387.

Caswell, H. 2001. *Matrix population models: construction, analysis, and interpretation.* Sinauer, Sunderland, Mass.

Christian, K. A., Griffiths, A. D., and Bedford, G. S. 1996. Physiological ecology of frillneck lizards in a seasonal tropical environment. *Oecologia* 106: 49–56.

Dyer, R., Jacklyn, P., Partridge, I., Russell-Smith, J., and Williams, R. (Eds.). 2001. *Savanna burning: understanding and using fire in northern Australia*. Tropical Savannas CRC, Darwin. N.T.

Edwards, A., Hauser, P., Anderson, M., McCartney, J., Armstrong, M., Thackway, R., Allan, G., Hempel, C., and Russell-Smith, J. 2001. A tale of two parks: contemporary fire regimes of Litchfield and Nitmiluk National Parks, monsoonal northern Australia. *International Journal of Wildland Fire* 10: 79–89.

Gill, A. M., Ryan, P. G., Moore, P. H. R., and Gibson, M. 2000. Fire regimes of World Heritage Kakadu National Park, Australia. *Austral Ecology* 25: 616–625.

Griffiths, A. D. 1999. Demography and home range of the frillneck lizard, *Chlamydosaurus kingii* (Agamidae), in northern Australia. *Copeia* 1999: 1089–1096.

Griffiths, A. D., and Christian, K. A. 1996a. Diet and habitat use of frillneck lizards in a seasonal tropical environment. *Oecologia* 106: 39–48.

Griffiths, A. D., and Christian, K. A. 1996b. The effects of fire on the frillneck lizard (*Chlamydosaurus kingii*) in northern Australia. *Australian Journal of Ecology* 21: 386–398.

Lacy, R. C. 2000. Structure of the VORTEX simulation model for population viability analysis. *Ecological Bulletins* 48: 191–203.

McCarthy, M. A., and Thompson, C. 2001. Expected minimum population size as a measure of threat. *Animal Conservation* 4: 351–355.

Miller, P. M., and Lacy, R. C. 1999. *VORTEX: a stochastic simulation of the extinction process. Version 8 User's Manual*. Conservation Breeding Specialist Group (SSC/IUCN), Apple Valley, Minn.

Pardon, L. G., Brook, B. W., Griffiths, A. D., and Braithwaite, R. W. 2002. Determinants of survival for the northern brown bandicoot under a landscape-scale fire experiment. *Journal of Animal Ecology* 72: 106–115.

Press, A. J. 1988. Comparisons of the extent of fire in different land management systems in the Top End of the Northern Territory. *Proceedings of the Ecological Society of Australia* 15: 167–175.

Press, A. J., Lea, D., Webb, A., and Graham, A. 1995. *Kakadu: natural and cultural heritage and management*. Australian Nature Conservation Agency and North Australian Research Unit, Australian National University, Darwin.

Russell-Smith, J., Ryan, P. G., and DuRieu, R. 1997. A Landsat MSS-derived fire history of Kakadu National Park, monsoonal northern Australia, 1980–1994: seasonal extent, frequency and patchiness. *Journal of Applied Ecology* 34: 748–766.

Shine, R. 1990. Function and evolution of the frill of the frillneck lizard, *Chlamydosaurus kingii* (Sauria: Agamidae). *Biological Journal of the Linnean Society* 40: 11–20.

Shine, R., and Lambeck, R. 1989. The ecology of frillneck lizards, *Chlamydosaurus kingii* (Agamidae), in tropical Australia. *Australian Wildlife Research* 16: 491–500.

Taylor, J. A., and Tulloch, D. 1985. Rainfall in the wet-dry tropics: extreme events at Darwin and similarities between years during the period 1870–1983 inclusive. *Australian Journal of Ecology* 10: 281–295.

Trainor, C. R., and Woinarski, J. C. Z. 1994. Responses of lizards to three experimental fires in the savanna forests of Kakadu National Park. *Wildlife Research* 21: 131–148.

White, G. C., and Burnham, K. P. 1999. Program MARK: survival estimation from populations of marked animals. *Bird Study* 46: 120–139.

Williams, R. J., Gill, A. M., and Moore, P. H. R. 1998. Seasonal changes in fire behaviour in a tropical savanna in northern Australia. *International Journal of Wildland Fire* 8: 227–239.

Williams, R. J., Griffiths, A. D., and Allan, G. E. 2001. Fire regimes and biodiversity in the savannas of northern Australia. Pages 281–304 in R. A. Bradstock, J. E. Williams, and A. M. Gill (eds.), *Flammable Australia: the fire regimes and biodiversity of a continent*. Cambridge University Press, Cambridge.

Woinarski, J. C. Z., Brock, C., Fisher, A., Milne, D., and Oliver, B. 1999. Response of birds and reptiles to fire regimes on pastoral land in the Victoria River District, Northern Territory. *Rangeland Journal* 21: 24–38.

Woinarski, J. C. Z., Milne, D. J., and Wanganeen, G. 2001. Changes in mammal populations in relatively intact landscapes of Kakadu National Park, Northern Territory, Australia. *Austral Ecology* 26: 360–370.

29

Sand Lizard (*Lacerta agilis*) in Central Sweden

Modeling Juvenile Reintroduction and Spatial Management Strategies for Metapopulation Establishment

SVEN-ÅKE BERGLIND

A common consequence of habitat loss and fragmentation is the creation of nonequilibrium metapopulations in which the local populations have become so isolated that among-population dispersal no longer occurs (Harrison 1994, Gilpin 1996). Metapopulations that have reached this state are destined to extinction unless the loss and fragmentation of habitat is reversed (Hanski 1997). In this situation, restoration and reintroduction at the scale of metapopulations should be considered (Huxel and Hastings 1999). Since introduced populations in general are small—and thus highly susceptible to environmental fluctuations, catastrophes, demographic stochasticity, and inbreeding depression—it is crucial to maximize the efficiency of initial introductions in order to shorten the period during which the populations would be exposed to these risks (Komers and Curman 2000).

In central Sweden a small number of nonequilibrium metapopulations of the sand lizard (*Lacerta agilis* L.) occur on "biotope islands" of large, glaciofluvial sand deposits covered by pine (*Pinus sylvestris*) heath forests. These populations are some of the northernmost in the world (Gasc et al. 1997) and represent genetically differentiated relicts from an ancient continuous distribution during the postglacial warm period ca. 7000–500 B.C. (Gislén and Kauri 1959, Gullberg et al. 1998). The sand lizard is a Eurasian species that has declined in most of northwestern Europe during the last decades, mainly due to loss of open, sandy, early successional habitats (Glandt and Bishoff 1988), and it is red-listed as "Vulnerable" in Sweden (Gärdenfors 2000). It is a ground-dwelling, medium-sized lizard that feeds mainly on insects and spiders.

The past forest structure on the sandy areas in central Sweden was strongly shaped by frequent fires that may have created a spatiotemporal mosaic of open patches with exposed sand used for egg-laying by the sand lizard and a rich field layer of heather (*Calluna vulgaris*) used for foraging and shelter (Berglind 1988, 1999). However since the beginning of the twenthieth century, effective forest fire suppression and afforestation have resulted in a dramatic increase in tree canopy formation, with subsequent loss of such open patches. It is probable that the number and size of sand lizard populations have decreased in parallel and that the present-day areas of occupancy are minute in comparison to the past. In 2001, only four local populations remained on the 11,000-ha nature reserve Brattforsheden, which is one of the largest of these sandy areas in central Sweden (Figure 29.1). All populations were completely isolated from each other by commercially managed pine forest at a distance of 2.5 to 10 km, and the number of adult females per population was estimated at < 5, 5, 8, and 13, respectively (Berglind 2000 and unpublished data). Each of the two largest populations was analyzed independently (Berglind 2000) by constructing age-structured, stochastic, single-population models using RAMAS Metapop 3.0 (Akçakaya 1998). These analyses predict that combined habitat and demographic management could dramatically improve the otherwise slim chances for population persistence (Berglind 2000; Sjögren-Gulve, Chapter 24, this volume).

However, beyond rescuing the present small sand lizard populations on Brattforsheden from imminent extinction, there is a need to transform this highly vulnerable metapopulation into a metapopulation of viable and more-or-less connected local popula-

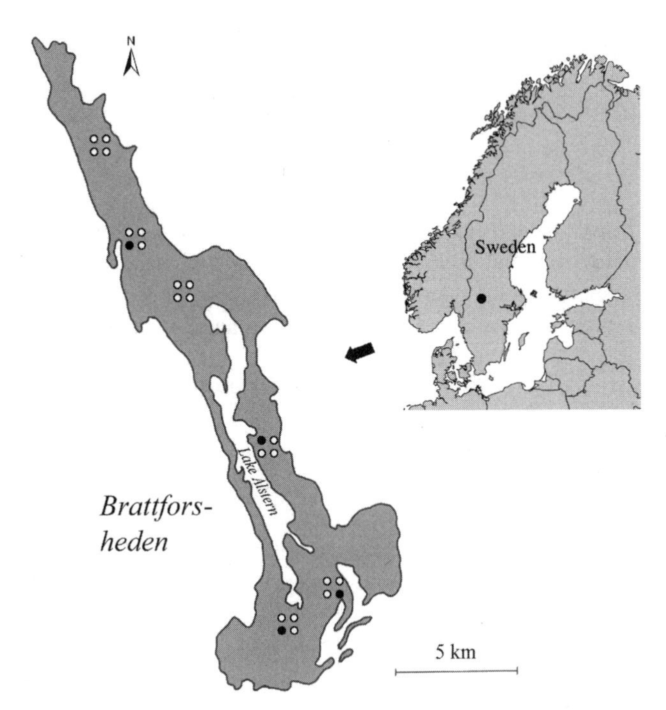

Figure 29.1 The nature reserve Brattforsheden in southcentral Sweden: the spatial distribution of inhabited patches by the sand lizard *Lacerta agilis* (*filled circles*) and the hypothetical future patches to be restored for reintroduction of sand lizards (*empty circles*). In this example there are six metapopulation networks, with four patches per network. The actual location of future restored patches will depend on a combination of suitable aspect, soil texture, potential for among-patch dispersal, and negotiation success with the landowners.

tions that require little demographic management for persistence. Because most of the area is subject to conventional forestry and because the past natural disturbance regimes are no longer operating (and will not do so in the foreseeable future), it is necessary to create networks of managed habitat for long-term conservation (Figure 29.1). After habitat restoration, one option is simply to let the few present populations colonize adjacent sites through natural dispersal. Another option, which potentially could create metapopulations much faster and not necessarily close to the existing populations, is to introduce lizards to restored sites. Sand lizards are comparatively easy to breed, and by introducing captive-born juveniles to new sites there are good chances for population establishment (Corbett 1988, Moulton and Corbett 1999, Berglind 2000).

This chapter explores the potential for metapopulation establishment of the sand lizard by reintroduction, by modeling the effect of differences in (1) patch size, (2) number of patches, (3) number of introduced juveniles, and (4) among-patch dispersal. The study is part of the planning for a conservation action program for the sand lizard on Brattforsheden and is directed at determining optimum reintroduction strategies. My basic model is a spatially explicit, age-structured metapopulation model developed using RAMAS Metapop 4.0 (Akçakaya 2002). The computer files are available on the CD that accompanies this volume.

Methods

Models

Models were constructed for females only, and the matrices were based on postbreeding census (see Akçakaya 2000, Caswell 2001). The basic model included 15 age classes (ranging from 0- to 14-year-olds); demographic and environmental stochasticity; one type of catastrophe (regional cold summers); contest-type density dependence; and a hypothetical metapopulation system with one, two, four, and eight local populations, respectively, all of the same initial size and structure, and connected by dispersal. Comparisons were also made without dispersal. In addition, a sensitivity analysis was made to compare extinction risks of individual patches of different size.

All models used a simulation that ran for 50 years with 1,000 replicates. The basic model used survival and fecundity data from "Scenario II, site FL" in Appendix 1 in Berglind (2000), corresponding to a relatively conservative deterministic effect on the population growth rate ($\lambda = 1.03$) after "optimal" habitat management (cutting of dense tree stands, excavation of new sand patches, and enhancement of heather growth). Details of parameterizations, except those of dispersal and density dependence, are given in Berglind (2000).

Demography

Annual survival rates were estimated from a 10-year mark-recapture study during 1988–1998 undertaken for the two largest sand lizard populations (site FL and SB) on Brattforsheden, central Sweden (Berglind 2000). Additional survival rate data for juveniles were taken from a 7-year mark-recapture study by Strijbosch and Creemers (1988) of a stable sand lizard population in a similar habitat in The Netherlands. In all, this gave the following annual survival rates: juveniles (0-year-olds) = 38.4 %, subadults (pre-reproductive ages; 1– to 2–year-olds) = 61.4%, and adults (that have hibernated 3 times or more) = 69.4%. Following Strijbosch and Creemers (1988), the age-specific probability of reproduction was set to 52.3% for 3-year-olds, 82.4% for 4-year-olds, and 100% for 5+-year-olds. The sex ratio among the offspring was set to 50% females, also in accordance with Strijbosch and Creemers (1988). Maternity (the average number of female offspring hatched per year per reproductive female) was measured as 3.113 over a 9-year period for site FL. Age-specific fecundities (*fx*) were calculated as the product of adult survival rate, probability of reproduction, and maternity.

Environmental Stochasticity and Catastrophes

Environmental stochasticity was modeled as randomly drawn values from lognormal distributions. The environmental stochasticity for yearly survival was calculated as the standard deviation (SD) of the series of annual recapture rates and yielded the following SDs: 0-year-olds = 0.176, 1- to 2-year-olds = 0.155, 3+-year-olds = 0.109 (Berglind 2000). These estimates are crude and also include effects of demographic stochasticity and sample error (but see Brook 2000).

The measure of environmental stochasticity for fecundity was calculated using the coefficient of variation of numbers of hatched eggs per clutch per year at site FL during 1988–1997 (CV = 0.18) as in Berglind (2000). Catastrophic events were ignored in these variance estimates. Catastrophes were incorporated as 0% survival for 0-year-olds every tenth year (Berglind 2000).

There was no within-population correlation between yearly adult survival and fecundity during year 1989–1996 for site FL (Pearson's $R = -0.36$; $p = 0.38$). Thus, no correlation between adult survival and fecundity was used in my models.

Correlation in environmental variation between local populations was established using the RAMAS correlation-distance function, with $a = c = 1$ and $b = 8000$, such that there was 100% correlation for a 0 m distance and 73% correlation for a 2500 m distance between populations. The latter value is equivalent to the correlation for adult female population sizes between site FL and SB (situated 2500 m apart) during the 14-year period 1988–2001 (Pearson's $R = 0.73$; $p = 0.0029$). In the hypothetical metapopulation systems modeled here, the distance between the most adjacent local populations was 750 m (from center to center), which gives a 91% correlation using this function.

Density Dependence

Density-dependent phenomena are suspected to be prominent in lizards, since population sizes are thought to fluctuate moderately compared to other vertebrates (Clobert et al. 1994). In this study I first compared the effects of exponential (no density dependence), ceiling, and contest types of density dependence, because the type of density dependence can strongly influence extinction risks. For the contest model, the equilibrium population size (K) was set at 60 females (including hatchlings)/ha, which is the average population size for site FL during 1992–2001, when the amount of habitat was restricted to ca. 1 ha. The maximal rate of population increase (R_{max}) was set to 1.10 (as compared to $\lambda = 1.03$ of the basic model's yearly population growth rate), which assumes an increase in juvenile survival to the mean between the above given estimates for juvenile and subadult survival (cf. scenario III in Berglind 2000). For the ceiling model, carrying capacity (K) was set at 120 females/ha, which is simply twice the observed equilibrium population density.

In the basic model I used contest type density dependence, for which there is evidence in populations of the common lizard (*Lacerta vivipara*) (Massot et al. 1992, Lena et al. 1998), a close and often sympatric relative to the sand lizard. Since my two study populations have fluctuated moderately under constant habitat size conditions during the 10-year period 1992–2001, and individuals have been homogenously distributed within habitat patches (Berglind, unpublished data), it is reasonable to assume that contest type density dependence is operating also in these sand lizard populations. Note, however, that neither the sand lizard nor the common lizard is territorial; instead, intraspecific competition probably occurs by preemptive use of (micro-)sites that differ in suitability (Olsson et al. 1997, Ronce et al. 1998), which is one type of contest competition (Rodenhouse et al. 1997, 2000). In the common lizard, juvenile mortality is positively related to population density (Massot et al. 1992), and adult female density is a major factor promoting juvenile dispersal (Lena et al. 1998). In the sand lizard, young individuals occur rather frequently in places uninhabited by adults (Yablokov et al. 1980, Nature Conservancy Council 1983), and intraspecific competition also occurs to varying degrees by adult predation on juveniles (Corbett and Tamarind 1979).

Dispersal and Population Geometry

Adult sand lizards are usually highly sedentary (Nature Conservancy Council 1983, Olsson et al. 1997, Berglind 1999). Most dispersal probably occurs among immatures (Yablokov et al. 1980, Nature Conservancy Council 1983), with considerable variation in dispersal tendencies (Olsson et al. 1996). No data exist on among-population dispersal rates, and there are few data on dispersal capacities. The longest dispersal distances reported seem to be of one individual found 2 km away from an isolated population (Strijbosch and van Gelder 1997) and of one subadult moving 500 m along a forest road within one season (Berglind 2000). Within forested areas, dispersal seems to be very limited outside open patches, probably occurring mostly along sun-exposed forest roads (Dent and Spellerberg 1988). In my models, it was assumed that suitably open dispersal corridors had been created between restored patches for introduction.

I modeled hypothetical (meta-)populations composed of one, two, four, and eight local populations, each corresponding to a 5–ha patch (equilibrium population size K = 300) and connected to the most adjacent patches by a distance of 750 m (center to center). Dispersal among populations was modeled using the RAMAS dispersal-distance function, with $a = 0.40$, $b = 220$, and $c = 1$. Maximum dispersal, D_{max}, was set to 1200 m per year. This function gives an annual dispersal rate of 1.3% per population in the two-patch system, and 3% and 6% for the least and most connected patches, respectively, in the eight-patch system. I used relative dispersal weightings of 1.0 for immatures (0-, 1-, and 2-year-olds), and 0 (i.e., no dispersal) for adults. Dispersal was assumed to be density dependent, and the above rates represent maximum dispersal at $K = 300$. (Density-dependent dispersal as a function of source population size (slope), under the Populations dialog box, was set at stable age distribution, such that the curve crossed the origin and the total rate of dispersal, given as the sum under each population in the Dispersal matrix, was adjusted to the equilibrium population size $K = 300$.)

Initial Population Structures

For the models used to test effects of different patch size and types of density dependence, initial population size was based on the observed population density of 60 females/ha, with a stable age distribution. The smallest patch size, 0.1 ha, is the smallest observed on Brattforsheden (a now extinct population). Although populations, not patches per se, are modeled in RAMAS Metapop, I here prefer to present the effects of differences in patch size (with the same initial population density) since "patch size" is the relevant term used by forest managers.

For the basic model for juvenile introduction scenarios, I set $K = 300$ for each local population. The starting population (year 0) was composed of 10, 20, or 40 introduced juveniles per patch. For some scenarios, propagules of 10, 20, or 40 additional juveniles per patch were introduced for up to 2 subsequent years. The chosen propagule sizes and introduction time periods were a compromise between expected population establishment success, costs, and administrative continuity (cf. Snyder et al. 1996). In all, 36 introduction combinations (scenarios) were analyzed.

For different alternatives to breed and raise juveniles for release, and a discussion of their potential survival rates, see Berglind (2000). Here I assumed that juveniles were introduced to restored habitat patches empty of sand lizards. It is probable that this will result in increased juvenile survival and higher individual growth rates during population buildup, compared to what occurs in stable populations, since survivorship of juvenile lizards is negatively related to density of conspecific lizards (cf. Massot et al. 1992, Tinkle et al 1993). Such an effect is allowed for in my basic model by the incorporation of contest density dependence with a maximal yearly population growth rate of 10%.

Results

Effects of Patch Size and Type of Density Dependence

The quasi-extinction risk for populations inhabiting patches of six different sizes, each of which has three types of density dependence, is shown in Figure 29.2. The risk of decline was > 56% for patches ≤ 1 ha, with relatively small differences between the three density dependence types. For patches 5 to 10 ha, the relative difference between contest-type density dependence, on one hand, and exponential and ceiling, on the other, is larger, with the first (probably most realistic) type showing a risk of decline from 6% to 1%. Thus, suitable habitat patches between 5 and 10 ha seem to offer "acceptable" chances of population persistence over a 50-year period. Larger patches contribute only a little more to persistence.

Effects of Number of Patches and Introduced Juveniles

Since several of my introduction populations started very small, and with small introductions in following years, interval risk of decline is not so informative. Instead, I used expected minimum abundance, extinction risk, terminal risk of decline (threshold of 10 females per population), and interval probability of increase (threshold of 300 females per population) as risk-based outputs (Table 29.1).

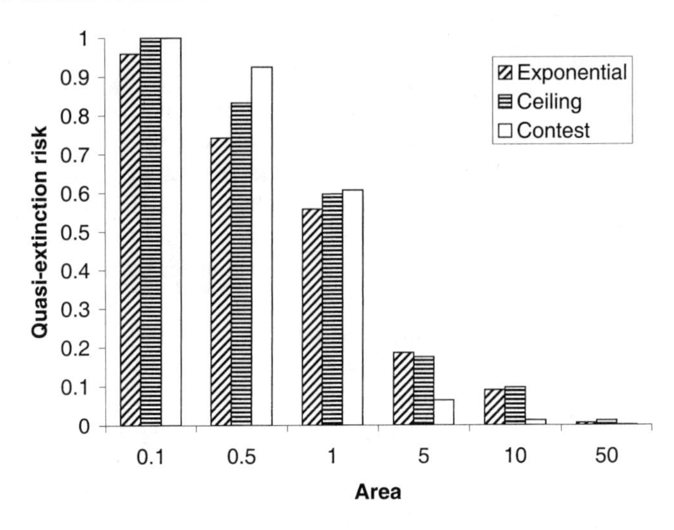

Figure 29.2 Quasi-extinction risks in relation to patch area and density dependence for female sand lizards on Brattforsheden. The quasi-extinction threshold is set to ≤10 individuals (including hatchlings). For all models, initial abundance = 60 individuals/ha × patch area (ha), except patch size 0.1 ha, where initial abundance was set to 12 individuals. For the ceiling model, K = 120 individuals/ha × patch size; for the contest model, K = 60 individuals/ha × patch size. R_{max} = 1.10.

As expected, extinction risks decreased with larger propagule sizes of introduced juveniles and with increasing number of introductions. Table 29.1 shows that with 40 juveniles introduced per patch for 3 years (years 0–2) the extinction risk is ≤ 5% (given in bold) for all patch systems, including only one patch (scenario 9). Also low risk are the two-, four-, and eight-patch systems with 40 juveniles introduced per patch for 2 years. A propagule size as low as 10 juveniles per patch also gives < 5% extinction risk for a four-patch system given three introductions (scenario 21) or an eight-patch system given at least two introductions (scenarios 29–30), but with such small propagule sizes the interval probability of increase is low (Table 29.1).

It seems that the same total number of introduced juveniles gives roughly the same extinction risk irrespective of whether these individuals are placed in one patch or divided into several, as in scenarios 9, 15, and 21. This result is also corroborated by the similar values for expected minimum abundance (Table 29.1). However, the probability of increase is lower for multipatch systems. Thus, given that the supply of juveniles is limited, it seems that in the highly correlated metapopulation networks simulated here (see also next section), a wise strategy may be to introduce juveniles for three successive years into one patch in each of several metapopulation networks to minimize correlation among the initial populations.

Effects of Among-Population Dispersal

Table 29.2 shows that introduced metapopulation networks with and without low and high rates of dispersal, respectively, had similar abundances and extinction risks, irrespective of the number of local populations in each network (scenarios a–c). Hence, dispersal does not seem to be important over a 50-year horizon to reduce extinction risks for these types of highly correlated, managed metapopulation networks of the sand lizard. This was also confirmed by testing the effects of absence of among-population correlation in environmental variation, which increased expected minimum abundance and reduced extinction risks relatively more (Table 29.2, scenarios d–e).

The potential for substantial dispersal is of course still necessary for metapopulation establishment if we assume that juveniles are not introduced into all patches in each network. An example of metapopulation development after introducing 40 juveniles for three successive years into one patch in a connected four-patch network is given in Figure 29.3. The expected minimum abundance for this metapopulation was 24 and the extinction risk was 7.5%, which are figures similar to those for the corresponding one-patch introduction scenario 9 in Table 29.1.

Furthermore, simulations over 500 years showed that dispersal is important on longer (evolutionary) time scales. The four-patch metapopulation scenarios 24 f–g in Table 29.2 reveal substantially higher expected minimum abundance and metapopulation occupancy, and lower extinction risk, for the dispersal scenario.

Discussion

Habitat restoration and reintroduction techniques have been employed for the sand lizard in England (Corbett 1988). To establish viable populations in England, three annual releases each of around 50 juveniles (both sexes) are recommended (Moulton and Corbett 1999). These numbers and introduction periods are similar to the "acceptable" one-patch scenario 9 in my simulations (Table 29.1). However, evaluation data from the English reintroductions are scarce. No doubt there is a need to maximize efficiency of sand lizard introductions to overcome demographic and stochastic problems during the buildup phase, especially since the supply of captive-born juveniles will normally be limited.

This study indicates that 5- to 10-ha suitable habitat patches have acceptably low extinction risks over a 50-year horizon. Dispersal among local populations had negligible effects on metapopulation persistence over the same time frame. Dispersal is generally expected to have little effect on local population persistence when growth rates of local populations in the system are highly correlated (Burgman et al. 1993, Stacey et al. 1997), like the ones in my basic metapopulation systems (91% correlation between adjacent local populations). However, in the past landscape of Brattforsheden and other Scandinavian sandy pine forests, dispersal was probably a much more important component for metapopulation survival than is implied here. In this landscape, with a natural forest fire regime and extensive human activities like forest grazing by cattle, sand lizard colonizations and extinctions might have occurred in a shifting spatial mosaic of habitat, with lizards tracking early successional habitats. Thus, continuity of suitable habitat within dispersal distance must have been critical to persistence (cf. Thomas 1994, 1996). In such a landscape it is likely that there was spatiotemporal variation in growth rates within sand lizard metapopulation networks, due to differences in successional stage, patch size, local topography (affecting microclimate and egg-hatching success), catastrophic short-term effects of forest fires, and so on.

Today and in the foreseeable future, however, we probably have to rely on habitat management to "freeze" patches in suitably early successional stages. Such manage-

Table 29.1 Juvenile sand lizard introduction scenarios and their effect on female metapopulation abundance and persistence over the next 50 years

Scenario	No. of Populations	Propagule size	Introduction Year	Final Abundance ± 1 SD	Expected Minimum Abundance	Extinction Risk	Terminal Quasi-extinction Risk	Interval Probability of Increase	Terminal Probability of Increase	Median Time to Quasi-extinction	Metapopulation Occupancy ± 1SD
1	1	10	0	43 ± 88	1.3	0.583	0.61	0.05	0.026	0.5	0.4 ± 0.5
2	1	10	0–1	62 ± 98	3	0.41	0.456	0.08	0.026	1.5	0.6 ± 0.5
3	1	10	0–2	79 ± 100	5.4	0.297	0.347	0.146	0.044	2.8	0.7 ± 0.5
4	1	20	0	63 ± 93	3	0.406	0.448	0.088	0.031	0.6	0.6 ± 0.5
5	1	20	0–1	92 ±114	7.2	0.238	0.302	0.163	0.063	3.3	0.8 ± 0.4
6	1	20	0–2	112 ± 113	12.5	0.121	0.164	0.209	0.069	>50	0.9 ± 0.3
7	1	40	0	86 ± 105	6.7	0.274	0.323	0.168	0.046	1.7	0.7 ±0.5
8	1	40	0–1	109 ± 108	14.8	0.151	0.193	0.255	0.071	>50	0.8 ± 0.4
9	**1**	**40**	**0–2**	**137 ± 119**	**24.1**	**0.053**	**0.088**	**0.338**	**0.092**	**>50**	**0.9 ± 0.3**
10	2	10	0	82 ± 136	3	0.39	0.582	0.036	0.015	0.5	1.0 ± 0.9
11	2	10	0–1	125 ± 157	7.5	0.235	0.397	0.059	0.015	1.5	1.3 ± 0.9
12	2	10	0–2	153 ± 179	13.4	0.133	0.212	0.094	0.035	2.9	1.6 ± 0.7
13	2	20	0	126 ± 170	6.9	0.274	0.355	0.063	0.03	0.7	1.3 ± 0.8
14	2	20	0–1	188 ± 202	16.4	0.114	0.192	0.142	0.054	4.6	1.6 ± 0.7
15	2	20	0–2	215 ± 210	26.9	0.066	0.129	0.17	0.063	>50	1.8 ± 0.5
16	2	40	0	172 ± 195	14.5	0.191	0.255	0.14	0.043	1.7	1.5 ± 0.8
17	**2**	**40**	**0–1**	**234 ± 213**	**32.7**	**0.054**	**0.095**	**0.227**	**0.068**	**>50**	**1.8 ± 0.5**
18	**2**	**40**	**0–2**	**273 ± 229**	**52**	**0.026**	**0.057**	**0.311**	**0.088**	**>50**	**1.9 ± 0.4**

19	4	10	0	169 ± 223	7.2	0.264	0.41	0.019	0.01	0.5	2.2 ± 1.7
20	4	10	0–1	278 ± 338	17.4	0.097	0.225	0.072	0.03	1.5	3.0 ± 1.4
21	**4**	**10**	**0–2**	**352 ± 372**	**29.2**	**0.04**	**0.142**	**0.093**	**0.04**	**3**	**3.4 ± 1.1**
22	4	20	0	284 ± 395	16.6	0.169	0.268	0.071	0.028	0.7	2.9 ± 1.5
23	**4**	**20**	**0–1**	**372 ± 371**	**37.1**	**0.054**	**0.122**	**0.122**	**0.041**	**8.7**	**3.4 ± 1.2**
24	**4**	**20**	**0–2**	**453 ± 401**	**58.9**	**0.019**	**0.074**	**0.181**	**0.069**	**>50**	**3.7 ± 0.8**
25	4	40	0	357 ± 385	31.3	0.139	0.206	0.129	0.038	1.7	3.1 ± 1.5
26	**4**	**40**	**0–1**	**483 ± 455**	**72.8**	**0.024**	**0.072**	**0.218**	**0.07**	**>50**	**3.7 ± 0.8**
27	**4**	**40**	**0–2**	**535 ± 425**	**110**	**0.006**	**0.025**	**0.294**	**0.076**	**>50**	**3.9 ± 0.5**
28	8	10	0	380 ± 574	16	0.178	0.335	0.018	0.011	0.5	4.8 ± 3.0
29	**8**	**10**	**0–1**	**583 ± 660**	**38.3**	**0.04**	**0.158**	**0.061**	**0.022**	**1.5**	**6.6 ± 2.0**
30	**8**	**10**	**0–2**	**711 ± 678**	**62.9**	**0.015**	**0.094**	**0.089**	**0.031**	**3**	**6.9 ± 1.9**
31	8	20	0	527 ± 597	33.1	0.124	0.237	0.056	0.017	0.6	5.8 ± 2.9
32	**8**	**20**	**0–1**	**820 ± 788**	**78**	**0.027**	**0.087**	**0.131**	**0.059**	**19**	**7.1 ± 1.8**
33	**8**	**20**	**0–2**	**944 ± 833**	**126.8**	**0.003**	**0.054**	**0.169**	**0.063**	**>50**	**7.5 ± 1.3**
34	8	40	0	699 ± 695	65.3	0.108	0.172	0.106	0.024	1.8	6.4 ± 2.7
35	**8**	**40**	**0–1**	**991 ± 817**	**148.4**	**0.011**	**0.04**	**0.214**	**0.06**	**>50**	**7.5 ± 1.4**
36	**8**	**40**	**0–2**	**1101 ± 870**	**232**	**0.001**	**0.018**	**0.299**	**0.09**	**>50**	**7.8 ± 0.8**

Note: *Propagule size* = number of introduced juvenile females per patch. *Expected minimum abundance* = the average of the smallest metapopulation size that is expected to occur (cf. McCarthy and Thompson 2001). *Extinction risk* = the probability that metapopulation abundance will fall to zero. *Terminal quasi-extinction risk* = the probability that metapopulation abundance will be less than 10 individuals × number of populations at the end of the 50-year period. *Interval probability of increase* = the probability that metapopulation abundance will exceed 300 individuals × number of populations, at least once within the 50-year period. *Terminal probability of increase* = the probability that metapopulation abundance will end up above 300 individuals × number of populations, at the end of the 50-year period. *Median time to quasi-extinction* = the median time it takes the metapopulation size to fall below 10 individuals × number of populations. *Metapopulation occupancy* = average ± 1 standard deviation of the number of extant populations (i.e., occupied patches) during the 50-year period. Boldface = scenarios with ≤ 5% extinction risk.

Table 29.2 Juvenile sand lizard introduction scenarios with and without among-population dispersal and correlation, respectively, and the effects on metapopulation abundance and persistence over the next 50 years

Scenario	Dispersal	Correlation	No. of Populations	Propagule Size	Introduction Year	Final Abundance ± 1 SD	Expected Minimum Abundance	Extinction Risk	Terminal Quasi-extinction Risk	Interval Probability of Increase	Median Time to Quasi-extinction	Metapopulation Occupancy ± 1SD
15a	Yes	Yes	2	20	0–2	215 ± 210	26.9	0.066	0.129	0.17	>50	1.8 ± 0.5
15b	Yes 10×	Yes	2	20	0–2	234 ± 217	27.6	0.063	0.120	0.210	>50	1.8 ± 0.5
15c	No	Yes	2	20	0–2	217 ± 201	27.7	0.042	0.098	0.169	>50	1.7 ± 0.6
15d	Yes	No	2	20	0–2	231 ± 178	31.4	0.020	0.049	0.114	>50	1.9 ± 0.4
15e	No	No	2	20	0–2	216 ± 166	32.0	0.026	0.061	0.124	>50	1.7 ± 0.5
24a	Yes	Yes	4	20	0–2	453 ± 401	58.9	0.019	0.074	0.181	>50	3.7 ± 0.8
24b	Yes 10×	Yes	4	20	0–2	486 ± 437	59.9	0.019	0.066	0.206	>50	3.8 ± 0.7
24c	No	Yes	4	20	0–2	430 ± 395	59.8	0.018	0.074	0.156	>50	3.4 ± 1.0
24d	Yes	No	4	20	0–2	495 ± 270	74.2	0.003	0.009	0.053	>50	3.9 ± 0.4
24e	No	No	4	20	0–2	444 ± 257	74.1	0.003	0.010	0.038	>50	3.4 ± 0.8
24f	Yes 500 yrs	Yes	4	20	0–2	545 ± 436	36.0	0.08	0.090	0.952	298	3.6 ± 1.1
24g	No 500 yrs	Yes	4	20	0-2	326 ± 389	30.2	0.14	0.205	0.740	210	2.1 ± 1.4
33a	Yes	Yes	8	20	0–2	944 ± 833	126.8	0.003	0.054	0.169	>50	7.5 ± 1.3
33b	Yes 10×	Yes	8	20	0–2	1002 ± 912	124.0	0.005	0.050	0.192	>50	7.7 ± 1.1
33c	No	Yes	8	20	0–2	841 ± 740	120.2	0.003	0.048	0.129	>50	6.8 ± 1.6
33d	Yes	No	8	20	0–2	1030 ± 442	157.5	0.000	0.001	0.013	>50	7.9 ± 0.4
33e	No	No	8	20	0–2	876 ± 377	157.8	0.000	0.000	0.002	>50	6.8 ± 1.1

Note: The dispersal scenarios b (= "l0x") refer to about a 10 times higher annual dispersal rate than that used in the basic model (see Methods), and scenarios 24f–g (= "500 yrs") refer to simulations over 500 years. For further details see Table 29.1.

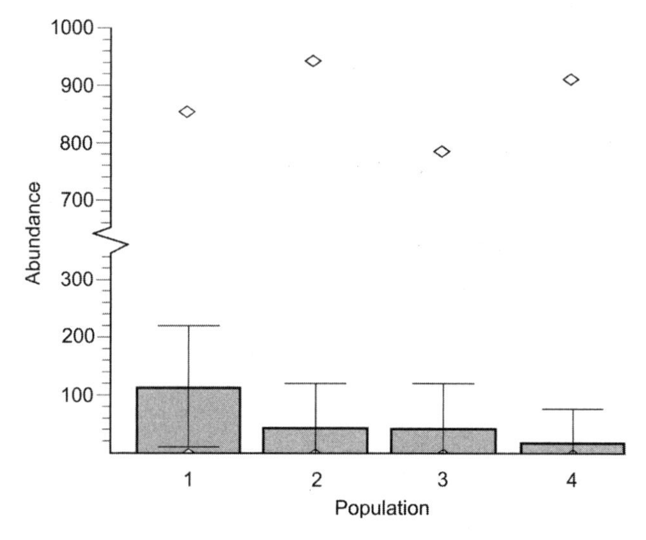

Figure 29.3 Simulated metapopulation structure of female sand lizards 50 years after the start of introduction of 40 juveniles for three successive years into one patch (patch 1) in a connected, empty four-patch network. The histogram shows the average ±1 SD and the minimum and maximum number of individuals in each population. The annual dispersal rate was 3% per patch for 0-, 1-, and 2-year-olds, respectively. See "Methods" for details of dispersal parameterization.

ment is now undertaken on Brattforsheden, with felling of up to 10-ha large pine stands, and within these excavation of, on average, one sand patch ≥100 m^2 per ha for egg laying by the sand lizard. At the moment, ca. 20 new habitat patches (divided into six networks) are planned to be restored within the coming years. The lesson from this modeling exercise is that if a limited number of juveniles per year are available for introduction into these patches, it seems wise to introduce them into one patch in a restored network at a time. If fairly large numbers of juveniles are available, it may be a good strategy to divide them into several patches belonging to different, moderately correlated metapopulation networks in order to reduce the extinction risk for the entire regional metapopulation. When self-sustaining, viable local populations have been established, it may also be a plausible strategy to translocate eggs and juveniles from these populations to other restored, empty patches.

Acknowledgments Thanks to Reşit Akçakaya and Per Sjögren-Gulve for inviting me to contribute to this volume. Barry W. Brook and one anonymous reviewer gave much appreciated comments on the manuscript. Thanks also to Jan Bengtsson and Lars Furuholm for assistance in the practical and administrative sand lizard conservation work, and to the timber company Stora Enso for good cooperation with the habitat restoration measures undertaken so far. Financial support for the demographic studies was obtained from the Swedish World Wildlife Fund (WWF), Swedish Biodiversity Centre (CBM), County Administrative Board of Värmland, Swedish Environmental Protection Agency, Carl Tryggers foundation, and Oskar och Lili Lamms foundation.

References

Akçakaya, H. R. 1998. *RAMAS GIS: linking landscape data with population viability analysis (version 3.0)*. Applied Biomathematics, Setauket, N.Y.

Akçakaya, H. R. 2000. Population viability analysis with demographically and spatially structured models. *Ecological Bulletins* 48: 23–38.

Akçakaya, H. R. 2002. RAMAS *Metapop: viability analyses for stage-structured metapopulations (version 4.0)*. Applied Biomathematics, Setauket, N.Y.

Berglind, S.-Å. 1988. The sand lizard, *Lacerta agilis* L., on Brattforsheden, south central Sweden: habitat, threats and conservation. *Fauna och flora* (Stockholm) 83: 241–255 (in Swedish with English summary).

Berglind, S.-Å. 1999. Conservation of relict sand lizard (*Lacerta agilis* L.) populations on inland dune areas in central Sweden. Ph.Lic. thesis, Uppsala University.

Berglind, S.-Å. 2000. Demography and management of relict sand lizard *Lacerta agilis* populations on the edge of extinction. *Ecological Bulletins* 48: 123–142.

Brook, B. W. 2000. Pessimistic and optimistic bias in population viability analysis. *Conservation Biology* 14: 564–566.

Burgman, M. A., Ferson, S., and Akçakaya, H. R. 1993. *Risk assessment in conservation biology*. Chapman and Hall, London.

Caswell, H. 2001. *Matrix population models: construction, analysis, and interpretation*. 2nd ed. Sinauer Associates, Sunderland, Mass.

Clobert, J., Massot, M., Lecomte, J., Sorci, G., de Fraipont, M., and R. Barbault. 1994. Determinants of dispersal behavior: the common lizard as a case study. Pages 183–206 in L. J. Vitt and E. R. Pianka (eds.), *Lizard ecology: historical and experimental perspectives*. Princeton University Press, Princeton, N.J.

Corbett, K. F. 1988. Conservation strategy for the sand lizard (*Lacerta agilis agilis*) in Britain. *Mertensiella* 1: 101–109.

Corbett, K. F., and Tamarind, D. L. 1979. Conservation of the sand lizard, *Lacerta agilis*, by habitat management. *British Journal of Herpetology* 5: 799–823.

Dent, S., and Spellerberg, I. F. 1988. Use of forest ride verges in southern England for the conservation of the sand lizard *Lacerta agilis* L. *Biological Conservation* 45: 267–277.

Gärdenfors, U. (Ed.). 2000. *The 2000 Red List of Swedish species*. Threatened Species Unit, Swedish University of Agricultural Sciences, Uppsala.

Gasc, J.-P., Cabela, A., Crnobrnja-Isailovic, J., Dolmen, D., Grossenbacher, K., Haffner, P., Lescure, J., Martens, H., Martínez Rica, J. P., Maruin, H., Oliveira, M. E., Sofianidou, T. S., Veith, M., and Zuiderwijk, A. (Eds.). 1997. *Atlas of amphibians and reptiles in Europe*. Societas Europaea Herpetologica and Muséum National d'Histoire Naturelle, Paris.

Gilpin, M. 1996. Metapopulations and wildlife conservation: approaches to modeling spatial structure. Pages 11–27 in D. R McCullough (ed.), *Metapopulations and wildlife conservation*. Island Press, Washington, D.C.

Gislén, T., and Kauri, H. 1959. Zoogeography of the Swedish amphibians and reptiles. *Acta Vertebratica* 1(3): 193–397 (Special issue).

Glandt, D., and Bishoff, W. (Eds.). 1988. Biologie und Schutz der Zauneidechse (*Lacerta agilis*). *Mertensiella* 1: 1–257.

Gullberg, A., Olsson, M., and Tegelström, H. 1998. Colonization, genetic diversity, and evolution in the Swedish sand lizard, *Lacerta agilis* (Reptilia, Squamata). *Biological Journal of the Linnaean Society* 65: 257–277.

Hanski, I. 1997. Metapopulation dynamics: from concepts and observations to predictive models. Pages 69–91 in I. A.Hanski and M. E. Gilpin (eds.), *Metapopulation biology: ecology, genetics, and evolution*. Academic Press, San Diego.

Harrison, S. 1994. Metapopulations and conservation. Pages 111–128 in P. J. Edwards, R. M. May, and N. R. Webb (eds.), *Large-scale ecology and conservation biology*. Blackwell Scientific Publications, Oxford.

Huxel, G. R., and Hastings, A. 1999. Habitat loss, fragmentation, and restoration. *Restoration Ecology* 7: 309–315.

Komers, P. E., and Curman, G. P. 2000. The effect of demographic characteristics on the success of ungulate re-introductions. *Biological Conservation* 93: 187–193.

Lena, J., Clobert, J., de Fraipont, M., Lecomte, J., and Guyot, G. 1998. The relative influence of density and kinship on dispersal in the common lizard. *Behavioral Ecology* 9: 500–507.

Massot, M., Clobert, J., Pilorge, T., Lecomte, J., and Barbault, R. 1992. Density dependence in the common lizard: demographic consequences of a density manipulation. *Ecology* 73: 1742–1756.

McCarthy, M. A., and Thompson, C. 2001. Expected minimum population size as a measure of threat. *Animal Conservation* 4: 351–355.

Moulton, N., and Corbett, K. 1999. *The sand lizard conservation handbook*. English Nature, Peterborough, U.K.

Nature Conservancy Council. 1983. *The ecology and conservation of amphibian and reptile species endangered in Britain*. Nature Conservancy Council, London.

Olsson, M., Gullberg, A., and Tegelström, H. 1996. Malformed offspring, sibling matings, and selection against inbreeding in the sand lizard (*Lacerta agilis*). *Journal of Evolutionary Biology* 9: 229–242.

Olsson, M., Gullberg, A., and Tegelström, H. 1997. Determinants of breeding dispersal in the sand lizard, *Lacerta agilis* (Reptilia, Squamata). *Biological Journal of the Linnaean Society* 60: 243–256.

Rodenhouse, N. L., Sherry, T. W., and Holmes, R. 1997. Site-dependent regulation of population size: a new synthesis. *Ecology* 78: 2025–2042.

Rodenhouse, N. L., Sherry, T. W., and Holmes, R. T. 2000. Site-dependent regulation of population size: a reply. *Ecology* 81: 1168–1171.

Ronce, O., Clobert, J., and Massot, M. 1998. Natal dispersal and senescence. *Proceedings of the National Academy of Sciences USA* 95: 600–605.

Snyder, N. F. R., Derrickson, S. R., Beissinger, S. R., Wiley, J. W., Smith, T. B., Toone, W. D., and Miller, B 1996. Limitations of captive breeding in endangered species recovery. *Conservation Biology* 10: 338–348.

Stacey, P. B., Johnson, V. A., and Taper, M. L. 1997. Migration within metapopulations: the impact upon local population dynamics. Pages 267–291 in I. A. Hanski and M. E. Gilpin (eds.), *Metapopulation biology: ecology, genetics, and evolution*. Academic Press, San Diego.

Strijbosch, H., and Creemers, R. C. M.1988. Comparative demography of sympatric populations of *Lacerta vivipara* and *Lacerta agilis*. *Oecologia* 76: 20–26.

Strijbosch, H., and van Gelder, J. J. 1997. Population structure of lizards in fragmented landscapes and causes of their decline. Pages 347–351 in W. Böhme, W. Bishoff, and T. Ziegler (eds.), *Herpetologia bonnensis*. Societas Europaea Herpetologica, Bonn.

Thomas, C .D. 1994. Extinction, colonization, and metapopulations: environmental tracking by rare species. *Conservation Biology* 8: 373–378.

Thomas, C. D. 1996. Essential ingredients of real metapopulations, exemplified by the butterfly *Plebejus argus*. Pages 292–307 in M. E. Hochberg, J. Clobert, and R. Barbault (eds.), *Aspects of the genesis and maintenance of biological diversity*. Oxford University Press, Oxford.

Tinkle, D. W., Dunham, A. E., and Congdon, J. D. 1993. Life history and demographic variation in the lizard *Sceloporus graciosus:* a long-term study. *Ecology* 74: 2413–2429.

Yablokov, A.V., Baranov, A. S., and Rozanov, A. S. 1980. Population structure, geographic variation, and microphylogenesis of the sand lizard (*Lacerta agilis*). Pages 91–127 in M. K. Hecht, W. C. Steere, and B. Wallace (eds.), *Evolutionary biology*, vol. 12. Plenum Press, New York.

30

Southern Great Barrier Reef Green Sea Turtle (*Chelonia mydas*) Stock

Consequences of Local Sex-Biased Harvesting

MILANI CHALOUPKA

The green turtle (*Chelonia mydas*) is a marine species with a broad pantropical distribution (Bowen et al. 1992) and is the most abundant large herbivore in marine ecosystems (Bjorndal et al. 2000). The southern Great Barrier Reef (sGBR) green turtle stock is one of the most important breeding populations of green turtles in the western South Pacific region (FitzSimmons et al. 1997) with benthic foraging grounds in Australian, Indonesian, Papua New Guinean, Coral Sea, and New Caledonian waters (Limpus et al. 1992).

The Great Barrier Reef and coastal Queensland benthic habitat component of this stock includes a spatially disjunct metapopulation with foraging grounds that span 12° latitude and 1800 km with foraging grounds that range from tropical waters in the northern Great Barrier Reef (nGBR) to warm temperate waters in southern coastal Queensland (Figure 30.1) (Chaloupka 2002a). Pelagic juveniles recruit to these benthic habitat foraging grounds after pelagic development in the western South Pacific Ocean (Limpus and Chaloupka 1997).

Adult turtles from this stock resident in all foraging grounds then migrate every few years to breed at one regional rookery in sGBR waters, and the females nest on the nearby coral cays (Figure 30.1). There is strong philopatry by adults to the foraging grounds (Limpus et al. 1992), but this does not necessarily occur with the immature green turtles from this stock. The age class–specific dispersal behavior of green turtles is poorly understood (Chaloupka 2002a).

The sGBR green turtle metapopulation abundance is stable, with some subpopulations apparently increasing in abundance since the mid-1980s (Figure 30.2) (see Chaloupka and Limpus 2001, Chaloupka 2002a). Nonetheless, this stock is subject to increasing anthropogenic hazards, including foraging habitat change, coastal trawl fisheries, and indigenous harvesting (Chaloupka 2002a).

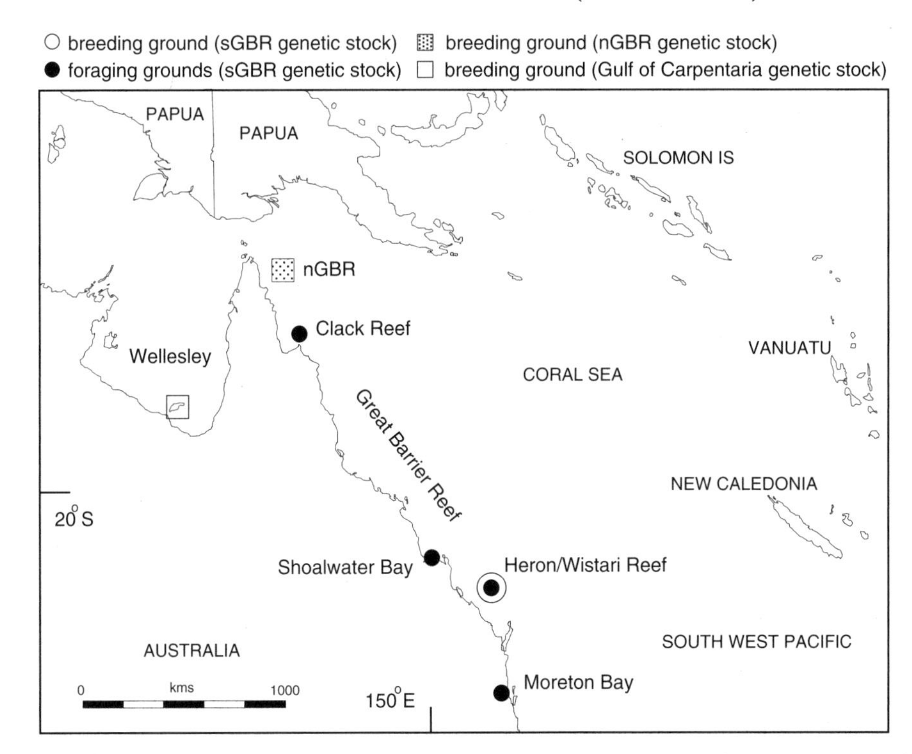

Figure 30.1 Location of the four foraging grounds used here to represent the sGBR green turtle stock resident in GBR and southern coastal Queensland waters (*solid dot*). The foraging grounds comprising this metapopulation are Clack Reef, Shoalwater Bay, Heron/Wistari Reef, and Moreton Bay, representing the estimated 15,675 km^2 of reefal (algae and/or seagrass) and coastal seagrass habitat occupied by the sGBR benthic habitat metapopulation. It was assumed that Clack Reef residents represent the nGBR reefal habitat component of the benthic metapopulation, Shoalwater Bay represents a central coastal Queensland seagrass habitat component, Heron/Wistari Reef represents the sGBR reefal component, and Moreton Bay represents a southern coastal Queensland seagrass habitat component. The regional rookery for this benthic habitat metapopulation is located on the coral cays in the sGBR region shown by open dot. The major rookeries of the other two genetic stocks of Australian green turtles in the same region are also shown (nGBR, Wellesley Island group).

The population dynamics of this stock was explored here using RAMAS Metapop, which extends a single habitat stochastic simulation model (Chaloupka 2002b) by taking into account some aspects of the spatial structure of the sGBR benthic habitat metapopulation (Chaloupka 2002a) and the assumed immature dispersal behavior. The metapopulation model was then used to explore the potential impacts of turtle harvesting on local and metapopulation abundance, given the assumed demographic structure.

Chaloupka and Musick (1997) provide a comprehensive review of sea turtle population dynamics modeling, while Chaloupka (2002a) provides a more demographically detailed model of sGBR green turtle population dynamics based on dynamically coupled systems of differential equations than what is presented here or in Chaloupka (2002b).

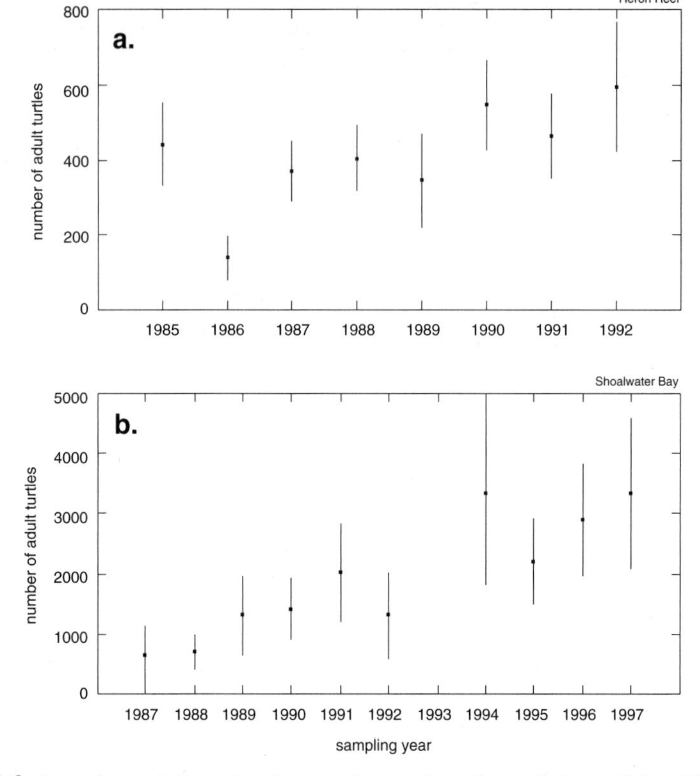

Figure 30.2 Annual population abundance estimates for subpopulations of the sGBR benthic habitat metapopulation that were resident in the (a) Heron/Wistari Reef and (b) Shoalwater Bay foraging grounds during the 1980s and 1990s. Solid square = mean Horvitz-Thompson type abundance estimate; vertical bar = 95% confidence interval for Horvitz-Thompson estimate. See Chaloupka and Limpus (2001, 2002) and Chaloupka (2002a) for details.

Methods

Metapopulation Model

Stage Structure

A stochastic sex- and stage-structured model of sGBR green turtle metapopulation dynamics was developed using RAMAS Metapop; see Chaloupka (2002b) for more details on the demographic basis for this model and on fractional factorial sampling approaches to model sensitivity evaluation. The model was based on six stages and four subpopulations comprising the Great Barrier Reef (GBR) benthic metapopulation using eight sex- and subpopulation or habitat-specific birth-pulse and postbreeding census projection matrices (Table 30.1). The six stages were (1) first-year neonates dispersed southward in the east Australian current, (2) pelagic juveniles resident in an oceanic habitat in the western South Pacific Ocean, (3) benthic juveniles that recruit at variable size

Table 30.1 Summary of the annual sex-specific stage-structured projection matrices for the four habitat-specific subpopulations comprising the sGBR benthic metapopulation of the sGBR green sea turtle stock

		Stage					
Subpopulation	Stage	S1	S2	S3	S4	S5	S6
Northern Great Barrier Reef	S1 (neonate: female)	0	0	0	0.2488	40.592	59.065
	S1 (neonate: male)	0	0	0	0.1296	18.808	31.804
	S2 (pelagic juvenile)	0.4394	0.5690	0	0	0	0
	S3 (benthic juvenile)	0	0.0755	0.8552	0	0	0
	S4 (subadult)	0	0	0.0252	0.8348	0	0
	S5 (maturing adult)	0	0	0	0.0126	0.8646	0
	S6 (adult)	0	0	0	0	0.0836	0.9482
Southern Great Barrier Reef	S1 (neonate: female)	0	0	0	0.2488	40.592	59.065
	S1 (neonate: male)	0	0	0	0.1296	18.808	31.804
	S2 (pelagic juvenile)	0.4394	0.5690	0	0	0	0
	S3 (benthic juvenile)	0	0.0755	0.8391	0	0	0
	S4 (subadult)	0	0	0.0413	0.8397	0	0
	S5 (maturing adult)	0	0	0	0.0077	0.7749	0
	S6 (adult)	0	0	0	0	0.1733	0.9482
Central coastal Queensland	S1 (neonate: female)	0	0	0	0.2488	40.592	59.065
	S1 (neonate: male)	0	0	0	0.1296	18.808	31.804
	S2 (pelagic juvenile)	0.4394	0.5690	0	0	0	0
	S3 (benthic juvenile)	0	0.0755	0.8507	0	0	0
	S4 (subadult)	0	0	0.0297	0.8374	0	0
	S5 (maturing adult)	0	0	0	0.0100	0.7636	0
	S6 (adult)	0	0	0	0	0.1846	0.9482
Southern coastal Queensland	S1 (neonate: female)	0	0	0	0.2488	40.592	59.065
	S1 (neonate: male)	0	0	0	0.1296	18.808	31.804
	S2 (pelagic juvenile)	0.4394	0.5690	0	0	0	0
	S3 (benthic juvenile)	0	0.0755	0.8604	0	0	0
	S4 (subadult)	0	0	0.0200	0.8324	0	0
	S5 (maturing adult)	0	0	0	0.0150	0.8239	0
	S6 (adult)	0	0	0	0	0.1243	0.9482

Note: Matrix values for stages 2–6 are the same for females and males for each subpopulation.

and ages from the pelagic habitat to settle in shallow coastal or coral reef habitats along the northeastern Australian coast (Figure 30.1), (4) subadults or large immature turtles, (5) maturing adults or large turtles that are approaching the onset of both physical and sexual maturity, and (6) mature adults. The somatic growth behavior of the various subpopulations that comprise the sGBR metapopulation is different (Figure 30.3a; see Chaloupka 2002a for details), and this was taken into account in the construction of the stage matrix for each subpopulation (Table 30.1). Age class–specific maturity functions derived from size-specific functions for the subpopulations (Figure 30.4) were then used to define the maturing and mature (adult) stages for the stage-based demographic structure. The same matrix parameter values ($P_{1,2}$; $G_{1,2}$) were used for each subpopulation for the pelagic stages since pelagic demography is independent of the benthic habitat.

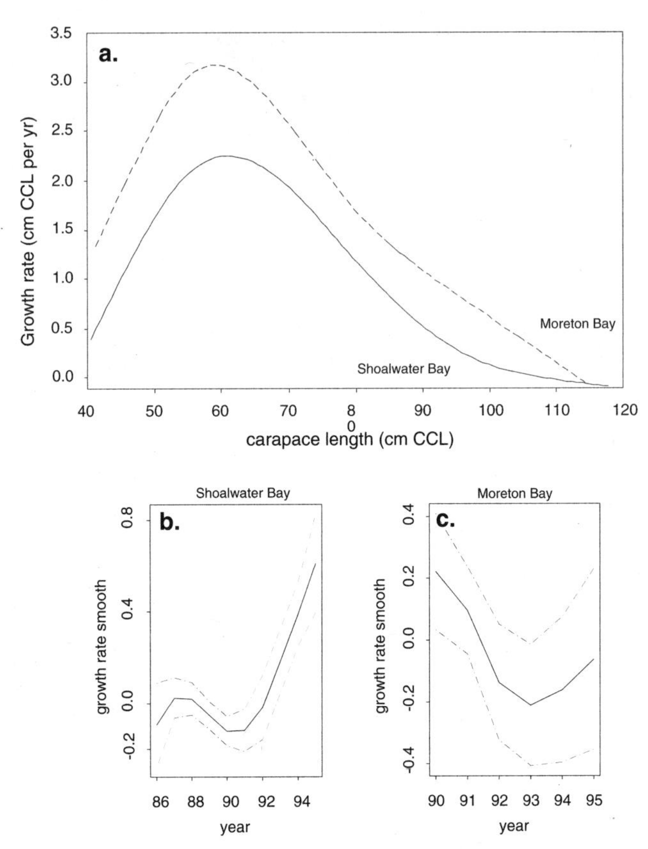

Figure 30.3 Estimated size-specific somatic growth functions for two foraging ground populations shown in **(a)** derived from nonparametric regression modeling of sex- and size-specific growth rates (Limpus and Chaloupka 1997, Chaloupka 2002a). Panels **(b)** and **(c)** show the year-specific growth rate functions derived from nonparametric regression modeling for the foraging ground populations summarized in (a). Year-effects reflect environmental stochasticity in size-specific growth behavior; the depressed growth in these foraging grounds during the early 1990s was due to anomalous rainfall and coastal flooding resulting in loss of inshore seagrass habitat (see Preen et al. 1997).

These eight sex- and habitat-specific projection matrices were derived using lower-level parameter inputs in a spreadsheet linked to robust MATHEMATICA (Wolfram 1996) code provided in Chaloupka (2002b) for deterministic models or for stochastic models drawing from a wide range of probability density functions for the matrix parameters. The model is also sex structured with the same stages for each sex within each of the four subpopulations (Table 30.1). The mating system was assumed here to be polygynous with males able to mate with at least five females. Multiple paternity is known for this stock, but it is apparently at a very low level (FitzSimmons 1998). The stage structure, spatially synchronized breeding behavior (Chaloupka 2001a), and sex- and subpopulation-specific maturation and survival probabilities were based on Limpus et al. (1994), Limpus and Chaloupka (1997), Chaloupka (2001b), Chaloupka and Limpus

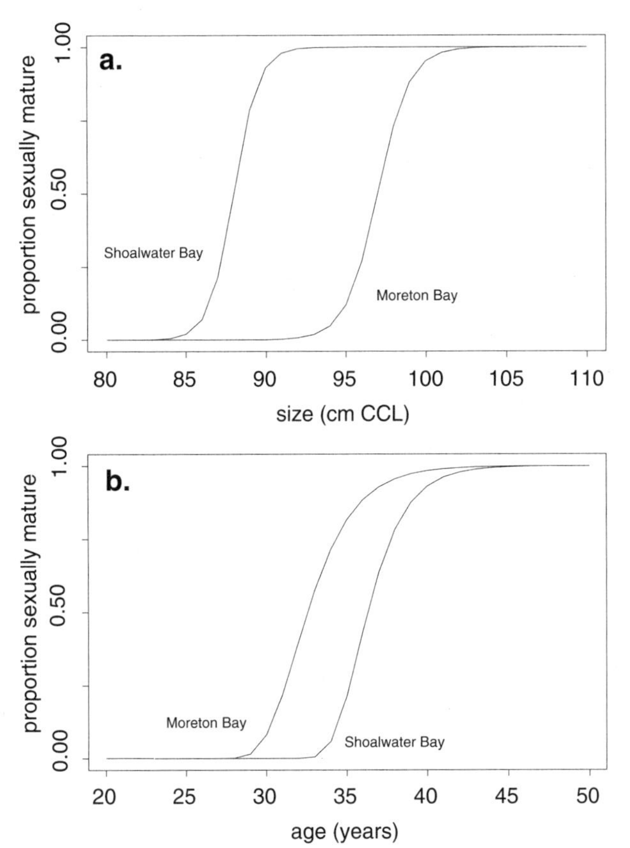

Figure 30.4 **(a)** Expected size-at-maturity functions estimated at two foraging grounds, based on maturity estimates, mean size at nesting of first-time breeders, and courtship size for males (Limpus et al. 1994) and size-specific growth functions (Limpus and Chaloupka 1997, Chaloupka 2002a). **(b)** Estimated age-specific functions based on the size-specific functions in (a) and Weibull-type age class-specific growth functions (Chaloupka 2001b) fitted to numerically integrated forms of the size-specific growth rate functions shown in Figure 30.3a.

(2001, 2002), and Chaloupka (2002a, 2002b). Subpopulation demographic parameters were drawn from comprehensive capture-mark-recapture programs (Limpus et al. 1992, 1994; Limpus and Chaloupka 1997; Chaloupka and Limpus 2002).

Stochasticity

Demographic stochasticity was implemented here by sampling the abundance and associated matrix parameters from a binomial probability mass function embedded in RAMAS Metapop (Akçakaya 1991). Environmental stochasticity is evident in both somatic growth (Figure 30.3b and c) and subpopulation-specific breeding behavior (Figure 30.5) and was accounted for here using sampling variability in the sex-specific matrix fertility parameters based on sampling from lognormal probability density functions. It was assumed that there was within-subpopulation correlation for stochastically

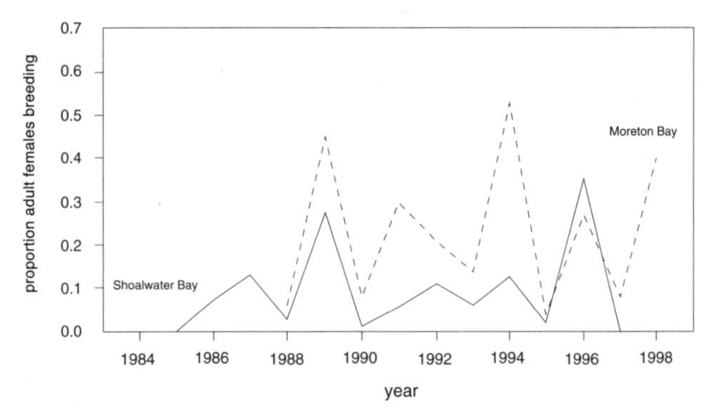

Figure 30.5 Temporal variability in proportion of female green turtles preparing to breed each year at 2 sGBR metapopulation foraging grounds. Males show similar temporal patterns. Breeding status derived from visual examination of gonads using laparoscopy (see Limpus et al. 1985, 1994). See Chaloupka (2002a) for more details relating to other foraging ground subpopulations of this stock.

varying fecundity, survival, and carrying capacity parameters, with a negative correlation imposed to reduce sampling truncation as some survival probabilities were high with considerable sampling variability. The metapopulation abundance was also sampled with error (CV = 0.1) to simulate the effect of imperfect information in abundance estimation for harvesting policies. Catastrophes or very rare but extreme environmental fluctuations were not included here, but such events are known to occur in some sGBR metapopulation foraging grounds (see Preen et al. 1995).

Spatial Structure and Dispersal

Spatial variation in the demographic processes was accounted for explicitly by including four habitat-specific subpopulations. The Great Barrier Reef (GBR) benthic metapopulation habitat comprises 15,675 km^2 of coral reef (74.5%) and coastal seagrass (25.5%) habitats distributed along the Queensland coast and offshore GBR region (Figure 30.1). The four habitat-specific subpopulations were (1) northern GBR coral reef habitat (nGBR) comprising 4203 km^2, or 26.8% of the benthic metapopulation habitat (see, for example, Clack Reef as representative of this habitat; Figure 30.1); (2) central and southern GBR coral reef habitat comprising 7472 km^2, or 47.7% (Heron/Wistari Reef; Figure 30.1); (3) central coastal Queensland seagrass habitat comprising 2560 km^2, or 16.3% (Shoalwater Bay); and (4) southern coastal Queensland seagrass habitat comprising 1440 km^2, or 9.2% (Moreton Bay). These subpopulations were linked in the model by using a four-parameter nonlinear distance-dependent dispersal function incorporated in RAMAS Metapop to account explicitly for spatial variation in local demographic processes. This function was used to derive a symmetrical matrix of dispersal probabilities between the four habitats that are assumed to be very low between habitats further apart. Rogers (1995) provides a useful introduction to construction and use of multistate matrix projection models to account for local demography within a regional or metapopulation context given intraregional dispersal. Significant synchronous fluctuations in the breeding behavior of the subpopulations are well known for this stock (Figure 30.5; see also Chaloupka 2002a), so that the stochastically varying demographic probabilities in the model were assumed to be correlated between the four habitats based on a three-parameter nonlinear distance-dependent function in RAMAS Metapop. Spatial synchrony in breeding and nesting behavior is also well known for other green turtle stocks; see Chaloupka (2001a) for more details on regional Moran effects for green sea turtle breeding in the southeast Asian region.

Density Dependence

Compensatory density-dependent processes were assumed in the model to be of the scramble type, as suggested for a population of immature green turtles resident in seagrass habitats (Bjorndal et al. 2000) with no sex-specific difference. Compensatory density dependence was assumed in the model to affect fecundity within each subpopulation and to be a function of subpopulation-specific benthic stage abundance (stages 3–6). Depensatory density dependence was also included in the model by assuming an Allee effect whenever the population dropped to below 500 individuals. It was assumed that, at this population level, the probability of a female finding at least one male to mate with would be extremely low.

Initial Abundance

Initial sex- and stage-specific subpopulation abundances were based on sex-specific and age class–specific Horwitz-Thompson type abundance estimates for the benthic component (stages 2–6) of the sGBR metapopulation (Chaloupka and Limpus 2001). The model was initialized with ca. 3.2 million individuals, assuming a stable stage structure comprising ca. 65% female and 23% of turtles in the benthic component (see Chaloupka 2002b). Green sea turtle subpopulations from the sGBR metapopulation are usually female-biased (Chaloupka 2002a) due to temperature-dependent sex determination and the warm nesting beaches at the regional rookery (Limpus et al. 1985). Initial abundances for the four subpopulations were based on proportion of habitat type as follows (Chaloupka 2002a, 2002b): nGBR (41.3%), offshore sGBR (28.2%), inshore sGBR (19.4%), and southern coastal Queensland (11.1%).

Simulation Scenarios

The metapopulation dynamics of the sGBR green turtle stock were evaluated in relation to the possible impact of a local or subpopulation-specific harvest by indigenous communities on the long-term viability of the metapopulation. The local effect was implemented here by harvesting only from the nGBR subpopulation, which is the region in close proximity to most coastal indigenous communities in eastern coastal Queensland. It is also the region with the largest habitat suitable for green turtles and probably supports the greatest proportion of the benthic component of the sGBR metapopulation. Moreover, given the distance-dependent dispersal function in the model, it is assumed that the nGBR subpopulation will function as a sink for the influx of immature turtles from the sGBR subpopulations as the nGBR subpopulation declines due to sustained harvesting.

Two harvesting strategies were evaluated:

Scenario 1: (1) at-risk stock was the nGBR subpopulation, (2) at-risk stages were maturing female and mature female adults (female stages 5–6: Table 30.1), (3) harvest strategy was a constant annual offtake of 100 turtles comprising 35 maturing female and 65 mature adult females per annum, (4) harvest duration was from simulation year = 75 and lasted for 50 years, and (5) metapopulation abundance was assessed each year with substantial error reflecting stock assessment error or an inability to fulfill the quota (see "Methods/Stochasticity.")

Scenario 2: same as for scenario 1, but now the harvest strategy was a constant annual offtake of 500 turtles comprising 175 maturing female and 325 mature adult females per annum.

The rationale behind these two scenarios was to take the largest available turtles (stages 5–6), which are usually females (see Limpus and Chaloupka 1997) using a simple-to-implement offtake quota for, say, indigenous harvesting in Great Barrier Reef or coastal Queensland waters. Taking a fixed number of larger turtles is the most easily implemented policy to manage indigenous harvesting in the Great Barrier Reef region. More sophisticated policies—such as proportional threshold-based harvesting—are possible but would not be easily implemented (Chaloupka 2002a). The harvest duration of 50 years over the 200-year simulation period was adopted to ensure that there was sufficient simulation time to observe any response due to the low reproductive capacity of green turtles resident in Great Barrier Reef or southern Queensland waters. The 200-year simulation period approximates ca. four generations and is needed to enable sufficient simulation time to elapse to observe any potential density-dependent metapopulation response for this low reproductive stock.

Metapopulation model response to the harvesting scenarios compared with a no-harvest base model was then assessed by using (1) simulated local and metapopulation abundances and (2) interval extinction probabilities derived from 1,000 Monte Carlo trials over a 200-year simulation period. The interval extinction risk in RAMAS Metapop is a measure of the probability that metapopulation abundance falls below a specific metapopulation threshold at least once during the 200-year simulation period (see Burgman et al. 1993 for details).

Results and Discussion

Model Estimation

The stochastic sex- and stage-specific metapopulation model described here was estimated using RAMAS Metapop on a Wintel-PC (Windows 2000) and on an Apple Macintosh (Mac OS 10.2) using Virtual PC 5.0 (Windows 98/2000). Model output was platform independent, but the model runs slower on the Macintosh as the Wintel PC environment is implemented using software emulation. Hence RAMAS Metapop can be run on a range of platforms using a wide range of operating systems. All RAMAS Metapop output was then saved to disk, modified in a word processor, and input to S-PLUS (MathSoft 1999) to create high-resolution graphical output suitable for publication.

Deterministic Metapopulation Growth Behavior

Deterministic subpopulation growth behavior can be derived easily in RAMAS Metapop with all subpopulations being in deterministic steady-state or perhaps slightly increasing (λ = ca. 1), which is consistent with robust MATHEMATICA code using eigensystem analysis (Chaloupka 2002b) and empirical estimates (Chaloupka and Limpus 2001; see Figure 30.2). Hence, RAMAS Metapop numerical precision is sound for deterministic matrix projection model estimation.

Stochastic Metapopulation Growth Behavior

Figure 30.6a shows that the model produces significant temporal variability in projected metapopulation abundance, given the assumed demographic processes and environmental and demographic stochasticity. The mean of 1,000 runs shows no significant pathological trend in metapopulation abundance over the 200-year period (Figure 30.6b). Autoregressive spectral density analysis of individual runs (Figure 30.6a) shows evidence of reddened spectral behavior (low-frequency fluctuations), as expected for stochastic model behavior of green turtle populations dynamics (Chaloupka 2002b), which is due to the strong environmental stochasticity that affects breeding behavior (Figure 30.5).

Evaluation of Local Sex-Biased Harvesting

Figure 30.6b shows the mean subpopulation abundance for the nGBR habitat component of the sGBR benthic population subject to the two subpopulation-specific harvesting scenarios: (1) at-risk stock was the nGBR subpopulation, (2) at-risk stages were maturing females and mature female adults (female stages 5–6), and (3) harvest strategy was a constant offtake strategy of either 100 or 500 female turtles per annum over a 50-year period starting in simulation year = 75. These harvest scenarios are compared to the no-harvest base model where a significant rebound capacity is readily apparent following cessation of the harvest, which reflects the level of density dependence that was assumed in the model for each subpopulation (Figure 30.6b). Confidence intervals have not been included here simply to avoid visual clutter, but they are readily derived from the model output.

The nGBR subpopulation takes ca. 75 years to recover or rebound after cessation of the sex-biased constant offtake harvesting scenario of 100 female turtles in simulation year =125 (Figure 30.6b). The nGBR subpopulation had not recovered by the end of the 200-year simulation period following an annual offtake of 500 female turtles, although it would appear to be well on the way to recovery by simulation year = 200. Figure 30.6c shows the interval extinction risk for the (1) base model (no harvest) compared to the interval extinction risk for the entire metapopulation subject to the sex-biased harvest scenarios of (2) 100 female turtles per year and (3) 500 female turtles per year over a 50-year harvesting period starting in simulation year = 75.

Given model assumptions, it is apparent that harvesting 100 or even 500 large females each year over the 50-year period from the nGBR subpopulation has little chance of causing the extinction of the sGBR metapopulation. Nonetheless, both harvest levels appear to have a significantly higher interval extinction risk than did the no-harvest base model (Figure 30.6c). Furthermore, the nGBR subpopulation would be seriously de-

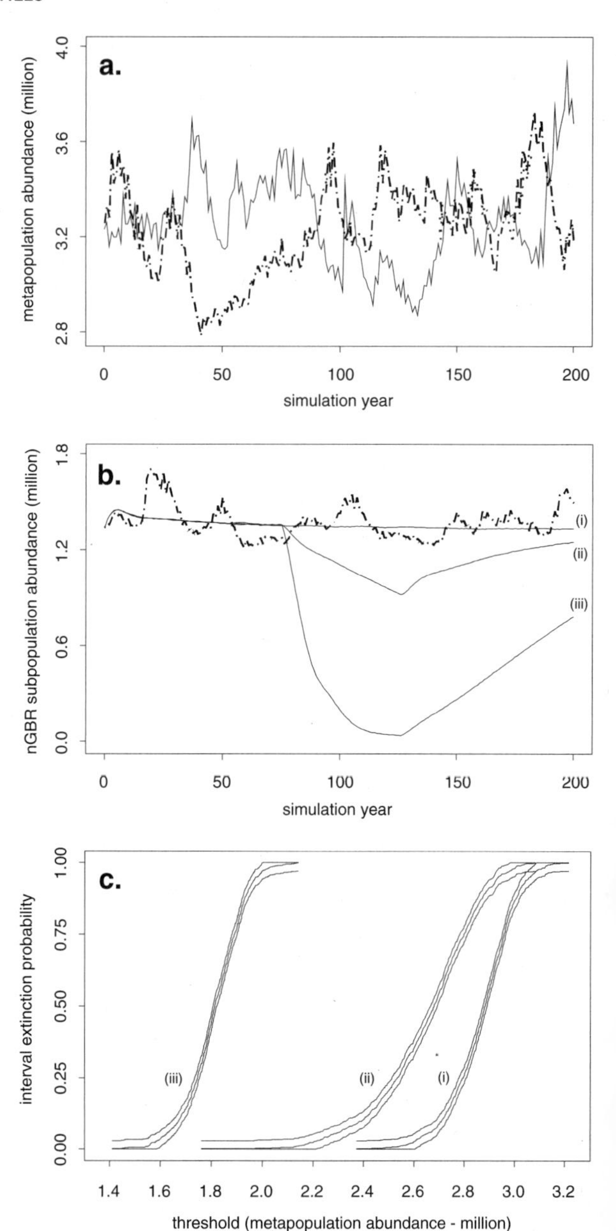

Figure 30.6 Simulated abundance for the sGBR green turtle stock. **(a)** Metapopulation abundance trajectories for two randomly selected runs drawn from 1,000 Monte Carlo trials of the four-habitat model in the absence of any anthropogenic hazards. **(b)** The nGBR subpopulation abundance for one run drawn from the 1,000 trials in the absence of any anthropogenic hazards (*dashed curve*) compared with the expected nGBR subpopulation abundance from the 1,000 trials when subject to (i) no annual offtake, (ii) an annual offtake of 100 female turtles over a 50–year period, and (iii) an annual offtake of 500 female turtles over a 50–year period. The annual 100–female offtake comprised 35 maturing females and 65 adult females from the nGBR subpopulation only; the annual 500–female offtake comprised 175 maturing females and 325 adult females from the nGBR subpopulation only. The constant offtake for each harvest scenario started in simulation year = 75 and concluded in year = 125. **(c)** The expected interval extinction risk and 95% confidence band for the metapopulation abundance associated with the three harvest scenarios: (i) no offtake, (ii) annual offtake of 100 turtles, and (iii) annual offtake of 500 turtles.

pleted by cessation of the harvest if it were exposed to an annual offtake of 500 large females (Figure 30.6b). It is apparent, then, that either level of harvest could decrease the long-term viability of local subpopulations (Figure 30.6b), as well as the metapopulation (Figure 30.6c).

Of course, if harvesting continued for longer than 50 years or if the harvest level were greater than 500 females per year, then more serious depletion of the metapopulation would occur. The model could be used more extensively than here to design a local sustainable harvest level that might minimize the risk to long-term viability of the metapopulation. However, extensive model evaluation would be required before application to policy design would be warranted (see Chaloupka 2002a, 2002b). The results given are exploratory only and are intended to help start developing a better understanding of green sea turtle metapopulation dynamics.

Model Sensitivity to Distance-Dependent Dispersal Functional Form

Figure 30.7a shows that the distance-dependent dispersal functional form assumed in the model with parameter $c = 1.25$—the dispersal-distance function in RAMAS Meta-

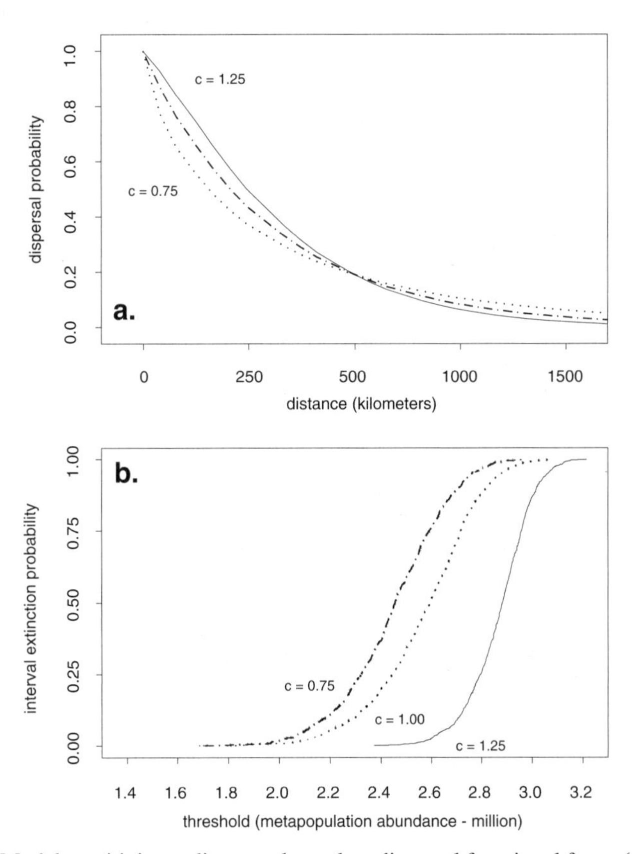

Figure 30.7 Model sensitivity to distance-dependent dispersal functional form. **(a)** Three variations of the dispersal-distance function used in the model derived by varying parameter "c" of the three-parameter function implemented in RAMAS Metapop. **(b)** The expected interval-specific extinction probability function for the metapopulation, given the constant sex-specific offtake harvest scenario summarized in Figure 30.6b that correspond with the three distance-dependent dispersal functions shown in (a).

pop—is a four-parameter function that is readily adjusted to evaluate the effect of various functional forms on model output. The dispersal between subpopulations is poorly known for the sGBR metapopulation, as it is for all sea turtle populations. There is a strong belief that there is substantial foraging ground fidelity for green sea turtles, but this cannot be correct or else there would never be any colonization of new or recolonization of changed habitats over evolutionary time if not ecological time. However, it is apparent from capture-mark-recapture programs that any such dispersal is limited for most sea turtle species, except perhaps for leatherbacks. Nonetheless, there is some empirical evidence based on capture-mark-recapture programs for local and regional dispersal.

Therefore, this function was varied here to evaluate the possible impact of small changes to dispersal-distance functional form assumed in the metapopulation model. Decreasing parameter c in the four-parameter dispersal-distance function invokes more dispersal between the various subpopulations that make up the sGBR benthic habitat metapopulation (see Figure 30.7a). The effect of increasing subpopulation dispersal was evaluated using the no-harvest base model but subject to the two small variations of the distance-dependent dispersal function shown in Figure 30.7a.

Figure 30.7b shows the expected interval extinction risk (confidence intervals excluded to avoid visual clutter) for the metapopulation given three functional forms: (1) the base model with $c = 1.25$, (2) $c = 1.00$, and (3) $c = 0.75$. The c-parameter in the four-parameter RAMAS Metapop function reflects the decline rate of the dispersal probability with distance between subpopulations (see Figure 30.7a).

It is clear from Figure 30.7b that the model is highly sensitive to variations in the functional form of the distance-dependent dispersal process: even a small decrease in this parameter has a significant effect on sGBR green turtle metapopulation dynamics. In other words, decreasing the dispersal of immature benthic green turtles (stages 3 and 4) between the foraging ground subpopulations would result in increased risk of both local and metapopulation decline.

It is possible that immature dispersal is also a function of local or subpopulation density, as well as being distance-dependent. If this is so, then the sGBR metapopulation dynamics could also comprise dynamic subpopulation source-sink interactions and lead to more complex behavior. This density-dependent dispersal function was not included here but is readily implemented in RAMAS Metapop.

Thus, it is apparent that metapopulation spatial structure and subpopulation dispersal is an important demographic process that needs to be better understood for sea turtle populations, which is a finding consistent with other metapopulation studies of marine (York et al. 1996) or maritime species (Spendelow et al. 1995). RAMAS Metapop provides a useful means for evaluating the effect of such processes on long-term viability of sea turtle metapopulations.

Important extensions of the metapopulation model presented here would include developmental or distributed age structure within the stages to account for delayed growth and maturation as shown for modeling of the population dynamics of stage-structured loggerhead sea turtle models (Chaloupka 2003). It would also be worthwhile to explore inclusion of density-dependent dispersal, despite the current lack of an empirical basis to derive the appropriate functional forms. The RAMAS/Metapop files for this model are available on the CD that accompanies this volume.

References

Akçakaya, H. R. 1991. A method for simulating demographic stochasticity. *Ecological Modelling* 54: 133–136.

Bjorndal, K. A., Bolten, A. B., and Chaloupka, M.Y. 2000. Green turtle somatic growth model: evidence for density-dependence. *Ecological Applications* 10: 269–282.

Bowen, B. W., Meylan, A. B., Ross, J. P., Limpus, C. J., Balazs, G. H., and Avise, J. C. 1992. Global population structure and natural history of the green turtle (*Chelonia mydas*) in terms of matriarchal phylogeny. *Evolution* 46: 865–881.

Burgman, M. A., Ferson, S., and Akçakaya, H. R. 1993. *Risk assessment in conservation biology*. Chapman and Hall, New York.

Chaloupka, M. 2001a. Historical trends, seasonality and spatial synchrony in green turtle egg production. *Biological Conservation* 101: 263–279.

Chaloupka, M. 2001b. System-of-equations growth function for southern Great Barrier Reef green sea turtles. *Chelonian Conservation and Biology* 4: 1–6.

Chaloupka, M. 2002a. *Assessment of suitability of Queensland Parks and Wildlife Service sea turtle data for use in models of the population dynamics of the southern Great Barrier Reef green turtle stock*. Research Publication 74. Great Barrier Reef Marine Park Authority, Townsville, Australia.

Chaloupka, M. 2002b. Stochastic simulation modelling of southern Great Barrier Reef green turtle population dynamics. *Ecological Modelling* 148: 79–109.

Chaloupka, M. 2003. Stochastic simulation modeling of loggerhead sea turtle population viability exposed to competing mortality risks in the western South Pacific region. Pages 274–295 in A. B. Bolten and B. E. Witherington (eds.), *The biology and conservation of loggerhead sea turtles*. Smithsonian Institution Press, Washington D.C.

Chaloupka, M. Y., and Limpus, C. J. 2001. Trends in the abundance of sea turtles resident in southern Great Barrier Reef waters. *Biological Conservation* 102: 235–249.

Chaloupka, M. Y., and Limpus, C. J. 2002. Estimates of survival probabilities for the endangered loggerhead sea turtle resident in southern Great Barrier Reef waters. *Marine Biology* 140: 267–277.

Chaloupka, M. Y., and Musick, J. A. 1997. Age, growth and population dynamics. Pages 233–276 in P. J. Lutz and J. A. Musick (eds.), *The biology of sea turtles*. CRC Marine Science Series. CRC Press, Boca Raton, Fla.

FitzSimmons, N. N. 1998. Single paternity of clutches and sperm storage in the promiscuous green turtle (*Chelonia mydas*). Molecular Ecology 7:575–584.

FitzSimmons, N. N., Moritz, C., Limpus, C. J., Pope, L., and Prince, R. 1997. Geographic structure of mitochondrial and nuclear gene polymorphisms in Australian green turtle populations and male-biased gene flow. *Genetics* 147: 1843–1854.

Limpus, C. J., and Chaloupka, M. 1997. Nonparametric regression modelling of green sea turtle growth rates (southern Great Barrier Reef). *Marine Ecology Progress Series* 149: 23–34.

Limpus, C. J., Reed, P. C., and Miller, J. D. 1985. The green turtle, *Chelonia mydas*, in Queensland: a preliminary description of the population structure in a coral reef feeding ground. Pages 47–52 in G. Grigg, R. Shine, and H. Ehmann (eds.), *Biology of Australasian frogs and reptiles*. Royal Society of New South Wales, Sydney, Australia.

Limpus, C. J., Miller, J. D., Parmenter, C. J., Reimer, D., McLachlan, N., and Webb, R. 1992. Migration of green (*Chelonia mydas*) and loggerhead (*Caretta caretta*) turtles to and from eastern Australian rookeries. *Wildlife Research* 19: 347–358.

Limpus, C. J., Couper, P. J., and Read, M. A. 1994. The green turtle, *Chelonia mydas*, in Queensland: population structure in a warm temperate feeding area. *Memoirs of the Queensland Museum* 35: 139–154.

MathSoft. 1999. *S-PLUS 2000 Guide to Statistics*. Vol 1. Data Analysis Products Division, MathSoft, Seattle.

Preen, A. R., Lee Long, W. J., and Coles, R. G. 1995. Flood and cyclone related loss and partial recovery of more than 1000 km^2 of seagrass in Hervey Bay, Queensland, Australia. *Aquatic Botany* 52: 3–17.

Rogers, A. 1995. *Multiregional demography: principles, methods and extensions*. Wiley, New York.

Spendelow, J. A., Nichols, J. D., Nisbet, I. C., Hays, H., Cormons, G. D. , Burger, J., Safina, C., Hines, J. E., and Gochfield, M. 1995. Estimating annual survival and movement rates of adults within a metapopulation of roseate terns. *Ecology* 76: 2415–2428.

Wolfram, S. 1996. *The MATHEMATICA book*. 3rd ed. Cambridge University Press, Cambridge.

York, A. E., Merrick, R. L., and Loughlin, T. R. 1996. An analysis of the Steller sea lion metapopulation in Alaska. Pages 259–292 in D. R. McCullough (ed.), *Metapopulations and wildlife conservation*. Island Press, Washington D.C.

PART V

BIRDS

31

Modeling Birds

An Overview

JEFF S. HATFIELD

Currently, there are estimated to be over 10,000 species of birds in the world, of which about 12% are considered to be threatened or endangered with extinction (BirdLife International 2000). The major causes for the decline of bird populations are thought to be habitat loss and habitat fragmentation (e.g., see Robbins et al. 1989), increased mortality or decreased productivity due to anthropogenic causes (e.g., overharvesting), introduction of exotic diseases or competition with exotic species (e.g., Hawaiian forest birds; see Scott et al. 2001), and various contaminant-related issues (Hoffman et al. 2003). Population viability analysis (PVA) has become a popular tool in recent years, due to its ability to synthesize field data and make objective projections about future population size and the probability of extinction for threatened and endangered species (Beissinger and McCullough 2002).

PVA makes use of estimates of a species' age-specific survival rates, fecundity or productivity, and dispersal rates among subpopulations (i.e., the probability of an individual moving permanently from one subpopulation to another) to produce estimates of the probability of extinction and expected population size over a specified time frame (usually 100 years into the future). If, under a particular management scenario, the probability of extinction is small (e.g., < 5% in 100 years) and under a different scenario the probability of extinction is larger (e.g., > 5% in 100 years) then we may conclude that the first scenario is better for the future management of this species than the second scenario. However, the accuracy and precision of estimates produced from a PVA depend on accurate and precise estimates of vital rates, as well as on the assumption that the demographic model chosen is appropriate for the species (Brook et al. 2000).

Many people and organizations have opinions concerning management actions that should be undertaken to recover a particular declining species, but PVA provides the only objective tool for using the available data to evaluate alternative management scenarios and to identify sensitive life stages (Morris et al. 2002). In many cases, especially for birds, capture-recapture studies have already been conducted, and their analysis can provide estimates of survival rates and temporal variances (Gould and Nichols 1998). Fecundity or productivity studies are also commonly undertaken for bird species, and temporal variances can be produced for these parameters (Akçakaya 2002a). Thus, for many species of birds being studied, survival and fecundity estimates are available, as well as temporal variances for these parameters.

Unfortunately, the one parameter (or set of parameters) that is often not estimated in field studies of birds is dispersal. This is also called "migration" in the PVA literature, but that term is used in the bird literature to describe the seasonal movements of many species of migratory birds. So, although migration may be related to dispersal, dispersal should not be confused with seasonal types of movement. Regardless, dispersal rates may be among the most important parameters to estimate, because dispersal among subpopulations will determine whether a metapopulation dynamic is occurring for any given species in which the available habitat is naturally patchy or has been fragmented due to anthropogenic causes.

Another problem related to dispersal is that, for species with high-dispersal abilities, the survival rates estimated from capture-recapture studies may be biased downward; this is because capture-recapture studies estimate apparent survival (the probability that a bird has survived and returned to the study area) and not actual survival. Actual survival may be higher because some of the missing study birds may have actually survived and dispersed elsewhere. Certain age classes may be more affected than others (i.e., hatch-year birds and other immature age classes that have not yet reached breeding age) because those age classes are more likely to disperse. If the bias is large, however, the survival rate can be increased somewhat, especially if data from other species corroborate this approach (e.g., see hatch-year survival in U.S. Fish and Wildlife Service 1996a).

Many bird species migrate over vast areas, so it is difficult to argue that dispersal is not important for these species, especially if the population is divided into subpopulations. However, at least in my own experience (U.S. Fish and Wildlife Service 1996a, 1996b; Alldredge et al., Chapter 33 in this volume), recaptures of birds dispersing from one study area to another are rare in field studies, so the estimation of dispersal rates is usually not possible. For species that are migratory or have high dispersal, this implies that we may never be able to provide accurate predictions of future population dynamics using PVA. Perhaps in the future, with higher levels of funding and more extensive field studies, better estimates of dispersal rates will become available for bird species for which it is deemed necessary.

Although other software packages are available (see Brook et al. 1999, 2000), RAMAS Metapop/GIS (Akçakaya 2002b) provides the researcher with a convenient, user-friendly, windows-based, spatially explicit software package for conducting PVA. It has been used for birds (Akçakaya and Atwood 1997, Akçakaya and Raphael 1998, Root 1998, Cox and Engstrom 2001, U.S. Fish and Wildlife Service 2001), as well as for many other species. In this section, we have examples of conducting PVA on birds that span the range of problems found in birds.

Rodríguez et al. (Chapter 32 in this volume) addressed the issue of overharvesting for the yellow-shouldered parrot (*Amazona barbadensis*) on islands in Venezuela.

Alldredge et al. (Chapter 33 in this volume) studied the golden-cheeked warbler (*Dendroica chrysoparia*)—in particular, whether a metapopulation dynamic could be occurring in the patchy landscape where this species breeds in central Texas. Sachot and Perrin (Chapter 34 in this volume) conducted a PVA to investigate management options for the capercaillie (*Tetrao urogallus*) in Switzerland, and Shriver and Gibbs (Chapter 35 in this volume) conducted a PVA to investigate the possible effects of rise in sea level on a metapopulation of seaside sparrows (*Ammodramus maritimus*). McCarthy et al. (Chapter 36 in this volume) have a somewhat different problem: they evaluated a reintroduction program for the helmeted honeyeater (*Lichenostomus melanops cassidix*) in which the captive population is considered one of the subpopulations in a metapopulation. Finally, Inchausti and Weimerskirch (Chapter 37 in this volume) report on a PVA for the wandering albatross (*Diomedea exulans chionoptera*), a pelagic seabird that is distributed patchily across the Southern Oceans.

The PVAs presented in this section should help the reader understand how the various authors used RAMAS Metapop/GIS to conduct PVA on these particular bird species. We hope readers will feel inspired to conduct PVAs on other species. With at least 1,200 threatened or endangered species of birds in the world today, a number that is likely to increase, surely the need to conduct more PVAs will also increase.

References

Akçakaya, H. R. 2002a. Estimating the variance of survival rates and fecundities. *Animal Conservation* 5: 333–336.

Akçakaya, H. R. 2002b. *RAMAS Metapop: viability analysis for stage-structured metapopulations (version 4.0).* Applied Biomathematics, Setauket, N.Y.

Akçakaya, H. R., and Atwood, J. L. 1997. A habitat-based metapopulation model of the California gnatcatcher. *Conservation Biology* 11: 422–434.

Akçakaya, H. R., and Raphael, M. G. 1998. Assessing human impact despite uncertainty: viability of the northern spotted owl metapopulation in the northwestern USA. *Biodiversity and Conservation* 7: 875–894.

Beissinger, S. R., and McCullough, D. R. (Eds.). 2002. *Population viability analysis.* University of Chicago Press, Chicago.

BirdLife International. 2000. *Threatened birds of the world.* Lynx Edicions and Birdlife International, Barcelona.

Brook, B. W., Cannon, J. R., Lacy, R. C.,Mirande, C., and Frankham, R.. 1999. Comparison of the population viability analysis packages GAPPS, INMAT, RAMAS, and VORTEX for the whooping crane (*Grus Americana*). *Animal Conservation* 2: 23–31.

Brook, B. W., O'Grady, J. J., Chapman, A. P., Burgman, M. A., Akçakaya, H. R., and Frankham, R. 2000. Predictive accuracy of population viability analysis in conservation biology. *Nature* 404: 385–387.

Cox, J., and Engstrom, R. T. 2001. Influence of the spatial pattern of conserved lands on the persistence of a large population of red-cockaded woodpeckers. *Biological Conservation* 100: 137–150.

Gould, W. R., and Nichols, J. D. 1998. Estimation of temporal variability of survival in animal populations. *Ecology* 79: 2531–2538.

Hoffman, D. J., Rattner, B. A., Burton Jr., G. A., and Cairns Jr., J. 2003. *Handbook of ecotoxicology*. 2nd ed. Lewis Publishers, Boca Raton, Fla.

Morris, W. F., Bloch, P. L., Hudgens, B. R., Moyle, L. C., and Stinchcombe, J. R. 2002. Population viability analysis in endangered species recovery plans: use and future improvements. *Ecological Applications* 12: 708–712.

Robbins, C. S., Dawson, D. K., and Dowell, B. A. 1989. Habitat area requirements of breeding forest birds of the Middle Atlantic States. *Wildlife Monographs* 103:1–34.

Root, K. V. 1998. Evaluating the effects of habitat quality, connectivity and catastrophes on a threatened species. *Ecological Applications* 8: 854–865.

Scott, J. M., Conant, S., and van Riper III, C. (Eds.). 2001. *Evolution, ecology, conservation, and management of Hawaiian birds: a vanishing avifauna.* Studies in Avian Biology 22, Cooper Ornithological Society. Allen Press, Lawrence, Kans.

U.S. Fish and Wildlife Service. 1996a. *Golden-cheeked warbler population and habitat viability assessment report.* Compiled and edited by Carol Beardmore, Jeff Hatfield, and Jim Lewis in conjunction with workshop participants. Report of an August 21–24, 1995, workshop arranged by the U.S. Fish and Wildlife Service in partial fulfillment of U.S. National Biological Survey Grant No. 80333–1423. Austin, Texas.

U.S. Fish and Wildlife Service. 1996b. *Black-capped vireo population and habitat viability assessment report.* Compiled and edited by Carol Beardmore, Jeff Hatfield, and Jim Lewis in conjunction with workshop participants. Report of a September 18–21, 1995, workshop arranged by the U.S. Fish and Wildlife Service in partial fulfillment of U.S. National Biological Survey Grant No. 80333–1423. Austin, Texas.

U.S. Fish and Wildlife Service. 2001. *Southwestern willow flycatcher recovery plan.* Albuquerque, N.M.

32

Yellow-Shouldered Parrot (*Amazona barbadensis*) on the Islands of Margarita and La Blanquilla, Venezuela

Poaching and the Survival of a Threatened Species

JON PAUL RODRÍGUEZ
LAURIE FAJARDO
ILEANA HERRERA
ADA SÁNCHEZ
ANNEDIS REYES

Approximately one-third of the 140 New World parrot species are considered at risk of extinction (Collar and Juniper 1992, Collar 1996, Snyder et al. 1999, BirdLife 2000). Venezuela is home to 14 genera and 49 species in the family Psittacidae, representing 35% of neotropical taxa (Desenne and Strahl 1991, 1994; Phelps and Meyer de Schauensee 1994). They occupy all habitat types in the country, from the dry scrub across the Caribbean and Atlantic coasts to the highland meadows (known locally as *páramos*) of the Andean cordillera in the west. Some 7 parrot taxa are considered nationally threatened, while 11 more are listed as facing a lower extinction risk (Rodríguez and Rojas-Suárez 1999).

The yellow-shouldered parrot (*Amazona barbadensis*) presents a fragmented population that includes sites along the mainland Venezuelan coast, the Venezuelan islands of Margarita and La Blanquilla, and the island nation of Bonaire. One subpopulation, which previously inhabited Aruba, is now considered extinct (Low 1981). The yellow-shouldered parrot is listed globally as "vulnerable" (BirdLife 2000), while nationally it is considered "endangered." It is one of the four most threatened parrots in the country (Rodríguez and Rojas-Suárez 1999).

In 1989, Provita (a Venezuelan nongovernmental organization), in collaboration with the Wildlife Conservation Society, Fundación para la Defensa de la Naturaleza, Ministry of the Environment, National Parks Institute, National Guard, and owners of the privately held ranch Fundo San Francisco, initiated a conservation program for the yellow-shouldered parrot on Margarita and La Blanquilla islands (Figure 32.1). At that time, the population on Margarita was 750 individuals, and approximately 100 parrots inhabited La Blanquilla (Rojas-Suárez 1994a, 1994b; Silvius 1997). On both islands, the capture of nestlings for the pet trade (both domestic and international) was the main

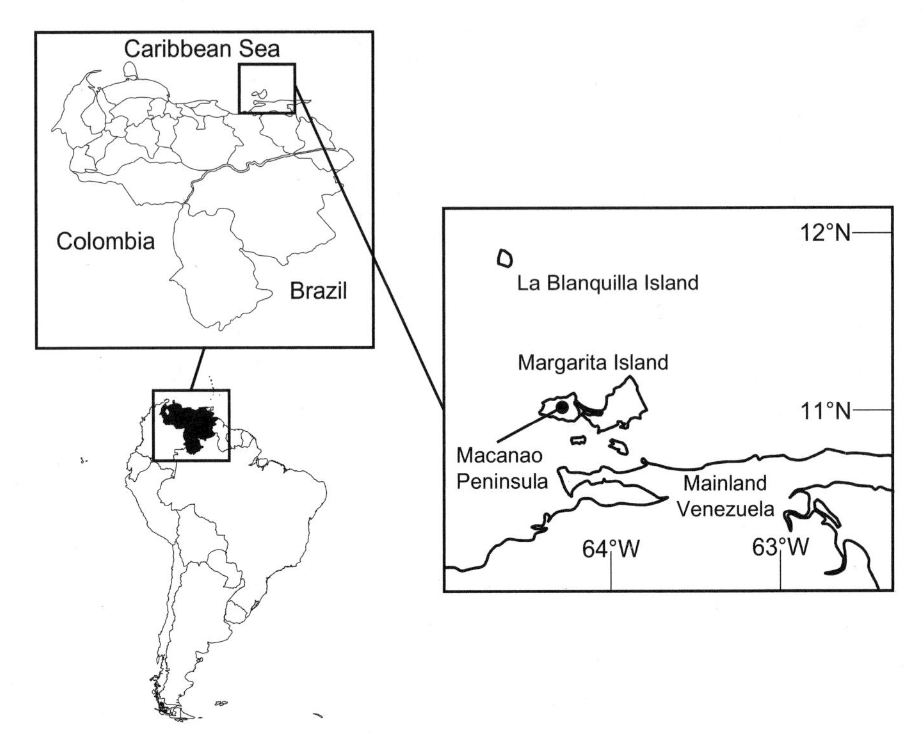

Figure 32.1 Location of La Blanquilla and Margarita islands. The insets show the location of Venezuela within South America (*bottom left*) and the position of our study site off the northeastern Venezuelan mainland coast (*top left*).

factor driving population declines (Rojas-Suárez 1994a, 1994b). Habitat conversion for tourism infrastructure developments and the uncontrolled mining for construction materials in Margarita were the principal factors destroying nesting habitat (Sanz 2001). A population viability analysis performed in the early 1990s estimated that the population on La Blanquilla could be considered "critically endangered" (sensu IUCN 2001), while the population on Margarita was "endangered" (Rodríguez and Rojas-Suárez 1994).

Provita's yellow-shouldered parrot conservation program has been built around the concept of "regional pride" (Butler 1992), combining scientific research with increasing awareness, environmental education, and active involvement of local stakeholders (e.g., see Albornoz et al. 1994, Rojas-Suárez 1994a, Rodríguez 1995, Sanz and Rojas-Suárez 1997, Silvius 1997, Sanz and Grajal 1998). Following the pioneering work of Kirsten M. Silvius and Patricia Márquez (then working with the Wildlife Conservation Society), a permanent field-based staff has monitored parrot reproductive biology and population size on Margarita, prevented poaching at major breeding sites, and raised awareness about the conservation of what is probably the best location for assuring the long-term survival of the species. No estimates exist for the population sizes of yellow-shouldered parrots along the Venezuelan mainland (Sanz 2001), whereas numbers on Bonaire are on the order of 360 individuals (BirdLife 2000).

The program has been highly successful: the yellow-shouldered parrot was officially named regional bird of Nueva Esparta state (where Margarita island is located), while the 2001 annual census revealed that the Margarita population increased from 750 birds in 1989 to 2,400 (Provita, unpublished data). Though Provita has been less active in the implementation of direct conservation actions in La Blanquilla, an informal census conducted in February 2002 estimated a minimum of 120 individuals (W. Rodríguez, A. Reyes, and J. P. Rodríguez, pers. obs.). Despite clear signs of population growth, mainly in Margarita, the threats continue. Parrot poaching is far from being controlled at either location, while habitat destruction in Margarita has increased as a consequence of the progressive expansion of open sky sand mining for the construction industry (Sanz 2001).

In this chapter, we update the prior population viability analysis of the yellow-shouldered parrot on Margarita and La Blanquilla (Rodríguez and Rojas-Suárez 1994), using demographic data collected over the last decade. First, we attempt to estimate the joint extinction risk for these populations, in order to classify them according to the World Conservation Union's Red List Criteria (known as "IUCN Red List Criteria"; see IUCN 2001). Second, we quantify the maximum allowable poaching intensity (MAPI) of nestlings that would permit the populations to persist for 100 years. As poaching cannot be completely eliminated, identifying the MAPI lets us define a management goal for our conservation program. In other words, if poaching intensity is successfully maintained at a level below MAPI, the populations should be able to persist into the foreseeable future. Naturally, the achievement of such a goal must be viewed in an adaptive management context. As several authors have pointed out (e.g., Boyce 1992, 1993; Possingham et al. 1993; Beissinger and Westphal 1998), the use of demographic models for threatened species conservation is best served when, as the quality of the data improves and models are strengthened, management goals are reevaluated and reassessed such that conservation actions can adjust to our better understanding of the population dynamics and to the consequences of our past management decisions.

Methods

Study Sites

La Blanquilla is a small (52.5 km^2), isolated offshore island located approximately 100 km northwest of Margarita island and 170 km from the continent (Figure 32.1). The climate is hot and dry, with a mean annual temperature of 25 to 26 °C and a mean annual precipitation of 300 to 600 mm. Though the vegetation is dominated by columnar cacti and thorny scrub, mangrove forests are found throughout the coastline, and gallery forests grow along seasonal creeks and depressions in the landscape. No permanent human settlements exist on La Blanquilla. The only residents inhabit a military base administered by the Venezuelan Navy, while temporary quarters are established by fishermen that work these waters for periods of 3 to 6 months at a time (Rojas-Suárez 1994b).

Although there are records of yellow-shouldered parrots throughout Margarita island, their current distribution is restricted to the western Macanao Peninsula (Figure 32.1). Spanning only 330 km^2, this location exhibits a pronounced topography that ranges from sea level to 745 m at Cerro Macanao. Mean annual temperature is 27 °C, and mean annual rainfall is 500 mm. As a consequence of the dominant trade winds from the northeast and the rugged topography, precipitation is highest on the northern portion of the peninsula. Plant communities are dominated by open cactus-chaparral with columnar cacti and legume trees. Seasonal riverbeds support permanent deciduous forests (Rojas-Suárez 1994a, Sanz and Grajal 1998).

Demographic Data and Parameter Values

The breeding season of the yellow-shouldered parrot begins at the end of the dry season (November to May), with egg laying in late March on Margarita and mid-May on La Blanquilla. Eggs are incubated for approximately 26 days, and nestlings are fledged in approximately 50 to 60 days (Rojas-Suárez 1994a).

Since Rojas-Suárez's (1994a) initial work on the reproductive biology of the yellow-shouldered parrot, Provita has performed annual estimates of population size and monitored reproductive performance on Margarita island. These results are mostly published as internal reports and are not available in the scientific literature. Table 32.1 summarizes data collected from 1990 to 1999.

Females enter the breeding pool at the end of their third year of life (Sanz and Grajal 1998). Mean clutch size is 3.4 eggs per nest (coefficient of variation = 24%), estimated from an examination of 17 to 47 nests monitored at Quebrada La Chica (Rojas-Suárez 1994a; Provita, unpublished data), one of the main nesting areas of the yellow-shouldered parrot on the Macanao Peninsula. Approximately one-third (33%) of adult females breed in any given year (Provita, unpublished data), thus we estimate that the per-female production of nestlings is 3.4 divided by 3, or 1.13. Because our census counts take place after reproduction, fecundity needs to include adult survival (Caswell 1989). Therefore, in the model, per-female production is reduced to 1.02 (1.13 * 0.9). As we have no evidence that sex ratio at birth is biased, half of this number, 0.51, is allocated equally to male and female nestlings.

By monitoring the nests in Quebrada La Chica from before eggs were laid to after the juveniles were fledged, it was also possible to estimate the main sources of mortal-

Table 32.1 Parameter values used to build the model (Rojas-Suárez 1994a, 1994b; Sanz 2001; and results from Provita's monitoring program)

Parameter	Value
Mean clutch size ± SD (number of eggs)	3.4 ± 0.8
Proportion of females breeding	0.33
Survival	
First year (nestlings)	0.50
Second year (juveniles)	0.80–0.95
Third year and onward (adults)	0.80–0.95
Coefficient of variation in survival (environmental stochasticity)	0.10–0.30
Initial abundance	
Macanao	2,400
La Blanquilla	120
Coefficient of variation of population size (sampling error)	0.10

ity (Rojas-Suárez 1994a; Sanz 2001; Provita, unpublished data). In the absence of poaching, roughly 20% of laid eggs do not hatch and 20% more die of natural causes during the nestling stage (half due to predation and the other half to unknown causes). This means that a conservative estimate of survival during the first year, which assumes that most mortality occurs while still in the nest, is 0.5. We also assume that survival increases sharply after the first year and remains high during the rest of their life (Table 32.1). Since Provita's field research does not include estimates of juvenile or adult survival directly, we use an interval ranging from 0.8 to 0.95. The extreme values of this interval predict average longevities of 7 and 22 years, respectively (calculated by RAMAS Metapop), which span the range of values characteristic of the larger parrots (Beissinger and Bucher 1992).

One catastrophe is modeled, which we call "extreme weather," with a probability of 0.01. This catastrophe is introduced to symbolize occasional widespread mortality of fledglings due to major environmental effects. Though Margarita and La Blanquilla are not in the traditional path of Caribbean hurricanes, occasional extreme weather events are known to occur. Precipitation records collected since 1949 show that this variable oscillates regularly between 200 and 1100 mm. As nesting takes place at the end of the dry season, extremely dry or wet years would mainly impact fledgling productivity.

The coefficient of variation of population size estimates was set to 10%. This is based on field measures of population size during annual censuses: variation in abundance estimates taken in consecutive days by the same observer and under similar weather conditions tend to be within 10% of each other (F. Rojas-Suárez, pers. comm.). The effect of this feature of the program is variation in the percentage of population poached each year.

Survival is assumed to vary as a consequence of the fluctuation in environmental conditions. Fledgling production is known to reflect the magnitude of precipitation during the previous year (Sanz 2001), but the degree of coupling between survival and environmental variation is unknown. To estimate environmental stochasticity, we used the observed variation in precipitation, which, from 1959 to 2001, had a coefficient of variation of 43%. We then assumed that variation in survival due to environmental stochasticity would vary between 5% (if coupling is loose) to 30% (if coupling is tight). Higher coefficients of variation in survival lead to overestimations in the standard deviation of survival rates (as indicated by RAMAS Metapop).

Carrying capacity was set to 3,000 birds for Margarita and 200 for La Blanquilla. These figures are rough estimates based on field observations of food availability and habitat use (Rodríguez and Rojas-Suárez 1994; F. Rojas-Suárez, pers. comm.). Density dependence is assumed to affect all vital rates following a simple ceiling model.

Simulation Scenarios

Using the RAMAS Metapop subprogram in RAMAS GIS, a baseline scenario was created for the yellow-shouldered parrot on La Blanquilla and Margarita islands (Figure 32.1). As these populations do not exchange individuals among themselves, or with other locations, they were simulated as a population composed of two isolated local populations.

The baseline scenario is a relatively optimistic scenario, where poaching reflects recently observed trends (see the next paragraph) and the variables modeled as intervals are assigned values that generate relatively lower predictions of extinction risk. In the baseline scenario, the coefficient of variation in survival is set to 10%, while the survival of juveniles and adults is 0.9 (other values are shown in Table 32.1).

As poaching is directed exclusively to nestlings, we simulated it as a "harvesting" population management action, impacting only birds during their first year of life. Male and female nestlings are equally exposed to poachers, as it is impossible to identify sex based on morphological characteristics. Poaching intensity can be very high: between 1990 and 1999, the average value was 30% of the clutch size, ranging between 20% in 1993 and 65% in 1999 (Sanz 2001). In 2002, due mainly to a decrease in the active surveillance of nests, 100% of the nests monitored by Provita were poached. In our reference scenario, poaching was assumed to be 60%. Note that in RAMAS Metapop this value applies to surviving fledglings and thus corresponds to approximately 30% additional mortality above the 50% that parrots experience during their first year due to natural causes (Table 32.1).

Since the recent increase of the parrot population on Margarita island, interest in harvesting and selling parrots in the illegal pet trade seems to have grown. Higher productivity in the wild reduces the effort that is necessary to capture a valuable load of parrots. Paradoxically, one of the unexpected consequences of this successful conservation program seems to be the improvement of the conditions for sustaining parrot trade. Therefore, in addition to the baseline scenario, we explored alternative scenarios with poaching intensities of 0% to 100%, in 10% increments. All simulations lasted 100 years and were replicated 1,000 times. Another set of alternative scenarios, where the coefficient of variation in survival and the survival of juveniles and adults were allowed to vary along the intervals defined in Table 32.1, were implemented to evaluate the sensitivity of the model outcome to these variables.

To classify the populations according to the IUCN Red List categories, we used the quantitative criteria listed under criterion "E" (IUCN 2001): a taxon is "critically endangered" if a quantitative analysis, such as a population viability analysis, shows that the probability of extinction in the wild is >50% in 10 years; "endangered" if the probability of extinction in the wild is estimated to be at least 20% within 20 years; and "vulnerable" if the probability of extinction in the wild is at least 10% within 100 years. As these populations are isolated, it is not necessary to adjust their category of threat in response to natural migrants from other portions of the range (Gärdenfors et al. 2001).

No estimates of habitat conversion, although known to be taking place (e.g., see Silvius 1997), are currently available. Our estimates of extinction risk, based mainly on the impact of poaching, are clear underestimates of the real extinction risk of this population.

Results

In the absence of poaching, the model suggests that the likelihood of persistence of the yellow-shouldered parrot on Margarita and La Blanquilla during the next 100 years is high, virtually 100%. At levels of poaching comparable to those observed in recent years (i.e., 70%–80%), extinction risk reaches and surpasses 10%, which would justify the designation of the La Blanquilla-Margarita population as "vulnerable" (Figure 32.2).

Under no circumstances should poaching intensity be allowed to exceed 60% in the wild. Shortly after this critical point in Figure 32.2, relatively small increases in poaching intensity lead to increasingly larger levels of extinction risk. Final mean population

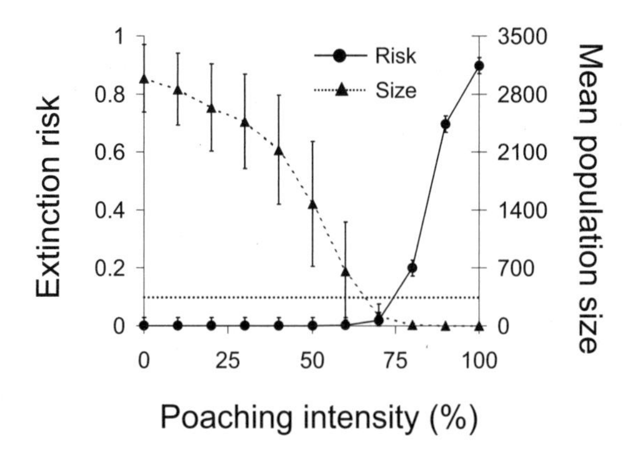

Figure 32.2 Simulated impact of nestling poaching on the extinction risk and mean final population size of the yellow-shouldered parrot on La Blanquilla and Margarita combined. Error bars of extinction risk estimates represent 95% confidence intervals. Error bars of mean final population size represent 1 SD. The horizontal dotted line indicates a 10% extinction risk.

size decreases relatively slowly when poaching is below 30% but declines sharply with increasing poaching intensity between 50% and 70%. The lag between the decline of final population size and extinction risk is a consequence of the parrot's longevity (which in the simulations of Figure 32.2 is 12 years), thus giving a false sense of safety when poaching is on the order of 50%. A small population of adult parrots is able to persist for relatively long periods of time when poaching is directed mainly at fledglings but will eventually collapse if productivity remains low. Though the model is relatively robust to uncertainties in the coefficient of variation in survival, sensitivity to juvenile and adult survival is high (Figure 32.3). Less optimistic scenarios, where juvenile and adult survival is less than 0.9, lead to sharp increases of extinction risk.

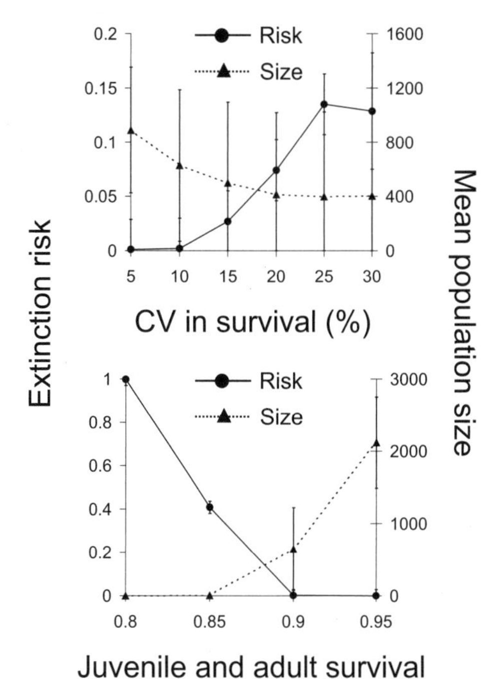

Figure 32.3 Sensitivity of the baseline simulation scenario to variations in the coefficient of variation (CV) in survival (*top*) and survival rates of juveniles and adults (*bottom*). Error bars of extinction risk estimates represent 95% confidence intervals. Error bars of mean final population size represent 1 SD.

According to these results, our conservation recommendation is the maintenance of the maximum allowable poaching intensity at a value no larger than 50%. Given the sensitivity of the model to juvenile and adult survival (a largely unknown parameter), as well as the implicit uncertainties related to regulating an illegal activity such as poaching, it is always critical to maintain a safety margin. This would follow the "precautionary principle," as mandated by the Convention on Biological Diversity, which has been signed and ratified by the Venezuelan government.

When the simulations are performed such that La Blanquilla and Margarita are modeled as individual populations, there is no perceptible difference in the extinction risk for Margarita. In the case of La Blanquilla, however, given its smaller size, the situation is quite different. For example, if the poaching level is set to the reference value of 60% (which is probably an underestimate, as there is no active surveillance and nests are easily found by poachers in this relatively flat and small island), extinction risk in 100 years is 0.24 [95% confidence interval = (0.22, 0.27)]. Therefore, in contrast to Macanao, if the La Blanquilla population were to persist indefinitely without translocation of individuals from other sites, MAPI must be maintained at levels well below 50%.

Discussion

Currently, there is no evidence of the existence of any yellow-shouldered parrot population anywhere in its range comparable in size to the population on Margarita island (Sanz 2001). However, after more than a decade of conservation efforts on Margarita, as well as a significant population recovery during that period, poaching levels are still within the range of those required to designate this population as vulnerable to extinction (Figure 32.2). These findings have two implications. First, until other large populations are identified and protected, potentially on the Venezuelan mainland (Desenne and Strahl 1994, Ridgely 1981), the conservation of the population on Margarita is key for the long-term survival of the species. Second, if poaching can threaten the largest, best-protected population of this species, it could easily represent a major threat throughout its range. In Venezuela, from the mid-1980s onward, the weakening economy has caused an increase in human pressure on fish and wildlife populations as sources of income and food (Rodríguez 2000). When one travels across the interior of the country it is common to see people selling wildlife—especially the larger parrot species—at the roadside. Under these circumstances, it is unlikely that the mainland populations of the yellow-shouldered parrot can be considered viable. We recommend that for the time being both the Margarita population and the species as a whole be classified vulnerable. Future field research should attempt to generate reliable estimates of juvenile and adult survival and thus improve quantification of extinction risk.

Although we did not explicitly model the risk posed by habitat conversion, this is also an important, increasing threat (Sanz 2001). The conservation goal that MAPI be maintained below 50% on Margarita can also be achieved by spatially segregating poachers. Instead of trying to protect all nests such that poachers impact less than 50% of nestlings, it may be more feasible to prevent poaching in 50% of the nests. Focusing on the protection of nesting habitat through the implementation of an effective reserve system on Margarita island would fulfill both the short-term goal of reducing the extinction risk of the yellow-shouldered parrot and the long-term goal of assuring sufficient nesting habitat for this and other threatened species on the island. In addition to strengthening the existing nature reserve network, we believe that a key component of such a process would be the creation of at least one private reserve in currently unprotected habitat.

Acknowledgments Since the establishment of the project for the conservation of the yellow-shouldered parrot more than a decade ago, many people and organizations have supported Provita's work or participated directly in the project's activities. We are very grateful to all of them. Major support has been provided by the Alcaldía del Municipio Península de Macanao, American Bird Conservancy, Armada de Venezuela, British Embassy in Venezuela, Fondo para el Desarrollo del Estado Nueva Esparta (Fondene), Fundación Museo Marino, Fundo San Francisco, Golden Ark Foundation, Guardia Nacional, IdeaWild, Instituto Nacional de Parques, Ministerio del Ambiente y de los Recursos Naturales, Palm Beach Zoo, Tulsa Zoological Park, Wildlife Conservation Society, Wildlife Trust, and World Parrot Trust. Our gratitude to Wallis Rodríguez, Franklin Rojas-Suárez, and Virginia Sanz for facilitating access to their unpublished observations. A previous version of the manuscript was critically reviewed and improved by H. Resit Akçakaya, Jeff S. Hatfield, and three anonymous reviewers.

References

Albornoz, M., Rojas-Suárez, F., and Sanz, V. 1994. Conservación y manejo de la cotorra cabeciamarilla (*Amazona barbadensis*) en la isla de Margarita, estado Nueva Esparta. Pages 197–207 in L. G. Morales, I. Novo, D. Bigio, A. Luy, and F. Rojas-Suárez (eds.), *Biología y conservación de los Psitácidos de Venezuela*. SCA, EBAFY, EcoNatura, SCAPNHP, Provita. Caracas, Venezuela.

Beissinger, S. R., and Bucher, E. H. 1992. Sustainable harvesting of parrots for conservation. Pages 73–115 in S. R. Beissinger and N. F. R. Snyder (eds.), *New World parrots in crisis: solutions from conservation biology*. Smithsonian Institution Press, Washington, D.C.

Beissinger, S. R., and Westphal, M. I. 1998. On the use of demographic models of population viability in endangered species management. *Journal of Wildlife Management* 62: 821–841.

BirdLife 2000. *Threatened birds of the world*. Lynx Edicions, Barcelona, Spain.

Boyce, M. S. 1992. Population viability analysis. *Annual Review of Ecology and Systematics* 23: 481–506.

Boyce, M. S. 1993. Population viability analysis: adaptive management for threatened and endangered species. *Transactions of the 58th North American Wildlife and Natural Resources Conferences* 520–527.

Butler, P. J. 1992. Parrots, pressures, people, and pride. Pages 25–46 in S. R. Beissinger and N. F. R. Snyder (eds.), *New World parrots in crisis: solutions from conservation biology*. Smithsonian Institution Press, Washington, D.C.

Caswell, H. 1989. *Matrix population models: construction, analysis, and interpretation*. Sinauer Associates, Sunderland, Mass.

Collar, N. J. 1996. Priorities for parrot conservation in the New World. *Cotinga* 5: 26–31.

Collar, N. J., and Juniper, A. T. 1992. Dimensions and causes of the parrot conservation crisis. Pages 1–24 in S. R. Beissinger and N. F. R. Snyder (eds.), *New World parrots in crisis: solutions from conservation biology*. Smithsonian Institution Press, Washington, D.C.

Desenne, P., and Strahl. P. 1991. Trade and the conservation status of the family Psittacidae in Venezuela. *Bird Conservation International* 1: 153–169.

Desenne, P., and Strahl, P. 1994. Situación poblacional y jerarquización de especies para la conservación de la familia Psittacidae en Venezuela. Pages 231–272 in L. G. Morales, I. Novo, D. Bigio, A. Luy, and F. Rojas-Suárez (eds.), *Biología y conservación de los Psitácidos de Venezuela*. SCA, EBAFY, EcoNatura, SCAPNHP, Provita, Caracas, Venezuela.

Gärdenfors, U., Hilton-Taylor, C., Mace, G. M., and Rodríguez, J. P. 2001. The application of IUCN Red List criteria at regional levels. *Conservation Biology* 15: 1206–1212.

IUCN 2001. *IUCN Red List categories and criteria: Version 3.1*. Species Survival Commission, World Conservation Union (IUCN), Gland, Switzerland.

Low, R. 1981. The yellow-shouldered amazon (*Amazona barbadensis*). Pages 215–225 in R. F. Pasquier (ed.), *Conservation of New World parrots*. Smithsonian Institution Press, Washington, D.C.

Phelps, W. H., Jr, and Meyer de Schauensee, R. 1994. *Una guía de las aves de Venezuela*. 2nd ed. Castellano. Editorial Ex Libris, Caracas, Venezuela.

Possingham, H. P., Lindenmayer, D. B., and Norton, T. W. 1993. A framework for the improved management of threatened species based on population viability analysis (PVA). *Pacific Conservation Biology* 1: 39–45.

Ridgely, R. S. 1981. The current distribution and status of mainland neotropical parrots. Pages 233–384 in R. F. Pasquier (ed.), *Conservation of New World parrots*. Smithsonian Institution Press, Washington, D.C.

Rodríguez, J. P. 1995. Investigación y conservación de la cotorra margariteña. Pages 100–103 in P. Noble (ed.), *Technology, the cnvironment, and social change*. Papers of the Thirty-Eighth Annual Meeting of the Seminar on the Acquisition of Latin American Materials (SALALM). SALALM Secretariat, General Library of the Universitiy of New Mexico, Albuquerque, N.M.

Rodríguez, J. P. 2000. Impact of the Venezuelan economic crisis on wild populations of animals and plants. *Biological Conservation* 96: 151–159.

Rodríguez, J. P., and Rojas-Suárez, F. 1994. Análisis de viabilidad poblacional de tres poblaciones de psitácidos insulares de Venezuela. Pages 97–113 in G. Morales, I. Novo, D. Bigio, A. Luy, and F. Rojas-Suárez (eds.), *Biología y conservación de los Psitácidos de Venezuela*. SCAV, EBAFY, EcoNatura, SCAPNHP, Provita, Caracas, Venezuela.

Rodríguez, J. P., and Rojas-Suárez, F. 1999. *Libro rojo de la fauna venezolana*. 2nd ed. PROVITA, Fundación Polar, Caracas, Venezuela.

Rojas-Suárez, F. 1994a. Biología reproductiva de la cotorra, *Amazona barbadensis* (Aves: Psitaciformes), en la península de Macanao, estado Nueva Esparta. Pages 89–96 in L. G. Morales, I. Novo, D. Bigio, A. Luy, and F. Rojas-Suárez (eds.), *Biología y conservación de los Psitácidos de Venezuela*. SCA, EBAFY, EcoNatura, SCAPNHP, Provita, Caracas, Venezuela.

Rojas-Suárez, F. 1994b. Evaluación preliminar de la población de cotorra (*Amazona barbadensis*) en la isla La Blanquilla, Venezuela. Pages 89–96 in L. G. Morales, I. Novo, D. Bigio, A. Luy, and F. Rojas-Suárez (eds.), Biología y conservación de los Psitácidos de Venezuela. SCA, EBAFY, EcoNatura, SCAPNHP, Provita, Caracas, Venezuela.

Sanz, V. 2001. Ecología de *Amazona barbadensis* (Aves: psittacidae) en la península de Macanao (isla de Margarita): Un estudio de las características del hábitat, dinámica de uso y selección. Seminario de Avance de la Tesis Doctoral, Postgrado en Ecología, Facultad de Ciencias, Universidad Central de Venezuela, Caracas.

Sanz, V., and Grajal, A. 1998. Successful reintroduction of captive-raised yellow-shouldered amazon parrots on Margarita island, Venezuela. *Conservation Biology* 12: 430–441.

Sanz, V., and Rojas-Suárez, F. 1997. Los nidos nodriza como técnica para incrementar el reclutamiento de la cotorra cabeciamarilla (*Amazona barbadensis*, Aves: Psittacidae). *Vida Silvestre Neotropical* 6: 8–14.

Silvius, K. M. 1997. What it takes to save a parrot. *Wildlife Conservation* 100: 52–57, 66.

Snyder, N., McGowan, P., Gilardi, J., and Grajal, A. (Eds.). 1999. *Parrots: status survey and conservation action plan*. World Conservation Union (IUCN), Gland, Switzerland.

33

Golden-Cheeked Warbler (*Dendroica chrysoparia*) in Texas

Importance of Dispersal toward Persistence in a Metapopulation

MATHEW W. ALLDREDGE
JEFF S. HATFIELD
DAVID D. DIAMOND
C. DIANE TRUE

The golden-cheeked warbler (*Dendroica chrysoparia*) is a neotropical migrant songbird whose breeding range is currently restricted to fewer than 35 counties in the Edwards Plateau region of central Texas (U.S. Fish and Wildlife Service 1996). Breeding habitat consists of Ashe juniper (*Juniperus ashei*) along with a variety of deciduous tree species (Wahl et al. 1990). Wintering range consists of high-elevation pine (*Pinus*) and oak (*Quercus*) woodlands located in Honduras, Guatemala, and Mexico (Pulich 1976, Vidal et al. 1994, Rappole et al. 2000). The golden-cheeked warbler was listed as endangered in 1990 by the U.S. Fish and Wildlife Service (USFWS), primarily because of the high rate of breeding habitat loss (USFWS 1992) due to clearing of Ashe juniper for range improvement and commercial harvest (Kroll 1980) and for urbanization (U.S. Fish and Wildlife Service 1996). Currently, efforts are being made by public and private organizations to acquire or protect golden-cheeked warbler breeding habitat for the continued survival of the species. However, it is unclear how these lands should be acquired in relation to lands already managed for this species in terms of spatial distribution and size.

The fragmented landscape of golden-cheeked warbler breeding habitat creates a potential metapopulation dynamic. Understanding how different demographic parameters and environmental variables influence warbler population dynamics will aid in effective management, help guide research, and direct future land acquisitions. USFWS (1996) performed a population viability analysis (PVA) for the golden-cheeked warbler and "tentatively recommended" that to assure a probability of extinction less than 5% over 100 years for an isolated patch, a carrying capacity of 3,000 breeding pairs would be necessary. However, this analysis was limited by information available at the time and did not examine the effects of metapopulation structure.

This chapter describes the potential metapopulation dynamic that exists for golden-cheeked warblers and outlines a PVA that was conducted, using current demographic parameters and an assumed metapopulation structure.

Methods

With the limited information available for golden-cheeked warblers, we chose to derive a few, readily comparable models to identify critical population parameters and to define future research needs. Following this approach we constructed three basic models that are examined on two different landscape scenarios (explained in the following discussion). Demographic stochasticity was used in all models, and environmental stochasticity was modeled with a lognormal distribution. All simulations were run for 100 years.

Stage Matrix

Because the majority of the information about golden-cheeked warblers is on males and we know that males establish breeding territories, all models were run for males only. This approach assumes that males will mate with only one female and that sex ratios of young are 50:50. The first assumption can be relaxed slightly as some of the fecundity estimates (see the following discussion) account for double brooding. It also assumes that the dynamics of female golden-cheeked warblers (survival and dispersal) are similar to males and that territorial males have mating opportunities and produce offspring.

Age classes used were hatch year (HY), second year (SY), and after-second year (ASY). Since SY and ASY birds have the same survival rates, the matrix included only two stages, HY and AHY. The matrix was based on postbreeding census, with HY fecundity as the product of survival and fledging rate for each stage. Equation 5 of Goodman (1960) was used for estimation of the variance of the product of fecundity and survival.

Estimates of fledging rate (Table 33.1) and corresponding temporal variances were obtained from several different studies conducted at the Department of Defense's Fort Hood military base (U.S. Fish and Wildlife Service 1996, Anders 2000). However, some of these estimates may be low because golden-cheeked warblers are known to split broods, and thus the number of young per territorial male may only represent half of the young produced (A. Anders, The Nature Conservancy, pers. comm.). Furthermore, not all territorial males actually pair (Anders 2000). The 2000 and 2001 estimates are comparable to those found for black-throated blue warblers (Holmes et al. 1992, 1996).

Estimates of survival (Table 33.2) were obtained from Fort Hood for 1997–2001 (unpublished data, The Nature Conservancy, Fort Hood project), and three other sources for golden-cheeked warblers (see USFWS 1996). Survival rates reported here are not actual survival but apparent survival, or those birds that survived and returned to the study area. Fort Hood data from 1997 to 2001 were analyzed using mark-resight methods, Jolly-Seber estimation, in program MARK (White and Burnham 1999). Although data were insufficient to warrant the increased number of parameters involved for group and time models according to the Akaike information criterion (AIC; Burnham and Anderson 1998), the time model is still valid and was used to produce yearly estimates

Table 33.1 Golden-cheeked warbler fledging rates from various studies (see text for references)

Study and Age (Fort Hood)	Fecundity Measure	Units	SE	Temporal Variance
92–94 SY	0.7535	Male fledge/singing male		0.024
92–94 ASY	1.075	Male fledge/singing male		0.0056
97 AHY	0.77	Young/territorial male	0.11	
98 AHY	1.30	Young/territorial male	0.14	
99 AHY	0.76	Young/territorial male	0.12	
97–99 AHY	0.943	Young/territorial male		0.0493
2000 SY	1.44	Female young/SY female		
2000 ASY	1.56	Female young/ASY female		
2001 SY	1.41	Female young/SY female		
2001 ASY	1.74	Female young/ASY female		

with corresponding variances and covariances. This information was then used to separate the model variance into process and sampling variance, following procedures outlined by Gould and Nichols (1998). Process variance is an estimate of temporal variation in the survival rates. The survival rates obtained are comparable to those of other warbler species (USFWS 1996, Chase et al. 1997, Sillet and Holmes 2002).

We developed three models based on varying demographic rates (Table 33.3). Model 1 represents the highest survival and fecundity rates documented for golden-cheeked warblers, and Model 3 represents the lowest documented rates. Model 2 uses parameter estimates that fall between the other models.

Survival and fecundity for a patch can be affected by local management strategies. Currently, Fort Hood has a relatively aggressive strategy that protects golden-cheeked warbler habitat and controls nest predators and parasites (specifically, the brown-headed cowbird, *Molothrus ater*). To simulate the possibility that different areas may have dif-

Table 33.2 Golden-cheeked warbler male apparent survival rates from various study areas (see text for references)

Species, Study, and Year	Survival Rate	SE	Temporal Variance
Fort Hood 1997–2001			
HY	0.302	0.110	0.058
AHY	0.563	0.044	0.007
Fort Hood 1991–1995			
HY	0.30		
AHY	0.57		0.0119
Pulich			
HY	0.42		
AHY	0.69		
Balcones Canyonlands NWR			
HY	0.00		
AHY	0.61		

Table 33.3 Parameters used in the age matrix assuming a post-breeding-season census

	Fecundity		Survival		Dispersal (%)		
Model	HY	AHY	HY	AHY	None	Low	High
1	0.72 (0.36)	0.94 (0.17)	0.5 (0.24)	0.57 (0.10)	0	15	30
2	0.48 (0.30)	0.74 (0.14)	0.4 (0.24)	0.57 (0.10)	0	15	30
3	0.23 (0.19)	0.62 (0.12)	0.3 (0.24)	0.57 (0.10)	0	15	30

Note: Fecundity is the product of fledging rate for the age class times the HY survival rate. Dispersal indicates the number of surviving birds that disperse and are then divided among the number of patches. Temporal standard deviations are in parentheses.

ferent vital rates, we parameterized Fort Hood patches with Model 1 and the Balcones Canyonlands National Wildlife Refuge (NWR) with either Model 2 or Model 3.

Density Dependence and Carrying Capacity

We used the "ceiling" model for density dependence, which allows the population to grow as long as it is below carrying capacity. There are many factors that influence carrying capacity (K), but for golden-cheeked warblers territoriality is probably the most important; territory size will limit the number of males that can establish a territory in a given area. Since territorial males attract mates and thus contribute to yearly productivity, carrying capacity of the patches was determined by the number of territories that fit within patch boundaries. Estimated territory size for high-density populations on Fort Hood for 1991 to 1995 were 4.3 ha per territory (USFWS 1996). Golden-cheeked warbler territory size on Fort Hood for the year 2000 was estimated at 4.7 ha per territory, and the number of territories has increased since the earlier estimates (Anders 2000). It appears that territory size is relatively constant because the size did not change with increasing number of territories.

Landscape Scenarios

We simulated two different landscape scenarios. The first included all currently available habitats on public and private lands. This extreme, where all current habitat is maintained indefinitely into the future, is unrealistic but presents a convenient starting point for management and land acquisition decisions. The second scenario included only habitat occurring on public land and ignored all other areas. Also an extreme, this scenario is based on land that has a reasonable chance of remaining as suitable habitat, even if at some future time suburban sprawl removes all other available habitat.

Golden-cheeked warbler habitat has been classified and is available on a GIS system (Diamond and True 1999). For simplicity, we chose to restrict our analyses to the area between Fort Hood and the city of Austin (Figure 33.1) and created a coverage for each landscape scenario, with 28.8-m pixel size. In order to create a reasonable number of

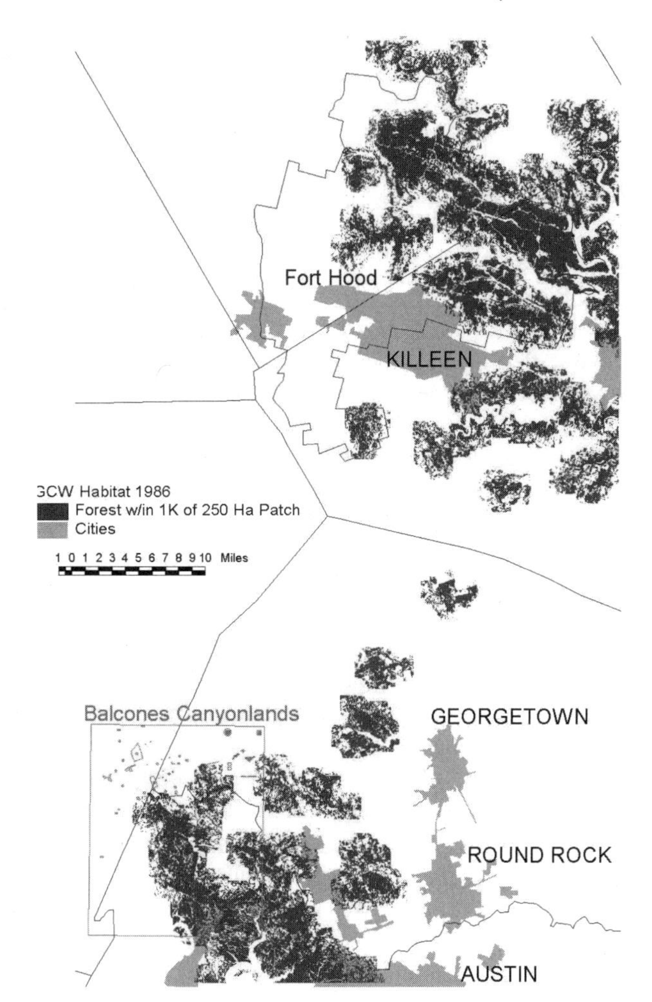

Figure 33.1 Study area (from Fort Hood to the city of Austin) depicting habitat suitable for golden-cheeked warblers during the summer breeding season in central Texas (see Diamond and True 1999 for details of habitat classification).

patches in the landscape, a cell was included in a patch if it was suitable habitat occurring within 1 km of another suitable habitat. Requiring continuity between pixels would have created 100s of patches, thereby rendering modeling cumbersome and interpretation difficult. After delineating patches, we used total patch area to determine the carrying capacity of each patch by dividing total area by average male territory size. Since some of these territories may not truly be potential territories, we reduced carrying capacity of each patch by 10%. All simulations, except those specifically noted, were started with initial population size of each patch at one-half the patch carrying capacity, and distributed between HY and AHY stage classes according to the stable age distribution for the model being used for the given patch.

Dispersal

Most of the information on dispersal is from three small study areas on Fort Hood, all located within the perimeter of the largest patch in our metapopulation. Adults exhibit strong site fidelity and adult dispersal between patches is probably negligible, so only HY dispersal is allowed in the model. To represent the extreme cases, we compared the scenario of no HY dispersal among patches to the scenario of equal HY dispersal among all patches. Dispersal rates were set at 15% and 30% from each patch, distributed evenly among all other patches. Because survival estimates are actually apparent survival (incorporating dispersal from the study area into mortality), we eliminated Model 3 from the dispersal analysis, as it would be unrealistic to have 15% or 30% dispersal of birds removed from the 30% survival of HY birds.

Recolonization is often an important aspect of metapopulation dynamic. Recolonization was simulated for the all-available-habitat scenario with the Model 1 parameterization incorporating dispersal. We started a simulation with a randomly selected patch set to 0 population size. In this way we determined the possibility of a patch being recolonized by migrants from other patches with specified demographic and dispersal rates. In doing this, we assumed that whatever process rendered the patch extinct did not alter the habitat suitability and that the patch could be immediately reoccupied.

Results

Four patches were delineated in the public-land-only scenario, and eight patches were determined in the scenario incorporating all available habitat (Table 33.4). Total carrying capacity for public land within the study area was 9,731 territories: the carrying capacity of the largest patch was 6,609 territories, and the smallest was 46 territories. Spatial layout of these patches was one large patch (>2,500 territories/patch) and one small patch (<300 territories/patch) for both Fort Hood and Balcones Canyonlands NWR (Figure 33.2a). Total area for the four patches combined was 450 km^2. Considering all currently available habitat, the total carrying capacity was 20,964 territories; the carrying capacity of the largest

Table 33.4 Delineation of patches and corresponding area, carrying capacity, and initial abundance

	Area (km^2)	Carrying Capacity	Initial Abundance
Public land only			
Patch 1	305.7	6,609	3,305
Patch 2	13.1	284	142
Patch 3	2.1	46	23
Patch 4	129.1	2,792	1,396
Total	450.0	9,731	4,866
All available land			
Patch 1	15	330	165
Patch 2	572	12,371	6,186
Patch 3	11	238	119
Patch 4	12	255	128
Patch 5	44	959	479
Patch 6	34	742	371
Patch 7	245	5,296	2,648
Patch 8	36	773	387
Total	970	20,964	10,483

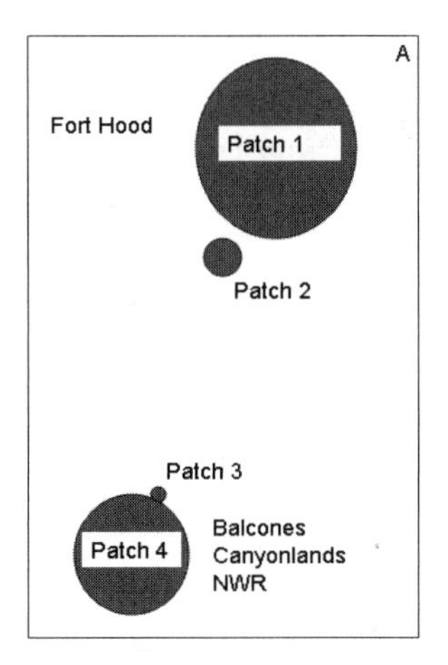

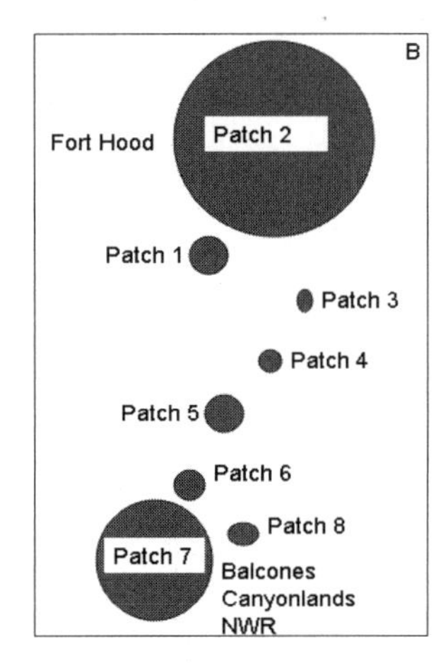

Figure 33.2 Patch layout as constructed by the Landscape Data procedure of RAMAS GIS. Schematic A represents only publicly owned lands, and B represents all habitat currently available in the area.

patch was 12,371 territories, and of the smallest it was 238 territories. The two largest patches consisted of the same two large public-land patches with an approximate doubling in size as surrounding lands were incorporated into the patch. The remaining six patches were scattered between the large patches, indicating a possibility of patch connectivity, depending on dispersal (Figure 33.2b). The combined area of all eight patches was 970 km^2.

Patch Trajectories

To determine the general behavior of the system in the absence of dispersal, we first examined individual patch trajectories without a metapopulation dynamic. This scenario also represents a potential extreme case since actual dispersal is not known and could be close to 0. Trajectories of the four patches from the public-land scenario represent similar population trajectories to those observed for all possible patch sizes using the three model parameterizations (see files PUBM1.mp, PUBM2.mp, and PUBM3.mp on the CD that accompanies this volume). Model 1 resulted in stable populations for all patches. These populations quickly increased to carrying capacity and remained there for the duration of the 100-year period, with an expected minimum population size of 4,817 (Table 33.5). The terminal extinction probability that the total population would fall below 200 or 500 territorial males during the 100-year period was 0.0 for the parameterization in Model 1 (Table 33.6).

The population trajectories for Model 2 consistently trended downward over the 100-year period (see files PUBM1.mp, PUBM2.mp, and PUBM3.mp). Smaller patches (carrying capacity < 2,000 or initial population size < 1,000) declined sufficiently so that temporal variation (i.e., poor years) would actually cause the populations to go extinct. Larger patches did not decrease to the point that extinction was probable after 100 years. The expected minimum population size after 100 years was 764 (Table 33.5). The terminal extinction probability that the population would fall below 500 territorial males in 100 years was 0.179 (95% CI, 0.151–0.207) and below 200 was 0.074 (95% CI, 0.045–

Table 33.5 Expected minimum population size after 100 years

Model	Dispersal Rate (%)		
	0	15	30
Public land only			
Model 1	4,817	4,536	3,997
Model 2	764	631	310
Model 3	0	NA	NA
All available land			
Model 1	11,710	11,159	9,701
Model 2	2,228	2,371	1,864
Model 3	0	NA	NA
Mixed models			
Model 1–2	NA	3,592	2,947
Model 1–3	NA	2,797	1,982
Catastrophe			
Model 1 (patch 2 extinct)	NA	5,352	5,306

Note: NA indicates simulations that were not run.

0.101) (Table 33.6). Extinction probabilities decrease for the scenario that considers all available habitat (Table 33.6), but this is attributable to the largest patch with an initial population size of over 6,000 territorial male golden-cheeked warblers. The maximum difference between the terminal (end of 100-year period) extinction risk for Model 1 and Model 2 was 0.89, which indicates the risk curves are significantly different (two-sample Kolmogorov-Smirnov test; $p < 0.001$) (see files PUBM1.mp, PUBM2.mp, and PUBM3.mp).

Population trajectories for Model 3 indicate almost certain extinction in less than 50 years (see files PUBM1.mp, PUBM2.mp, and PUBM3.mp), even for the largest local population in the all-available-habitat scenario starting at more than 6,000 territorial male golden-cheeked warblers. Extinction probabilities for all patches using this parameterization were 1.00 (Table 33.6). Since even the largest population cannot survive under these model conditions, Model 3 was omitted from dispersal scenarios examining the metapopulation dynamic.

Metapopulation Trajectories

Incorporating 15% and 30% dispersal from each patch (distributed evenly among the other three patches) under the public-land scenario demonstrated stable population trajectories for all four patches using Model 1 (see files PUBM1_15.mp and PUBM1_30.mp). The largest patch maintained approximately 6,000 males without dispersal, 5,350 with 15% dispersal, and 4,400 with 30% dispersal. Patch 4, which represented about one-third of the total metapopulation size, maintained between 2,400 and 2,500 territorial males, with or without dispersal. Both small populations increased in size to their patch carrying capacity at either dispersal rate. Probability of terminal extinction below 200 or 500 territorial males was 0.0 for all dispersal rates using Model 1 for the public-land scenario. The maximum distance between risk curves was significantly different from the no-dispersal curve (two-sample Kolmogorov-Smirnov test; $D = 0.19$ at 15% dispersal,

Table 33.6 Terminal extinction risk to fall below 200 or 500 birds in 100 years for scenarios considering public land only and considering all available habitat with either 0, 15%, or 30% dispersal from each patch distributed evenly among all other patches for three models

	Model 1	Model 2	Model 3
Public land only			
	Terminal extinction risk below 200		
No dispersal	0.0	0.074 (0.045, 0.101)[a]	1.00
15% dispersal	0.0	0.048 (0.020, 0.076)[a]	NA
30% dispersal	0.0	0.116 (0.088, 0.144)[a]	NA
	Terminal extinction risk below 500		
No dispersal	0.0	0.179 (0.151, 0.207)[a]	1.00
15% dispersal	0.0	0.210 (0.182, 0.238)[a]	NA
30% dispersal	0.0	0.470 (0.444, 0.500)[a]	NA
All available habitat			
	Terminal extinction risk below 200		
No dispersal	0.0	0.0[a]	1.00
15% dispersal	0.0	0.0	NA
30% dispersal	0.0	0.0	NA
	Terminal extinction risk below 500		
No dispersal	0.0	0.017 (0.0, 0.045)[a]	1.00
15% dispersal	0.0	0.0	NA
30% dispersal	0.0	0.0	NA

Note: NA indicates simulations that were not run.

[a]Models where local populations go extinct when extinction risk is <1 including 0.

$p < 0.001$, and $D = 0.27$ at 30% dispersal, $p < 0.001$) (see files PUBM1_15.mp and PUBM1_30.mp); however, biological significance is doubtful.

Dispersal for the public-land scenario with Model 2 resulted in a faster total population decline than in the absence of dispersal. The expected minimum population size after 100 years decreased from 764 with no dispersal to 631 at 15% dispersal and 310 at 30% dispersal (Table 33.5). The population size for the largest patch was about 1,000 fewer territorial male golden-cheeked warblers for 15% dispersal after 100 years and 1,500 fewer territorial males for 30% dispersal (see files PUBM2_15.mp and PUBM2_30.mp). Patch 4, at one-third of the total metapopulation size, showed only a small decrease at 15% dispersal after 100 years and a much larger decrease at 30% dispersal. Terminal extinction risk below 500 territorial males was 0.210 and 0.470 for 15% and 30% dispersal, respectively, and was 0.048 and 0.116 for 15% and 30% dispersal, respectively, at a level of 200 males (Table 33.5). Again, the terminal extinction risk curves are significantly different from the no-dispersal scenario (two-sample Kolmogorov-Smirnov test: $D = 0.25$ at 15% dispersal, $p < 0.001$ and $D = 0.51$ at 30% dispersal, $p < 0.001$), which probably is biologically significant in the short term (see files PUBM2_15.mp and PUBM2_30.mp).

Examination of the scenario considering all available habitat using Model 1 with 15% and 30% dispersal from each patch (distributed evenly among all other patches) demonstrated a similar effect as with the public-land-only scenario. All patches stabilized at population levels near their respective carrying capacity. The probability of extinction was 0.0 (Table 33.6). Population trajectories for Model 2 were similar, considering all available habitat to those observed using only public land with similar parameterization. Large patches steadily declined in population size over the 100-year period, with faster declines being associated with increasing dispersal rates. Smaller patches had an initial population increase followed by a decline as populations in larger patches declined. Even though population trajectories were similar between the two landscape scenarios, extinction probabilities were lower when all available land was considered due to the large population found in and around Fort Hood. The probability of metapopulation extinction below 500 and 200 territorial males was 0.0 at 15% and 30% dispersal, respectively (Table 33.6).

Mixed Models and Catastrophes

Modeling public land patches with Model 1 parameterization for the Fort Hood patches and Model 2 or Model 3 parameterization for Balcones Canyonlands NWR resulted in stable populations (see files MIXEDM1M2_15.mp, MIXEDM1M2_30.mp, MIXEDM1M3_15.mp, and MIXEDM1M3_30.mp). The largest population at Fort Hood maintained over 3,800 territorial male birds with the Model 2 and Model 3 parameterization at Balcones Canyonlands NWR. This is less than the stable population levels when all patches were parameterized with Model 1, but populations still appear stable. When Balcones Canyonlands NWR patches are parameterized with Model 3, the largest patch does decline to less than one-third of its carrying capacity and is at higher risk of extinction. The probability of the metapopulation going extinct is 0.0, but this is attributable to the Fort Hood populations, especially when Model 3 is used for Balcones Canyonlands NWR. Looking at the final population structure of each patch shows that the differences in the scenarios for Fort Hood patches are caused by dispersal rates, but the differences for Balcones Canyonlands NWR patches are attributable to the difference between Model 2 and Model 3 parameterizations (see files MIXEDM1M2_15.mp, MIXEDM1M2_30.mp, MIXEDM1M3_15.mp, and MIXEDM1M3_30.mp).

Simulating patch catastrophes by initially setting a patch population size to 0 demonstrated the possibility of patches to be recolonized under the assumption of dispersal and Model 1 parameterization. Examination of the worst-case scenario of a catastrophe for Patch 2, which represents more than one-half the total metapopulation size in the study area, showed that the stable population size of Patch 2 can be reached in < 40 years (see files CATAST_15.mp and CATEST_30.mp). Examining Patch 7 (the largest remaining population) shows that the overall size of this population remains stable.

Discussion

Public land comprises less than one-half of the current habitat suitable for golden-cheeked warblers within our study area (Figure 33.1). Fortunately, the two largest patches within the study area do occur within and around Fort Hood and Balcones Canyonlands NWR, which means substantial portions of golden-cheeked warbler habitat in this area are relatively secure from human-caused destruction. This suggests that it may be possible to maintain two reasonably large (> 1,000 male territories) populations. However, concentrating efforts on only these two populations may be risky because they are geographically distant, and it is not known if golden-cheeked warblers disperse between these areas. The risk is that one patch would undergo a population crash and could not be recolonized by the remaining population. Until better information is obtained for the golden-cheeked warbler, it would be wise to maintain as many patches as possible.

There are only two patches that have carrying capacities sufficient (>2,700 male territories) to meet the recommendations made from the previous PVA (>3,000 breeding pairs; USFWS 1996) for either the public-land-only or all-habitat scenarios. Based on the conclusions by the USFWS, the remaining patches would not be viable over 100 years unless a metapopulation dynamic exists. Even the large Balcones Canyonlands patch (carrying capacity of 2,792 territories) may not be viable in the absence of dispersal unless adjacent lands are maintained.

The geographic distribution of remaining patches does indicate the potential for a metapopulation dynamic even if dispersal distances are small compared to the total size of the study area. Moderate dispersal distances could act to minimize population fluctuations in adjacent patches, and this increased connectivity would moderate negative stochastic events on local populations. This approach is appealing, given the lack of specific information for the golden-cheeked warbler.

Simulations for the three different model parameterizations indicate that current populations do not experience rapid growth, except under the best conditions, and stochastic events that depress populations may have long-term effects. Any event that reduces survival or fecundity for multiple years could drive populations to very low numbers. Prudent management would involve careful monitoring of populations across the entire geographic range in order to detect population trends before declines become serious. Detecting reduced survival or fecundity in a local population while other populations do not change may indicate that active management is necessary.

It is important to note that the parameters for Model 1 (the stable population model) come from Fort Hood populations, which were being actively managed while data were being collected. One of the management actions was cowbird trapping, which was implemented to increase nest success of golden-cheeked warblers and other species (Summers et al. 2000). Given the source of our data for model parameterization, it may not be realistic to assume that habitats outside of Fort Hood will have similar survival and fecundity rates; this indicates that Model 1 may be overly optimistic.

Given the importance of nest success and the current effort to remove cowbirds at Fort Hood, our Model 1 parameterization using recent data from Fort Hood may represent a best-case scenario for current conditions. Nest parasitism by brown-headed cowbirds has been documented for the golden-cheeked warbler (Summers et al. 2000) and many other bird species (Robinson et al. 1995a, 1995b; Braden et al. 1997; DeGroot et al. 1999). Cowbird trapping has proven to be an effective management technique in reducing nest parasitism for golden-cheeked warblers and other bird species (Braden et al. 1997).

The previous PVA emphasized that, for a single patch to be viable for 100 years, it would need to have a carrying capacity of 3,000 breeding pairs (male territories) (USFWS 1996). Our analysis of individual patch trajectories supports such a number. A large population would buffer against catastrophes and environmental influences that may depress the population. In other words, a large population under Model 2 is sustainable for several years and could even survive under Model 3 for a few years.

This population would then be able to recover as conditions improved or management actions were implemented to increase fecundity and HY survival. Without accurate information on dispersal, it would be wise to maintain multiple large patches and not rely on small patches under the best-case scenario of Model 1.

We also demonstrated that, given our dispersal assumptions, patches could be successfully recolonized after extinction. What our simulations demonstrate is that, given the ability to recolonize, Model 1 parameterization would support such an event. For example, survival and fecundity rates are sufficient to maintain a stable population and produce enough migrants to establish a new population.

Given the patch sizes that occur outside of public lands, it seems imperative that actions to acquire or protect these areas begin while large areas of suitable habitat still exist. This analysis also points out the need for regional studies and management to determine what the population dynamic truly is and how populations across the region, and not just within our study area, should be managed to maintain such a dynamic. Studies on local populations are not sufficient to understand the entire system. Some key parameters that need to be determined for the entire breeding range are natal dispersal among and within habitat patches, as well as survival and fecundity rates (and their temporal variability) across the region in association with habitat quality; an assessment of the effects of winter range on breeding populations is also important. Such information would greatly improve our ability to conduct PVA, determine population persistence, and analyze the metapopulation dynamic. Such studies are generally prohibitive due to the cost and time required to complete such an effort. However, the golden-cheeked warbler provides a unique opportunity for such an investigation because of its limited breeding range, making a regional study more feasible than for many other neotropical migrants.

Acknowledgments This research was partially funded by a challenge grant from the U.S. Fish and Wildlife Service to J. S. H. and D. D. D. (FWS Agreement No. 1448–20181–00–J605). The data from Fort Hood were collected by The Nature Conservancy through cooperative agreement DPW-ENV 97-A-0001 between the Department of the Army and The Nature Conservancy. Information contained in this report does not necessarily reflect the position or the policy of the government, and no official endorsement should be inferred.

References

Akçakaya, H. R. 2002. *RAMAS Metapop: Viability analysis for stage-structured metapopulations (version 4.0)*. Applied Biomathematics, Setauket, N.Y.

Anders, A. D. 2000. Demography of golden-cheeked warblers on Fort Hood, Texas in 2000: productivity, age structure, territory density, and adult return rates. In *Endangered species monitoring and management at Fort Hood, Texas: 2000 annual report*. Fort Hood project, The Nature Conservancy of Texas, Fort Hood, Texas.

Braden, G. T., McKernan, R. L., and Powell, S. M. 1997. Effects of nest parasitism by the brown-headed cowbird on nesting success of the California gnatcatcher. *Condor* 99: 858–865.

Burnham, K. P., and Anderson, D. R. 1998. *Model selection and inference: a practical information-theoretic approach*. Springer-Verlag, New York.

Chase, M. K., Nur, N., and Geupel, G. R. 1997. Survival, productivity, and abundance in a Wilson's warbler population. *Auk* 114: 354–366.

DeGroot, K. L., Smith, J. N. M., and Taitt, M. J. 1999. Cowbird removal programs as ecological experiments: measuring community-wide impacts of nest parasitism and predation. *Studies in Avian Biology* 18: 229–234.

Diamond, D. D., and True, C. D. 1999. *Golden-cheeked warbler habitat area, habitat distribution, and change and a brief analysis of land cover within the Edwards Aquifer recharge zone*. Final report. U.S. Fish and Wildlife Service, Austin, Texas.

Goodman, L. A. 1960. On the exact variance of products. *Journal of the American Statistical Association* 57: 708–713.

Gould, W. R., and Nichols, J. D. 1998. Estimation of temporal variability of survival in animal populations. *Ecology* 79: 2531–2538.

Holmes, R. T., Sherry, T. W., Marra, P. P., and Petit, K. E. 1992. Multiple brooding and annual productivity of a Neotropical migrant, the black-throated blue warbler (*Dendroica caerulescens*), in an unfragmented temperate forest. *Auk* 109: 321–333.

Holmes, R. T., Marra, P. P., and Sherry, T. W. 1996. Habitat-specific demography of breeding black-throated blue warblers (*Dendroica caerulescens*): implications for population dynamics. *Journal of Animal Ecology* 65: 183–195.

Kroll, J. C. 1980. Habitat requirements of the golden-cheeked warbler: management implications. *Journal of Range Management* 33: 60–65.

Pulich, W. M. 1976. *The golden-cheeked warbler: a bioecological study*. Texas Parks and Wildlife Department, Austin.

Rappole, J. H., King, D. I., and Leimgruber, P. 2000. Winter habitat and distribution of the endangered golden-cheeked warbler (*Dendroica chrysoparia*). *Animal Conservation* 2: 45–59.

Robinson, S. K., Rothstein, S. I., Brittingham, M. C., Petit, L. J., and Grzybowski, J. A. 1995a. Ecology and behavior of cowbirds and their impact on host populations. Pages 428–460 in T. E. Martin and D. M. Finch (eds.), *Ecology and management of neotropical migratory birds*. Oxford University Press, New York.

Robinson, S. K., Thompson III, F. R., Donovan, T. M., Whitehead, D. R., and Faaborg, J. 1995b. Regional forest fragmentation and the nesting success of migratory birds. *Science* 267: 1987–1990.

Sillet, T. S., and Holmes, R. T. 2002. Variation in survivorship of a migratory songbird throughout its annual cycle. *Journal of Animal Ecology* 71: 296–308.

Summers, S. G., Eckrich, G. H., and Cavanagh, P. M. 2000. Brown-headed cowbird control program on Fort Hood, Texas, 1999–2000. In *Endangered species monitoring and management at Fort Hood, Texas: 2000 annual report*. Fort Hood Project, Nature Conservancy of Texas, Fort Hood.

U.S. Fish and Wildlife Service. 1992. *Golden-cheeked warbler (Dendroica chrysoparia) recovery plan*. U.S. Fish and Wildlife Service, Albuquerque, N.M.

U.S. Fish and Wildlife Service. 1996. *Golden-cheeked warbler population and habitat viability assessment report*. Compiled and edited by Carol Beardmore, Jeff Hatfield, and Jim Lewis in conjunction with workshop participants. Report of an August 21–24, 1995, workshop arranged by the U.S. Fish and Wildlife Service in partial fulfillment of U.S. National Biological Survey Grant No. 80333–1423. Austin, Texas.

Vidal, R. M., Macias-Caballero, C., and Duncan, C. D. 1994. The occurrence and ecology of the golden-cheeked warbler in the highlands of northern Chiapas, Mexico. *Condor* 96: 684–691.

Wahl, R., Diamond, D. D., and Shaw, D.1990. *The golden-cheeked warbler: a status review*. U.S. Fish and Wildlife Service, Albuquerque, N.M.

White, G. C., and Burnham, K. P. 1999. Program MARK: survival estimation from populations of marked animals. *Bird Study* 46(Suppl.): 120–138.

34

Capercaillie (*Tetrao urogallus*) in Western Switzerland

Viability and Management of an Endangered Grouse Metapopulation

SÉBASTIEN SACHOT
NICOLAS PERRIN

Capercaillie, the largest European grouse, is listed as Threatened in western, central, and southeastern Europe (IUCN 1996). Though the species still occupies most of its original range throughout northern Europe and Asia, western and central European populations have experienced important declines and local extinctions over the past decades (Storch 2000), owing to habitat losses and human disturbance (Ménoni et al. 1994, Storch 1994). Rainy conditions during egg incubation and young rearing, as well as predation (Storch 1991), can locally amplify population decline. Remnant populations are restricted to mountain regions where habitats are often isolated and fragmented (Blair et al. 1997).

In Switzerland, concerns raised by capercaillie's rapid decline resulted in financial support for silvicultural practices that would improve its habitat conditions. Political pressure to invest money in a cost-effective way called for a comprehensive approach aimed at identifying crucial life history stages, key populations parameters, and potential effects of habitat improvements.

In this chapter, we present a viability analysis for the endangered capercaillie metapopulation of the Swiss Jura mountains. Field studies on this focal population started in 1976. Its specific life history parameters and decline are thus well documented. This analysis is based on a series of computer simulations performed by the software RAMAS GIS. Our specific goals were (1) to build a habitat suitability map and identify landscape parameters explaining capercaillie's distribution, (2) to define crucial life history stages, and (3) to explore the relative benefits of several management scenarios.

Study Area and Species

The Jura mountains (47°25′ N; 6°42′ E), at the border between France and Switzerland, spread over 250 km length and 65 km width at the widest point. The landscape is characterized by a succession of forests, pastures, rocky crests, and valleys (the latter that never exceed 3 km width). Elevation ranges from 500 to 1718 m, and the treeline is around 1550 m. Coniferous forests are dominant above 1000 m and are mainly represented by the fir-beech mountain association (*Abieti albae–Fagetum sylvaticae typicum*) (Vittoz 1998). Forests are dominated by spruce (*Picea excelsa*) mixed with beech (*Fagus sylvatica*) and fir. The climate is harsh, with a mean annual temperature of 5.5 °C and an average of 180 frost days per year at 1200 m. Mean annual rainfall is 2000 mm, with 4 m of total snowfall. Snow cover is present from November to May and reaches 0.5 to 2 m in January. Spring and summer are mild and damp, with a growing period that does not exceed 4 months. Human density is low (50–100 people km^{-2}), although the area is frequently visited for various types of leisure activities. Cattle farming is the main economic activity, and it occurs in pastures and valley bottoms.

Capercaillie is strictly associated with boreal or montane forests with highly specific habitat requirements in terms of structure and composition that are, in most cases, best met by old stands (Leclercq 1987a, Picozzi et al. 1992). Bilberry (*Vaccinium myrtillus*) represents the major part of its summer diet, while in central Europe, fir (*Abies alba*) is required as winter food. Home range varies from 100 to 1000 ha (Gjerde and Wegge 1989, Ménoni 1991) for adults of both sexes according to local conditions. These characteristics, together with a high sensitivity to human perturbations (especially during the winter period and the reproductive season), make capercaillie an indicator of undisturbed forest ecosystems (Storch 1993b).

As a result of cattle farming and silvicultural practices started as early as the fourteenth century (Leclercq 1987a), the distribution of capercaillie populations in the Jura mountains has long been patchy. It has drastically regressed since the 1960s, however, being now restricted to altitudes between 1100 and 1550 m. The remaining populations are distributed over 24 forest patches in France and 15 in Switzerland, covering 542 km^2 in total and with an overall population of ca. 760 breeding birds (Montadert and Chamouton 1997).

Model Parameters

The PVA analysis was developed in three major steps. First, a habitat suitability (HS) map was built on capercaillie observations and on ecogeographical variables within RAMAS GIS Spatial Data module. This module also allowed conversion of the HS map to a patch structure. Second, the patch structure was exported to the Metapopulation Model module to perform population viability analysis. Third, this module was used to explore management options.

Spatial Data

The habitat suitability map was restricted to the area between the French border at the northwest as ecogeographical variability was not available in France, and to the 500-m isohypse at the southeast as historical data indicate that the species never occurred below this altitude (Leclercq 1987a). We investigated 1-km^2 squares ($N = 150$) between 1100 and 1600 m elevation twice between 1995 and 2000. Each square was explored radially from its center toward the border for 1 hour maximum. Signs of capercaillie presence (droppings, footprints, and bird sightings) were located with an accuracy of ±10 m, using compass, altimeter, and global positioning system (GPS), and marked on 1:25,000 vectorial topographical maps (Federal Office of Topography). Moreover in each of these squares, eight random plots were carefully sampled (15 min each) for capercaillie signs. None of these 1,200 random plots provided any sign of capercaillie occupancy. Presence signs ($N = 1{,}161$) were rasterized to 1 ha (100 × 100 m) cell resolution and occurred in 783 cells. An equal number of cells ($N = 783$) were randomly chosen as absence sites from the 1,200 random plots.

Topographical characteristics (altitude, slope, and exposure) and land-use data for each hectare in the study area were obtained from the Geostat database (Swiss Federal Office of Statistics). From the 74 land-use categories defined in this database, 52 could be dropped because they were either not represented in the study area or irrelevant for assessing capercaillie habitat (e.g. vineyards, orchards). As several studies suggested disturbance effects of human activities on capercaillie (Leclercq 1987a, Ménoni and Bougerol 1993), we included information on the road network from the Vector25 database (Federal Office of Topography). Data on nordic ski trails, alpine ski runs, skilifts, snowshoe trails, and sled dog trails were calculated from digitalized maps of leisure activities edited by the Association Vaudoise du Tourisme Pédestre and the tourist offices from the Vallée de Joux, St-Cergue, and Les Rasses. These boolean data were transformed into quantitative variables by calculating either frequencies (of occurrences within a 1-km^2 circle, corresponding to capercaillie home range) or distances (from the focal plot to the closest occurrence of a given disturbance category). Finally, as some influence of winter wild boar hunting on capercaillie distribution was expected (Lefranc 1987), we added a layer derived from the mean number of wild boars killed per 100-ha forest (Service Forêts-Faune-Nature, unpublished data). The complete data set thus comprised 37 layers in raster format of 1-ha pixel size, referenced to the plane projection of the Swiss coordinate system. Data conversion was performed using Mapinfo 6.0 (Mapinfo Corp.), and rasterization was done with Idrisi32 (Clarks Lab.).

We used logistic regression to estimate the habitat suitability (HS) function (Table 34.1). Following Hosmer and Lemeshow (1989), we used a first-order polynomial as linear predictor, a binomial error distribution, and a logit link function. Multicolinearity was limited by computing pairwise comparisons. Whenever a correlation exceeded 0.5, the variable with lower biological meaning was dropped. Of the initial variable set, 34 variables were then used to fit the models. Untransformed variables were used, as factor normality is not a prerequisite and error terms may have non-Gaussian distributions (Guisan and Zimmermann 2000). A stepwise forward and backward selection procedure was used to select explanatory variables. The final model retained only uncorrelated variables significant at the 5% level.

By inverting the logit link function, we obtain the following HS function:

$$HS = \frac{1}{1+e^{(43.776\mathrm{IN}+263.471\mathrm{HU}+6.184\mathrm{FA}+4.342\mathrm{SM}-2.713\mathrm{OP}-0.466\mathrm{FO}+4.37\mathrm{RO}+0.913\mathrm{RA}-0.107\mathrm{EL}+0.456\mathrm{SL}-0.154\mathrm{DI})}}$$

Table 34.1 GLM model for capercaillie habitat suitability with binomial error distribution and logit link function

Variable	Code	Estimate	SE	t-Ratio	p-Value
Intercept	IN	–43.776	3.254	–13.454	<0.001
Elevation	EL	0.107	0.008	12.745	<0.001
Slope	SL	–0.456	0.122	–3.745	<0.001
Open forest (tree cover < 60%) (%)	OP	2.713	0.457	5.933	<0.001
Normal forest (tree cover > 60%) (%)	FO	0.466	0.046	10.114	<0.001
Small woods (size < 0.25 ha) (%)	SM	–4.342	1.037	–4.186	<0.001
Rocks (%)	RO	–4.370	1.107	–3.947	<0.001
Distance to roads (all categories pooled)	RA	–0.913	0.376	–2.429	0.015
Distance to farms	FA	–6.184	3.068	–2.015	0.044
Distance to alpine ski trails	DI	0.154	0.039	3.956	<0.001
Wild boar hunting index	HU	–263.471	84.400	–3.122	0.002

Note: The 10 retained variables explained 75.6% of the differences in habitat selection.

This function was used to assess capercaillie habitat suitability for each pixel of the map. We introduced a habitat quality threshold of 0.5 and a habitat quantity threshold of 100 ha, which corresponds to the minimum home range size for a female capercaillie (Storch 1993b). In central Europe, hens do not use distinct seasonal ranges and mean movement distance between seasons was estimated to 0.8 km ± 0.3 ($N = 7$) (Storch 1995). We thus set the neighborhood distance to eight cells. This corresponds to the assumption that suitable cells within 800 m of each other belong to the same habitat patch.

A linear relationship was assumed between the average HS value for a patch and home range size. In Europe, recorded home range sizes were around 100 ha in very good habitats and 400 ha in medium-quality habitats (Storch 1993b, Wegge and Rolstad 1986). Home ranges overlap more than 50%, so we used a 50–ha spring home range for HS = 1 and 200 ha for HS = 0.5. The mean home range size for each population (h_i) was obtained by

$$h_i = -300 \times \frac{\sum_{i=1}^{n} \mathrm{HS}}{n_i} + 350$$

where n_i = the number of pixels in population i. Population carrying capacity (K_i) was given by $K_i = n_i / h_i$.

Patch Delineation

The patch recognition algorithm identified 15 capercaillie patches in western Switzerland (Figure 34.1). The largest patches (nos. 14 and 15) are located at the southwestern part of the study area. Moderate- and small-sized populations are distributed at the northeast. Three patches (nos. 2, 10, and 12) identified by the algorithm are presently empty. The absence of birds at those locations could be explained by parameters that were locally important for these populations but were not retained by the logistic regression or by local extinction through the demographic stochasticity process.

Initial abundance of males is the total number of males counted at the leks in each population during spring 2000 (Table 34.2). We assume a 1:1 sex ratio in each popula-

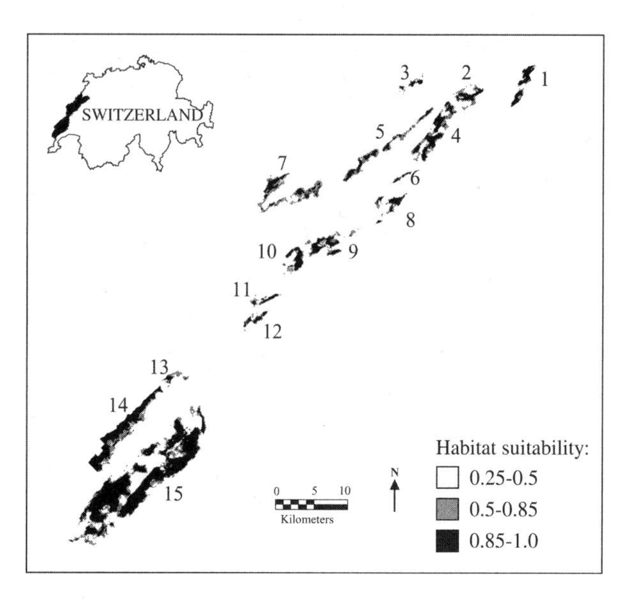

Figure 34.1 Habitat suitability map encompassing locations of 15 capercaillie patches in western Switzerland. Delineation of patches by the patch-recognition algorithm was made with a habitat suitability threshold of 0.5, a habitat quantity threshold of 100 ha, and an 800-m neighborhood distance.

tion and extrapolate female initial abundance. In four patches (nos. 1, 3, 5, and 12), no female was recorded since 1995, thus initial female abundance was set to 0.

Metapopulation Model

Baseline demographic parameter estimations (Table 34.3) were obtained from route censuses performed in a neighboring French Jura population during July 15–30 (i.e., 3 months after egg laying) from 1976 to 1986 (Leclercq 1987c). A line of three to eight persons disposed at 25-m intervals walked simultaneously through entire forest patches and counted flushed capercaillie. Age- and sex-undetermined birds were omitted from the calculations of demographic parameters.

Fertility was estimated as the mean number of eggs in all clutches discovered between 1976 and 1999 ($N = 28$). The proportion of reproductive females is the average proportion of females observed with a brood during a census. Juvenile survival during the first 3 months of life was estimated from brood composition in July, assuming a 1:1

Table 34.2 Capercaillie initial abundance (adults only) and carrying capacity (K_i) for the 15 patches delineated in western Switzerland

	Patch number														
Parameter	1	2	3	4	5	6	7	8	9	10	11	12	13	14	15
Number of males	1	0	1	1	1	1	2	1	3	3	0	1	1	9	43
Number of females	0	0	0	1	0	1	2	1	3	3	1	0	1	9	43
K_i	6	2	5	13	9	1	11	4	8	6	2	2	2	27	101

Note: See text for parameter estimations.

Table 34.3 Baseline parameter estimates used in the simulations of capercaillie persistence in western Switzerland

Parameter	Mean	SD
Sex ratio (proportion of female hatchlings)	0.5	0
Fertility (no. of eggs per female)	7.54	0.60
Proportion of reproductive females (%)	0.34	0.25
Female chick survival (0–3 months)	0.55	0.30
Female juvenile survival (3–15 months)	0.19	0.15
Female adult survival (annual survival after 15 months)	0.82	0.24
Female juvenile survival from census to breeding (9 months)	0.288	0.227
Female adult survival from census to breeding (9 months)	0.862	0.252
Male chick survival (0–3 months)	0.45	0.27
Male survival (3–15 months)	0.2	0.15
Male survival (annual survival 15–27 months)	0.7	0.1
Male survival (annual survival after 27 months)	0.92	0.10

Note: Derived from Leclercq (1987a).

sex ratio at hatching. Annual survival rates of yearlings and adults were estimated by Leclercq (1987c) from a linear regression of $S(t)$ on $R(t)$:

$$S(t) = Sa - R(t) \times (Sa - Si)$$

where $S(t)$ is the population average survival from year t to $t + 1$, calculated by dividing the number of adults at year $t + 1$ by the total number of individuals (adults plus yearlings) at year t, and $R(t)$ is the proportion of yearlings at year t, calculated as the ratio of the number of yearlings to the total number of individuals (adults plus yearlings). The parameters Sa and Si represent adult and yearling survival rates, respectively. Leclercq obtained annual survival rates by fitting linear regressions for the two sexes separately.

The transition matrix (Figure 34.2) included sex structure with two female stages and five male stages. Matrix fecundities were calculated as a product of five variables. The corresponding fecundity for the first female stage was

$$\text{Fecundity}(3) = Sfc(0\text{–}3) \times SR\ F \times PF \times Sfj(3\text{–}9)$$

where $Sfc(0\text{–}3)$ is the female chick survival rate for the first 3 months of life, SR the sex ratio, F the fertility, PF the proportion of reproductive females, and $Sfj(3\text{–}9)$ the juvenile female survival rate from the census to the next breeding season. Fecundity for the first male stage was calculated by replacing $Sfc(0\text{–}3)$ with juvenile male survival $Smj(0\text{–}3)$. Fecundity for the second female stage was calculated as:

$$\text{Fecundity}(15+) = Sfc(0\text{–}3) \times SR \times F \times PF \times Sfa(3\text{–}9)$$

where $Sfa(3\text{–}9)$ is the adult female survival rate from the census to the next breeding season. Fecundity for the second male stage was calculated by replacing $Sfc(0\text{–}3)$ with juvenile male survival $Smj(0\text{–}3)$.

Standard deviations of transition rates are shown in Figure 34.3. The variance of the product of two random numbers was a function of their means, variances, and covariance (for an example, see Akçakaya and Raphael 1998). Our calculations assume a 0 covariance.

		Female 3	Female 15+	Male 3	Male 15	Male 27	Male 39	Male 51+
Female	3	0.202	0.607	0	0	0	0	0
	15+	0.19	0.82	0	0	0	0	0
Male	3	0.166	0.497	0	0	0	0	0
	15	0	0	0.2	0	0	0	0
	27	0	0	0	0.7	0	0	0
	39	0	0	0	0	0.92	0	0
	51+	0	0	0	0	0	0.92	0.92

Figure 34.2 Stage matrix including two female stages and five male stages.

Juveniles disperse away from the natal territory after fledging (during autumn and early winter). Females reach maturity at age 1 and then establish a breeding territory. Because no local information was available on dispersal, we used values from the literature on other West European populations. Average values of about 8 km and a maximum value of 75 km (Koivisto 1963) documented for juvenile females were used to parameterize the dispersal-distance function (see Chapter 1 in this volume) with $a = 1$, $b = 8$ km, and $D_{max} = 75$ km.

Breeding females defend territories and additional females have a highly reduced breeding success, if any (Ménoni 1991). Therefore, we chose a ceiling type of density

		Female 3	Female 15+	Male 3	Male 15	Male 27	Male 39	Male 51+
Female	3	0.306	0.662	0	0	0	0	0
	15+	0.15	0.24	0	0	0	0	0
Male	3	0.258	0.566	0	0	0	0	0
	15	0	0	0.15	0	0	0	0
	27	0	0	0	0.1	0	0	0
	39	0	0	0	0	0.1	0	0
	51+	0	0	0	0	0	0.1	0.1

Figure 34.3 Matrix of standard deviations including two female stages and five male stages.

dependence, the number of adult females on each patch being limited by the local carrying capacity, itself a function of patch habitat suitability.

Simulation Scenarios

One simulation was run on the baseline demographic parameter values (Table 34.3) and initial population sizes (Table 34.2). Further simulations were then conducted with one parameter changed at a time and all other parameters held to their baseline value. In this uncertainty analysis, the modified parameter takes a low or high value (Table 34.4) that represents the extreme estimates derived from a compilation of all known capercaillie data reported in published papers for other European populations (Höglund 1952; Koivisto 1963; Wegge et al. 1981, 1990; Moss and Oswald 1985; Leclercq 1987a, 1987c; Moss and Weir 1987).

Finally, we tested a series of management scenarios corresponding to improvements in habitat availability and quality. The current political context and financial resources make habitat enhancement of one-half of the available area a reasonable and realistic measure. Habitat availability improvements consisted of a 50% increase of the surface of (1) all 15 patches, (2) the two largest patches only, and (3) the 13 smallest patches only. All simulations consisted of 1,000 replicates run over a 50-year time horizon.

Results

The populations of capercaillie were moderately endangered according to the baseline life history traits: extinction risk was estimated to be 7.77%, and the median time to extinction was more than 50 years. All simulations have a median time to extinction greater than 50 years, but the uncertainty analysis gave contrasting results in terms of

Table 34.4 Parameter uncertainty and PVA analysis for the capercaillie in western Switzerland

	Value		Extinction Risk(%)	
Parameter	Low	High	Low	High
Female chick survival (0–3 months)	0.1	0.8	23.4	2.5
Sex ratio (proportion of female hatchlings)	0.22	0.75	0.5	5.1
Fertility (no. of eggs per female)	6	8	14.5	8.2
Proportion of reproductive females (%)	0.16	0.55	26.7	0.5
Female juvenile survival (3–5 months)	0.1	0.6	8.8	2.9
Female adult survival (annual, after 15 months)	0.55	0.9	14.5	7.1
Male chick survival (0–3 months)	0.1	0.5	19.9	8
Male survival (3–15 months)	0.15	0.75	10.6	2.3
Male survival annual survival 15–27 months)	0.2	0.8	20.2	6.9
Male survival (annual survival 27–39 months)	0.84	0.96	9.6	8.2
Male survival (39–51 months)	0.84	0.96	9.6	8.7
Male survival (51+ months)	0.84	0.96	41.7	0.1

Note: Parameters values are extreme estimates derived from various sources (see text).

extinction risk (Table 34.4). The variation in risk estimates for low parameter values was high: from 0.5% to 41.7%. The decrease of reproductive male survival had the strongest impact on extinction risk. A reduction in the proportion of reproductive females, chick survival, and survival of juvenile males between 3 and 27 months also markedly increased extinction risk.

Risk estimates for high parameter values were less variable and range from 0.5% to 8.7%. An increase of reproductive male survival or of the proportion of reproductive females markedly decreased extinction risk down to 0.1% and 0.5%, respectively. High values for sex ratio, fertility, and juvenile male stages (15–51 months) were insufficient to completely eliminate extinction risk.

An increase in available suitable habitat had a very small impact on estimated capercaillie persistence. Extinction risk ranged from 7.3% for scenario 3 to 8.5% for scenario 2, while scenario 1 had an intermediate value (7.5%). All simulations have a median time to extinction greater than 50 years.

Discussion

Habitat Suitability

Capercaillie preferred high elevations and moderate slopes (Table 34.1), consistent with Storch (1993a) who found strong preferences for high altitudes and slopes between 10° and 20° in the Bavarian Alps. From historical information on the decline process in the Jura mountains, low-altitude localities were abandoned first (Leclercq 1987b). This pattern may apply to most central, western, and southeastern Europe, where remaining capercaillie populations are presently restricted to mountain forests (Storch 2000). This trend may result from changes in the structure or composition of low-altitude forests, but also from increasing perturbation levels resulting from their proximity to human settlements. Preference for moderate slopes certainly relates to the topography of the Jura mountains and potential capercaillie habitat availability. Unfavorable habitats are located below 1100 m and above 1550 m and correspond to steep areas with either too dense tree cover or no trees at all.

Capercaillie showed a strong preference for open forests (20%–60% canopy cover) or normal forests (>60% canopy cover). Open forests in the study area showed abundant and diversified ground vegetation with bilberry and other ericaceous shrubs in nutrient-poor soils (Vittoz 1998). Bilberry, an essential component of capercaillie diet in summertime, optimally grows in forests with canopy cover around 50% (Storch 1993a).

Small woods had a negative influence on capercaillie habitat suitability. Forest patches less than 50 ha in size do not sustain permanent leks, and patches below 25 ha are not colonized (Ménoni 1994). Small woods arise from habitat fragmentation by human-induced processes and are more prone to predation by generalist predators and to human disturbance (Angelstam 1986).

Sources of human disturbance had a clear negative impact on capercaillie habitat (Table 34.2). Wild boar hunting largely decreased habitat suitability, as did farms, roads, and alpine ski trails. High wild boar density in Jura capercaillie sites is a recent phenomenon (<20 years) and induces predation on nests and chicks (Lefranc 1987). The negative coefficient attributed to this factor indicates that capercaillie prefer areas with lower wild boar hunting success rates. This can be explained by selection of sites with moderate predator pressure and by avoidance of disturbances induced by hunters and trained dogs.

In the Jura mountains, farms are patchily distributed, according to meadow availability. Farms resulted in habitat fragmentation beginning in the early fourteenth century, when cattle grazing and farming practices were very intensive (Leclercq 1987b). Farms also represent a major disturbance in potential capercaillie breeding and wintering sites. Similarly, roads facilitate human access into these sensitive sites for logging or leisure activities. Repeated disturbances force hens to abandon their clutches, reduce chick survival, and drastically increase winter mortality (Ménoni et al. 1994, Storch 1991).

The negative effect of alpine ski-lifts has been shown in previous studies reporting breeding and wintering site desertion in the vicinity of ski trails (Leclercq 1987a, Ménoni et al. 1994). During winter, capercaillie feeds essentially on coniferous needles, mainly firs with a high amount of nitrogen content (Leclercq 1987a). This superabundant and continuous supply of resources allows a reduced activity of the birds (Gjerde and Wegge 1987), and several birds may forage in the same high-quality tree (Leclercq 1987a). Low habitat disturbance is essential during this highly sensitive season (Storch 1993c).

Population Viability Analysis

Simulations with baseline parameter values suggested that the studied capercaillie populations would be moderately endangered within the next 50 years. This prediction is conservative as we did not include in our model any factors that could further accelerate the extinction process, such as complex density dependence regulation or Allee effects (Burgman et al. 1993). However, the most appropriate use of PVAs does not lie in the absolute values of extinction risks it provides but in the ranking of threats and management options (Akçakaya and Altwood 1997). In this respect, we do not expect the main limitations outlined by the model to affect the ranking of threats and management options delineated in our simulations.

The results of the uncertainty analysis suggest that the most important parameters decreasing the viability of these capercaillie populations are reproductive (51+ months) and juvenile (3–27 months) male survival, proportion of reproductive females, and chick survival.

In long-lived birds, adult survival has often been seen as being the main parameter on which conservation actions should be taken (Trouvilliez et al. 1988). Our results confirm this view. In capercaillie, juvenile mortality is high during summer and late winter, then mortality largely decreases after the first year of life (Storch 1993b). Several factors affect the breeding success, which exhibits large annual and site-to-site variances (Leclercq 1987a). Habitat structure; food availability; and quality, predator number, and weather conditions could all play a role (Moss et al. 2001, Storch 1993b).

Building the capercaillie model linking landscape data on habitat quality to spatial structure for a PVA analysis required some simplifying assumptions, including ceiling type of density dependence, a simple link between habitat patches and carrying capacity, and spatial extent restricted to western Switzerland. Analyses at a larger geographic scale including the 24 capercaillie patches of the French Jura mountains should provide a more reliable estimate about the future of capercaillie.

Management Scenarios

Management scenarios did not strongly reduce extinction risk, and small differences among them were probably due to stochastic events. Our results suggest that managing forested habitat on the two largest patches has the same impact as managing the 15 patches. Increasing habitat size without increasing habitat quality thus seems ineffective. However, the simulated management scenarios were not directly linked to vital rates because of insufficient data. Linking these characteristics will probably give more sensitive results on the effect of habitat management.

Grouse-adapted forest management positively influences capercaillie and could act on food availability and predator numbers (Kurki et al. 1998, Storch 1993b). A major constraint of habitat enhancement is the long time between logging and the effective impact on capercaillie populations. The bird has large area requirements that call for wood and rejuvenation cutting on a large extent and this over several years.

Conclusion

The spatial PVA performed in this chapter improved our understanding of the system in two ways. First, the logistic regression identified key landscape parameters. At this spatial level, land managers should maintain the present forest distribution, encourage grouse-adapted silvicultural practices for high-altitude sites, and reduce human disturbance. Preliminary essays suggest that forest structure and composition could be improved at reasonable costs (S. Sachot et al., unpublished data). Second, the PVA and uncertainty analysis showed that survival of not only females but also males and chicks is an essential parameter for population maintenance. To increase these life history parameters, one option would be to limit human disturbance by restricting access into breeding and wintering sites. Our PVA could be developed in several complementary directions to explore the effects of simplifying assumptions about density dependence and carrying capacity. Future investigations on capercaillie distribution in the French Jura must be carried out to perform a PVA for the whole metapopulation.

References

Akçakaya, H. R., and Altwood, J. L. 1997. A habitat-based metapopulation model of the California Gnatcatcher. *Conservation Biology* 11: 422–434.

Akçakaya, H. R., and Raphael, M. G. 1998. Assessing human impact despite uncertainty: viability of the northern spotted owl metapopulation in the northwestern USA. *Biodiversity and Conservation* 7: 875–894.

Angelstam, P. 1986. Predation on ground nesting birds' nests in relation to predator densities and habitat edge. *Oikos* 47: 365–373.

Blair, M., Bijlsma, R., and Hagemeijer, W. 1997. *The EBCC atlas of European breeding birds*. T&AD Poyser, London.

Burgman, M. A., Ferson, S., and Akçakaya, H. R.1993. *Risk assessment in conservation biology*. Chapman and Hall, London.

Gjerde, I., and Wegge, P. 1987. Activity patterns of capercaillie during winter. *Holarctic Ecology* 10: 286–293.

Gjerde, I., and Wegge, P. 1989. Spacing pattern, habitat use and survival of capercaillie in a fragmented winter habitat. *Ornis Scandinavica* 20: 219–225.

Guisan, A., and Zimmermann, N. E. 2000. Predictive habitat distribution models in ecology. *Ecological Modelling* 135: 147–186.

Höglund, N., 1952. Capercaillie reproduction and climate. *Papers in Game Research* 8: 78–86.

Hosmer, D. W., and Lemeshow, S. 1989. *Applied logistic regression*. Wiley, New York.

IUCN, 1996. *Red List of threatened animals*. IUCN, Gland, Switzerland.

Koivisto, L. 1963. Über den Ortwechsel der Geschlechter beim Auerhuhn nach Markierungergebnissen. *Volgelwarte* 22: 75–79.

Kurki, S., Nikula, A., Helle, P., and Linden, H. 1998. Abundances of red fox and pine marten in relation to the composition of boreal forest landscapes. *Journal of Animal Ecology* 67: 874–886.

Leclercq, B. 1987a. Ecologie et dynamique des populations du grand tétras (*Tetrao urogallus major* L.) dans le Jura français. Thesis, Faculté des Sciences de la Vie, Université de Bourgogne, Dijon.

Leclercq, B. 1987b. Influence des modes de gestion forestière passées sur la gestion actuelle et la structure des forêts de montagne ainsi que sur leurs peuplements en grand tétras. Pages 265–282 in *Actes du colloque Galliformes de montagne*. Gibier Faune Sauvage, Grenoble, France.

Leclercq, B. 1987c. Premières données sur la comparaison de la dynamique des populations de grand tétras (*Tetrao urogallus*) et de gélinotte des bois (*Bonasa bonasia*) d'un même massif forestier du Haut-Jura. Pages 21–36 in *Actes du colloque Galliformes de montagne*. Gibier Faune Sauvage, Grenoble, France.

Lefranc, N. 1987. La situation du Grand Tétras (*Tetrao urogallus*) dans le massif vosgien. *Bulletin mensual de l'office national de la chasse* 112: 5–18.

Ménoni, E. 1991. Ecologie et dynamique des populations du grand tétras dans les Pyrénées, avec des références spéciales à la biologie de la reproduction chez les poules: quelques applications à sa conservation. Thesis, Université Paul Sabatier, Toulouse.

Ménoni, E. 1994. Plan de restauration du grand tétras (*Tetrao urogallus*) en France. *Gibier Faune Sauvage* 11: 159–202.

Ménoni, E., and Bougerol, J. 1993. Capercaillie populations in forests fragmented by topography and human activities in the French Pyrenees. Pages 148–159 in *Twenty-first Congress of the International Union of Game Biologists*. I. D. Thompson, Halifax, Canada.

Ménoni, E., Brenot, J. F., and Catusse, M. 1994. Grand tétras et ski de fond. *Bulletin mensuel de l'ONC* 190: 12–21.

Montadert, M., and Chamouton, A. 1997. Statut des tétraonidés dans le massif jurassien. Pages 73–95 in *Secondes rencontres iurassiennes*. Parc Naturel du Haut-Jura, Lajoux, France.

Moss, R., and Oswald, J. 1985. Population dynamics of Capercaillie in a north-east Scottish glen. *Ornis Scandinavica* 16: 229–238.

Moss, R., and Weir, D. N. 1987. Demography of capercaillie *Tetrao urogallus* in north-east Scotland. II. Age and sex distribution. *Ornis Scandinavica* 18: 135–140.

Moss, R., Oswald, J., and Baines, D. 2001. Climate change and breeding success: decline of the capercaillie in Scotland. *Journal of Animal Ecology* 70: 47–61.

Picozzi, N., Catt, D. C., and Moss, R.1992. Evaluation of capercaillie habitat. *Journal of Applied Ecology* 29: 751–762.

Storch, I. 1991. Habitat fragmentation, nest site selection, and nest predation risk in Capercaillie. *Ornis Scandinavica* 22: 213–217.

Storch, I. 1993a. Habitat selection by capercaillie in summer and autumn: is bilberry important? *Oecologia* 95: 257–265.

Storch, I. 1993b. Habitat use and spacing of capercaillie in relation to forest fragmentation patterns. Thesis, Faculty of Biology, Université Ludwig-Maximilian, Munich.

Storch, I. 1993c. Patterns and strategies of winter habitat selection in alpine capercaillie. *Ecography* 16: 351–359.

Storch, I. 1994. Habitat and survival of capercaillie *Tetrao urogallus* nests and broods in the Bavarian Alps. *Biological Conservation* 70: 237–243.

Storch, I. 1995. Annual home ranges and spacing patterns of capercaillie in central Europe. *Journal of Wildlife Management* 59: 392–400.

Storch, I. 2000. *Grouse status survey and conservation action plan 2000–2004*. WPA/BirdLife/

SSC Grouse Specialist Group. IUCN Publications, Gland, Switzerland, and the World Pheasant Association, Reading, U.K.

Trouvilliez, J., Gaillard, J. M., Allaine, D., and Pontier, D. 1988. Stratégies démographiques et gestion des populations chez les oiseaux: particularités des Galliformes. *Gibier Faune Sauvage* 5: 27–41.

Vittoz, P. 1998. Flore et végétation du Parc jurassien vaudois: typologie, écologie et dynamique des milieux. Thesis, Faculté des Sciences, Université de Lausanne, Lausanne.

Wegge, P., and Rolstad, J. 1986. Size and spacing of capercaillie leks in relation to social behavior and habitat. *Behavioural Ecology and Sociobiology* 19: 401–408.

Wegge, P., Larsen, B., and Storaas, T. 1981. Dispersion printanière d'une population de coqs de bruyère dans le sud-est de la Norvège. Pages 138–153 in C. Kempf (ed.), Actes du Colloque International Grand Tétras. Parc Naturel des Ballons des Vosges, France.

Wegge, P., Gjerde, I., Kastdalen, L., Rolstad, J., and Storaas, J. 1990. Natural mortality and predation of adult capercaillie in south-east Norway. Pages 49–56 in *Proceeding of the International Symposium on Grouse*. World Pheasant Association, Reading, U.K.

35

Seaside Sparrows (*Ammodramus maritimus*) in Connecticut

Projected Effects of Sea-Level Rise

W. GREGORY SHRIVER
JAMES P. GIBBS

Global average surface temperature is projected to increase by 1.4 to 5.80 °C between 1900 and 2100 (International Panel on Climate Change 2001), largely due to human industrial activities, such as the burning of fossil fuels (coal, oil, and gas). Rising sea levels predominantly associated with oceanic thermal expansion are one consequence of changes in surface temperatures (Titus et al. 1991). Although according to some estimates, average sea-level rise is predicted to be 50 cm by the year 2100 (Houghton 1999, International Panel on Climate Change 2001), there have been few attempts to estimate the magnitude of the adverse effects that these coastal ecosystems and, in particular, the species these unique habitats support may experience (Donnelly and Bertness 2001).

Salt marshes are coastal ecosystems dominated by halophytic plants and exposed to daily inundation of saline water. Because salt marsh ecosystems are transitional between terrestrial and marine environments, they are particularly susceptible to sea-level rise at their outer boundaries and lowest elevations (Simas et al. 2001). Historically, salt marshes could expand and retreat with changes in sea level, but presently in many areas there is little opportunity for marshes to migrate inland as the advance is blocked by dense human settlements along the coastal fringe (Alexander et al. 1986, Chabreck 1988, Bildstein et al. 1991). Titus et al. (1991) estimated that, with no coastal protection (e.g., bulkheads, seawalls), a 50-cm increase in sea-level rise would result in a 30% loss of coastal wetlands and a 100-cm rise in sea level by 2100 could result in a 50% to 80% loss in coastal wetlands in the United States. The effects of these potential losses on obligate salt marsh species are unknown and have not been estimated.

Seaside sparrows (*Ammodramus maritimus*) are habitat specialists in tidal and brackish marshes of North America (Post and Greenlaw 1994). Seaside sparrows are monogamous, are patchily distributed, and have been shown to have greater abundance on larger marshes (Greenlaw 1992, Benoit 1997). Near their range limit in southern New England (Greenlaw 1992), populations are small and relatively uncommon, occurring in the wettest portions of the marsh where *Spartina alterniflora* is abundant (Woolfenden 1956, Marshall and Reinart 1990). Because of the salt marsh specificity for seaside sparrow, the species' distributional pattern indicates coastal wetland integrity (Post and Greenlaw 1994) and has already proven to be sensitive to habitat modification caused by water-level manipulation in Florida (Baker 1974, 1978; Walters 1992).

Our goal was to examine whether sea-level rise poses risks to the long-term persistence of seaside sparrow. Specifically, we examined the problem for sparrows in Connecticut salt marshes and (1) estimated through field surveys the distribution and abundance of seaside sparrow populations within the present marsh mosaic, (2) determined the relationship between seaside sparrow patch occupancy and marsh area, and (3) estimated the viability of seaside sparrow populations in response to three scenarios of sea-level rise during the twenty-first century (0-, 50-, and 100-cm rise in sea level).

Methods

To estimate seaside sparrow distribution and abundance, we surveyed 142 fixed radius points (100 m) in 75 Connecticut salt marshes during May 15 to August 15, 1999 (mean = 1.9 points per marsh). Because the survey was designed as an inventory of salt marsh breeding birds in New England, marshes were chosen to include as many as logistically possible to visit. Therefore, we limited sizes of marshes surveyed to all marshes >2.5 ha. We visited each sampling point twice, with at least 10 days between visits (Ralph et al. 1995). All point centers were greater than 300 m from any other point center and at least 50 m from upland (non-salt marsh vegetation). At each point, we observed for 10 minutes and recorded all birds seen and heard within the 100-m radius point. We surveyed from dawn to noon on days with low wind and, at most, light rain. We used logistic regression to develop an incidence function for seaside sparrow occupancy in relation to marsh size to better estimate patch selection criteria in the metapopulation model (Hosmer and Lemeshow 1989).

We developed a stage-structured spatially explicit metapopulation model for seaside sparrows in Connecticut using RAMAS GIS version 4.0 (Akçakaya 2002). The analysis involved describing the habitat mosaic and then developing a demographic model for birds residing within that habitat mosaic.

Salt Marsh Mosaics

We used a raster-based GIS (Idrisi; Clark Labs, Clark University, Worcester, Mass.) to simulate the effects of sea-level rise on the Connecticut coast. The base data were tidal marsh habitats digitized from 1995 aerial photographs by the Connecticut Department of Environmental Protection's Office of Long Island Sound. We converted this polygon coverage to a grid with 50- × 50-m cell sizes. We then estimated the amount and distribution of salt marsh habitat that would remain in response to three different sea-level rise scenarios (Titus et al. 1991). The first scenario was the current distribution of salt marsh habitat subjected to no rise in sea level or any associated salt marsh loss. The second scenario was a reduction of salt marsh habitat by 30% to approximate the effects of a 50-cm increase in sea level by 2100, as projected by Titus et al. (1991). The third scenario was a 50% reduction in salt marsh habitat to approximate the effects of a 100-cm increase in sea level by 2100 (Titus et al. 1991). To generate spatially explicit marsh mosaics for scenarios 2 and 3, we used Idrisi to progressively erode salt marshes from their seaward edges to a given level of habitat loss (Figure 35.1). We assumed that sea-level rise would occur equally across the Connecticut coast and that subsidence and

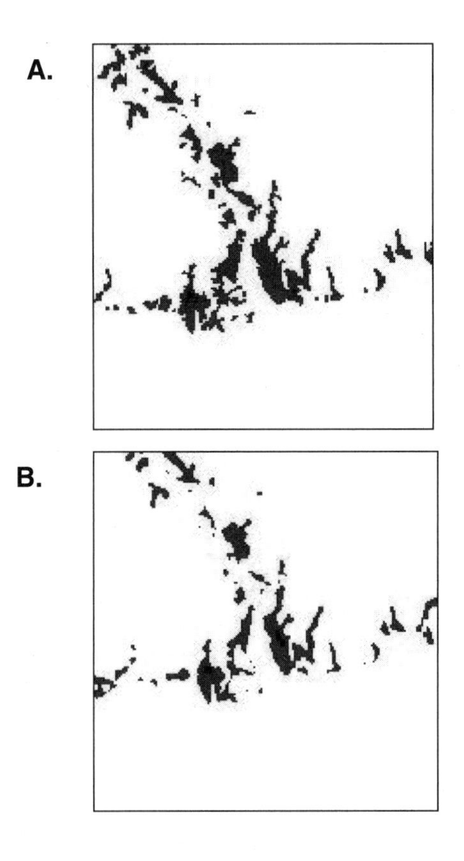

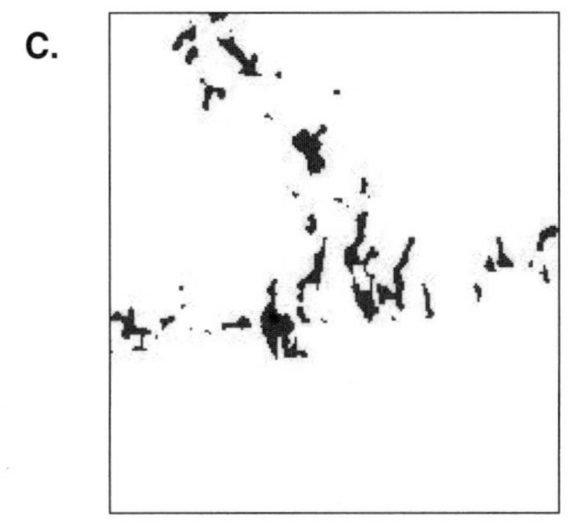

Figure 35.1 Example of the "nibbling" function used to progressively erode salt marshes from their seaward edges in Connecticut: **(A)** present salt marsh mosaic; **(B)** 30% reduced mosaic; **(C)** 50% reduced mosaic.

coastal development would limit marsh migration inland. Thus only salt marsh cells abutting ocean cells were eroded. The three scenarios were imported into the spatial data module of RAMAS GIS 4.0, which was then used to identify patches on each of the three landscapes. We incorporated habitat-area relationships for seaside sparrows into patch selection based on a nearest neighbor distance that limited minimum patch size to 50 ha (Figure 35.2) (see Benoit 1997).

Initial Abundance

To initiate population viability analyses, we used field survey data to populate marshes that were consistent with seaside sparrow occurrence and densities during the 1999 survey. We estimated population density based on the average number of males detected per point and marsh area. We estimated initial population abundance for each patch by taking the mean number of males detected per point, dividing by the area of the point (3.14 ha for 100-m radius points) to estimate the number of territories per ha, and multiplying by marsh area. We then multiplied this abundance estimate by the reciprocal of 0.8, an estimate of the detection probability to account for incomplete detection of sparrows at each point (Dawson et al. 1995; W. G. Shriver, unpublished data).

Stage Structure

We developed a stage-structured model with two life stages: adults and juveniles. A bird was considered an adult after surviving its first winter (see Post and Greenlaw 1994). We chose to model males because there is no reason to suspect that sex ratios are strongly imbalanced (Post and Greenlaw 1982), and our survey data were based primarily on the detection of males. We assumed that all breeding took place during a defined breeding season (birth pulse model, Caswell 1989). The census was assumed to be a postreproductive census, and we assumed that no mortality took place between breeding and the census. We used estimates of survival from a New York population of seaside sparrows based on cumulative return rates (Post and Greenlaw 1994; Table 35.1). We calculated a juvenile fecundity rate by multiplying the juvenile survival rate by the average number of male offspring produced (0.35 × 1.98) and adult fecundity rate as the adult survival rate multiplied by the number of male offspring (0.57 × 1.98). We used estimates of the number of males fledged per female per year from a Massachusetts population of seaside sparrows and assumed a 1:1 sex ratio (Marshall and Reinert 1990).

Carrying Capacity

We calculated carrying capacity for each patch based on patch size. We assumed a conservative estimate of seaside sparrow territory density as one territory per hectare of marsh (Woolfenden 1956, Post and Greenlaw 1994) and set carrying capacity for each patch equal to patch area (ha).

Density Dependence and Allee Effects

Density dependence has been demonstrated with song sparrows (Smith et al. 1996) and incorporated into PVA models for other sparrows (Wells 1997). Because density de-

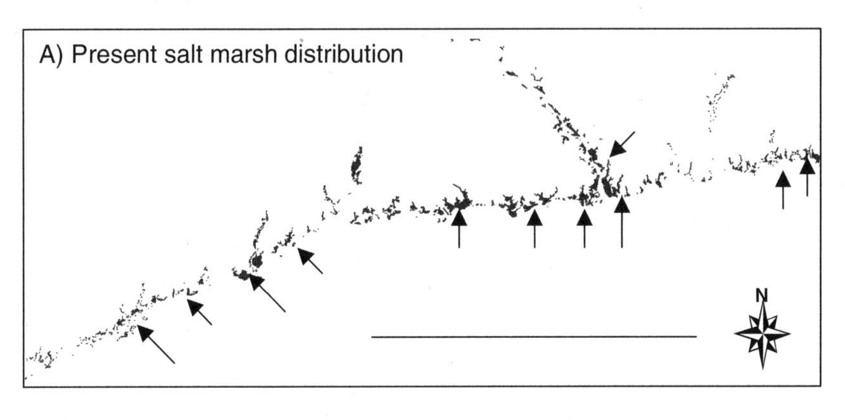

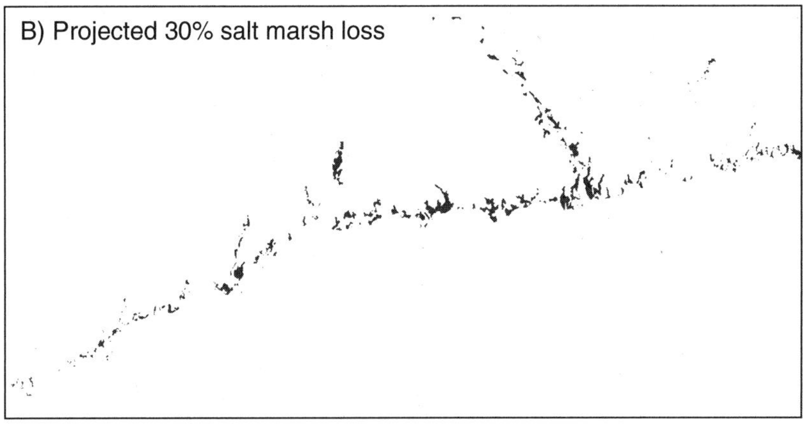

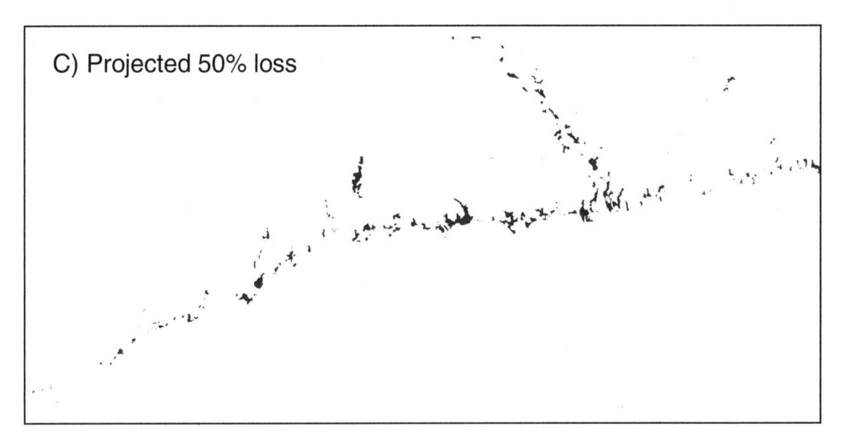

Figure 35.2 Salt marsh mosaics in Connecticut given three scenarios of sea-level rise during the twenty-first century: **(A)** present salt marsh mosaic; **(B)** 30% reduced mosaic; **(C)** 50% reduced mosaic. Arrows denote locations of seaside sparrow populations used to populate the metapopulation.

Table 35.1 Stage matrix of survival and fecundity values used in the seaside sparrow metapopulation model for the three sea-level rise scenarios in Connecticut

Parameter	Juvenile	Adult
Fecundity	0.694	1.131
Survival	0.350	0.570

Note: See text for sources.

pendence is likely to occur at high population levels, we incorporated it into our PVA with a ceiling model as have others for PVA models with passerines (Akçakaya and Atwood 1997, Wells 1997). We also included Allee effects to reduce vital statistics by 50% when populations reached fewer than five males. Allee effects are difficult to document and separate from dispersal events; however, we included Allee effects in our model to simulate small population vulnerability associated with stochastic events to which seaside sparrows can be susceptible (Kale 1983, Walters 1992). Last, we incorporated a metapopulation threshold of 100 males to determine the probability that the metapopulation would go below that threshold and reach "quasi extinction" in a given time period.

Dispersal

We assumed juveniles were twice as likely to disperse as adults because adults are philopatric (Post and Greenlaw 1994). To ensure that the majority of the Connecticut coast was a potential habitat for sparrows returning from the wintering habitat, we set a maximum dispersal distance of 20 km (with coefficient of variation for dispersal = 0.2).

Stochasticity

Seaside sparrows can be susceptible to storm surges that flood nests and limit reproduction in a given year (Post and Greenlaw 1994); therefore, we incorporated environmental fluctuations into our simulations. For this we generated a standard deviation for carrying capacities (±0.10) and estimated a standard deviation matrix for survival and fecundity rates (±0.17). Variation in vital rates for birds is poorly known (Burgman et al. 1993), and this value is similar to those estimated for other passerines (Akçakaya and Atwood 1997). Changes in the magnitude of variation will likely affect the magnitude of model estimates but should not change the relative relationships among the different models. For all scenarios we altered the standard deviation matrix by 10% for high and low estimates of environmental fluctuations and also incorporated a 10% probability of a catastrophe that would reduce fecundity by 5% to simulate rare storm flooding events, fires, or predation.

Simulations

Simulations (n = 5,000) were run over a 100-year period for consistency with the sea-level rise projections of Titus et al. (1991) on which our scenarios are based. At each time step the number of juveniles and adults were projected, using a set of vital rates drawn from a random normal distribution with mean values taken from the stage matrix and standard deviations taken from the standard deviation matrix. To determine the effects of habitat loss and fragmentation on population viability, we held all model parameters and initial abundances constant for the three sea-level rise scenarios. To conduct a sensitivity analysis we varied each model parameter by 10% while holding others constant to examine their relative influences on the average number of populations occupied after 100 years and the risk of the metapopulation falling below the extinction threshold of 100 adult males.

Results

We detected seaside sparrows on 11 of 75 (15%) surveyed salt marshes in Connecticut in 1999 (Figure 35.2). We found a significant relationship between seaside sparrow occurrence and marsh size (Figure 35.3, model $\chi^2 = 24.932$, d.f. = 1, $p < 0.001$; Hosmer-Lemeshow $\chi^2 = 3.599$, d.f. = 7, $p = 0.825$; overall classification rate = 92%).

Comparisons of Sea-Level Rise Scenarios

Salt marsh patch structure differed among the three sea-level rise scenarios. The number of patches increased and patch size decreased in both reduced salt marsh mosaics over those in the contemporary mosaic (Figure 35.4): 23 patches occurred on the contemporary marsh mosaic (mean size = 125.62 ha, ±58.27 SE), 29 patches on the 30% reduced marsh mosaic (mean size = 67.65 ha, ±27.69 SE), and 37 patches on the 50% reduced marsh mosaic (mean size = 39.95 ha, ±13.63 SE; Figure 35.4). As large patches were fragmented into smaller patches, dispersal connections increased with reduced levels of salt marsh habitat. In the contemporary marsh mosaic, 194 of the possible 506 connections (38%) were possible given the dispersal function. This increased to 338 of the 812 possible dispersal connections (42%) in the 30% reduced marsh mosaic and

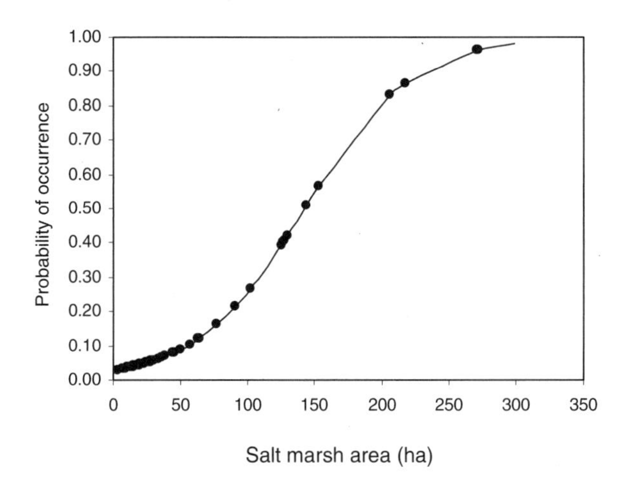

Figure 35.3 Incidence function for seaside sparrow in Connecticut in relation to marsh area based on 75 surveyed marshes, 1999. Points indicate the probability of seaside sparrow occupancy given a specific marsh size based on logistic regression.

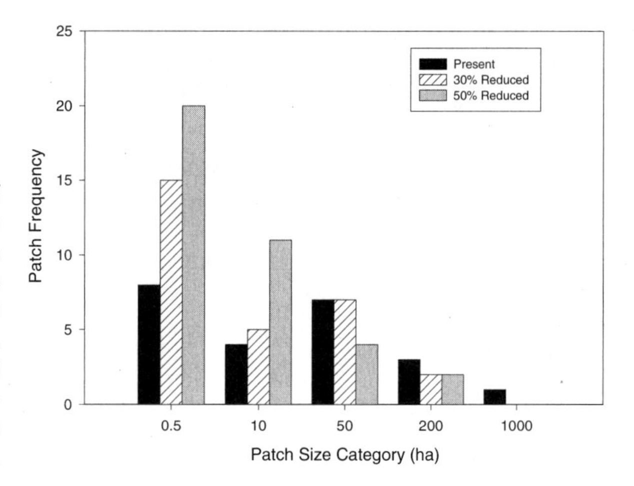

Figure 35.4 Distribution of salt marsh patch sizes in Connecticut given three hypothetical landscapes where salt marsh habitat was unchanged (*solid bars*), reduced by 30% (*hatched bars*), and reduced by 50% (*gray bars*) in response to sea-level rise.

542 of 1,332 possible connections (41%) in the 50% reduced marsh mosaic. Changes in patch structure also lowered the metapopulation carrying capacity as salt marsh habitat patches became smaller in response to sea-level rise. Carrying capacity in the contemporary marsh mosaic was 2,889 sparrows, which declined to 1,748 in the 30% reduction scenario and 1,231 in the 50% reduction scenario.

Patch structure, representing salt marsh loss associated with sea-level rise, had profound effects on the population persistence of seaside sparrows. There was <1% chance of metapopulation quasi extinction in the contemporary marsh mosaic, whereas the 30% and 50% reduced marsh mosaics had 12% and 33% chances of quasi extinction, respectively (Figure 35.5A). The probability of decline as a function of the magnitude of the decline differed substantially between the contemporary marsh mosaic and the two reduced mosaics (Figure 35.5B). There was a <1% chance of 100% seaside sparrow population decline in the contemporary marsh mosaic, a 7% chance in the 30% reduced mosaic, and a 22% chance in the 50% reduced marsh mosaic (Figure 35.5B). The average number of occupied patches after 100 years declined from 20 in the contemporary marsh mosaic to 15 in the 30% reduced mosaic and 14 in the 50% reduced mosaic.

Sensitivity

The contemporary marsh mosaic model was most sensitive to survival and fecundity rates for adults and juveniles (Table 35.2). A 10% increase in the stage matrix values increased the number of populations occupied by 103% and the probability of quasi extinction by 13,650% (Table 35.2). Increasing the standard deviation matrix by 10% reduced the number of populations occupied by 17% and increased the probability of quasi extinction by 360% (Table 35.2). The contemporary marsh mosaic model was also sensitive to Allee effects. Reducing Allee effects by 10% increased the number of occupied populations by 10% and decreased the probability of quasi extinction by 100% (Table 35.2). To further examine the effects of the Allee parameter in the model on the number of populations occupied and the probability of quasi extinction after 100 years, we ran the model with a two alternate Allee thresholds: 1 and 3 (initial Allee value = 5). The lowest Allee value of 1 increased the number of occupied populations by 16% and

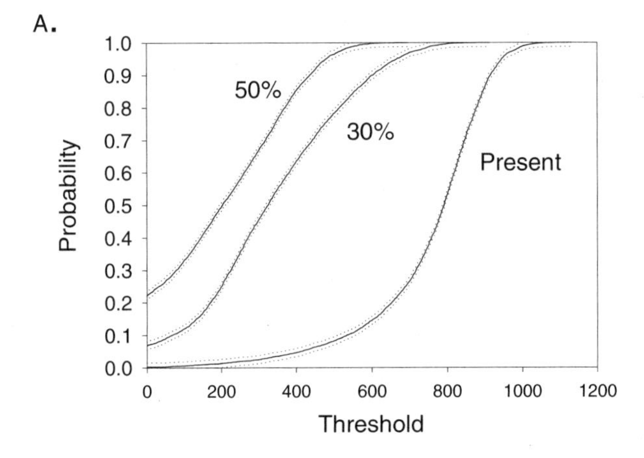

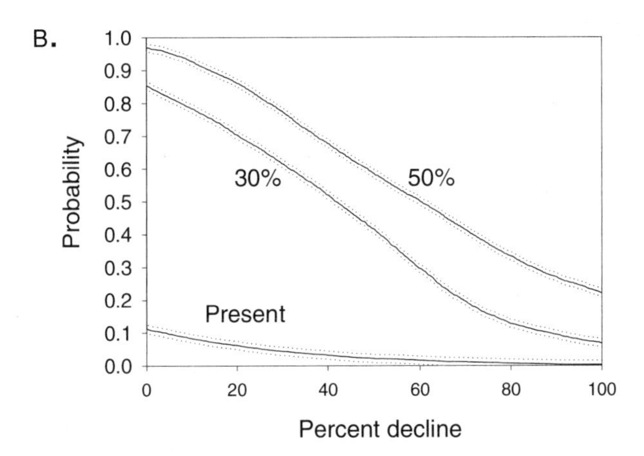

Figure 35.5 Results of seaside sparrow metapopulation analyses on the three hypothetical salt marsh landscapes in Connecticut: (**A**) predicted risk of the total abundance falling lower than a threshold population; (**B**) probability of a population decline of a given magnitude.

reduced the probability of quasi extinction by 100% compared to the initial value of five individuals (Table 35.2). Increasing dispersal rates by 10% reduced the number of occupied populations by 4% and the probability of quasi extinction by 83% (Table 35.2). Increasing the initial abundance by 10% had a positive effect on metapopulation persistence and increased the number of occupied populations by 2% and reduced the probability of quasi extinction by 33% (Table 35.2).

Discussion

Sea-level rise can disrupt coastal wetlands by increasing inundation times, changing vegetation composition, increasing erosion, or converting marshes to open water (Titus 1994). In our modeling approach, we assumed that sea-level rise would primarily affect portions of marshes adjacent to the ocean, tidal creek, or ditch (any "water" cell in the

Table 35.2 Sensitivity of the seaside sparrow metapopulation as indicated by the difference in the number of occupied populations and the probability of decline between models with high and low estimates of each parameter

		No. of Populations after 100 Years		Probability of Decline below 100 Individuals after 100 Years	
Parameter	Effect[a]	Absolute[b]	Percent[c]	Absolute[b]	Percent[c]
Initial abundance	+	0.40	2.02	0.002	33.33
Dispersal rates	–	0.80	4.04	0.005	83.33
Allee effects (±10%)	–	1.90	9.60	0.006	100.00
Allee = 1	–	3.20	16.16	0.006	100.00
Allee = 3	–	2.60	13.13	0.006	100.00
Stage matrix means	+	20.30	102.50	0.819	13,650.00
Standard deviation matrix	+	3.30	16.67	0.022	360.00
Catastrophe	–	0.00	0.00	0.000	0.00
Duration	–	0.40	0.02	0.006	100.00

[a]The effect of increasing the parameter on metapopulation viability. Indicates whether the high parameter increased (+) or decreased (–) the persistence of the metapopulation.
[b]The absolute difference between low and high estimates of the parameter.
[c]Difference as a percentage of the result with the initial value of the parameter.

raster GIS), assuming these areas were at lowest elevation and hence at highest risk with increasing sea level. This is, however, a simplistic model of how sea-level rise may actually alter the available salt marsh habitat. We did not account for potential salt marsh accretion or creation of new salt marshes inland, which may limit the effects of sea-level rise (Stevenson et al. 1986, Roman et al. 1997, van Wijnen and Bakker 2001). Studies conducted during the 1970s and 1980s in New England generally reported that rates of accretion exceeded rates of sea-level rise (Roman et al. 1997). However, most recent IPCC assessments of sea-level rise indicate that model estimates of the twentieth century were low compared to observations and there was a wide range of observational estimates (Church 2001, Intergovernmental Panel on Climate Change 2001). Estimates of sea-level rise over the next century range from 0 to 3.5 m (Titus and Narayanan 1995, Simas et al. 2001). The IPCC Special Report on Emission Scenarios estimated that sea levels would increase from 0.09 to 0.88 m between 1900 and 2100 (Intergovernmental Panel on Climate Change 2001). Thus, the ranges that we invoked in our model (0 m, 0.5 m, and 1.0 m) represent reasonable scenarios of sea-level rise over the next 100 years, with our 1.0–m scenario a "worst case." In light of these caveats, we think that sea-level rise has the potential to reduce salt marsh area and that it is important to understand how these habitat losses may affect the persistence of salt marsh–dependent species. Warren and Niering (1993) observed changes in vegetation in a Connecticut marsh over the past half century that could be a response to wetland submergence (Roman et al. 1997), and, generally, the amount of newly created wetlands will be much smaller than the area of wetlands lost (U.S. Environmental Protection Agency 1999).

Declines in populations of many passerines have been associated with fragmentation of formerly contiguous habitats (Robbins et al. 1989, Askins et al. 1990). Metapopulation dynamics, where extinction-prone populations exist in discrete habitat patches (Hanski 1998), have been suggested as the process driving the patterns of response by sensitive species to habitat fragmentation (Hames et al. 2001). The status of seaside sparrows in Connecticut has been stable, but local declines have been observed in response to destruction of particular marshes (Proctor 1994). Based on this analysis, seaside sparrows were not likely to go extinct during the twenty-first century under the contemporary marsh mosaic. Our model was realistic in the spatial distribution of populated sites as we only populated sites in the models where we actually detected seaside sparrows during the 1999 survey. Because nondetection does not always indicate species absence, our estimate of 11 occupied patches may be lower than the true number of occupied patches. Nevertheless, our survey results identified the same 11 patches occupied by seaside sparrows during the survey for the Connecticut breeding bird atlas (Bevier 1994), which provides corroboration that our projections were based on the true contemporary distribution of these birds.

The increase in sea-level rise we modeled fragmented the contemporary marsh mosaic by reducing patch size, increasing the number of patches, and increasing the distance between patches. This reduction in habitat area and increase in patch isolation had negative effects on the persistence of seaside sparrows. Both hypothetical sea-level rise scenarios in our model reduced the probability of seaside sparrow population persistence over the next 100 years. Because all demographic parameters were equal among the three models, changes in population persistence can be attributed to changes in landscape structure—specifically, the distribution and abundance of salt marsh habitat.

Long-term persistence of this species has been linked to the conservation of extensive salt marsh areas as small marshes do not provide adequate nesting habitat (Proctor 1994, Benoit 1997). Habitat loss has been suggested as the main threat to population persistence (Post and Greenlaw 1994). To date, habitat loss has occurred through ditching, filling, draining, and diking salt marshes, all of which could negatively affect seaside sparrows. Oil spills, storms, and fires also displace populations (Baker 1973). These perturbations usually occur at local scales, and seaside sparrows may have the opportunity to emigrate to other, nonimpacted sites, thus maintaining the metapopulation. The effects of sea level rise on the distribution and abundance of salt marsh habitat would be an omnipresent perturbation adversely impacting seaside sparrow population persistence by fragmenting the salt marsh mosaic and thereby limiting the potential for populations to be rescued (Brown and Kodric-Brown 1977, Harrison 1991) or to redistribute themselves in response to local perturbations. Current conservation efforts for salt marsh habitats and the unique species they support, such as seaside sparrows, may soon become obsolete if scenarios such as those we have projected indeed transpire (Hannah et al. 2002). Conservation land acquisition must begin to account for the potential inland migration of coastal wetlands and work to protect remaining upland habitat or nearshore freshwater wetlands to provide species such as seaside sparrows with places to live in the future.

Acknowledgments J. Hatfield, R. Akçakaya, and two anonymous reviewers provided valuable comments on earlier versions of this manuscript and are warmly acknowledged. We thank H. Brawley and P. Pendergast for assistance in data collection. R. Schauffler provided greatly appreciated GIS support. This project was funded in part by the National Fish and Wildlife Foundation, the Massachusetts Audubon Society, the State University of New York College of Environmental Science and Forestry, the Long Island Sound License Plate Fund, the Sounds Conservancy, and a National Estuarine Research Reserve Graduate Fellowship awarded to W.G.S.

References

Akçakaya, H. R. 2002. RAMAS Metapop 4.0. *Applied biomathematics*. Setauket, N.Y.
Akçakaya, H. R., and Atwood, J. L. 1997. A habitat-based metapopulation model of the California Gnatcatcher. *Conservation Biology* 11: 422–434.
Alexander, C. E., Boutman, M. A., and Field, D. W. 1986. *An inventory of coastal wetlands of the USA*. U.S. Department of Commerce, Washington, D.C.
Askins, R. A., Lynch, J. F., and Greenberg, R. 1990. Population declines in migratory birds in eastern North America. *Current Ornithology* 7: 1–57.
Baker, J. L. 1974. Preliminary studies of the dusky seaside sparrow on the St. Johns National Wildlife Refuge. Southeastern Association of Game and Fish Commissioners. *Proceedings of the Annual Conference* 27: 207–214.
Baker, J. L. 1978. *Status of the dusky seaside sparrow*. Technical Bulletin. Georgia Department of Natural Resources.
Benoit, L. K. 1997. Impact of the spread of *Phragmites* on populations of tidal marsh birds in Connecticut. M.S. thesis, Connecticut College, New London.
Bevier, L. R. 1994. *The atlas of breeding birds of Connecticut*. State Geological and Natural History Survey of Connecticut, Hartford.
Bildstein, K. L., Bancroft, G. T., Dugan, P. J., Gordon, D. H., Erwin, R. M., Nol, E., Payne, L. X., and Senner, S. E. 1991. Approaches to the conservation of coastal wetlands in the Western Hemisphere. *Wilson Bulletin* 103: 218–254.
Brown, J. H., and Kodric-Brown, A. 1977. Turnover rates in insular biogeography: effect of immigration on extinction. *Ecology* 58: 445–449.
Burgman, M. A., Ferson, S., and Akcakaya, H. R. 1993. *Risk assessment in conservation biology*. Chapman and Hall, London.
Caswell, H. 1989. *Matrix population models: construction, analysis, and interpretation*. Sinauer Associates, Sunderland, Mass.
Chabreck, R. H. 1998. *Coastal marshes: ecology and wildlife management*. University of Minnesota Press, Minneapolis.
Church, J. A. 2001. How fast are sea levels rising? *Science* 294: 802–803.
Dawson, D. K., Smith, D. R., and Robbins, C. S. 1995. Point count length and detection of forest neotropical migrant birds. Pages 35–44 in C. J. Ralph, J. R. Sauer, and S. Droege (eds.), *Monitoring bird populations by point counts*. Gen. Tech. Rep. PSW-GTR-149. Pacific Southwest Research Station, Forest Service, U.S. Department of Agriculture, Albany, Ca.
Donnelly, J. P., and Bertness, M. D. 2001. Rapid shoreward encroachment of salt marsh cordgrass in response to accelerated sea-level rise. *Proceedings of the National Academy of Sciences* 98: 14218–14223.
Greenlaw, J. S. 1992. Seaside sparrow, *Ammdramus maritimus*. Pages 211–232 in K. J. Schneider and D. M. Pence (eds.), *Migratory nongame birds of management concern in the northeast*. U.S. Department of the Interior, Fish and Wildlife Service, Newton Corner, Mass.
Hames, R. S., Rosenberg, K. V., Lowe, J. D., and Dhondt, A. A. 2001. Site reoccupation in fragmented landscapes: testing predictions of metapopulation theory. *Journal of Animal Ecology* 70: 182–190.
Hannah, L., Midgley, G. F., Lovejoy, T., Bond, W. J., Bush, M., Lovett, J. C., Scott, D., and Woodward, F. I. 2002. Conservation of biodiversity in a changing climate. *Conservation Biology* 16: 264–268.
Hanski, I. 1998. Metapopulation dynamics. *Nature* 396: 41–49.
Harrison, S. 1991. Local extinction in a metapopulation context: an empirical evaluation. *Biological Journal of the Linnean Society* 42: 73–88.
Hosmer, D. W., and Lemeshow, S. 1989. *Applied logistic regression*. Wiley, New York.
Houghton, J. 1999. *Global warming: the complete briefing*. 2nd ed. Cambridge University Press, New York.
Intergovernmental Panel on Climate Change (IPCC). 2001. *Climate change 2001*. Cambridge University Press, New York.
Kale, H. W., II. 1983. Distribution, habitat, and status of breeding seaside sparrows in Florida. Pages 41–48 in T. L. Quay, J. B. Funderburg Jr., D. S. Lee, E. F. Potter, and C. S. Robbins

(eds.), *The seaside sparrow, its biology and management*. North Carolina Biological Survey, Raleigh.

Marshall, R. M., and Reinart, S. E. 1990. Breeding ecology of seaside sparrows in a Massachusetts salt marsh. *Wilson Bulletin* 102: 501–513.

Post, W., and Greenlaw, J. S. 1982. Comparative costs of promiscuity and monogamy: a test of reproductive effort theory. *Behavioral Ecology and Sociobiology* 10: 101–107.

Post, W., and Greenlaw, J. S. 1994. Seaside sparrow (*Ammodramus maritimus*). No. 127 in A. Poole and F. Gill (eds.), *The birds of North America*. Academy of Natural Sciences, Philadelphia, and the American Ornithologists' Union, Washington, D.C.

Proctor, N. S. 1994. Seaside sparrow, *Ammodramus maritimus*. Pages 370–371 in L. R. Bevier (ed.). *The atlas of breeding birds of Connecticut*. State Geological and Natural History Survey of Connecticut, Hartford.

Ralph, C. J., Sroege, S., and Sauer, J. R. 1995. Managing and monitoring birds using point counts: standards and applications. Pages 161–168 in C. J. Ralph, J. R. Sauer, and S. Droege (eds.), *Monitoring bird populations by point counts*. Gen. Tech. Report PSW-GTR-149. Pacific Southwest Research Station, Forest Service, U.S. Department of Agriculture, Albany, Ca.

Robbins, C. S., Dawson, D. K., and Dowell, B. A. 1989. *Habitat area requirements of breeding forest birds of the middle Atlantic states*. Wildlife Monographs No. 103. Wildlife Society, Bethesda, Md.

Roman, C. T., Pack, J. A., Allen, J. R., King, J. W., and Appleby, P. S. 1997. Accretion of a New England salt marsh in response to inlet migration, storms and sea level rise. *Estuarine, Coastal, and Shelf Science* 45: 717–727.

Simas, T., Nunes, J. P., and Ferreira, J. G. 2001. Effects of global climate change on coastal salt marshes. *Ecological Modeling* 139: 1–15.

Smith, J. N. M., Taitt, M. J., Rogers, C. M., Arcese, P., Keller, L. K., Cassidy, A. L. E. V., and Hochachka, W. M. 1996. A metapopulation approach to the biology of the song sparrow *Melospiza melodia*. *Ibis* 138: 120–128.

Stevenson, J. C., Ward, L. G., and Kearney, M. S. 1986. Vertical accretion in marshes with varying rates of sea level rise. Pages 241–236 in D. A. Wolfe (ed.), *Estuarine variability*. Academic Press, New York.

Titus, J. G. 1994. *Greenhouse effects and coastal wetland policy: how Americans could abandon an area the size of Massachusetts at minimum cost*. U.S. Environmental Protection Agency, Washington, D.C.

Titus, J. G., and Narayanan, V. K. 1995. *The probability of sea-level rise*. Available at http:/// users.erols.com/jtitus/holding/NRJ.html. U.S. Environmental Protection Agency.

Titus, J. G., Park, R. A., Leatherman, S. P., Weggel, J. R., Greene, M. S., Mausel, P. W., Brown, S., Gaunt, C., Trehan, M., and Yohe, G. 1991. Greenhouse effect and sea level rise: the cost of holding back the sea. *Coastal Management* 19: 171–210.

van Wijnen, H. J., and Bakker, J. P. 2001. Long-term surface elevation change in salt marshes: a prediction of marsh response to future sea-level rise. *Estuarine and Coastal Shelf Science* 52: 381–390.

Walters, M. J. 1992. *A shadow and a song*. Chelsea Green, Post Hills, Vt.

Warren, R. S., and W. A. Niering. 1993. Vegetation change on a northeast tidal marsh: interaction of sea level and marsh accretion. *Ecology* 74: 96–103.

Wells, J. V. 1997. Population viability analysis for Maine grasshopper sparrows. In P. D. Vickery and P. W. Dunwiddie (eds.), *Grasslands of Northeastern North America*. Massachusetts Audubon Society, Lincoln.

Woolfenden, G. E. 1956. Comparative breeding behavior of *Ammospiza caudacuta* and *A. maritima*. *University of Kansas Museum of National History* 10: 45–75.

36

Helmeted Honeyeater (*Lichenostomus melanops cassidix*) in Southern Australia

Assessing Options for Establishing a New Wild Population

MICHAEL A. McCARTHY
PETER W. MENKHORST
BRUCE R. QUIN
IAN J. SMALES
MARK A. BURGMAN

Population viability analysis (PVA) was originally developed to estimate the risk of extinction faced by species (Shaffer 1981). However, because of uncertainties associated with these predictions, determining appropriate management strategies is often more useful (Possingham et al. 1993, Akçakaya and Raphael 1998). One conservation strategy for endangered species management is captive breeding and release (Beck et al. 1994), which can be evaluated using PVA (McCarthy 1995, Saltz 1998). In this chapter, we develop a population model to assess options for establishing a new population of helmeted honeyeaters (*Lichenostomus melanops cassidix*) from a captive population.

The helmeted honeyeater is endemic to remnant riparian forests in southern Victoria, Australia. Its population ecology has been studied by Wykes (1985) and more recently by several researchers working under the auspices of the Helmeted Honeyeater Recovery team (Menkhorst et al. 1999). Over the last decade, the latter researchers have conducted an extensive field study (Franklin et al. 1995, 1999; see also McCarthy et al. 1994, Akçakaya et al. 1995, McCarthy 1996). Helmeted honeyeaters are predominately sedentary, with local movements by nonbreeding birds and by a minority of breeding females during the nonbreeding season (Runciman et al. 1995). Pairs occupy exclusive territories that are used for foraging and nesting. Territories are clustered into colonies, and strong social bonds exist within each colony. The main food sources are manna (a sugary exudate from *Eucalyptus* trees) and arthropods.

The helmeted honeyeater is one of the rarest birds in the world. Backhouse (1987), Smales et al. (1990), Menkhorst and Middleton (1991), and Menkhorst et al. (1999) have documented its decline and management. Extensive clearing of its habitat in the nineteenth century led to a dramatic decline in abundance and distribution. By the early 1980s only one small population remained, centered on Yellingbo Nature Conservation Reserve. This population declined to about 60 birds, including just 15 or 16 breeding pairs in 1990. A recovery program was initiated in 1989, with the population growing to more than 100 individuals and 24 breeding pairs by the mid-1990s.

Long-term conservation plans for the helmeted honeyeater have focused on strategies for reintroduction to one or more locations where suitable habitat exists. As part of the recovery program, a captive colony has been established to support the wild population, assist research to aid recovery, and provide individuals to establish populations in new areas (Smales et al. 2000). Because of uncertainties about the fates of individuals and the difficulty of integrating the available information from numerous different sources, the optimal release strategy is not immediately apparent. In the following sections, we develop a population model to achieve this integration and use it to evaluate alternative management strategies.

Data Collection

Banding of the population of helmeted honeyeaters at Yellingbo began in 1984, with the entire known population banded by 1990. All breeding birds and most if not all nonbreeding birds were banded until 1995. During this time, an intensive monitoring program provided an almost complete census of the population throughout the year. In particular, the territories of breeding birds were mapped accurately, and their production of offspring was closely monitored. Since 1995, monitoring has focused on surveying breeding pairs, and banding has been mostly discontinued because of leg injuries caused by the bands and a need to prioritize the use of limited resources. The only birds that have been banded since 1995 have been those associated with the release program. The first unbanded birds entered the breeding population in the 1996/97 season, and 14 of the 41 breeding birds were still banded in 2001/02. The revised surveys still provide an accurate assessment of the number of breeding pairs at Yellingbo and the number of fledglings produced by each pair. They also allow breeding territories to be accurately mapped (Helmeted Honeyeater Recovery Team, unpublished data; Figure 36.1).

Analysis

Annual survival rates of adults and 1-year-old birds, and survival rates of fledglings until they are 1 year old, were analyzed using the beta-binomial model of Kendall (1998). Kendall's method allows the variation in survival rates due to demographic stochasticity to be separated from that of environmental stochasticity. Confidence intervals can be constructed for the estimates by analyzing the profile likelihoods (Kendall 1998). Due to the relatively consistent population size over the last decade, these data analyses were conducted without accounting for density dependence, although density dependence was included in the model (see the section "Model Structure"). Adults and 1-year-old birds appear to have similar survival rates, so these two groups were pooled for the analysis. The estimated annual survival rate of 1-year-old males and adults was 0.82 (95% confidence interval of 0.75–0.87), with the coefficient of variation due to environmental stochasticity being estimated as 0.0 with 95% confidence intervals of 0.0–0.14. The

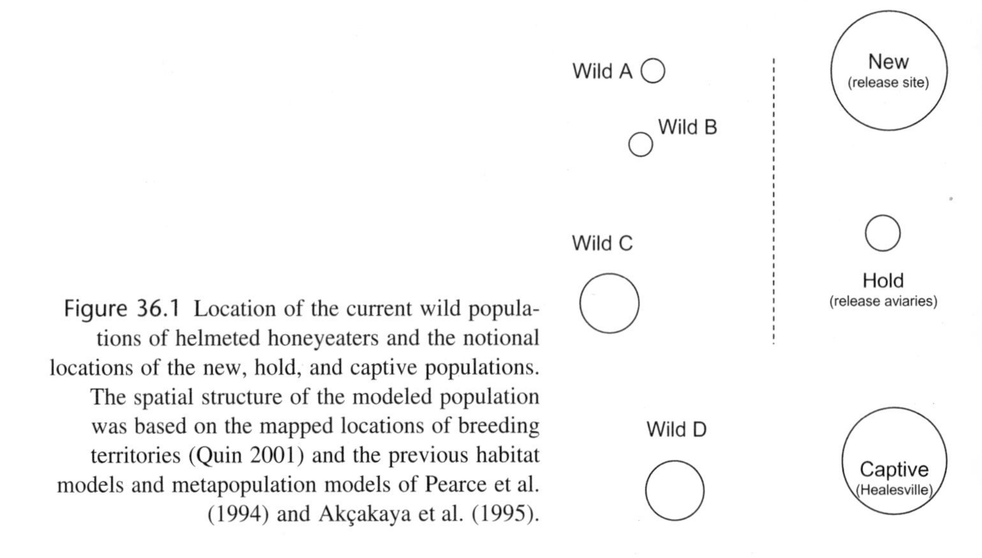

Figure 36.1 Location of the current wild populations of helmeted honeyeaters and the notional locations of the new, hold, and captive populations. The spatial structure of the modeled population was based on the mapped locations of breeding territories (Quin 2001) and the previous habitat models and metapopulation models of Pearce et al. (1994) and Akçakaya et al. (1995).

corresponding value for survival of fledglings to 1 year of age was 0.48 (0.42–0.54), with the coefficient of variation being 0.0 (0.0–0.24).

Similar analyses were conducted to remove the influence of demographic stochasticity from the estimate of environmental stochasticity in fledgling production. In this case, demographic stochasticity was modeled using a Poisson distribution and environmental stochasticity using a gamma distribution, resulting in a negative binomial distribution for fecundity (Ludwig 1996). These analyses led to an estimated mean fecundity rate of 1.63 fledglings per pair, with 95% confidence interval of 1.45–1.82. The coefficient of variation was estimated as 0.0 (0.0–0.14).

Model Structure

The model included two age classes: first years (juveniles—that is, those born in the previous breeding season) and birds in their second year or older (adults), with the census in the model occurring immediately prior to breeding. Only males were modeled because their dispersal behavior better matched that simulated by RAMAS Metapop, and males establish and defend a single breeding territory so their abundance limits population growth. Most males breed in their natal colonies, while most females disperse away from their natal sites but toward available vacancies. The sex structure option in RAMAS Metapop was not used because we were unable to simulate appropriate dispersal rates by females.

It was assumed that the fecundity of the juvenile age class was 0. The fecundity of adults was determined by multiplying the estimated fledgling production (1.63) by their subsequent survival rate to 1 year of age (0.48) and the assumed sex ratio of 0.5. This led to an estimate of 0.39 1-year-old males produced per adult male. The survival rate of juveniles (survival from 1 year of age to 2 years of age) and adults was estimated as 0.82 (see the preceding discussion), leading to the stage transition matrix

$$\begin{bmatrix} 0.0 & 0.39 \\ 0.82 & 0.82 \end{bmatrix}$$

This matrix has a finite rate of increase of approximately 1.11, which was used as the estimate of the population growth rate when all adults were breeding. The vital rates in this matrix were reduced under the influence of density dependence such that the finite rate of increase was equal to 1 when the population was at carrying capacity.

Given the strong territorial behavior of helmeted honeyeaters, density dependence in the wild population was modeled using contest competition (the Beverton-Holt function). Density dependence was assumed to affect fecundity rates, due to limitations in the number of breeding territories. Survival rates in the model were also density-dependent (see the following discussion).

In the years when the population was intensively monitored, the ratio of breeding birds to nonbreeding birds was approximately 2:1 (Akçakaya et al. 1995, McCarthy 1996). Thus, the carrying capacity of each population in the model was set as 1.5 times the number of breeding birds (Akçakaya et al. 1995). Preliminary analyses with RAMAS Metapop indicated that the average age structure (under the influence of density dependence) was approximately 4 juveniles to 11 adults. Assuming the ratio of breeding to nonbreeding birds is 10:5, the average reproductive rate was approximately 0.4 one-year-old males per breeding male, which is close to the observed rate of 0.39. There is uncertainty about whether the survival rate of adult birds is density-dependent. However, a model without density dependence in survival rates was not able to generate the observed ratio of breeding to nonbreeding birds with the correct age structure (as discussed). These analyses suggest that the Beverton-Holt model provided a reasonable representation of density dependence in the populations. To account for declining habitat at site C and the apparent increases in habitat at site E, linear changes in carrying capacity of 0.2 birds per year were assumed for these two populations.

Environmental stochasticity was incorporated by specifying a coefficient of variation in the vital rates. Rather than using the maximum likelihood estimates of the coefficient of variation (0.0 for all vital rates), we chose the midpoints of the 95% confidence intervals for the base case scenario. We did this because we believe a priori that there is likely to be at least some annual variation in the vital rates, given the environmental variability of southeastern Australia. Sensitivity analyses were conducted to examine the influence of parameter uncertainty on the results by varying the parameters across the range of the 95% confidence intervals. We used 0.12 as the coefficient of variation for fecundity, which was the midpoint of the estimate for fledgling production. We assumed that environmental variation was perfectly correlated within populations (among vital rates) and, due to the proximity of the populations, perfectly correlated among wild populations at Yellingbo. Because the candidate release site for birds is tens of kilometers away in a different catchment, we assumed that this new population would be uncorrelated with the original population and that neither of these populations was correlated with the captive population. To assess the importance of our assumptions about the correlations, we also analyzed a model in which all correlations equaled 0.0, and a second alternative in which all correlations were 1.0.

The probability of a 1-year-old male moving from its natal colony (i) to another colony (j) was modeled using the formula $p_{i,j} = 0.5 * e^{-0.6 * d_{i,j}}$, where $d_{i,j}$ is the distance (km) between the colonies (McCarthy 1996). Because movement from one colony to another by breeding males is rare, the dispersal rate of adults was one-tenth that of juveniles. It was assumed that there was no dispersal between the proposed release site and the current population.

The captive colony was modeled as an additional population. It was assumed to have ceiling density dependence of 24 males because the breeding capacity is limited by the 14 breeding aviaries and other aviaries that can house up to 20 nonbreeding birds of both sexes. A slightly modified matrix of vital rates was used to reflect the apparently small differences between the captive and wild populations:

$$\begin{bmatrix} 0.0 & 0.4 \\ 0.8 & 0.8 \end{bmatrix}$$

To reflect the more stable environment in captivity, the coefficient of variation in vital rates was assumed to be half that of the wild.

In addition to the captive population at Healesville, there are seven breeding aviaries that can be constructed at the proposed release site. The aim is to have breeding pairs in these aviaries and to release the offspring that are produced. These extra aviaries were modeled as a separate population (the "hold"). The release of birds to the wild was modeled by translocating all offspring at hold to the release population (the "new") before dispersal, and all surviving adults at hold were translocated to the Healesville population (the "captive") at the same time. Then, depending on the total abundance of adults in the captive population, a specified number of adults were moved back to the hold population using the postdispersal translocation of Metapop. Although in reality birds will not necessarily be moved back and forth between the release aviaries and Healesville, the strategy that was modeled meant that the number of birds in the hold aviaries (and the rate of release) could be based on the total abundance in captivity.

Simulations

Simulations were conducted to determine how the probability of success of the reintroduction over 20 years was influenced by the rate of release from the captive population of helmeted honeyeaters. Ten thousand iterations were used for each scenario. The strategies were implemented in Metapop by varying the proportion of adult males moved from the captive population to the release aviaries (hold). This was achieved by varying the threshold captive population sizes at which these rates occurred. The release strategy was implemented using the "Linearly Change" option of the Conditionals screen in the Population Management part of Metapop. The maximum number of adults moved from captive to hold was held constant at seven, which is the number of release aviaries that are available. The actual rate of release depended on the upper threshold. Values used for the upper threshold were such that the maximum proportion of the adult captive population being moved to the hold population was 10%, 20%, 30.4%, 38.9%, or 50%. If the lower population threshold was 0, the number of translocated adults was proportional to the number of adults in captivity. When the lower threshold was greater than 0, a threshold was established below which all birds were held in captivity at Healesville to help reduce the probability of decline in the captive population. This threshold was set at zero, two, or four adult males.

The sensitivity of the results to variation in the parameter estimates for the wild population was analyzed by conducting simulations in which parameter values were changed to extreme values (based on the 95% confidence intervals of the parameter estimates). Scenarios examined were high and low values for survival rates, fecundity rates, and environmental stochasticity. Because the helmeted honeyeater is colonial and breeding activity may decline at low population sizes, a model with an Allee effect (Allee parameter equal to 1) was also analyzed (Allee 1938). Uncertainty about the parameters of the captive colony was assessed by reducing the survival rate of adults from 0.8 to 0.75. It was assumed that changes in other parameters would have a qualitatively similar effect on the results.

The best overall release strategy was chosen (subjectively) as the one that provided a good chance of persistence of the new population without contributing substantially to the risks faced by the captive population. Finally, to assess the overall benefit of the release of captive birds to the total wild population size, the entire population was simulated and the results were compared to the case without the release of captive birds.

A deterministic analysis of the population growth rate of the captive colony was also used to evaluate release strategies. Assuming the population is at a stable age distribution and experiences deterministic population dynamics without density dependence, the population growth rate would be found as the solution to the equation

$$\lambda^2 - s\lambda - s(1 - r)f = 0$$

where s is the survival rate, f is the fecundity rate, r is the proportion of 1-year-olds released each year, and λ is the population growth rate of the captive colony (McCarthy et al. 1994). The maximum sustainable yield from the captive colony can be found by determining the value of r when λ is equal to 1. Some simple algebra leads to

$$r = 1 - (1 - s)/fs \qquad (1)$$

This equation was analyzed to help determine the sensitivity of the release strategy to changes in the population growth rate of the captive population.

Results

For the new population after 20 years, the mean population size was predicted to increase and the risk of being at or below 10 individuals was predicted to decline as the maximum proportion of males that were moved from the captive population to the hold population increased (Figure 36.2). Increases in the population size and decreases in the risks tended to be small once the proportion was greater than 0.3. Additionally, establishing a threshold captive population size below which releases did not occur reduced the expected population size and increased the risks of decline, although the differences tended to be unimportant when the maximum proportion released was 0.3 or more. The results for the other scenarios are not given here but were qualitatively similar for all scenarios in which the parameters of the wild population were modified.

Risks of falling to or below 10 individuals in captivity increased steadily as the maximum proportion of released birds increased (Figure 36.3). Similarly, the expected minimum population size of the captive colony declined steadily over the same parameter range (Figure 36.3). Establishing a threshold below which releases did not occur led to

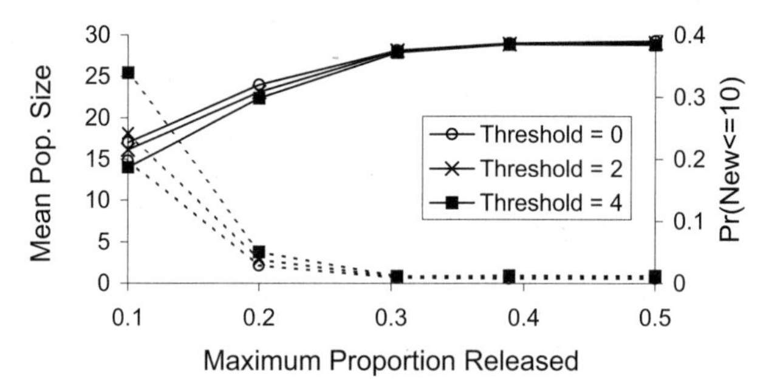

Figure 36.2 Mean population size of the new population (*solid line*) and the probability of the new population being at or below 10 individuals (*dotted line*) after 20 years for different release rates and different captive population thresholds, assuming the standard set of parameters. The threshold is the population size of the captive colony below which releases did not occur and which was established in an attempt to reduce risks faced by the captive population. The fate of the new population is only marginally improved once more than 30% of the juveniles are released from captivity.

small but measurable benefits to the captive population. Given the relatively uniform changes in the risks faced by the captive colony with changes in the release strategy, and the moderately small changes in the fate of the new population once release rates increased above 0.3, this value (0.3) was chosen as the optimal release rate. Additionally, given the small but measurable benefit to the captive colony and indiscernible impact on the new population, a threshold of four adult males was chosen as optimal.

Reducing the annual survival rate of birds in captivity from 0.8 to 0.75 resulted in some qualitative changes to the results. In this case, establishing a threshold below which birds should not be released had a greater effect on the final abundance of the new population (Figure 36.4), presumably because the threshold was reached more frequently.

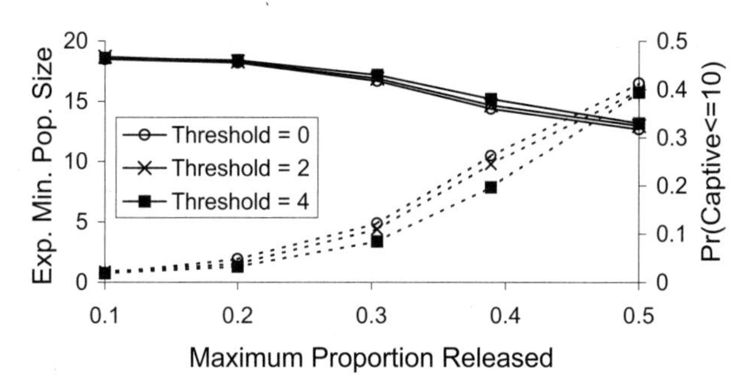

Figure 36.3 Expected minimum population size (*solid lines*) and probability of the captive population declining to or below 10 individuals at some time within the next 20 years (*dotted lines*) for different release rates and different captive population thresholds, assuming the standard set of parameters. The threshold is the population size of the captive colony below which releases did not occur and which was established in an attempt to reduce risks faced by the captive population. The risk of decline faced by the captive population increases relatively uniformly with the proportion of juveniles that are released.

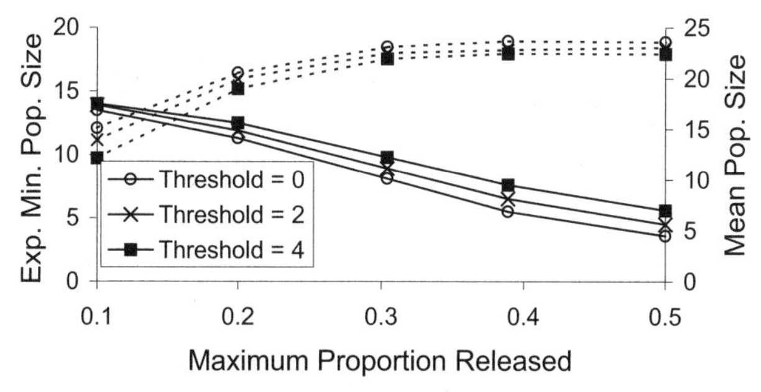

Figure 36.4 Mean population size of the new population after 20 years (*dotted lines*) and the expected minimum population size of the captive population within the next 20 years (*solid lines*) for different release rates and different captive population thresholds, assuming annual survival in captivity is 0.75. The threshold is the population size of the captive colony below which releases did not occur and which was established in an attempt to reduce risks faced by the captive population.

When the survival rate in captivity was 0.75, there was an appreciable risk of decline in the captive colony, with risks increasing steadily as the rate of release increased from 0.1 to 0.5 (Figure 36.4). When there are two aims (maintaining a healthy captive colony and maximizing the chance of success of the reintroduction), it is not straightforward to decide on the optimal release strategy.

Without the release of birds from captivity to the new population, the probability of the total wild population falling below 15 males (half the initial population size) within the next 20 years is 0.266, and the expected minimum population size is 19.4 (using the standard set of parameters). With the release of birds to the new population under the optimal release strategy, the risk of falling to 15 or fewer males in the wild is 0.012, and the expected minimum population size is 27.2, which is only slightly below the initial wild population size of 30 males. In the case where vital rates are perfectly correlated (within and among populations), the expected minimum population size under the optimal strategy is reduced slightly to 26.7. In the opposite case, where the correlations are all 0, the expected minimum population size is increased slightly to 27.9.

The deterministic analysis of the captive colony (eq. (1), $s = 0.8, f = 0.4$), indicated that the maximum release rate is $r = 0.375$. This is slightly greater than the value obtained from the stochastic analyses (0.3), but it is largely comparable. The value for the stochastic analyses is lower because the small size of the captive colony exposes it to risks of decline and because density dependence is not accounted for in the deterministic analysis.

Whereas the analyses indicated that the optimal release strategy was relatively insensitive to the parameter values assumed for the release site, the optimal release strategy is likely to be sensitive to the parameter estimates assumed for the captive colony. For example, if s equals 0.75 rather than 0.8 in captivity, r in the deterministic analysis (eq. (1)) more than halves from 0.375 to 0.166. This helps explain the simulation results that demonstrated a steady increase in risks to the captive colony as the release rate increased from 0.1 to 0.5 when the survival rate was assumed to be 0.75. Additionally, if the survival rate in captivity is 0.75 rather than 0.8, it is unlikely that a release rate greater than 0.1 will be sustainable in the long term. Reductions in the population growth rate due to reduced fecundity would have a similar influence on the ability of the captive colony to produce birds for release.

Discussion

The simulations suggest that the chance of success of the proposed reintroduction is moderately good, with little chance that the new population will have fewer than 10 males after 20 years. This gives reason for optimism, but this hope is contingent on the released birds remaining where they are released, establishing the same population behavior, and having the same vital rates as the current wild population. Release attempts so far have resulted in released birds breeding in the wild and raising offspring that have survived to adulthood (Smales et al. 2000). Furthermore, it is hoped that increased experience with different release protocols will be beneficial.

The results of the simulations indicated that the optimal release strategy was largely insensitive to changes in parameter values for the population at the release site, even though the actual fate of this population was sensitive. This suggests that despite uncertainty about the true fate of the population, it is still possible to be relatively confident about the optimal number of birds that should be released each year. This illustrates the contention that despite uncertainty about the predictions of PVA, these models can be used to help determine effective and efficient management strategies (Possingham et al. 1993, Akçakaya and Raphael 1998).

The ability of the captive colony to contribute birds to the wild is most sensitive to changes in the annual adult survival. While this survival is approximately 0.8 in captivity, in the past there have been two events with high mortality occurring over a few weeks. The analyses conducted here have assumed that such events will not happen in the future. If catastrophic losses of birds occurred approximately 1 year in 10, the average adult survival rate would be equal to 0.72, making the captive colony barely viable ($\lambda \approx 1$ with $r \approx 0$). If catastrophes were more frequent, the captive colony would act as a population sink on the wild. Therefore, these analyses and the apparent benefit to the wild population are contingent on there being no catastrophes in the captive population.

The simulations conducted here have not included the translocation of birds from the wild to captivity. Such a strategy might be beneficial in some circumstances, especially when the captive population is small and the wild population is large, although such removals may affect the existing population (McCarthy 1995). These risks may be small if birds are taken into captivity as eggs because they survive better in captivity compared to the wild, and helmeted honeyeaters readily re-nest following the loss of nests. Nevertheless, there is a measurable impact of egg removal and the possible benefits (including greater overall viability and improved genetic structure in the captive population) need to be weighed against the risks (McCarthy 1995).

The analyses conducted in this study are based on a similar framework to that of Saltz (1998), in which a population model is used as a tool for rigorous and systematic assessment of the likely benefits of different management strategies. These analyses help managers determine the parameters that are most critical for success and the strategy that is likely to be most effective at achieving the goal. The model can be used to explore the trade-off between the establishment of a wild population and the maintenance of the captive population as insurance against declines in the wild. The results of the analyses conducted here suggest that the optimal strategy is most sensitive to the demographic performance of the captive population.

Acknowledgments We are grateful for the comments of two anonymous referees and regret that some of them could not be fully addressed within the word limits.

References

Akçakaya, H. R., and Raphael, M. G. 1998. Assessing human impact despite uncertainty: viability of the northern spotted owl metapopulation in the northwestern USA. *Biodiversity and Conservation* 7: 875–894.

Akçakaya, H. R., McCarthy, M. A., and Pearce, J. L. 1995. Linking landscape data with population viability analysis: management options for the helmeted honeyeater. *Biological Conservation* 73: 169–176.

Allee, W. C. 1938. *The social life of animals*. Heinemann, London.

Backhouse, G. N. 1987. Management of remnant habitat for conservation of the helmeted honeyeater *Lichenostomus melanops cassidix*. Pages 287–294 in D. A. Saunders, G. W. Arnold, A. A. Burbidge, and A. J. M. Hopkins (eds.), *Nature conservation: the role of remnants of native vegetation*. Surrey Beatty, Sydney.

Beck, B. B., Rapaport, L. G., Stanley Price, M. R., and Wilson, A. C. 1994. Reintroduction of captive-born animals. Pages 265–286 in P. J. S. Olney, G. M. Mace, and A. T. C. Feistner (eds.), *Creative conservation: interactive management of wild and captive animals*. Chapman and Hall, London.

Franklin, D. C., Smales, I. J., Miller, M. A., and Menkhorst, P. W. 1995. The reproductive-biology of the helmeted honeyeater, *Lichenostomus melanops cassidix*. *Wildlife Research* 22: 173–191.

Franklin D. C., Smales, I. J., Quin, B. R., and Menkhorst, P. W. 1999. Annual cycle of the helmeted honeyeater *Lichenostomus melanops cassidix*, a sedentary inhabitant of a predictable environment. *Ibis* 141: 256–268.

Kendall, B. E. 1998. Estimating the magnitude of environmental stochasticity in survivorship data. *Ecological Applications* 8: 184–193.

Ludwig, D. 1996. The distribution of population survival times. *American Naturalist* 147: 506–526.

McCarthy, M. A. 1995. Population viability analysis of the helmeted honeyeater: risk assessment of captive management and reintroduction. Pages 21–25 in M. Serena (ed.), *Reintroduction biology of Australian and New Zealand fauna*. Surrey Beatty, Chipping Norton, New South Wales, Australia.

McCarthy, M. A. 1996. Extinction dynamics of the helmeted honeyeater: effects of demography, stochasticity, inbreeding and spatial structure. *Ecological Modelling* 85: 151–163.

McCarthy, M. A., Franklin, D. C., and Burgman, M. A. 1994. The importance of demographic uncertainty: an example from the helmeted honeyeater. *Biological Conservation* 67: 135–142.

Menkhorst, P., and Middleton, D. 1991. *Helmeted honeyeater recovery plan 1989–1993*. Department of Conservation and Environment, Melbourne, Victoria.

Menkhorst, P., Smales, I., and Quin, B. 1999. *Helmeted honeyeater recovery plan, 1999–2003*. Department of Natural Resources and Environment, Melbourne, Victoria.

Pearce, J. L., Burgman, M. A., and Franklin, D. C. 1994. Habitat selection by helmeted honeyeaters. *Wildlife Research* 21: 53–63.

Possingham, H. P., Lindenmayer, D. B., and Norton, T. W. 1993. A framework for improved threatened species management using population viability analysis. *Pacific Conservation Biology* 1: 39–45.

Quin, B. 2001. Wild helmeted honeyeaters: the 2000/2001 breeding season. Unpublished report to the Helmeted Honeyeater Recovery Team, Melbourne.

Runciman D., Franklin, D. C., and Menkhorst, P. W. 1995. Movements of helmeted-honeyeaters during the non-breeding season. *Emu* 95: 111–118.

Saltz, D. 1998. A long-term systematic approach to planning reintroductions: the Persian fallow deer and the Arabian oryx in Israel. *Animal Conservation* 1: 245–252.

Shaffer, M. L. 1981. Minimum population sizes for species conservation. *Bioscience* 31: 131–134.

Smales, I. J., Craig, S. A., Williams, G. A., and Dunn, R. W. 1990. The helmeted honeyeater: decline, conservation and recent initiatives for recovery. Pages 225–238 in T. W. Clark and J. H. Seebeck (eds.), *Management and conservation of small populations*. Chicago Zoological Society, Chicago.

Smales, I., Quin, B., Krake, D., Dobrozczyk, D., and Menkhorst, P. 2000. Re-introduction of helmeted honeyeaters, Australia. *Re-introduction News* (Newsletter of the Re-introduction Specialist Group of the IUCN's Species Survival Commission) 19: 34–36.

Wykes, B. J. 1985. The helmeted honeyeater and related honeyeaters of Victorian woodlands. Pages 205–217 in A. Keast, H. F. Recher, H. Ford, and D. Saunders (eds.), *Birds of eucalypt forests and woodlands: ecology, conservation and management*. Royal Australasian Ornithologists Union and Surrey Beatty, Sydney.

37

Wandering Albatross (*Diomedea exulans chionoptera*) in the Southern Oceans

Effects of Dispersal and Density Dependence on the Persistence of an Island Metapopulation

PABLO INCHAUSTI
HENRI WEIMERSKIRCH

Spatially explicit metapopulation models, in which local population dynamics result from the interplay between local and regional environmental variability and dispersal from nearby populations, have become an important research direction in conservation biology. By allowing populations that experienced declines to recover through dispersal from other nearby populations, metapopulation models explicitly assess the impact of the degradation, destruction, and fragmentation of natural habitats, and this is probably the single most important cause of the current biodiversity crisis (Heywood and Watson 1995). While studies of dispersal have focused on terrestrial animals, very little information exists for marine animals for obvious logistical reasons. Pelagic seabirds breed on islands scattered over open waters—that is, on highly fragmented breeding habitats—and, while they are thought to have a high philopatry (Warham 1990), there is scant evidence of actual dispersal rates among colonies or islands (Ainley et al. 1990, Pyle 2001, Russell 1999), especially for those breeding on remote scattered oceanic islands like pelagic petrels and albatrosses.

The wandering albatross (*Diomedea exulans chionoptera*) is a pelagic seabird that forages on patchily and unpredictably distributed resources over wide areas of the Southern Oceans, while breeding in highly fragmented habitat patches (oceanic islands) scattered from South America to Australia (Figure 37.1). Due to recent declines in many of its extant populations, this species is currently listed as Vulnerable by the International Union for the Conservation of Nature (IUCN) (2000). The life history of the wandering albatross is typical of the order Procellariiformes: it is a large-bodied (adults weigh between 7 and 12 kg), long-lived species (estimated maximum life expectancy up to 80 years, average 32 years) that has low fecundity (a single egg is laid at most every 2 years, and the chick has a long fledging period) and high adult annual survival (Tickell 1968, Weimerskirch and Jouventin 1997).

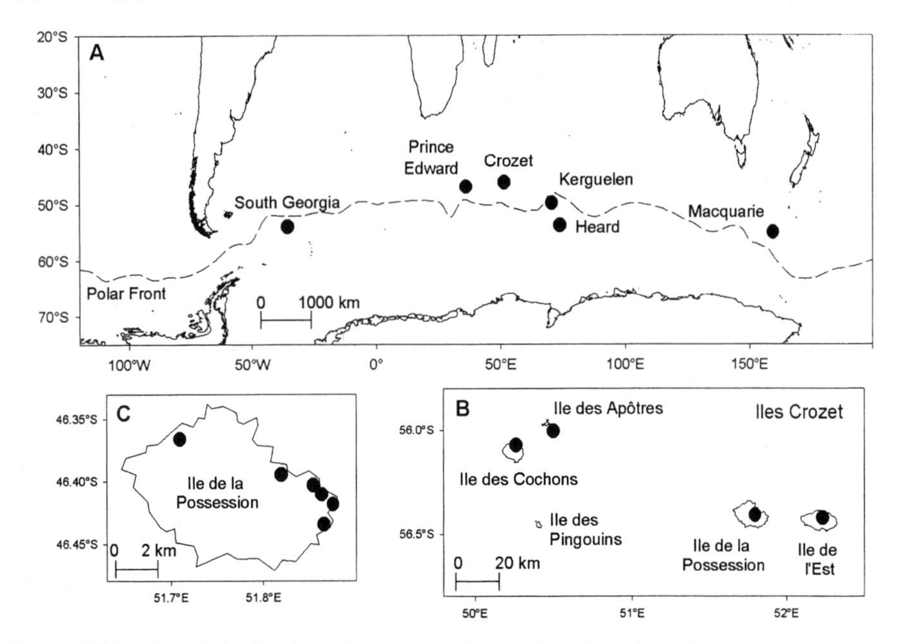

Figure 37.1 Map of the Southern Oceans showing the breeding sites of wandering albatrosses at different scale: **(A)** in the Southern Ocean; **(B)** in the Crozet Islands; **(C)** on Ile de la Possession. Banding of birds was carried out between 1966 and 1996.

In this chapter we show that the wandering albatross population over the Southern Oceans functions as a metapopulation: that is, dispersal events can influence the dynamics and degree of persistence of other island populations (Inchausti and Weimerskirch 2002). Then we assess whether the occurrence and type of density dependence at the island level alters the metapopulation functioning of the wandering albatross. In particular, we investigate the effect of the increase of the age at first reproduction with population abundance observed for the wandering albatross.

Methods

Long-term records of population parameters including population size, survival, and breeding performances are available for the Possession Island, Crozet, Indian Ocean (Weimerskirch et al. 1997); Macquarie, Pacific Ocean (de la Mare and Kerry 1994); and South Georgia, Atlantic Ocean (Croxall et al. 1990). The age-structured, stochastic metapopulation model consists of three postbreeding, age-structured matrices for each of these three breeding areas. There are six groups of islands in which wandering albatrosses breed (from the west to the east, see Figure 37.1)—South Georgia, Prince Edward (including Marion and Prince Edward islands), Crozet (including Cochons, Apôtres, Posession, and Est), Kerguelen, Heard, and Macquarie Islands—making 10 populations. Since demographic data are not available for the Prince Edward group, Kerguelen, and Heard we assumed the same demographic rates as the Crozet group. This is reasonable since population sizes of these island groups have changed in parallel to those recorded at Crozet (Weimerskirch et al. 1997, Nel et al. 2002). Due to the restricted exchange of breeding adults between colonies (see the discussion below), all colonies within an island were considered part of a single breeding population, and thus the model describes the population dynamics at the level of the island groups: Crozet, South Georgia, and Macquarie. Because wandering albatrosses aged 1 to 5 years remain at sea, the precise initial age structure of each population cannot be known. In the absence of detailed censuses for all age classes in each island, we assumed the initial age structure of all populations to be in proportion to their expected stable age structure. The life cycle of the wandering albatross was subdivided into 14 age classes: chicks, juveniles (2- to 5-year-olds), immatures (6- to 8-year-olds), and adults (9 years and older). As with other southern oceanic birds, juvenile wandering albatross disperse after fledging and remain scattered over extensive areas in the oceans (e.g., Weimerskirch et al. 1985). We accounted for the observed longevity of Procellariiformes by setting a non-zero probability of remaining at the last age class.

Annual survival rates based on capture-mark-recapture protocols for Crozet, Macquarie, and South Georgia were obtained from Weimerskirch et al. (1997), de la Mare and Kerry (1994), and Croxall et al. (1990). Fecundity rates were calculated as the product of breeding frequency (proportion of breeding adults), breeding success (proportion of eggs resulting in chicks), clutch size (equal to one), and sex ratio (a 1:1 sex ratio was assumed). Age-specific fecundity rates were obtained as the proportion of breeding adults in each age class (data obtained from Weimerskirch et al. 1997), and the estimate of fecundity was obtained as explained here (details in Inchausti and Weimerskirch 2002). Dispersal rates were based on recapture of 6,037 chicks banded before fledging at Ile de la Possession between 1966 and 1996, as well as on the monitoring programs on other islands (Weimerskirch et al. 1985, 1997). Dispersal rates of juveniles and adults were considered separately. All observed dispersal events of adults occurred between colonies of the same island (Inchausti and Weimerskirch 2002), but adult dispersal was not considered in the model because it took part within the same breeding population. In the case of juveniles, interisland dispersal rates were estimated as proportion of juveniles migrating = $a*\exp(b*\text{distance}^c)$ (see Akçakaya 2002 for details), from which we calculated the juvenile interisland dispersal rate. Juvenile dispersal rate was related to the size of the source population (Inchausti and Weimerskirch 2002; slope = 0.0016, SE = 0.00029, $p < 0.001$, $n = 19$, $R^2 = 0.699$), a feature that was incorporated into the metapopulation model.

Environmental stochasticity was modeled by drawing the values of the age-specific survival and fecundity rates from a set of lognormal distributions whose parameters (means and standard deviations) reflect the average value and the temporal variability of each age-specific demographic rate in each population. Due to the long rearing period of the young in this species (Croxall et al. 1997, Weimerskirch et al. 1997), we assumed that the values of fecundity and adult survival for a given year were positively correlated. The model also included demographic stochasticity, which was modeled using the method of Akçakaya (1991). Environmental correlations (i.e., the synchronicity of the temporal variation of demographic rates between island groups) were estimated using the Spearman correlation of adult survival rates, the demographic rate that has the largest influence on the population growth rate (Weimerskirch et al. 1997). We assumed that the demographic rates of island populations within an island group were perfectly synchronous (i.e., environmental correlation equal to 1) due to their geographic proximity compared to the distance separating island groups.

We considered a density-dependent variant of the basic model that assumed the observed positive relationship between the average age at first reproduction and the number of breeding pairs at Ile de la Possession (Figure 37.2). The age at first reproduction is a key parameter of a species life history that determines the "speed" of its life cycle and strongly influences the population growth rate of long-lived species (e.g., Stearns 1992, Roff 1993). Thus, for the purpose of this chapter, we take the observed positive association between age at first breeding and population abundance (Figure 37.2) as a given and use a modeling approach to explore its consequences on the estimated extinction risk. Modeling the relation between the age at first reproduction and the size of the breeding population required the consideration of an ensemble of age-transition matrices whose number of breeding age classes depended on the local population size of each island. This density dependence function was implemented as a user-specified file for density dependence (Albatros.DLL; to see the source code, open the file Albatros.DPR on the CD that accompanies this volume; see Figure 37.2). Carrying capacities were set as 20% above the current total abundance, based on the observation at Ile de la Possession where present population size is about 20% lower than its stable level during the 1960s (Weimerskirch et al. 1997). For the two smallest populations (Heard and Macquarie), we assumed that carrying capacity was equal to the current abundance of the largest population of the Indian Ocean (Marion). Contest type of density dependence, which we also considered, requires specifying the maximum annual growth rate, which was set to the largest value ever observed (1.04; Weimerskirch et al. 1997). All models simulated the dynamics of the ensemble of the 10 populations of wandering albatross over a time horizon of 200 years using 2,000 replications.

Results

Although the large majority of juvenile individuals returned to either their hatching colonies or nearby colonies, the remaining fraction (less than 1%) dispersed to breed in colonies of other island groups up to 1450 km away. The probability of dispersing to the farthest colonies at 6050 km was estimated to be 0.0001. Once birds start breeding in a colony, 97.1 % of adults return to the same colony, while the remaining individuals stayed in colonies on the same island: no wandering albatross has ever been found breeding outside the island where it has bred before. While the dispersal rate of juveniles from the hatching colony changed over time and was positively and significantly related to the size of population at Ile de la Possession, that of adults was unrelated to the size of the source population (slope = –0.006, SE = 0.006, p = 0.350; n = 21).

The dispersal of juvenile wandering albatross altered the persistence of populations of other islands. Dispersal had a variety of effects on the risk of decline of an island's population, depending on the distance to neighboring islands and on their relative population size (Figure 37.3). For populations that are very distant from other islands, dispersal has predictably little effect on their risk of decline (Figure 37.3A, F). However, dispersal decreased the probability of decline in abundance for small- or medium-sized populations with close larger populations that function as a source of migrants (Figure 37.3D, E). Finally, dispersal increased the risk of decline of large populations having as their closest neighbors either small populations (Figure 37.3C) or isolated medium-sized

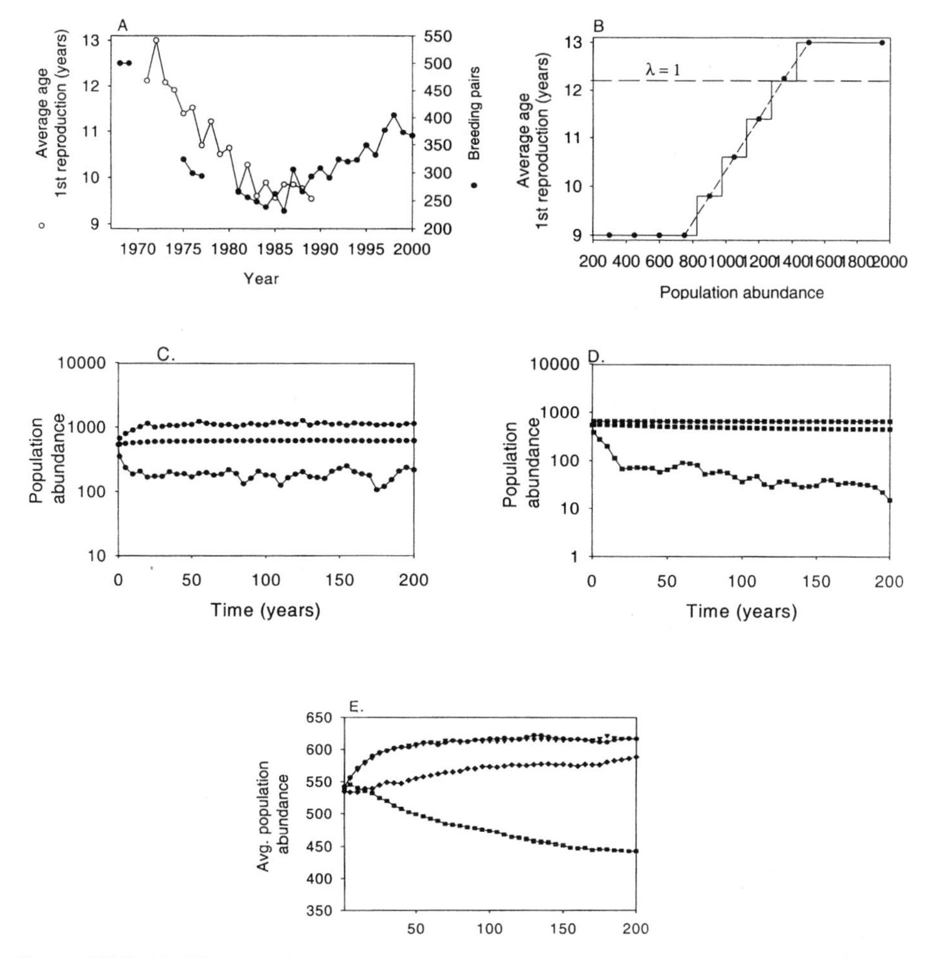

Figure 37.2 **(A)** Changes of the average age at first reproduction (*empty circles, left*) and number of breeding pairs (*filled circles, right*) at Ile de la Possession over time. **(B)** Density-dependent function depicting the relationship between the average age at first reproduction and population density (the horizontal line corresponds to age at first reproduction giving the equilibrium growth rate $\lambda = 1$). **(C)** The predicted maximum, average, and minimum abundance at Apotres over time. **(D)** and **(E)** The maximum, average, and minimum abundance at Apotres over time predicted by contest or undercompensation and by ceiling density dependence.

populations (Figure 37.3B). At the level of the entire metapopulation, the compounding of these different effects at the island level suggests that dispersal had a small effect of the risk of overall decline of the metapopulation (Figure 37.3G).

Much like dispersal, the influence of density dependence on the risk of decline depended on the spatial scale (local population or metapopulation) at which its effects were assessed (Figure 37.4). At the metapopulation level, density dependence reduced the risk of decline more than did density independence, and dispersal only slightly lowered

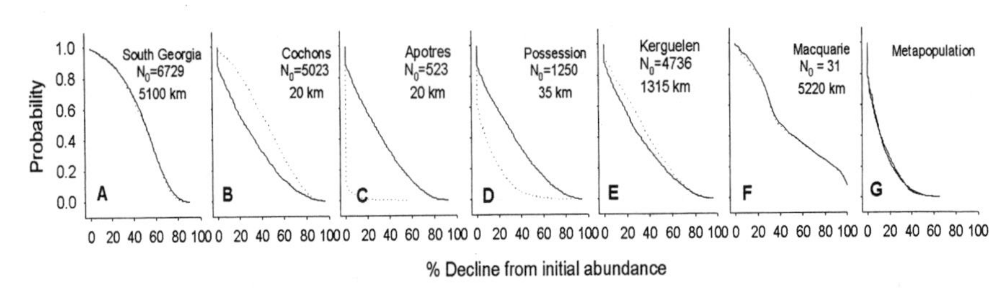

Figure 37.3 Probability of observing a decline from initial abundance during the next 200 years for **(A–F)** populations of six islands from west to east of the current geographic range of the wandering albatross, and **(G)** for the entire metapopulation, with (*dotted line*) and without (*continuous line*) juvenile migration without considering density dependence. The distance to the closest island and the initial population size of each island are indicated above each panel (see Inchausti and Weimerskirch 2002).

the risk of decline for density-dependent models (Figure 37.4D). At the level of the local population, density dependence typically increased the risk of decline regardless of the occurrence of dispersal (Figure 37.4A,B, and C). Dispersal led to an increase of the risk of decline of large populations that functioned as source of migrants (Figure 37.4C) while lowering the risk of decline of smaller populations (Figure 37.4A) that functioned as sinks.

Discussion

Our results suggest that the populations of wandering albatross in the Southern Oceans may have a metapopulation dynamic since the dispersal of juvenile birds can alter the degree of persistence of populations of other islands' populations. These findings would counter the view that albatrosses are strictly philopatric species (Warham 1990) whose populations can be considered closed to dispersal. Nevertheless, adult wandering albatross did show a strong fidelity to the breeding island after the first reproduction as only short-range movements between colonies of the same island have been observed. This pattern, while common to many seabird species (e.g., Cassin's Auklets in Pyle 2001; Cory's shearwater in Rabouam et al. 1998; South polar skua in Ainley et al. 1990; western gulls in Spears et al. 1988; sooty terns in Feare and Lesperance 2002; review in Coulson 2001), is not without exceptions (e.g., roseate terns in Spendelow et al. 1995). An important feature in the metapopulation functioning of the wandering albatross in the Southern Ocean was that juveniles were more philopatric to the island of birth when local population abundance was lower. It is this feature that largely determines whether a local population functions either as a sink or a source of juvenile migrants in the metapopulation. This type of source-sink metapopulation dynamic does not result from habitat heterogeneity on the breeding sites (as may be the case for some terrestrial animals; Holt 1985, Pulliam 1988). The dynamics of pelagic seabird populations are largely determined by the quality of their feeding habitats at sea and by their interaction with long-line fisheries (Robertson and Gales 1997, Russell 1999) and essentially independent from the nesting habitats—that is, oceanic islands—provided that the latter are free

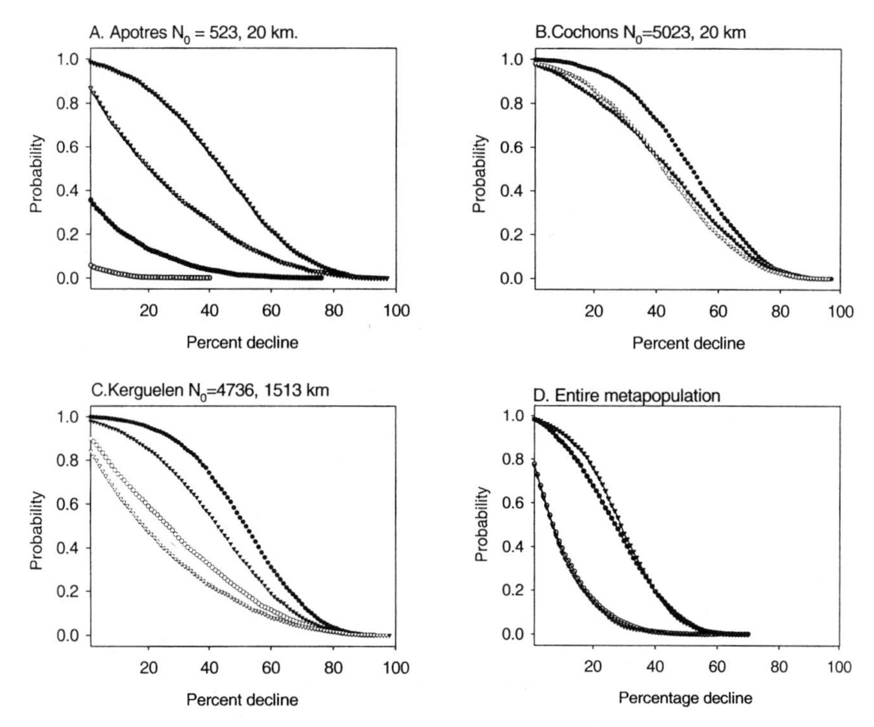

Figure 37.4 Probability of observing a decline from initial abundance during the next 200 years for the populations of **(A)** Apotres, **(B)** Cochons, **(C)** Kerguelen, and **(D)** the entire metapopulation, illustrating the diversity of effects of migration and density dependence in the age of first reproduction. The four curves in each panel correspond to the models that have density dependence in the age at first reproduction with (*filled triangle*) or without (*filled circle*) juvenile migration, and to density-independent models with (*empty triangle*) or without (*empty circle*) juvenile migration. The distance to the closest island and the initial population size of each island are indicated above each panel.

of ice or of land predators. Rather, it is the specific characteristics of the wandering albatross whose populations have a discrete spatial distribution with density- and distance-dependent dispersal that results in the asymmetric dispersal between islands, thus producing the observed source-sink metapopulation dynamic.

Considering density dependence in the age of first reproduction generally increased the risk of decline of an island population compared with density-independent models with the same demographic rates. Nevertheless, the magnitude of the difference between the density dependence–density independence risk curves and the actual probability of decline varied among local populations, depending on the proximity and the size of the closest population that govern the source-sink dynamics of the metapopulation. At the metapopulation level, the risk curve for the density-dependent models was always higher than that of the density-independent ones, suggesting that the latter can serve to estimate a lower bound of the risk of decline of the metapopulation and thus allow a conservative assessment (Ginzburg et al. 1990). Because density-dependent growth stabilizes population trajectory, it decreases its temporal variability and lowers the probability of pronounced population declines (Burgman et al. 1993). However, models of common mechanisms of density dependence constrain population growth in different ways and thus have distinct influences on the predicted extinction risk despite their similar effects on the average population abundance (Figure 37.2). For instance, ceiling density dependence (sensu Strong 1986) implies a sharp boundary to the maximum population abundance, effectively preventing gains during favorable years when population abundance is close to the carrying capacity, and it is the model of density dependence that yields the most conservative risk assessment (Figure 37.2). The density dependence in the age at first reproduction rendered risk curves within the envelope defined by the density-independent and ceiling density dependence models that defined the lower and upper bounds of the probable extinction risk. Decreases in the age at first reproduction with lower breeding abundance have been observed for royal (Robertson 1993) and wandering albatrosses (Croxall et al. 1990, Weimerskirch et al. 1997), herring gulls (Coulson et al. 1982), and Atlantic puffins (Kress and Nettleship 1988). Explanations for the positive relationship between age at first breeding and population abundance have included reduced competition for nest sites and food resources and a greater availability of experienced mates (Pyle 2001), as it could occur when human actions (e.g., incidental mortality during fishery operations) induce a decrease in adult survival. In the case of albatrosses, it is extremely difficult to link variation in demographic rates with changes in resource abundance (as recently shown by Lewis et al. 2001 for the northern gannet around Britain) because these species forage over very large areas of pelagic waters (Weimerskirch 1997). The observed decrease in age at first breeding in the wandering albatross as population sizes became smaller (Croxall et al. 1990, Weimerskirch et al. 1997, review in Russell 1999) suggests that density dependence occurs, although the underlying mechanism remains unclear at present. Our consideration of density dependence should then be strictly considered as a sensitivity analysis to examine whether the main features of the metapopulation function of the wandering albatross are affected by the occurrence of density dependence.

The metapopulation function of wandering albatross could have important consequences when analyzing the effects of incidental by-catch related to long-line fishery activities (Brothers 1991, Inchausti and Weimerskirch 2002), probably the main threat to the viability of the wandering albatross (Croxall et al. 1997, Weimerskirch et al. 1997). Although fisheries have probably affected albatross populations throughout the Southern Ocean, their impact on local populations is rather variable and difficult to document because of the extensive spatial and temporal variation of the fishing effort and the paucity of the available data (Croxall et al. 1997). The interrelated dynamics of local populations mediated by distance- and density-dependent juvenile dispersal produces a source-sink type of metapopulation that permits the occurrence and propagation of a diversity of effects through the network of island populations. For instance, a local decrease in population abundance caused by long-line fisheries may negatively impact the persistence of other island populations that are not directly affected by the fishery activities; this is because fisheries could shift the sink-source relation between pairs of populations in nearby islands. When operating at a regional level (i.e., affecting the population of more than one island), long-line fishery activities would have an augmented negative effect: a direct negative impact on local populations and an indirect regional effect through enhancing local philopatry, thereby decreasing the potential of "rescue" from nearby populations (Inchausti and Weimerskirch 2002).

References

Ainley, D., Ribic, C., and Wood, R. 1990. A demographic study of the south polar skua *Catharacta maccormiki* at Cape Crozier. *Journal of Animal Ecology* 59: 1–20.

Akçakaya , H. R. 1991. A method for simulating demographic stochasticity. *Ecological Modelling* 54: 133–136.

Akçakaya H. R. 2002. *RAMAS Metapop: viability analysis for stage-structured metapopulations.* Version 4.0. Applied Biomathematics, Setauket, N.Y.

Brothers, N. 1991. Albatross mortality and associated bait losses in the Japanese long-line fishery in the Southern Ocean. *Biological Conservation* 55: 255–268.

Burgman, M., Ferson, S., and Akçakaya, H. 1993. *Risk assessment in conservation biology.* Chapman and Hall, New York.

Coulson, J. 2001. Colonial breeding in seabirds. Pages 87–113 in E. Schreiber and J. Burger (eds.), *Biology of marine birds.* CRC Press, Boca Raton, Fla.

Coulson, J., Duncan, N., and Thomas, C. 1982. Changes in the breeding biology of the herring gull *Larus argentatus* induced by the reduction in the size and density of the colony. *Journal of Animal Ecology* 51: 739–756.

Croxall, J., Rothery, P., and Pickering, S. 1990. Reproductive performance, recruitment and survival of wandering albatrosses *Diomedea exulans* at Bird Island, South Georgia. *Journal of Animal Ecology* 59: 775–795.

Croxall, J., Prince, P., Rothery, P., and Wood, A. 1997. Population changes in albatrosses at South Georgia. Pages 69–83 in G. Robertson and R. Gales (eds.), *Albatross biology and conservation.* Surrey Beatty, Sydney.

de la Mare, W., and Kerry, K. 1994. Population dynamics of the wandering albatross *Diomedea exulans* on Macquarie Island and the effects of mortality from long-line fishing. *Polar Biology* 14: 231–241.

Feare, C., and Lesperance, C. 2002, Intra- and inter-colony movements of breeding adult sooty terns in Seychelles. *Colonial Waterbirds* 25: 52–55.

Ginzburg, L., Ferson, S., and Akçakaya, H. 1990. Reconstructibility of density dependence and the conservative assessment of extinction risks. *Conservation Biology* 4: 63–70.

Heywood, R., and Watson, R. (Eds.) 1995. *Global biodiversity asssessment.* Cambridge University Press, Cambridge.

Holt, R. 1985. Population dynamics in two-patch environment: some anomalous consequences of an optimal habitat distribution. *Theoretical Population Biology* 28: 181–208.

Inchausti, P., and Weimerskirch, H. 2002. Dispersal and metapopulation dynamics of an oceanic seabird, the wandering albatross, and its consequences for its reponse to long-line fisheries. *Journal of Animal Ecology* 71: 765–770.

International Union for the Conservation of Nature (IUCN). 2000. *Report 2000.* International Union for the Conservation of Nature, Geneva, Switzerland.

Kress, S., and Nettleship, D. 1988. Re-establishment of Atlantic puffins (*Fratercula arctica*) at a former breeding site in the Gulf of Maine. *Journal of Field Ornithology* 59: 161–170.

Lewis, S., Sherrat, T., Hamer, K., and Wanless, S. 2001. Evidence of intra-specific competition for food in a pelagic seabird. *Nature* 412: 816–818.

Nel, D., Ryan, P., Crawford, R., Cooper, J., and Huyser, O. 2002. Population trends of albatrosses and petrels at sub-Antarctic Marion Island. *Polar Biology* 25: 81–89.

Pulliam, H. 1988. Sources, sinks and population regulation. *American Naturalist* 132: 652–661

Pyle P. 2001. Age at breeding and natal dispersal in a declining population of Cassin's auklet. *Auk* 118: 996–1007.

Rabouam, C., Thibault, J-C., and Bretagnolle, V. 1988. Natal philopatry and close inbreeding in the Cory's shearwater *Calonectris diomedea. Auk* 115: 483–486.

Robertson, C. 1993. Survival and longevity of the Northern Royal Albatross *Diomedea epomophora sanfordi* at Taioara Head 1937–1993. *Emu* 93: 269–276.

Robertson, G., and Gales, R. (eds.). 1997. *Albatross biology and conservation.* Surrey Beatty, Sydney.

Roff, D. 1993. *The evolution of life histories.* Chapman and Hall, New York.

Russell, R. 1999 Comparative demography and life history tactics of seabirds: implications for conservation and marine monitoring. *American Fisheries Society Symposium* 23: 51–76.

Spears, L., Pyle, P., and Nur, N. 1998. Natal dispersal in the western gull: proximal factors and fitness consequences. *Journal of Animal Ecology* 67: 165–179.

Spendelow, J., Nichols, J., Nisbet, I., Hays, H., Cormons, G., Burger, J., Safina, C., Hines, J., and Gochfeld, M. 1995. Estimating annual survival and movements rates of adults within a metapopulation of roseate terns. *Ecology* 76: 2415–2428

Stearns, S. 1992. *The evolution of life histories*. Oxford University Press, New York.

Strong, D. 1986. Density-vague population change. *Trends in Ecology and Evolution* 1: 39–42.

Tickell, J. 1968. The biology of the great albatrosses *Diomedea exulans* and *Diomedea epomorpha*. *Antarctic Research Series* 12: 1–55.

Warham, J. 1990 *The petrels, their ecology and breeding systems*. Academic Press, London.

Weimerskirch, H. 1997. Foraging strategies of southern albatrosses and their relationship with fisheries. Pages 168–179 in G. Robertson and R. Gales (eds.), *Albatross biology and conservation*. Surrey Beatty, Sydney.

Weimerskirch, H., and Jouventin, P. 1997. Changes in population sizes and demographic parameters of six albatross species breeding on the French sub-Antarctic islands. Pages 84–91 in G. Robertson and R. Gales (eds.), *Albatross biology and conservation*. Surrey Beatty, Sydney.

Weimerskirch, H., Jouventin, P., and Mougin, J. 1985. Banding recoveries and the dispersion of seabirds breeding in the French Austral and Antarctic territories. *Emu* 85: 22–33.

Weimerskirch H., Brothers, N., and Jouventin, P. 1997. Population dynamics of wandering albatrosses *Diomedea exulans* and Amsterdam albatross *D. amsterdamensis* in the Indian Ocean and their relationships with long-line fisheries: conservation implications. *Biological Conservation* 79: 257–270.

PART VI

MAMMALS

38

Mammal Population Viability Modeling

An Overview

MICHAEL A. McCARTHY

Research on mammals has had an important role in our understanding of population ecology and the development of population models. Important areas of research have included studies on density dependence, predator-prey theory, disturbance dynamics, the use of models for synthesis of knowledge and as management tools, and approaches to validating models of population dynamics. This overview briefly discusses these areas of research and notes the relevant contributions of the following chapters in this volume. A common feature of these chapters is the use of models for assessing management strategies, a task for which stochastic population models such as RAMAS Metapop seem most suited. One area of research that requires further work is the development of efficient methods for analyzing these models so that optimal management strategies can be determined.

Development of Theory and Methods

Research on the population ecology of mammals has played an important part in many developments in population ecology. Chapter 39 on snowshoe hares by Griffin and Mills reminds us of the snowshoe hare–lynx cycle in Canada. It is perhaps one of the most widely known examples of predator-prey dynamics, has one of the longest time series, and has contributed substantially to our understanding of predator-prey dynamics (Krebs et al. 2001). The potential influence of predation (or other trophic interactions) is either ignored in most models of population viability or subsumed within the vital rates. Ignoring predation might be suitable when the main predators are absent from the system or are unlikely to contribute substantially to the total mortality (Yamada et al., Chapter

42 in this volume). Subsuming the effects of predation into the vital rates might be suitable when predation does not cause a deterministic pattern in the vital rates—for example, in population cycles (Griffin and Mills, Chapter 39 in this volume). The use of temporal change files in Metapop, in which the relevant vital rates are adjusted to ensure that a suitable cycle occurs, provides a possible mechanism for those cases where trophic interactions lead to deterministic cycles in population abundance. However, with this method it is not possible to develop a dynamic trophic model in which the vital rates varied as a function of other trophic levels. An alternative may be to include the second trophic level as one or more additional stages in the model and then model the dynamics with a user-defined density dependence function.

As with the snowshoe hare–lynx system, the economic and social interest in mammals means that we often have long time series to help us address fundamental questions in population ecology. One such question concerns the relative importance of density-dependent and density-independent processes in influencing the abundance of species (Coulson et al. 2000). RAMAS Metapop readily incorporates both environmental stochasticity (e.g., variation in weather) and deterministic processes (e.g., density dependence and trends in parameter values) in population models. It is likely that both these factors will be important for modeling mammal species, although the relative importance of each may vary. The following chapters on mammals include a range of types of density dependence. They also demonstrate various methods for estimating the level of environmental stochasticity when long time series are not available, such as using weather records (Sezen et al., Chapter 41 in this volume; Yamada et al., Chapter 42 in this volume) and substituting space for time (Griffin and Mills, Chapter 39 in this volume).

Disturbances such as fires, windstorms, and floods drive the dynamics of many ecosystems. The population dynamics of mammal species that inhabit such ecosystems often respond to these disturbances through changes in the abundance of predators and the availability of food, nesting sites, and dens (e.g., Lindenmayer and Possingham 1996, McCarthy and Lindenmayer 1999). RAMAS Metapop allows for such disturbance processes by simulating "catastrophes" as events that can lead to changes in survival, fecundity, and carrying capacity. Drought was simulated as a catastrophe that reduced fecundity to 0 in the ibex model of Yamada et al. (this volume). Although Griffin and Mills (this volume) did not simulate catastrophes per se, successional dynamics of the forest were incorporated as probabilistic changes in the state of the forest with concomitant changes in the vital rates of snowshoe hares. This was achieved by using an external program to generate the values for the temporal change files used by Metapop.

Applied Theoretical Ecology

The use of models as ecological management tools can be thought of as applied theoretical ecology. As demonstrated by the following chapters, models provide a framework for evaluating a range of management options more easily and quickly than can be done in the field. This may be achieved by explicitly modeling management strategies or scenarios (e.g., Griffin and Mills, Chapter 39 in this volume; Lopez, Chapter 40 in this volume; Root, Chapter 44 in this volume; Sezen et al., Chapter 41, this volume), or by sensitivity analysis in which the relative influence of different parameters is assessed (Gerber, Chapter 43 in this volume; Yamada et al., Chapter 42 in this volume).

When these models are used to suggest suitable management strategies that are evaluated by using "management experiments," the process is known as active adaptive management (Walters 1986, Walters and Holling 1990). One of the advantages of using models in this way is that expert opinion is documented explicitly in the form of model structures and parameters (e.g., Yamada et al.), providing predictions that can be tested. It is difficult to evaluate the reliability of expert opinion without formal documentation such as that provided by the construction of a model.

Mammals have proven to be important in testing the predictions of stochastic population models. McCarthy and Broome (2000) found that the mean and variance of the annual population growth rate of three populations of mountain pygmy possums were close to those predicted by a stochastic population model developed using a single population. Brook et al. (2000), examining a range of species including mammals, found that predicted risks of population decline appeared to be unbiased. Metapopulation models of mammals have also been evaluated: for example, average rates of extinction and recolonization by three shrew species inhabiting islands in Finland were close to those predicted by metapopulation models (Hanski, 1997). However, model predictions and field data do not always agree: in a patch system in southeastern Australia, predicted patterns of patch occupancy matched those observed for some arboreal marsupial species but not for others (McCarthy et al., 2001a). Such discrepancies should be expected because models are, by design, imperfect representations of reality. The aim of testing should be to discover important discrepancies, and this is an important part of using population models in decision making (McCarthy et al., 2001b). The range of output provided by RAMAS Metapop (e.g., abundance and occupancy) is suitable for such tests.

Methods for evaluating stochastic population models have tended to focus on predictions of risk rather than on predicted changes in risks in response to management and the relative benefits of different possible management strategies (McCarthy et al., 2003). There appears to be considerable scope for using models to suggest possible management strategies, to help design management experiments, and to assess the relative effectiveness of the different management strategies using both the results of the models and the experiments (Walters 1986, Possingham et al. 1993). The kinds of models developed in the following chapters could be readily used in this role. However, determining optimal management strategies is often computationally difficult, especially when those strategies depend on the state of the system and when different strategies have different costs and benefits. The difficulty occurs because of the large number of possible options and combination of options that can be employed. For example, determining the optimal arrangement of harvesting operations over a period of time in the model of Griffin and Mills (Chapter 39 in this volume) would be difficult given the vast number of possible combinations in their landscape of 484 patches. The difficulty is compounded when one trades the economic costs of different harvesting strategies against the possible environmental costs. Developing useful rules of thumb and methods for determining optimal strategies is an active area of research in population ecology (Possingham 2001).

Overview of Mammal Chapters

Each chapter in Part VI demonstrates how RAMAS Metapop can be used to address a range of different questions and how particular characteristics of the species can be incorporated into the models. Griffin and Mills (Chapter 39) use a model of snowshoe hares and forest succession to investigate possible effects of timber harvesting operations on the species. A novel feature of their model is the 3-month time step, corresponding to winter, spring, summer, and autumn. Abundance is modeled by using an expanded stage transition matrix that includes the two different age classes (juveniles and adults) in each of the four seasons. This allows different vital rates to apply to the different times of year. Griffin and Mills demonstrate that forest succession and timber harvesting are likely to have important impacts on snowshoe hare dynamics.

Yamada et al. (Chapter 42) also used a modified stage-transition matrix to model a particular feature of their species. They did not differentiate between males and females in the youngest age class of ibex to ensure that fecundity rates of male and female offspring were perfectly correlated and because at this age it is not necessary to distinguish between the two sexes. Yamada et al. demonstrate that the ibex is most vulnerable to changes in vital rates, reductions in carrying capacity and initial abundance, and encroachment by further human activity. The species appears to be relatively insensitive to harvest rates, perhaps because it is concentrated on males, which because of polygyny do not influence the population size to the same extent as females.

In contrast to Yamada et al., Sezen et al. (Chaptrer 41), in modeling the reintroduction and harvesting of mouflon, demonstrated that harvesting of males can considerably affect a species. A proposal to harvest up to 50 older males each year was unlikely to be sustainable, with 5 individuals being a more realistic figure. This illustrates one of the chief advantages of developing a population model: it demonstrates the consequences of what is believed to be true. It is difficult to determine suitable harvest rates by simply using intuition. Developing the population model forced a synthesis of available information and determined that a particular proposal was unrealistic.

Lopez (Chapter 40) demonstrates how RAMAS Metapop can be used to explore tradeoffs between human development and species conservation. In this context, the model is used as a decision support tool that again demonstrates the consequences of what is believed to be true. Lopez notes that the user can explore the importance of parameter uncertainty by modifying the model and investigating the consequences. Decisions can then be made with better knowledge about the magnitude and importance of uncertainty.

Using Florida panthers as an example, Root (Chapter 44) illustrates the three steps required to link habitat data with a metapopulation model. The steps are (1) identify the relationship between the species and its habitat, (2) locate discrete habitat patches, and (3) estimate population- and metapopulation-level parameters. These three steps are incorporated within RAMAS/GIS although the first must often be conducted partly with a separate statistical program. Root then explored the influence of translocation and natural dispersal on the recovery of the taxon. Her model demonstrates the importance of habitat to the Florida panther, with reductions in population size closely tracking any reductions in the amount of habitat. Further, recovery of the taxon depends on the availability and accessibility of suitable habitat.

Gerber (Chapter 43) used a population model of Steller sea lions to examine uncertainty regarding the conservation status of the species. A sensitivity analysis was used to explore the influence of parameter values on the viability of the species, thereby suggesting the suitability of different management actions. Gerber also notes that a comparison of the predictions of the model with observed data (e.g., population age structure) can suggest potential causes of population decline.

The following chapters illustrate a range of approaches to developing population models in RAMAS GIS, as well as the different methods for analysis of these models. They illustrate how models can be used to explore management actions and examine the importance of uncertainty.

Acknowledgments I thank Julian Fox, Darrell Land, David Maehr, Zeynep Sezen, Roel Lopez, Mark Maunder, and E. J. Milner-Gulland for help in reviewing the chapters.

References

Brook, B. W., O'Grady, J. J., Chapman, A. P., Burgman, M. A., Akçakaya, H. R., and Frankham, R. 2000. Predictive accuracy of population viability analysis in conservation biology. *Nature* 404: 385–387.

Coulson, T., Milner-Gulland, E. J., and Clutton-Brock, T. 2000. The relative roles of density and climatic variation on population dynamics and fecundity rates in three contrasting ungulate species. *Proceedings of the Royal Society of London* B 267: 1771–1779.

Hanski, I. 1997. Metapopulation dynamics: from concepts and observations to predictive models. Pages 69–91 in I. Hanski and M. E. Gilpin (eds.), *Metapopulation biology: ecology, genetics and evolution*. Academic Press, San Diego.

Krebs, C. J., Boonstra, R., Boutin, S., and Sinclair, A. R. E. 2001. What drives the 10-year cycle of snowshoe hares? *BioScience* 51(1): 25–35.

Lindenmayer, D. B., and Possingham, H. P. 1996. Ranking conservation and timber management options for Leadbeater's possum in southeastern Australia using population viability analysis. *Conservation Biology* 10: 235–251.

McCarthy, M. A., and Broome, L. S. 2000. A method for validating stochastic models of population viability: a case study of the mountain pygmy-possum (*Burramys parvus*). *Journal of Animal Ecology* 69: 599–607.

McCarthy, M. A., and Lindenmayer, D. B. 1999. Incorporating metapopulation dynamics of greater gliders into reserve design in disturbed landscapes. *Ecology* 80: 651–667.

McCarthy, M. A., Lindenmayer, D. B., and Possingham, H. P. 2001a. Assessing spatial PVA models of arboreal marsupials using significance tests and Bayesian statistics. *Biological Conservation* 98: 191–200.

McCarthy, M. A., Possingham, H. P., Day, J. R., and Tyre, A. J. 2001b. Testing the accuracy of population viability analysis. *Conservation Biology* 15: 1030–1038.

McCarthy, M. A., Andelman, S. A., and Possingham, H. P. 2003. Reliability of relative predictions in population viability analysis. *Conservation Biology* 17: 982–989.

Possingham, H. P. 2001. *The business of biodiversity: applying decision theory principles to nature conservation*. TELA series, no. 9, Australian Conservation Foundation (http://www.acfonline.org.au/docs/publications/tp009.pdf).

Possingham, H. P., Lindenmayer, D. B., and Norton, T. W. 1993. A framework for improved threatened species management using population viability analysis. *Pacific Conservation Biology* 1: 39–45.

Walters, C. J. 1986. *Adaptive management of renewable resources*. Macmillan, New York.

Walters, C. J., and Holling, C. S. 1990. Large-scale management experiments and learning by doing. *Ecology* 71: 2060–2068.

39

Snowshoe Hares (*Lepus americanus*) in the Western United States

Movement in a Dynamic Managed Landscape

PAUL C. GRIFFIN
L. SCOTT MILLS

Since the listing of the Canada lynx, *Lynx canadensis*, as threatened in the coterminous United States (U.S. Fish and Wildlife Service 2000), the effect of landscape-scale forest management on populations of snowshoe hares, *Lepus americanus*, has been a concern. Snowshoe hares are the central prey species of lynx (Aubry et al. 2000). Because snowshoe hare densities vary widely across different types of forested vegetation structure (Adams 1959, Dolbeer and Clark 1975, Wolff 1980, Wolfe et al. 1982), changes in forest vegetation structure through natural disturbance, succession, and silviculture can have dramatic impacts on populations. Snowshoe hare populations cycle with a regular 9- to 11-year period in the northern portion of the species range (Keith 1990), but evidence of a cycle to the south is mixed (Hodges 2000). Vegetative succession complicates population models (Johnson 2000). In western Montana, we have found highest snowshoe hare densities in regenerating forest stands with high sapling density and in uncut, late-seral-stage forest stands also with abundant saplings. This high-quality snowshoe hare habitat in Montana can be thought of as having a bimodal distribution relative to forest stand age, with only young and much older stands providing the "closed" understories with abundant cover and browse (Buskirk et al. 2000). The layer of abundant shrubs and saplings that regenerates following clearcuts or large fires later disappears as the lower limbs die on growing trees, but this layer may reappear when large trees die, creating canopy gaps (Oliver and Larson 1986). The potentially long period when the stand understory becomes and remains comparatively "open" is a time of lower habitat quality. Anthropogenic canopy gaps in partially harvested stands can also stimulate growth of a dense understory layer under an established canopy (DeBell et al. 1997, Tappeiner et al. 1997). Forest managers trying to maintain the prey base for lynx need to balance the maintenance of snowshoe hare and other prey habitats with other management goals (McKelvey et al. 2000). We used RAMAS GIS to explore the effects of timing and placement of one type of silvicultural treatment on a population of snowshoe hares that was highly connected by movement and distributed across many patches of varying quality.

In pre-commercial thinning of dense conifer regenerating stands in the western United States, live stem density is decreased from 3,000–6,000 per ha to 650–1,300 per ha to increase tree growth and yield, reduce future fuel load, and shift species composition (Seidel 1986, Johnstone 1995, Martin and Barber 1995). Through pre-commercial thinning, the sudden conversion of a "closed" understory to an "open" understory means the loss of cover and forage for hares. In preliminary results from a 3-year experiment, we have seen twofold to fourfold snowshoe hare density decreases during the 2 years after thinning (Griffin and Mills 2003). Because dense, young forest stands were known to hold high hare densities (Adams 1959, Wolfe et al. 1982, Hodges 2000), in 1999 the U.S. Forest Service halted pre-commercial thinning on lands defined as lynx habitat, although the practice continues on private and state-managed lands. Because pre-commercial thinning is a costly silvicultural investment, it may often be applied synchronously across large contiguous areas.

Current studies are addressing whether pre-commercial thinning could accelerate shrub and new seedling growth under the remaining trees, as well as what time scale is necessary for the regrowth of understory plants before a thinned stand is good habitat for snowshoe hares. Similarly, uncut mature stands may lose understory cover as a result of partial harvest operations. It is not clear how long a partially harvested mature stand must develop sufficient understory cover to consider it equivalent in hare habitat quality to uncut mature stands where shrubs and saplings are dense.

The montane forests of the study region are dominated by subalpine fir (*Abies lasiocarpa*), Douglas fir (*Pseudotsuga menziesii*), western larch (*Larix occcidentalis*), lodgepole pine (*Pinus contorta*), and Engelman spruce (*Picea engelmannii*), and they generally correspond to the "*Abies lasiocarpa* / *Menziesia ferruginea* warm phase" habitat type of Arno et al. (1985). Existing literature and successional models give a range of time necessary for development of a dense understory in our study region (Zamora 1982, Arno et al 1985, Moeur 1985).

We parameterized our demographic model with vital rate estimates (survival, birth, and movement) from ongoing studies in the Seeley-Swan region of western Montana, where we are evaluating snowshoe hare population dynamics in four forest vegetation structural types: uncut mature forest (henceforth referred to as "uncut mature"), partially harvested mature forest ("cut mature"), regenerating clearcut ("dense sapling"), and pre-commercially thinned or sparsely regenerating clearcut ("open sapling"). These structural types were determined in the field, based on sapling density, basal area, horizontal cover, and overhead canopy cover (Table 39.1). Both the cut mature and open sapling structural types generally had little understory cover. The preliminary data referred to in this chapter represent more than 4 years of intense work in 20 forest stands at five study sites (Griffin and Mills 2003).

Caution is appropriate in applying our data to larger spatial scales, for several reasons. First, by choosing four discrete vegetation structural types to study, we excluded many structure types of the forest matrix. For example, what we call the uncut mature structural type has dense understory, which should provide good cover and forage for hares; in reality, many mature stands have canopies of large trees without a dense layer

Table 39.1 A key to the vegetation structure attributes we used to delineate the four nominal structure types in this study

Habitat Type Name	Structural Attributes
Dense sapling	Dense stands of regenerating conifers with more than 5,600 saplings/ha
Open sapling	Open stands of regenerating conifers with fewer than 3,360 saplings/ha
Uncut mature	Uncut forest stands with more than 40% overhead canopy cover and a basal area greater than 13.8 m^2/ha (60 ft^2/acre)
Cut mature	Partially harvested forest with less than 30% overhead canopy cover and basal area greater than 13. 8 m^2/ha (60 ft^2/acre).

Note: These categories leave out a large number of vegetation structure types found in the study region. Saplings are defined here as trees over 0.5 m tall, but less than 10 cm diameter at breast height.

of shrubs or saplings. While our demographic research estimates vital rates in specific forest stands, hares can inhabit multiple structural types. Therefore, hares in our model occupy a "patchy metapopulation" (Harrison 1994) with a high degree of movement between contiguous patches, each of potentially different quality.

Our second concern for modeling our data in a RAMAS GIS framework was that the conditions and timing of succession that cause stands to change from one type to another can vary dramatically, depending on environmental factors, stand history, and site preparation (Arno et al. 1985).

Despite our considerable efforts in the field, the sample size and study duration are still not sufficient for us to evaluate carrying capacity and the type of density dependence. We also could not evaluate how dispersal varied according to patch type arrangement (Wiens et al. 1993). Finally, our preliminary data from a period of apparent decline give us no basis to evaluate whether vital rate estimates are close to average rates of the long term, especially for a species that may be cyclic.

Methods

Population Matrices

Our vital rate estimates are from the first 3 years of a 4-year study. Snowshoe hares born in the summer wean after 4 to 5 weeks (Keith 1990). Juveniles that survive to spring are reproductive adults. Adult females at our study area could have bred in each of three synchronous pulses per summer, with nursing taking place overall during ~15 weeks. Fertility, defined here as total number of young expected per summer per reproductive female, was estimated from observed values and ultrasonographic examinations (Griffin et al., in press). We detected no difference in fertility between vegetation structure types, although there were differences in litter size of newborns between the first, second, and third birth pulses.

Using known fates of 149 radio-collared hares, we estimated survival rates based on the candidate model with lowest AICc values computed in Program MARK (White and Burnham 1999). The 11 models for adult and juvenile survival ranged in complexity from simple models with one parameter to models with multiple parameters for season-specific and habitat-specific survival (Griffin and Mills 2003). The selected model structure had the lowest AICc value by 4.57 AICc units, indicating that it is more than 10 times as likely as all other models (after Burnham and Anderson 1998). Based on the selected model, survival rates were equal in all vegetation structure types in the summer and winter, but rates differed in fall and spring, when survival was lower in structure types with "open" canopies (the cut mature and open sapling structure types). Survival rates of adult and weaned juvenile hares were indistinguishable. There was no difference between male and female survival rates, so our matrix model reflects females only.

To reflect the seasonal differences in survival and movement, our matrix model has four time steps per year (Table 39.2). The 40-year simulations require 160 time steps. We account for juveniles starting in the first fall after birth, so the population vector includes three juvenile stages (fall, winter, and spring) and four adult stages (summer, fall, winter, spring). Because survival models for adult and juvenile hares were for 4-week time periods, mean seasonal survival rates were estimated by raising the 4-week rate to a power corresponding to the number of weeks in each season divided by 4 (Table 39.2). The fecundity term accounts for estimated survival of summer adults up to each birth pulse, fertility (number of female offspring per female per birth pulse), survival of newborns to weaning, and survival of weaned juveniles to the fall. To include all three 5-week nursing periods, our summer is 15 weeks and our spring is 11 weeks. Over 1 year (four time steps), population sizes decrease because of mortality then increase after each summer due to reproduction.

Movement rates were specified as a dispersal function in RAMAS Metapop, based on radio-collared hare movements. Observed maximum distances that hares moved every season were lognormally distributed with a mean of approximately 275 m per season (Griffin 2003). More individuals left natal home ranges during the fall, so we assigned highest relative dispersal weights to the fall stages.

Table 39.2 Our stage-based projection model for snowshoe hares makes explicit the seasonal differences in survival rates

	Juv (F)	Juv (W)	Juv (Sp)	Ad (Su)	Ad (F)	Ad (W)	Ad (Sp)
Juv (fall)	0	0	0	F[a]	0	0	0
Juv (winter)	S_{JFa}	0	0	0	0	0	0
Juv (spring)	0	S_{JW}	0	0	0	0	0
Adult (sum)	0	0	S_{JSp}[b]	0	0	0	S_{ASp}[b]
Adult (fall)	0	0	0	S_{ASu}[b]	0	0	0
Adult (wint)	0	0	0	0	S_{AFa}	0	0
Adult (spring)	0	0	0	0	0	S_{AW}	0

Note: Survival rates are subscripted with J for juveniles and A for adults; then with Fa, W, Sp, and Su for fall, winter, spring, and summer. The fecundity term, F, accounts for survival of mothers up to each enumerated birth pulse, fertility per mother in each birth pulse (f_i), survival of baby leverets up to weaning in each birth pulse (S_{LevBPi}), and survival of weaned juveniles to the fall. Fertility is for female offspring only: half the total number of offspring. Adult survival rates in the fecundity term are raised to a power to indicate the number of weeks survived at that rate, out of the 15 in the summer. Juvenile survival after weaning until the fall is the same as adult summer survival (S_{ASu}).

[a]Fecundity term here is $(f_1{*}S_{LevBP1}{*}S_{ASu}{}^{(10/15)}) + (S_{ASu}{}^{(5/15)}{*}f_2{*}S_{LevBP2}{*}S_{ASu}{}^{(5/15)}) + (S_{ASu}{}^{(10/15)}{*}f_3{*}S_{LevBP3})$

[b]Spring survival rates reflect an 11-week period; summer survival rates reflect a 15-week period.

In stochastic population models, variances of vital rates should reflect the temporal and spatial variation but should exclude sampling variance, which results from uncertainty in parameter estimates owing to incomplete sampling of the population (Burnham et al. 1987, Thompson et al. 1998, Gould and Nichols 1998, Ludwig 1999, White 2000). For survival rates, we could not partition out temporal variation from total variation because the most parsimonious survival model did not include parameters for year-to-year temporal variation. Instead, standard errors for survival rate estimates were taken from the survival model, and the corresponding standard deviations for seasonal survival rates used in RAMAS were found using the delta method (Agresti 1990). Similarly, because our fecundity data were limited in temporal scale, we could not partition temporal variance from total variance. We used the spatial variance in fecundity across all sites and vegetation structure types (0.75) as a proxy for temporal variance in fecundity. The implicit assumption that differences across sites are comparable to temporal variation at a single site is suspect because differences in parameter values at different sites can result from variation in abiotic and biotic factors (Tyre et al. 2000).

In the absence of dispersal, the population projection matrix corresponding to structure types with open understories ("open matrix" for cut mature and open sapling structure types) has a yearly rate of increase of $\lambda = 0.49$. The population projection matrix for structure types with more densely vegetated understories ("closed matrix" for uncut mature and dense sapling structure types) has a yearly rate of increase of $\lambda = 1.03$.

The data from 1998 to 2001 were from a period of overall decline (Mills et al. 2003). Average density estimates in summer 2001 were roughly 35% of summer 1998 values. Vital rate estimates for this period were probably influenced by whatever factor caused the declines. Because we do not know with certainty the cause of the declines, or how vital rates and movement would differ during population increases, we conducted simulations with the observed rates instead of increasing them arbitrarily to achieve stationary populations.

Model Landscapes and Succession

We limited model landscapes to four vegetation structure types (Table 39.1). We developed a single landscape, including a simple model of succession, and evaluated the population dynamics of snowshoe hares in three additional scenarios that varied only in the timing and placement of pre-commercial thinning. We used the Spatial Data module to create a single metapopulation map with 484 patches of 16 ha each, arranged in a square. Although snowshoe hare habitat is contiguous in reality, we included narrow 40-m boundaries so that RAMAS Metapop could distinguish patches.

Each patch was randomly assigned an initial forest vegetation structure type. The target proportions of patches in each structure type desired were approximately 60% uncut mature, 15% dense sapling, 10% cut mature, and 5% open sapling. Our model landscape is unlike managed landscapes in western Montana because few of those, if any, have 60% uncut mature forest. For example, in the Seeley Lake region the area of uncut mature forest is not more than 25%. When we initially allocated less uncut mature type, however, overall hare populations fell to 0 before the end of the 40-year simulations. This decline was unacceptable for meeting the objectives of this study because we wanted to explore the effects of timing and extent of pre-commercial thinning. By including so much uncut mature type in the model, we ensured that some good-quality hare habitat in each scenario would remain as a source of snowshoe hares after thinning.

Each patch was associated with snowshoe hare vital rates, depending on the vegetation structure type found on the patch in each year. Uncut mature and dense sapling structure types had the higher survival rates of the "closed matrix." Cut mature and open sapling structure types had the lower survival rates of the "open matrix." The matrices for each patch at each time step were recorded in a "temporal change file," thereby reflecting the starting vegetation structure type of each patch and any changes in structure type that resulted from simulated succession or pre-commercial thinning.

We devised a simple model for forest succession allowing for transitions over time between the different vegetation structure types, and we recorded the successional trajectories of each patch. The successional simulation rules were used to represent probabilistic changes between types. Transitions between types were based on mean rates with a stochastic component analogous to demographic stochasticity, whereby transition occurred if a random number drawn from a uniform 0–1 distribution was less than the mean transition probability.

Any patch assigned initially as uncut mature type stayed in that condition; this simplifying assumption ignores possible logging, catastrophic fire, or insect damage that could convert mature stands to younger structure types.

Reflecting potential tree crowding and crown lift, dense sapling type patches could change to a vegetation structure type with an open understory with an annual probability of 0.015. After such a transition we left these patches with an "open" matrix of vital rates for the remainder of the 40-year simulation. The 1.5% probability reflects the assumption of a long residence time in the dense sapling type with the expectation that half the patches would change to the open-understoried type within 45 years, given a binomial distribution. Current management guidelines for preserving dense cover on U.S. Forest Service lands in potential lynx habitats recommend not thinning in regenerating stands until 45 years or older.

To signify regrowth of understory plants to the point that open sapling type patches support a high density of shrub and sapling cover, each patch of the open sapling type changed to a vegetation structure type with dense understory with annual probability 0.033, based on the expectation that half the patches of open sapling type should change to a structure type with a denser understory after 20 years, given a binomial distribution. Arno et al. (1985) indicate that high sapling coverage occurs 12 to 33 years after stand initiation, so the 20-year horizon is reasonable for dense growth of shrubs and small saplings.

To represent the growth of shrubs and saplings in patches starting as the cut mature type, such patches could change to and stay in a vegetation structure type with a dense understory with annual probability of 0.02, based on an expectation that half the patches of cut mature type should develop adequate cover within 35 years, given a binomial distribution. The cut mature type had a lower probability of developing a dense understory than did the open sapling structure type because large standing trees can reduce the light available for understory growth. Tree crown area, which intercepts light, can be approximated as a linear function increasing with tree diameter and height (Moeur 1981). Shrub cover value is predicted to decline with increased basal area of a stand (Moeur 1985).

Four Scenarios of Landscape Change

We refer to the single configuration of initial vegetation structure types and the time series of changes for that configuration as the "succession" scenario (no. 1). We used this configuration and time series, with changes only to those stands that were pre-commercially thinned in the three other scenarios: "upper half thin" (no. 2), "year 10 random thin" (no. 3), and "random thin" (no. 4). If a dense sapling type patch was thinned in a scenario, hares in that patch then had the lower survival rates associated with structure types with low understory cover. In all three scenarios with thinning (nos. 2, 3, and 4), thinned patches could potentially convert back to the "closed" matrix with the higher survival rate as a result of succession, but only after 3 or more years. All scenarios were simulated 100 times.

In scenario no. 2, upper half thin, we simulated spatial autocorrelation that could arise from patterns of land ownership. The randomly assigned configuration of vegetation structure types that initialized all four scenarios included either 54 dense sapling type patches in the 242 patches of the model landscape's upper half or 68 such patches in the lower half. In the upper half thin scenario, all 54 dense sapling patches were thinned at year 10. All 68 dense sapling patches in the lower half of the landscape were not thinned and kept the same temporal change files as in the "succession" scenario.

In scenario no. 3, year 10 random thin, 54 dense sapling patches were again thinned at year 10, but the placement of those patches was chosen at random from all 122 dense sapling patches.

In scenario no. 4, random thin, an equal amount of thinning occurs on the landscape as in other scenarios, but it is spread out across the 40-year time period and the 122 dense sapling patches. For this, we calculated the annual risk of thinning to patches in the other scenarios by dividing the 54 thinned patches by the 122 total patches and by the 40-year duration of the simulations, giving 1.106% mean annual probability of thinning per patch. For every dense sapling patch in every year, we evaluated a random number against this quotient (0.01106) to determine whether or not it would be thinned. Fifty-five patches were thinned in this process over the 40-year simulation, while unthinned patches maintained the same temporal change files as in the succession scenario.

Other Model Parameters

We used the ceiling model for density dependence because of its simplicity and lack of evidence for a better alternative. We could not evaluate carrying capacity from snowshoe hare time series at our sites, so we assigned it at $K = 42$ for all 16-ha patches based on the highest density we observed (2.63 hares/ha). This is within the upper range observed in experimental plots in the Yukon (Hodges et al. 2001).

For all scenarios, we initialized all patches at a density of 0.6 hares/ha (10 individuals per patch), approximately the mean density observed in 1998. The initial distribution to stages was according to stable age distribution, with hares only in the fall juvenile and fall adult stage classes.

Model Predictions

Of the four scenarios, the random thin scenario maintained the highest average total snowshoe hare population size across the 40 years simulated (Figure 39.1). The succes-

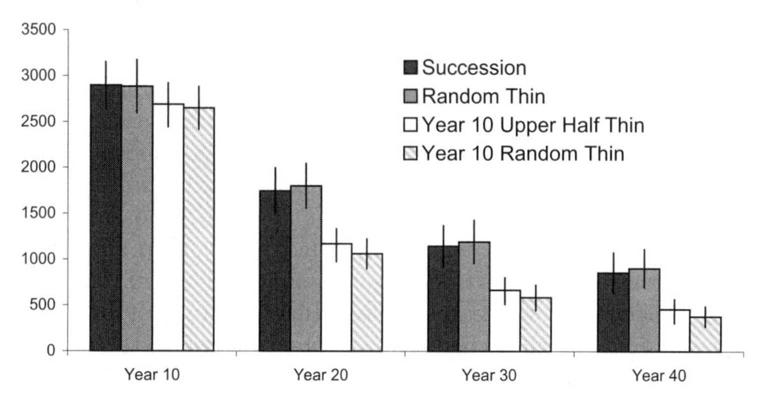

Figure 39.1 Overall abundance in four simulation scenarios for snowshoe hare populations at years 10, 20, 30, and 40. Scenarios in which thinning is applied in year 10 have lower trajectories than those with no thinning ("succession") or where thinning is spread out over time and space ("random thin"). Error bars are standard deviations from 100 simulations.

sion scenario without thinning had the next highest average number of hares. The predictions of these two scenarios overlap substantially for all years simulated.

Both scenarios in which thinning occurred at year 10 maintained a lower total number of snowshoe hares than the other scenarios. On average, the abundance was marginally higher in the upper half thin scenario than in the year 10 random thin scenario; their standard deviation bars overlap substantially at all times.

Discussion

It is not surprising that overall abundance of hares was higher in the succession scenario than in the upper half thin and year 10 random thin scenarios. Both of the latter scenarios include synchronous conversions of 54 out of 122 high-quality dense sapling structure type patches to low-quality open sapling type patches, with at least a 3-year lag before they could convert back to a vegetation structure type with dense understory and higher hare survival rates.

More interesting is the possibility that, using our model assumptions, landscapes with a low level of pre-commercial thinning (the random thinning scenario) may support as many hares in the long term as do landscapes without any thinning (the succession scenario). One critical assumption driving this result is that young thinned stands can eventually provide the understory cover that is necessary for snowshoe hares. A scenario without thinning may have supported higher hare densities, however, if stand-replacing processes such as catastrophic wildfire, insect damage, or clearcutting had been included in the model; these would have led to regeneration of young, dense stands. Some outstanding research questions for models such as ours relate to successional rates: time for shrub and sapling growth to provide ample cover for hares; subsequent time until trees in the canopy reduce understory light levels such that shrub and sapling cover becomes insufficient for hare survival; and time to breakup of closed canopy so that understory cover is again favorable. Also, we raise the vexing question of whether uncut mature stands can really be considered stable. For each of these successional rate questions, mean times will depend on many factors, including elevation, moisture, species composition, and disturbance history. For example, if the probability of changing from the open sapling type to a forest type with a dense understory were lower than the value we used, then negative effects of thinning could be more severe and longer lasting.

The simulation results also hinge on the initial proportions of structure types and the other rules for our simple model of thinning treatment allocation and succession. To maintain non-zero model populations over the 40-year simulated interval, we had to assign roughly 60% of patches as high-quality uncut mature structure type. Even though patches of the dense sapling structure type were converted to open habitats by thinning in the random thin scenario, these thinned patches were later available for dense understory growth. In contrast, during the 40-year simulation the 122 patches in the succession scenario starting in the dense sapling vegetation structure type could only change into a type with crowded trees and an open understory. After the creation of the starting arrangement of structure types in the landscape, the transition rates between structure types govern the overall availability of high-quality and low-quality patches. Furthermore, we simulated only one landscape arrangement in the succession scenario, which was then modified slightly under each of the thinning scenarios.

Our landscape contained only four of the many forest vegetation structure types in the region. Some other structural types may have higher vital rates than structure types with open understories. By assuming that pre-commercially thinned structure types had the same carrying capacity as all other structure types, we may have inflated the abundance that thinned patches could maintain.

Effects of our assumptions could be examined by linking randomized realistic models of succession to RAMAS GIS and by estimating demographic rates in more vegetation structure types. Successional rates, especially, are expected to have a large role in the long-term dynamics of populations living in successional landscapes (Johnson 2000). Before any model such as ours is used to make management recommendations, sensitivity analysis (as in Mills and Lindberg 2002) should be used to examine the effects of different landscape configurations and demographic and successional rates on projected hare population dynamics, perhaps with a program for generating large numbers of dynamic landscapes under different rules.

Populations declined over the 40-year duration of all four scenarios and remained at low but relatively steady levels in the latter decades of the simulations. This happened even though the proportion of uncut mature structure type started at a frequency not found in our study region. The overall decline reflects strong declines expected in the two vegetation structure types associated with the "open" stage matrix of vital rates and in the comparatively modest growth rates of the other two vegetation structure types associated with the "closed" matrix.

The field data leading to demographic rate estimates used here were from a period when the real snowshoe hare population declined in the study region. At this time it is not clear whether the apparent snowshoe hare declines in our study area occurred because of drought or other environmental factors, because of some underlying cyclicity, or because of habitat-specific demography linked to forest age and type composition in the real landscape. Evidence is mixed as to whether snowshoe hares cycle in the southern range (Hodges 2000, Malloy 2000). If (> 1 in all structure types during periods of increase, overall populations could periodically reach high levels despite pre-commercial thinning. Even during periods of overall population increase, we would not expect vegetation structure types with open understories to have higher snowshoe hare survival than do vegetation structure types with dense understories.

Despite the caution necessary in interpreting our results, the landscape modeling framework in RAMAS GIS allows us to make tentative suggestions about the influence of pre-commercial thinning on snowshoe hare populations, pending more refined information about vital rates and dispersal distances in many vegetation structure types and about rates of successional change between structure types of varying quality for hares. For this model landscape where uncut mature forest was prevalent and stable, results suggest that snowshoe hare populations stay higher when pre-commercial thinning is not applied all at once in a landscape. In the context of less stable landscapes with a lower prevalence of favorable patches, successional transition rates should largely determine the dynamics of snowshoe hares.

Acknowledgments We thank R. Akçakaya, D. Christian, J. Goodburn, M. McCarthy, K. McKelvey, D. Pletscher, and one anonymous reviewer for comments on the manuscript. We acknowledge financial support from the National Science Foundation (grant nos. DEB-9876054 to L.S.M. and DEB-0105123 to L.S.M. and P.C.G.), Rocky Mountain Research Station (U.S. Forest Service), Plum Creek Timber Co., and Universal Medical Systems. We also thank L. Bienen, D.V.M., and dozens of field assistants.

References

Adams, L. 1959. An analysis of a population of snowshoe hares in northwestern Montana. *Ecological Monographs* 29: 141–170.

Agresti, A. 1990. *Categorical data analysis*. Wiley, New York.

Arno, S. P., Simmerman, D. G., and Keane, R. E. 1985. *Forest succession on four habitat types in western Montana*. General Technical Report INT-177. U.S. Department of Agriculture, Forest Service, Intermountain Forest and Range Experiment Station, Ogden, Utah.

Aubry, K. B., Koehler, G. M., and Squires, J. R. 2000. Ecology of Canada lynx in southern boreal forests. Pages 373–396 in L. F. Ruggiero et al. (eds.), *The scientific basis for lynx conservation*. University Press of Colorado, Boulder.

Burnham, K. P., and Anderson, D. R. 1998. *Model selection and inference*. Springer-Verlag, New York.

Burnham, K. P., Anderson, D. R., White, G. C., Brownie, C., and Pollock, K. H. 1987. *Design and analysis methods for fish survival experiments based on release-recapture*. Monograph 5. American Fisheries Society, Bethesda, Md.

Buskirk, S. W., Ruggiero, L. F., Aubry, K. B., Pearson, D. E., Squires, J. R., and McKelvey K. S. 2000. Comparative ecology of lynx in North America. Pages 397–417 in L. F. Ruggiero et al. (eds.), *The scientific basis for lynx conservation*. University Press of Colorado, Boulder.

DeBell, D. S., Curtis, R. O., Harrington, C. A., and Tappeiner, J. C. 1997. Shaping stand development through silvicultural practices. Pages 141–149 in K. A. Kohm and J. F. Franklin (eds.), *Creating a forestry for the 21st century: the science of ecosystem management*. Island Press, Washington, D.C.

Dolbeer, R. A., and Clark, W. R. 1975. Population ecology of snowshoe hares in the central Rocky Mountains. *Journal of Wildlife Management* 39: 535–539.

Gould, W. R., and Nichols, J. D. 1998. Estimation of temporal variability of survival in animal populations. *Ecology* 79: 2531–2539.

Griffin, P. C. 2003. Landscape ecology of Snowshoe hares in Montana. Ph.D. diss., University of Montana.

Griffin, P. C., and Mills, L. S. 2003. Snowshoe hare population dynamics in fragmented forests of the Rocky Mountains. Unpublished ms.

Griffin, P. C., Bienen, L., Gillin, C., and Mills, L. S. In press. Estimating pregnancy rates and litter size in snowshoe hares using ultrasound. *Wildlife Society Bulletin*.

Harrison, S. 1994. Metapopulations and conservation. Pages 111–128 in P. J. Edwards et al. (eds.), *Large-scale ecology and conservation ecology*. Blackwell Science, Oxford.

Hodges, K. 2000. Ecology of snowshoe hares in southern boreal and montane forests. Pages 163–206 in L. F. Ruggiero et al. (eds.), *The scientific basis for lynx conservation*. University Press of Colorado, Boulder.

Hodges, K., Krebs, C. J., Hik, D. S., Stefan, C. I., Gillis, E. A. , and Doyle, C. E.. 2001. Snowshoe hare demography. Pages 141–178 in C. J. Krebs et al. (eds.), *Ecosystem dynamics of the boreal forest: the Kluane project*. Oxford University Press, New York.

Johnson, M. P. 2000. The influence of patch demographics on metapopulations, with particular reference to successional landscapes. *Oikos* 88: 67–74.

Johnstone, W. D. 1995. *Thinning lodgepole pine*. In U.S. Forest Service General Technical Report INT-319. U.S. Department of Agriculture, Forest Service, Intermountain Forest and Range Experiment Station, Ogden, Utah.

Keith, L. B. 1990. Dynamics of snowshoe hare populations. Pages 119–195 in H. H. Genoways (ed.), *Current mammalogy*. Vol. 2. Plenum, New York.

Ludwig, D. 1999. Is it meaningful to estimate a probability of extinction? *Ecology* 80: 298–310.

Malloy, J. C. 2000. Snowshoe hare, *Lepus americanus*, fecal pellet fluctuations in western Montana. *Canadian Field Naturalist* 114: 409–412.

Martin, F. C., and Barber, H. W. 1995. Pre-commercial thinning response in 7-year-old and 50-ear-old western larch: post-growth and future prognosis. In U.S. Forest Service General Technical Report INT-319. U.S. Department of Agriculture, Forest Service, Intermountain Forest and Range Experiment Station, Ogden, Utah.

McKelvey, K. S., Aubry, K. B., Agee, J. K., Buskirk, S. W., Ruggiero, L. F., and Koehler, G. M. 2000. Pages 419–441 in L. F. Ruggiero et al. (eds.), *The scientific basis for lynx conservation*. University Press of Colorado, Boulder.

Mills, L. S., and Lindberg, M. 2002. Sensitivity analysis to evaluate the consequences of conservation actions. Pages 338–366 in S. R. Beissinger and D. R. McCollough (eds.), *Population viability analysis*. University of Chicago Press, Chicago.

Mills, L. S., Griffin, P. C., Hodges, K. E., McKelvey, K., Henderson, C., and Ulizio, T. 2003. Pellet count indices compared to mark-recapture estimates for evaluating snowshoe hare density. Unpublished ms.

Moeur, M. 1981. *Crown width and foliage weight of Northern Rocky Mountain conifers*. Research Paper INT-283. U.S. Department of Agriculture. Forest Service. Intermountain Forest and Range Experiment Station, Ogden, Utah.

Moeur, M. 1985. *COVER: a user's guide to the CANOPY and SHRUBS extension of the stand prognosis model*. General Technical Report INT-190. U.S. Department of Agriculture, Forest Service, Intermountain Forest and Range Experiment Station, Ogden, Utah.

Oliver, C. D., and Larson, B. C. 1986. *Forest stand dynamics*. Wiley, New York.

Seidel, K. W. 1986. *Growth and yield of western larch in response to several density levels and two thinning methods: 15–year results*. Research Note PNW-RN-455. U.S. Forest Service, Portland, Ore.

Tappeiner, J. C., Lavender, D., Walstad, J., Curtis, R. O., and DeBell, D. S. 1997. Silvicultural systems and regeneration methods: current practices and new alternatives. Pages 151–164 in K. A. Kohm and J. F. Franklin (eds.), *Creating a forestry for the 21st century: the science of ecosystem management*. Island Press, Washington, D.C.

Thompson, W. L., White, G. C., and Gowan, C. 1998. *Monitoring vertebrate populations*. Academic Press, San Diego.

Tyre, A. J., Tenhumberg, B., McCarthy, M. A., and Possingham, H. P. 2000. Swapping space for time and unfair tests of ecological models. *Austral Ecology* 25: 327–331.

U.S. Fish and Wildlife Service. 2000. Determination of threatened status for the contiguous U.S. distinct population segment of the Canada lynx and related rule: final rule. *U.S. Federal Register* 65: 16051–16086.

White, G. C. 2000. Population viability analysis: data requirements and essential analyses. Pages 288–331 in L. Boitani and T. K. Fuller (eds.), *Research techniques in animal ecology*. Columbia University Press, New York.

White, G. C., and Burnham, K. P. 1999. Program MARK: Survival estimation from populations of marked animals. *Bird Study* 46(Suppl.): 120–138.

Wiens, J. A., Stenseth, N. C., Van Horne, B., and Ims, R. A. 1993. Ecological mechanisms and landscape ecology. *Oikos* 66: 369–380.

Wolfe, M. L., Debyle, N. V., Winchell, C. S., and McCabe, T. R. 1982. Snowshoe hare cover relationships in northern Utah. *Journal of Wildlife Management* 46: 662–670.

Wolff, J. O. 1980. The role of habitat patchiness in the population dynamics of snowshoe hares. *Ecological Monographs* 50: 111–130.

Zamora, B. A. 1982. Understorey development in forest succession: an example from the inland northwest. Pages 63–69 in J. E. Means (ed.), *Forest succession and stand development research in the northwest*. Oregon State University, Corvallis.

40

Florida Key Deer (*Odocoileus virginianus clavium*)

Effects of Urban Development and Road Mortality

ROEL R. LOPEZ

Key deer (*Odocoileus virginianus clavium*) are the smallest subspecies of white-tailed deer in the United States and are endemic to the Florida Keys on the southern end of peninsular Florida (Hardin et al. 1984). Due to uncontrolled hunting, Key deer numbers were estimated to be less than 50 by the 1940's (Hardin et al. 1984). As a result, increased law enforcement and the establishment of the National Key Deer Refuge (NKDR) in 1957 provided protection for the deer and its habitat. Currently, Key deer occupy 20 to 25 islands within the boundaries of the NKDR with approximately 75% of the deer population occupying two islands: Big Pine and No Name keys (Lopez 2001) (Figure 40.1).

Since 1960, urban development and habitat fragmentation have threatened the Key deer's recovery (Lopez 2001). For example, between 1968 and 1998 the human population on Big Pine and No Name increased nearly 10–fold (Monroe County Growth Management Division 1992). In addition to a loss of habitat, increase in urban development is of particular concern because highway mortality accounts for the majority of the total deer mortality (Hardin 1974, Folk 1991). For example, over half of all road mortalities occur on US 1, the only road linking the Keys to the mainland (Lopez 2001). Safe and expedient evacuation during hurricanes depends on the level of service or traffic on US 1. In 1995, Monroe County authorities imposed a building moratorium due to the inadequate level of service on the US 1 segment servicing Big Pine and No Name keys. In an effort to improve the level of service and lift the building moratorium, highway improvements such as intersection widening and adding a third lane northbound (traffic direction toward the mainland) have been proposed (Calvo 1996, Lopez 2001).

Because additional traffic and development on Big Pine and No Name keys might result in an *incidental take* of Key deer, highway improvements could be permitted with the initiation and approval of a habitat conservation plan (HCP). Legislators established the

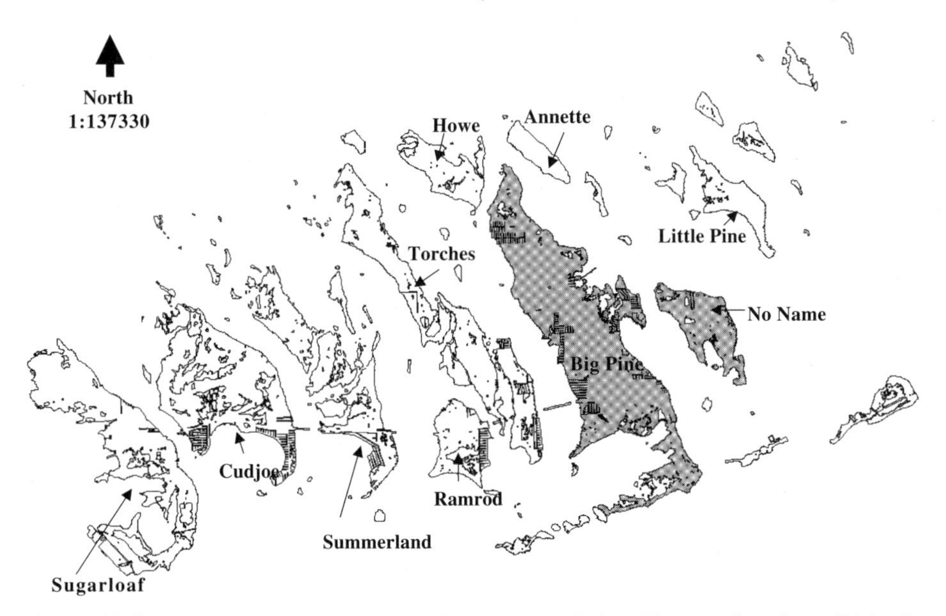

Figure 40.1 The endangered Florida Key deer metapopulation. They are found on all islands; however, the bulk of the population is on Big Pine and No Name keys.

Endangered Species Act (ESA) of 1973 to protect species and their habitats. In 1982, the ESA was amended to authorize incidental taking of an endangered species by landowners and non-federal entities provided they develop an HCP (ESA, Section 10a, 16 U.S.C. §1539a). A planning process in 1998 began with representatives from the Florida Department of Transportation (FDOT), Monroe County, and the Florida Department of Community Affairs (DCA) to draft and submit a regional Key deer HCP to biologists in the U.S. Fish and Wildlife Service (USFWS). A population viability analysis (PVA), a method used to evaluate population threats in terms of extinction risk or decline based on computer simulation models, is an integral part of this HCP (Boyce 1992, Burgman et al. 1993, Lopez 2001).

This chapter illustrates the use of RAMAS Metapop (Version 4.0) in the development of a demographic, stochastic model for Key deer. First, I review demographic data collected and used in model construction. Second, I demonstrate how these data were incorporated into a population viability analysis for Key deer, and how these models are being used in making conservation decisions for these two islands. Specifically, the Key deer models will provide wildlife managers and HCP stakeholders with an array of risk estimates under different management scenarios.

Methods

Study Area and Data Sources

Big Pine (2522 ha) and No Name (459 ha) Keys are within the boundaries of the NKDR, Monroe County, and support the majority of the Key deer population (Lopez 2001) (Figure 40.1). Data used to estimate model parameters were collected via radiotelemetry, survey or trend observations, and deer necropsies (Lopez 2001). The scope of the PVA was limited to Big Pine and No Name keys due to the lack of information for other island deer populations. No deer dispersal was assumed between outer islands (this does not include dispersal between Big Pine and No Name).

Stage Matrix

Each female population was modeled with a stage-structured, stochastic matrix model with three stages (fawn, yearling, and adult). In developing this model, it was assumed that (1) all reproduction occurred in a relatively short breeding season (1 April; Hardin 1974's "birth-pulse" population), (2) the deer population was surveyed after each breeding season (postreproductive survey), (3) all adults and yearlings breed (proportion of breeders determined by survival rate for adults and yearlings), and (4) the stage matrix was the same for both populations. With these assumptions, the stage matrix for the Key deer model was

$$\begin{bmatrix} 0 & F_y & F_a \\ S_f & 0 & 0 \\ 0 & S_y & Sa \end{bmatrix}$$

where S_f, S_y, and S_a are fawn, yearling, and adult survival, respectively, and F_y and F_a are yearling and adult fecundity estimates, respectively.

Survival

Annual female survival and variance estimates were determined by age-class using a known-fate model framework in Program MARK (White and Burnham 1999, Lopez 2001) (Table 40.1). Age categories used included fawns (assumed all deer were born on 1 April), yearlings (1–2 years old), and adults (> 2 years old) (Hardin 1974).

Fecundity

Fecundity was estimated for yearlings and adults from deer necropsy data. Hardin (1974) reported that annual deer maternity was 1.05 fawns per breeding female (yearlings and adults) (Hardin 1974, Folk and Klimstra 1991). Furthermore, Hardin (1974) reported fetal sex ratios were male-biased (59% males) in the Key deer population. From these data, fecundity estimates for yearlings (F_y) and adults (F_a) were determined by

$$F_y = R * M * S_y$$
$$F_a = R * M * S_a$$

where R is equal to the female fetal sex ratio, M is equal to maternity, and S_y and S_a are equal to yearling and adult survival, respectively (Table 40.1) (Lopez 2001). An approximate estimate of fecundity variances were determined as described by Akçakaya et al. (1999) where the variance of the product of two values (1 and 2) was given by

$$\text{var}_{1\times2} = \text{var}_1(\text{mean}_2)^2 + \text{var}_2(\text{mean}_1)^2 + 2\ \text{mean}_1\text{mean}_2\text{cov}_{12}$$

Mean, variance, and coefficient of variation estimates for maternities (1) and survivorships (2) were used in this formula.

Table 40.1 Model parameter estimates used in population viability analysis for Key deer on Big Pine (BPK) and No Name (NNK) Keys

Parameter	Format	Estimate (years)
Survival	Age = mean (SD)	Fawn = 0.470 (0.061)
		Yearling = 0.824 (0.071)
		Adult = 0.840 (0.030)
Fecundity	Age = maternity	Fawn = 0.000
		Yearling = 0.3548
		Adult = 0.3625
Fecundity SD	Age = fecundity SD	Yearling = 0.1016
		Adult = 0.0855
Maximal growth rate (r_{max})		1.05
Carrying capacity [a]		60 (NNK)
		332 (BPK = S1–S2)
		326 (BPK = S3)
		320 (BPK = S4)
		314 (BPK = S5)
Initial abundance	Adult, Yearling, Fawn	37, 7, 16 (NNK)
		203, 41, 88 (BPK)
Catastrophe 1 (Probability)		0.02 [b]
Catastrophe 1 (Impacts)	Age = survival rate	Fawn = 0.50
		Yearling = 0.75
		Adult = 0.85
Catastrophe 2 (Probability)		0.01[b]
Catastrophe 2 (Impacts)	Age = survival	Fawn = 0.10
		Yearling = 0.20
		Adult = 0.50
Dispersal		3.50%

[a]Carrying capacity changed for each management scenario depending on expected habitat loss. Changes applied to BPK only. Management scenarios: S1 (Scenario 1) = no change, S2 (Scenario 2) = road improvement, S3 (Scenario 3) = road improvement + 300 houses, S4 (Scenario 4) = road improvement + 600 houses, and S5 (Scenario 5) = road improvement + 900 houses.
[b]Storm probability (mean return period) of occurrence (catastrophe 1 = category 3–4 hurricane, catastrophe 2 = category 4–5 hurricane). Data from the National Weather Service.

Population Density

Initial abundances used in model simulations were determined from mark-resight estimates for the month of April (White and Garrot 1990, Lopez 2001). A stable age distribution was assumed for both populations, with 60 (16 fawns, 7 yearlings, and 37 adults) and 332 (81 fawns, 41 yearlings, and 203 adults) deer for Big Pine Key and No Name Key, respectively (Table 40.1).

Density Dependence

Density dependence was incorporated in the model using a contest-type density dependence (Beverton-Holt) with the assumption that populations grew 5% to 10% when the N was low and that there was 0% growth when $N = K$ (Akçakaya and Root 2002). It was assumed that available resources are shared unequally due to differences in range site selection among age classes of Key deer (Lopez 2001). Maximal growth rate (1.05% per year) was estimated from survey data during 1990–1999 (Lopez 2001).

Dispersal

Dispersal is an important mechanism in the persistence of a metapopulation (Akçakaya and Root 2002). Annual dispersal rates between Big Pine and No Name keys were estimated from radio-collared deer (Table 40.1) (Lopez 2001). The criterion for determining dispersal was the percentage of deer moving between islands during April (start date in simulations).

Urban Development

Urban development was assumed to result in (1) a change in carrying capacity and (2) an increase in mortality due to secondary impacts (e.g., traffic increases, entanglement in fences) in the Key deer population (Lopez 2001). Changes in these two parameters for a given scenario were estimated from survey and necropsy data and are discussed later in this chapter.

Carrying Capacity (K)

The hypothesized carrying capacity for these two islands was determined from survey data (Lopez 2001). The overall deer density estimate (0.131 deer/ha) from No Name Key was used as a baseline estimate because survey data and herd health indices suggest the island's deer population is at or near carrying capacity (Lopez 2001, Nettles et al. 2001). It was assumed that development would render the entire parcel unavailable to deer due to loss of habitat (house footprint) and possible fencing of the remaining area. Using the Monroe County tax roll database, the average area per house was determined to be 0.367 ha, and this was used in conjunction with the deer density estimate to determine the expected change in K due to development (Table 40.2).

Population "Harvest"

Secondary impacts due to development (e.g., traffic increases, entanglement in fences; Lopez 2001) were modeled using the harvest function in RAMAS Metapop. RAMAS Metapop allows users to determine the spatial extent and quantity (number or proportion of individuals) of animals to be "harvested" in a metapopulation (Akçakaya and Root 2002). Human-related mortality can be viewed as a form of "harvest" in modeling the population dynamics of the Key deer. For example, it is assumed the construction of 100 houses would increase traffic, thereby increasing the probability of deer dying. Data suggest that human-related mortality, particularly highway mortality, is density-dependent (Lopez 2001); therefore, an increase in development could be modeled as a form of harvest affecting a proportion of individuals on Big Pine Key. Spatial analyses suggested the majority of development would occur on Big Pine Key, thus, impacts were limited in model simulations to this island (Lopez 2001). Development in the Key deer models took two forms: transportation improvements and development of new homes or businesses. A review of each of these forms of development follows.

Table 40.2 Key deer population impacts on Big Pine Key (BPK) due to urban development by management scenario

Scenario	No. of Houses	Habitat Loss (K Decrease)	Harvest (Houses) (%)	Total Harvest (Houses + Road) (%)	New K (BPK)
S1	0	0	0.000	0.000	332
S2	0	0	0.000	0.409	332
S3	300	6	0.918	1.326	326
S4	600	12	1.836	2.244	320
S5	900	18	2.754	3.162	314

Note: Carrying capacity changed for each management scenario depending on expected habitat loss. Changes applied to BPK only. Management scenarios: S1 (Scenario 1) = no change, S2 (Scenario 2) = road improvement, S3 (Scenario 3) = road improvement + 300 houses, S4 (Scenario 4) = road improvement + 600 houses, and S5 (Scenario 5) = road improvement + 900 houses.

1. *Road improvements.* US 1 traffic problems have long plagued Big Pine and No Name key residents (Calvo 1996, Lopez 2001). A cross-island road allowing residents to avoid access to US 1 has been suggested as a corrective measure (Calvo 1996). Lyttons Way on Big Pine Key is an existing, unimproved road that parallels US 1; however, residents want the road to be improved or paved, which is expected to increase deer mortality. The percentage of highway mortality on Watson Boulevard, an improved road that is parallel to Lyttons Way, was determined and used to estimate expected mortality on Lyttons Way (Lopez 2001). The expected percentage of increase in mortality was adjusted, based on the length of the proposed development. It was assumed that a mortality increase of 0.409% would occur if Lyttons Way were to be paved (Table 40.2).
2. *Development.* It was assumed that additional development would increase deer mortality (Lopez 2001). The county tax roll was queried to determine the number of houses on these two islands. The USFWS mortality database also was searched to determine the average percentage of human-related deer mortality (highway mortality, dogs, entanglement) that occurred in the last 5 years (deer mortality divided by estimated deer population, 1996–2000; Lopez 2001). With this information, the proportion of deer that died due to human-caused mortality (8.35%) was estimated and divided by the average number of houses. The result was an estimate of the percentage of deer mortality per house (0.0031%; see Table 40.2).

Environmental and Demographic Stochasticity

Natural environments are known to fluctuate, causing changes in the population dynamics of species (Burgman et al. 1993). An attempt to model the population dynamics of Key deer included incorporating environmental fluctuations, demographic variability, and catastrophes. RAMAS Metapop allows model users to incorporate different types of stochasticity in a variety of ways. Environmental stochasticity was incorporated by sampling vital rates from random (lognormal) distributions with means taken from a mean stage matrix and standard deviations taken from a "standard deviation matrix" (Akçakaya 1991). Demographic stochasticity was incorporated in model simulations by sampling (1) the number of survivors from a binomial distribution, (2) the number of offspring from a Poisson distribution, and (3) the number of individuals dispersing between populations from a binomial distribution (Akçakaya 1991). With stochasticity incorporated in the Key deer models, repeated simulations in RAMAS Metapop could be used to generate a range of predictions.

Hurricanes have been hypothesized to have a negative impact on Key deer (Seal and Lacy 1990, Folk 1991). In the model, two levels of catastrophes were evaluated: catastrophe 1 (category 3–4 hurricane), and catastrophe 2 (category 4–5 hurricane). It was assumed hurricanes would impact Key deer population abundances regionally (all populations would be impacted). Hurricane probability information was obtained from the National Hurricane Center in Miami, Florida. The model HURISK was used to estimate the probability of the storm center striking within 25 nautical miles of Big Pine and No Name Keys (Neumann 1991).

Model Development and Scenarios

The program RAMAS Metapop was used to develop the Key deer metapopulation model (Akçakaya and Root 2002). Within each population, demographic changes were modeled using a stage-structured matrix model that allowed for annual changes in vital rates and incorporated impacts from regional catastrophes such as hurricanes. Furthermore, impacts due to urban development (i.e., changes in carrying capacity, increases in highway mortality) also were evaluated with RAMAS Metapop. Model results were summarized in terms of risk of decline, and alternative development scenarios were compared (Table 40.2). Only female Key deer were modeled in the PVA.

To illustrate the use of the Key deer model, five management scenarios were evaluated, with each scenario consisting of 10,000 simulations for 100 years (Table 40.2):

Scenario 1 = no change

Scenario 2 = road improvement

Scenario 3 = road improvement + 300 houses

Scenario 4 = road improvement + 600 houses

Scenario 5 = road improvement + 900 houses

where road improvement was defined as the paving of Lyttons Way and development referred to single resident homes. Two measures of evaluating the viability of the Key deer metapopulation were used: (1) risk of metapopulation going extinct in 50 and 100 years, and (2) risk of falling below metapopulation threshold (50 individuals) in 50 and 100 years. To analyze the sensitivity of model results to parameters, the low, medium, and upper estimates of each parameter could be varied while the medium estimates of all other parameters are held constant (Akçakaya 2000). For illustrating the use of RAMAS Metapop in this chapter, only medium estimates for each scenario was simulated (five models total; Tables 40.1–40.2).

Results

Models were developed to serve as a tool in the conservation of Key deer for Big Pine and No Name keys. A summary of model parameters used is outlined (Tables 40.1–40.2). In general, the model predicted a low risk of decline for the Key deer population increasing with more development (Figure 40.2). For example, the risk of the deer metapopulation falling below 50 individuals in 50 years ranged from 3% to 13%, depending on the management scenario. The level of risk for Key deer increased with a longer simulation period (100 years, 7%–43%).

Conservation Planning

Using RAMAS Metapop, demographic, stage-structured models were constructed and used in a PVA for the endangered Florida Key deer. These models provided a framework for the HCP in evaluating impacts from proposed development on Key deer population. In the last several years, conflicts between land owners, residents, and environmental groups have escalated due to land use issues, the building moratorium, and traffic congestion on these islands. It is proposed that the HCP will identify areas important to the conservation of Key deer while offering island residents some relief from building restrictions (Lopez 2001).

Deer Viability

Model results suggest the deer population has a relatively low risk of decline (Figure 40.2) and that a certain degree of development might occur on the islands, depending on the acceptable level of risk for HCP decision-makers. One advantage of the Key deer model and the proposed planning approach is the holistic evaluation of development in the core of the Key deer range rather than the evaluation of development by individual parcels. This approach can provide decision-makers with results that incorporate uncertainty and variability in risk projections for this endangered deer population. Finally, model users can understand the effect of uncertain input and make decisions with full knowledge of these uncertainties (Akçakaya and Sjogren-Gulve 2000).

Problems in conservation planning and wildlife management are increasing with changes in and demands on land shared with threatened or endangered species (McCullough 1996, Akçakaya and Sjogren-Gulve 2000). Use of PVAs provides conservationists and managers with a tool to determine future development in light of risk to threatened and endangered populations. Such was the case in the study of Florida Key deer. Cur-

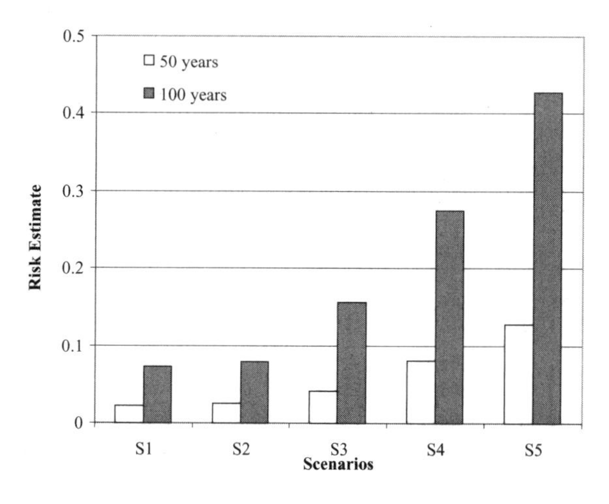

Figure 40.2 Risk estimate (probability of falling below 50 individuals at least once in 50 and 100 years) by management scenario for Big Pine and No Name Keys, 2001.

rently, RAMAS Metapop is being used by decision-makers in conservation planning in the Florida Keys. In addition, model users were provided with an array of risk estimates and other assessment end points to plan the future of this endangered deer herd. Finally, the Key deer model serves as an important management tool for the endangered Key deer and other populations of white-tailed deer. For example, the impact of translocations to overall metapopulation viability can be evaluated a priori using RAMAS Metapop.

References

Akçakaya, H. R. 1991. A method for simulating demographic stochasticity. *Ecological Modelling* 54: 133–136.

Akçakaya, H. R. 2000. Population viability analyses with demographically and spatially structured models. *Ecological Bulletins* 48: 23–38.

Akçakaya, H. R., and Root, W. 2002. *RAMAS Metapop: viability analysis for stage-structured metapopulations (version 4.0)*. Applied Biomathematics, Setauket, N.Y.

Akçakaya, H. R., and Sjogren-Gulve, P. 2000. Population viability analyses in conservation planning: an overview. *Ecological Bulletins* 48: 9–21.

Akçakaya, H. R., Burgman, M., and Ginzburg, L. 1999. *Applied population ecology: principles and computer exercises using RAMAS EcoLab 1.0*. Applied Biomathematics, Setauket, N.Y.

Boyce, M. S. 1992. Population viability analysis. *Annual Reviews Ecology and Systematics* 23: 481–506.

Burgman, M. A., Ferson, S., and Akcakaya, H. R. 1993. *Risk assessment in conservation biology*. Chapman and Hall, London.

Calvo, R. 1996. *US-1/SR-5 Key deer/motorist concept report*. Technical Report. Florida Department of Transportation, Miami.

Caswell, H. 2000. *Matrix population models: construction, analysis, and interpretation*. Sinauer Press, Sunderland, Mass.

Folk, M. L. 1991. Habitat of the Key deer. Ph.D. diss., Southern Illinois University, Carbondale.

Folk, M. L., and Klimstra, W. D. 1991. Reproductive performance of female Key deer. *Journal of Wildlife Management* 55: 386–390.

Hardin, J. W. 1974. Behavior, socio-biology, and reproductive life history of the Florida Key deer, *Odocoileus virginianus clavium*. Ph.D. diss., Southern Illinois University, Carbondale.

Hardin, J. W., Klimstra, W. D., and Silvy, N. J. 1984. Florida Keys. Pages 381–390 in L. K. Halls (ed.), *White-tailed deer: ecology and management*. Stackpole Books, Harrisburg, Pa.

Lopez, R. R. 2001. Population ecology of Florida Key deer. Ph.D. diss., Texas A&M University, College Station.

McCullough, D. R. 1996. Introduction. Pages 1–10 in D. R. McCullough (ed.), *Metapopulations and wildlife conservation*. Island Press, Washington, D.C.

Monroe County Growth Management Division. 1992. *Monroe County year 2010 comprehensive plan*. Monroe County Planning Department, Key West, Fla.

Nettles, V. F, Quist, C. F., Lopez, R. R., Wilmers, T. J., Frank, P., Roberts, W., Chitwood, S., and Davidson. W. R. 2002. Morbidity and mortality factors in Key deer, *Odocoileus virginianus clavium. Journal of Wildlife Diseases* 38: 685–692.

Neumann, C. J. 1991. *The National Hurricane Center risk analysis program (HURISK)*. Technical Memorandum, NWS NHC 38. National Oceanic and Atmospheric Administration (NOAA), Coral Gables, Fla.

Seal, U. S., and Lacy, R. C. 1990. *Florida Key deer population viability assessment*. Captive Breeding Specialist Group, IUCN, Apple Valley, Minn.

White, G. C., and Burnham, K. P. 1999. Program MARK: survival estimation from populations of marked animals. *Bird Study* 46: S120–S138.

White, G. C, and Garrott, R. A. 1990. *Analysis of wildlife radio-tracking data*. Academic Press, San Diego.

41

Turkish Mouflon (*Ovis gmelinii anatolica*) in Central Anatolia

Population Viability under Scenarios of Harvesting for Trophy

ZEYNEP SEZEN
H. REŞİT AKÇAKAYA
C. CAN BİLGİN

Turkish mouflon (*Ovis gmelinii anatolica*) is an endangered subspecies of mountain sheep that is endemic to Turkey. The range of the Turkish mouflon, which initially spanned the western and southern parts of the central Anatolian plain (Turan 1984), is presently confined to a single site in Bozdağ, Konya. The reasons for their decline are not well documented but may be due to the joint effect of predation, competition with domestic livestock, harsh weather conditions, and hunting.

This subspecies was first taken under protection in 1937 when it was listed as a protected species and its hunting was banned. The first action taken for the conservation of Turkish mouflon was the establishment of a 42,000-ha wildlife protection area in Konya, Bozdağ, in 1966 (the Bozdağ Wildlife Protection Area). In order to relieve the population from competition with domestic sheep and predation by wolves, a fence was put up in 1989, enclosing an area of 5000 ha. With this fence, which was later electrified in 1996, the only extant population of *O. g. anatolica* was artificially divided into two subpopulations. As there are no reliable data available on the small (sub)population outside the fence in the Bozdağ Wildlife Protection Area, this study takes into consideration only the fenced population, which is hereafter referred to as the Bozdağ population.

Due to the local extinction of Turkish mouflon in most of its former geographic range, probable sites for reintroduction have been considered by the wildlife authorities (O. Arıhan, pers. comm.). Although no one locality has been earmarked for this purpose, former habitats around Nallıhan (northwestern Ankara) are considered to be most suitable; therefore, Nallıhan was assumed to be the translocation site. In a previous study, the effect of reintroduction on the viability of the subspecies was investigated using the RAMAS Metapop Version 3.0 (Akçakaya 1998). Simulations were carried out to determine the size of the initial population to be reintroduced, as well as the age distribution of the reintroduced individuals. After a translocation, the newly formed population (hereafter the "New population") and the Bozdağ population were considered to be a metapopulation.

Another management option considered from time to time by the wildlife authorities in Turkey regarding the Turkish mouflon is trophy hunting (S. Tarhan, pers. comm.). Although trophy hunting is practiced under state control at a small number of selected sites within Turkey, Turkish mouflon is currently not among the species hunted. In 1998, a report was submitted to the Ministry of Forestry, General Directorate of National Parks, Game and Wildlife (İnal 1998) which proposed the harvesting of 50 to 60 mature rams (aged 5 and above) each year from the Bozdağ Protection Area.

In this study, RAMAS Metapop Version 4.0 was used to investigate the risk experienced by the Bozdağ population and the potential metapopulation when the males are harvested for trophy, given that a reintroduction was carried out. Each harvesting scenario was run by setting levels of trophy hunting of older males (ages 5 and above) at 0, 5, 15, and 50 males per year.

Methods

Metapopulation Model

The model is an age-structured, stochastic metapopulation model. Published and unpublished data on the demography and population trends of this species and closely related species were used in parameterizing the model. For the reintroduction scenarios, only the females were modeled because females are the limiting sex in polygynous species such as the Turkish mouflon. To investigate the effects of trophy hunting on the viability of the Bozdağ population and the potential Turkish mouflon metapopulation, both the females and males were included in the model as it is only the males (older rams) that would selectively be harvested for trophy. All simulations were carried out for 100 years and 10,000 replications.

Stage Matrix

Due to insufficient data on life table parameters of the Bozdağ population, data from closely related species were used to estimate the stage matrix elements. The survival rates were determined based on information and data from Cransac et al. (1997) and Festa-Bianchet (1988, 1989). The survival probabilities given in Table 41.1 were assumed to be the same for both males and females, again based on the information given in these sources.

There are six female and six male age classes (Table 41.2): 0, 1, 2, 3, 4, and 5+. The elements of the stage matrix are the vital rates of these 12 classes of individuals. The last age class (5+) is a composite age class that includes all individuals aged 5 and above.

Turkish mouflon exhibit a polygynous mating system, and an equal sex ratio is observed at birth (Kaya and Aksoylar 1992). It was assumed that each male could mate with up to 20 females in each breeding season. The model assumes a postreproductive census, therefore fecundities were calculated as the product of female survival rate, number of lambs per breeding female in that age group, sex ratio at birth, and percent-

Table 41.1 Survival probabilities of Turkish Mouflon, estimated from the literature

	Age Class					
	0	1	2	3	4	5+
Mouflon[a]	0.9	0.7	0.8	0.8	0.9	0.9[b]
Bighorn[c]	No data	0.67	0.82	0.94	0.91	0.85
This model	0.9	0.7	0.8	0.9	0.9	0.9

[a]From Cransac et al. (1997).
[b]This value was not given in the source and therefore was determined based on information presented in Cransac et al. (1997).
[c]From Festa-Bianchet (1989).

age of breeding females in that age group. These values given in Table 41.3 were determined with data and information from Arıhan (2000), Kaya and Aksoylar (1992), and Cransac et al. (1997).

Initial Abundances

The initial abundance of the Bozdağ population was calculated from two censuses carried out in August 1998 and November 1999 (Arıhan 2000). These two censuses were postreproductive transect counts in which more than 70% of the total area was surveyed. The number of individuals observed was then extrapolated to 100% of the total area, assuming an equal density in the nonsurveyed parts since the habitat within the 5000-ha area was equally available to the mouflon.

Table 41.2 Stage matrix for the Turkish Mouflon

	Females						Males					
	0	1	2	3	4	5+	0	1	2	3	4	5+
Females												
0	0.00	0.00	**0.31**	**0.61**	**0.65**	**0.65**	0.00	0.00	0.00	0.00	0.00	0.00
1	**0.90**	0.00	0.00	0.00	0.00	0.00	0.00	0.00	0.00	0.00	0.00	0.00
2	0.00	**0.70**	0.00	0.00	0.00	0.00	0.00	0.00	0.00	0.00	0.00	0.00
3	0.00	0.00	**0.80**	0.00	0.00	0.00	0.00	0.00	0.00	0.00	0.00	0.00
4	0.00	0.00	0.00	**0.90**	0.00	0.00	0.00	0.00	0.00	0.00	0.00	0.00
5+	0.00	0.00	0.00	0.00	**0.90**	**0.90**	0.00	0.00	0.00	0.00	0.00	0.00
Males												
0	0.00	0.00	**0.31**	**0.61**	**0.65**	**0.65**	0.00	0.00	0.00	0.00	0.00	0.00
1	0.00	0.00	0.00	0.00	0.00	0.00	**0.90**	0.00	0.00	0.00	0.00	0.00
2	0.00	0.00	0.00	0.00	0.00	0.00	0.00	**0.70**	0.00	0.00	0.00	0.00
3	0.00	0.00	0.00	0.00	0.00	0.00	0.00	0.00	**0.80**	0.00	0.00	0.00
4	0.00	0.00	0.00	0.00	0.00	0.00	0.00	0.00	0.00	**0.90**	0.00	0.00
5+	0.00	0.00	0.00	0.00	0.00	0.00	0.00	0.00	0.00	0.00	**0.90**	**0.90**

Note: The stage matrix gives the transition probabilities from one age class to the next for both males and females in separate stages. The elements of the first row of the first and third quadrants represent fecundities; the elements in the other rows represent survival rates.

Table 41.3 Parameter values used in the calculation of fecundities

	Age Class					
	0	1	2	3	4	5+
Survival rate	0.90	0.70	0.80	0.90	0.90	0.90
No. of lambs per breeding female[a]	0.00	0.00	1.10	1.70	1.80	1.80
Proportion of female lambs (%) [b]	0.00	0.00	0.50	0.50	0.50	0.50
Proportion of breeding females (%) [c]	0.00	0.00	0.70	0.80	0.80	0.80
Fecundity	0.00	0.00	0.31	0.61	0.65	0.65

[a]From Kaya and Aksoylar (1992) and Arıhan (2000).
[b]From Kaya and Aksoylar (1992).
[c]From Cransac et al. (1997) and Arıhan (2000).

The Bozdağ population size in the year 1999 was estimated as 560 individuals. The figure of 576 individuals was the abundance of the population in the year 2000, when extrapolated from the estimated figure of 560. The population size in 1999 was extrapolated to determine the abundance in the year 2000 when the reintroduction was to take place. The initial abundance of the Bozdağ population was determined as 576 individuals minus the number of individuals to be translocated. It is assumed that during any translocation a loss of 15% to 20% of transported individuals is inevitable. To ensure a successful translocation of 20 individuals to the New site, it was assumed that 24 individuals were taken from the Bozdağ population. A stable age distribution was assumed for the two populations, Bozdağ and New, whose initial abundances were set as 552 and 20, respectively.

Environmental Variation

There are no data available on the variation of vital rates of the Turkish mouflon population in Bozdağ. Data from Cransac et al. (1997) (Table 41.4) was used to determine the level of environmental variation (as standard deviations of the stage matrix elements given in Table 41.5).

In the absence of information on fecundity variability, a standard deviation of 0.1 was assumed for each age-specific fecundity value. This corresponds to a CV of about 32% for fecundity of 2-year-olds and about 15% for other reproductive age classes, consistent with the assumption that there is higher variation in the fecundity of younger age classes in ungulates (Gaillard et al. 2000). The average CV for all age classes was

Table 41.4 Environmental variation of survival rates

	Age Class				
	0	1	2	3	4
Average survival	0.95	0.74	0.79	0.83	0.93
SD	0.11	0.25	0.27	0.14	0.09
CV	0.12	0.34	0.35	0.17	0.10

From Cransac et al. (1997: 1871, Figure 3.

Table 41.5 Standard deviations matrix for the study populations

	Females						Males					
	0	1	2	3	4	5+	0	1	2	3	4	5+
Females												
0	0.00	0.00	**0.10**	**0.10**	**0.10**	**0.10**	0.00	0.00	0.00	0.00	0.00	0.00
1	**0.109**	0.00	0.00	0.00	0.00	0.00	0.00	0.00	0.00	0.00	0.00	0.00
2	0.00	**0.239**	0.00	0.00	0.00	0.00	0.00	0.00	0.00	0.00	0.00	0.00
3	0.00	0.00	**0.276**	0.00	0.00	0.00	0.00	0.00	0.00	0.00	0.00	0.00
4	0.00	0.00	0.00	**0.155**	0.00	0.00	0.00	0.00	0.00	0.00	0.00	0.00
5+	0.00	0.00	0.00	0.00	**0.09**	**0.09**	0.00	0.00	0.00	0.00	0.00	0.00
Males												
0	0.00	0.00	**0.10**	**0.10**	**0.10**	**0.10**	0.00	0.00	0.00	0.00	0.00	0.00
1	0.00	0.00	0.00	0.00	0.00	0.00	**0.109**	0.00	0.00	0.00	0.00	0.00
2	0.00	0.00	0.00	0.00	0.00	0.00	0.00	**0.239**	0.00	0.00	0.00	0.00
3	0.00	0.00	0.00	0.00	0.00	0.00	0.00	0.00	**0.276**	0.00	0.00	0.00
4	0.00	0.00	0.00	0.00	0.00	0.00	0.00	0.00	0.00	**0.155**	0.00	0.00
5+	0.00	0.00	0.00	0.00	0.00	0.00	0.00	0.00	0.00	0.00	**0.09**	**0.09**

about 20%, similar to the average CV for survival rates. A lognormal distribution was assumed for environmental variation.

To determine the correlation of environmental variation, data obtained for the years between 1985 and 1999, from Konya, Aksaray, Ankara, and Sivrihisar meteorological stations were analyzed. Correlation coefficients between Bozdağ and the New populations were computed from these meteorological data and were 0.9 for air temperature and 0.6 to 0.4 for precipitation and snow parameters, respectively. The correlation coefficient was set at 0.5.

Catastrophes

The two types of catastrophes included in the model were heavy snowfall that may lead to increased mortality due to predation, starvation, and freezing, and pneumonia which is one type of disease carried by domestic sheep that may affect wild sheep populations. It has been noted that wild sheep are especially vulnerable to diseases carried by domestic sheep (Toumazos and Hadjisterkotis 1997), and this disease is also a major problem in the management of bighorns (Jorgenson et al. 1993a). Both catastrophes—heavy snowfall and disease epidemics—would affect the abundances of the population.

The frequency of occurrence of heavy snowfall was determined from meteorological data (years 1985–1999 for Konya, Aksaray, Ankara, and Sivrihisar meteorological stations). "Heavy" snowfall was assumed to occur in years when either the total number of snow-covered days or the maximum number of consecutive days with full snow cover was higher than 1.8 times the average. For both the Bozdağ area and the probable location of reintroduction (for the New population), it was found to occur once in every 6 or 7 years, which gave a probability of 0.15. The stage-specific multipliers of heavy snowfall were based on the survivorship of individuals in the presence of snowfall after Bousses et al. (1994), with the exception of stage-specific multiplier for age 0, which was determined from Nahlik (2000). These multipliers are presented in Table 41.6. This

Table 41.6 Stage-specific multipliers for the effect of extreme snowfall

	Age Class					
	0	1	2	3	4	5+
Multiplier	0.67[a]	0.95	0.85	0.85	0.85	0.80
Survival for Bozdağ	0.60	0.66	0.68	0.76	0.76	0.72
Survival for "New"	0.48	0.53	0.54	0.61	0.61	0.58

Note: Survival refers to the overall annual survival rate for each age class in years with extreme snowfall (stage-specific multiplier times the survival rate for that age class).

[a]Determined from Nahlik (2000).

catastrophe is expected to affect the New population more severely, because heavy snowfall will result in increased predation only in the New population, whereas starvation and freezing will affect both populations. The local multipliers are 0.8 and 1.0 for the New and Bozdağ populations, respectively. The average annual rates of survival in a year with heavy snowfall are 0.6782 and 0.5438, for the Bozdağ and the New populations, respectively.

The effect of heavy snowfall was assumed to be regional in extent as the environmental correlation between Bozdağ and the probable translocation site was moderately high ($r = 0.5$) with respect to rainfall and snow cover, which are probably the most important factors.

The bighorn population at Sheep River in Alberta, experienced pneumonia epizootics twice in a 16-year study (Jorgenson et al. 1997). Similarly, the probability of pneumonia occurring at the New site was set as 0.125 (i.e., once every 8 years) while for the Bozdağ population a value of 0.05 was used. The local probability of the New population for this catastrophe is higher compared to that of the Bozdağ population, since the latter is isolated where the chances of coming into contact with domestic sheep are very low, while the New population may share grazing land with domestic livestock. The stage-specific multipliers of the disease pneumonia were based on research by Jorgenson et al. (1997) and Cransac et al. (1997). In a population of bighorn sheep, an outbreak of pneumonia caused an additional mortality of 20.5% in prime-age (2- to 7-year-olds) males and 18.1% in prime-age females during a single year (Jorgenson et al. 1997). This epizootic killed approximately 40% of the sheep in the Sheep River population, bringing overall survival after the epizootic to a very low 0.6 (Jorgenson et al. 1993b). Cransac et al. (1997) found a local multiplier of 0.77 for male mouflon under a keratoconjuctivitis epidemic. Therefore, a local multiplier of 0.8, which is also supported by Nahlik et al. (2000), was selected for the simulations. Combined with stage-specific multipliers (to model different vulnerabilities of age classes to epidemics), the annual overall survival rate in years with epidemics ranged from 0.532 to 0.72 for different age classes (see Table 41.7). (The overall annual survival rate during the epidemic was 0.6252).

Because of low dispersal ability of the species, and the distance between the current and any newly founded populations, no dispersal is assumed. A pneumonia outbreak therefore, would be local in extent.

Table 41.7 Stage-specific multipliers for the effect of pneumonia

	Age Class					
	0	1	2	3	4	5+
Multiplier	0.80	0.95	1.00	1.00	1.00	0.95
Survival	0.576	0.532	0.640	0.720	0.720	0.684

Note: These survival rates are valid for both populations as the local multiplier was assumed to be the same for both populations.

Density Dependence

The type of density dependence in wild sheep is scramble and it is assumed to affect the fecundities of the population. Reproduction in ruminants decreases as the population density increases (Geist 1971). This view is supported by studies on bighorn sheep by Jorgenson et al. (1997) and Festa-Bianchet et al. (1998). In a bighorn population study on Ram Mountain, it was clearly suggested that survival of adults of either sex was little affected by changes in population density but that population limitation should mostly depend on changes in either the production or survival of lambs (Jorgenson et al. 1997). Festa-Bianchet et al. (1998) state that at high density, heavy ewes had higher reproductive success than light ewes, and the reproductive cost and somatic costs of reproduction increased.

Maximum Growth Rate (R_{max})

The overall growth rate of the Turkish population would be 1.09 if the final population size is taken as 620 (1998) or as 733 (1999) given an initial population size of 35 (1966) (Arıhan 2000). The average annual growth rate before fencing (1966 to 1989) was 1.125 (from 35 to 520 in 23 years). For a single pair of years (1998–1999) for which census data exist, a maximum value of 1.20 was computed for the fenced Bozdağ population.

The R_{max} value was fit by running a 23-year retrospective simulation and selecting the value of R_{max} that matched the observed population growth. Both catastrophes were kept in the model used in the retrospective simulation, because the extreme snowfall catastrophe occurs relatively frequently and there is no evidence that an epidemic did not occur during the last 23 years. During the 23-year period (for which an R_{max} of 1.25 was fitted) the Bozdağ population was not fenced. Because of the recently built fence, a slightly higher figure of 1.30 was used for this population. Because the habitat conditions of the reintroduced population will be similar to the situation in Bozdağ before the electric fence was established, an R_{max} value of 1.25 is assumed appropriate for the New population.

Carrying Capacity

Current figures for Bozdağ are 673 individuals per 50 sq km (5000 ha), giving a density of 13.46 per sq km (July 1999; Arıhan, 2000). However, a slowdown in population growth from 1989 to 1998 suggests that carrying capacity may soon be reached. Therefore, the carrying capacity is assumed to be 1,000 individuals. For the New population, a carrying capacity of 600 individuals is selected, because the probable release site is about 40% smaller than the Bozdağ Protection Area.

Sensitivity Analyses

To account for uncertainty about the demography of the species, a sensitivity analysis was carried out. For some parameters this was done by running simulations for best-case and worst-case scenarios, but for the stage matrix and standard deviations matrix parameters, the Sensitivity Analysis program of RAMAS GIS was used.

The most sensitive parameter in the model was determined to be R_{max} for both the Bozdağ and New populations. One of the most uncertain parameters was the carrying capacity of the New population; therefore, this parameter was tested with a change of ±50% in the sensitivity analyses. Despite this, the model did not prove to be sensitive to this parameter.

Results and Discussion

To evaluate the possible consequences of opening the Bozdağ population to trophy hunting once a metapopulation has been established, the effect of a yearly harvest of 0, 5, 15, and 50 older males in the Bozdağ population was investigated. The results of these harvesting scenarios are summarized in Table 41.8. The scenarios are compared in terms of the probability of at least 1 year with 10 or less older males, as well as the risk of decline of the Bozdağ population and the metapopulation by 80% at least once during the next 100 years.

In the scenarios for harvesting of 0 and 5 older males per year, the risk that the abundance of older males will fall to 0 at least once during the next 100 years is 4% and 24%, respectively. In the extreme case of 50 older males being hunted, there exist no older males in the population for 81.5% of the time (Table 41.8). During years in which there are no older males present, the program does not harvest any males.

The longest consecutive years with no older males in the population averaged over the 10,000 iterations is 0.4, 1.5, 6.3, and 35 for yearly harvesting scenarios of 0, 5, 15, and 50 older males, respectively (Table 41.8). This may be an underestimate, however, because the social interactions among mouflon were not incorporated into the analyses and it was assumed that younger males that have just reached sexual maturity (age 3) are able to actively participate in rut. Not incorporating social interactions may also lead to an underestimate for the risk of extinction in the metapopulation that appears to be unaffected by harvesting.

The risk of decline of the Bozdağ population and the metapopulation does not appear to be affected by a yearly harvest of up to five males. In contrast, a yearly harvest of 50 males increases the risk of decline of the Bozdağ population by nearly twofold. This may also be an underestimate due to the reasons previously mentioned here.

The major limitation of the model was the lack of long-term demographic data on the Bozdağ population. Several assumptions had to be made through inferences from

Table 41.8 Results of harvesting scenarios

	No. of Older Males[a] Harvested per Year			
	0	5	15	50
Average no. of older males[a] in year 100	167.4	137.1	83.8	5.7
No. of years (out of 100) with at least one older male[a]	99.6	98.1	89.1	18.5
No. of years (out of 100) with no older males[a]	0.4	1.9	10.9	81.5
Longest consecutive no. of years with no older males[a]	0.4	1.5	6.3	35.0
Probability of at least 1 year with no older males[a]	0.04	0.24	0.87	1.00
Probability of at least 1 year with 10 or fewer older males[a]	0.11	0.39	0.93	1.00
Probability that the no. of females at the Bozdağ population will decline by 80% or more within the next 100 years[b]	0.12	0.12	0.17	0.21
Probability that the total no. of females in the metapopulation will decline by 80% or more within the next 100 years[b]	0.05	0.05	0.07	0.08

[a]Older male = male 5 years old or older.
[b]Expressed in terms of total females.

other mouflon or wild sheep populations. However, the two sources from which data were mostly drawn had quite similar values, despite the fact that one was a study on mouflon (Cransac et al. 1997) and the other on bighorns (Jorgenson et al. 1997).

Another limitation was regarding heavy snowfall. This catastrophe would potentially affect all vital rates of the populations (Bousses et al. 1994, Nahlik, 2000), but in the model it is set to affect only "abundances." The fecundities are thought to be lowered after heavy snow, as a result of poor nutrition of pregnant females or delayed maturity of younger females. These are not expected to occur in either Bozdağ or the New site since supplementary feeding is available, especially during harsh winters.

The model did not incorporate inbreeding as there were no available data, but this was not considered a major limitation since several mouflon populations elsewhere were established from as few as a single pair of individuals (Bousses et al. 1992) and yet showed no sign of decrease in vital rates.

According to the results of simulations, only harvesting of no more than 5 older males per year at Bozdağ is tolerable for trophy harvesting. Even then, simulation results suggest that there may not be any older males in the population in 1 out of 4 years. Harvesting of 50 to 60 older males as suggested previously by Inal (1998) is equivalent to extinction at the hands of legal hunters. Even if harvesting of 50 older males is carried out only once, the exact consequences would be unpredictable as it may also lead to the breakdown of social structure within the population.

Natural mortality levels of older males at Bozdağ envisage about six to eight natural deaths annually. The simulation assumes trophy harvests are additional to those dying such a natural death. Therefore, if hunted males include any individuals that will have died that year anyway, the associated risks calculated are somewhat conservative.

The results suggest a controlled harvesting regime of only a few older males per year once the legislature allows trophy hunting of mouflons. However, the population should be continuously monitored and harvesting plans reviewed if necessary.

References

Akçakaya, H. R. 1998. *RAMAS Metapop: viability analysis for stage-structured metapopulations.* Version 3.0. Applied Biomathematics, Setauket, N.Y.

Arıhan, O. 2000. Population biology, spatial distribution and grouping patterns of the Anatolian mouflon *Ovis gmelinii anatolica* Valenciennes 1856. M.S. thesis, Middle East Technical University, Ankara.

Bousses, P., Barbanson, B., and Chapuis, J. L. 1992. The Corsican mouflon (*Ovis ammon musimon*) on Kerguelen archipelago: structure and dynamics of the population. Pages 317–20 in F. Spitz, G. Joneau, G. Gonzales, and S. Auglanier (eds.), *Symposium Ongulés/Ungulates 91', Toulouse.* Société Française d'Etude et Protection de Mammifères, Institut de la Recherche sur les Grands Mammifères, Imprimerie des Escartons, Briançon.

Bousses, P., Reale, D., and Chapuis J. L. 1994. Mortalite hivernale massive dans la population de mouflons de Corse (*Ovis musimon*) de l'archipel subantarctique de Kerguelen. *Mammalia* 58(2): 211–223.

Cransac, N., Hewison, A. J. M., Gaillard, J. M., Cugnasse, J. M., and Maublanc, M. L. 1997. Patterns of mouflon (*Ovis gmelinii*) survival under moderate environmental conditions: effects of sex, age and epizootics. *Canadian Journal of Zoology* 75: 1867–1875.

Festa-Bianchet, M. 1988. Birthdate and survival in bighorn lambs (*Ovis canadensis*). *Journal of Zoology* (London) 214: 653–661.

Festa-Bianchet, M. 1989. Survival of male bighorn sheep in Southwestern Alberta. *Journal of Wildlife Management* 53(1): 259–263.

Festa-Bianchet, M., Gaillard, J. M., and Jorgenson, J. T. 1998. Mass and density-dependent reproductive success and reproductive costs in a capital breeder. *American Naturalist* 152: 367–379.

Gaillard, J. M., Festa-Bianchet, M., Yoccoz, N. G., Loison, A., and Toigo, C. 2000. Temporal variation in fitness components and population dynamics of large herbivores. *Annual Review of Ecology and Systematics* 31: 367–393.

Geist, V. 1971. *Mountain sheep: a study in behavior and evolution.* University of Chicago Press, Chicago.

Inal, S. 1998. *Anadolu Yaban Koyununun Geleceği.* Report to the Ministry of Forestry, General Directorate of National Parks, Game and Wildlife, Ankara, Turkey.

Jorgenson, J. T., Festa-Bianchet, M., and Wishart, W. D. 1993a. Harvesting bighorn ewes: consequences for population size and trophy ram production. *Journal of Wildlife Management* 57(3): 429–435.

Jorgenson, J. T., Festa-Bianchet, M., Lucherini, M., and Wishart, W. D., 1993b. Effects of body size, population density, and maternal characteristics on age at first reproduction in bighorn ewes. *Canadian Journal of Zoology* 71: 2509–2517.

Jorgenson, J. T., Festa-Bianchet, M., Gaillard, J-M., and Wishart, W. D. 1997. Effects of age, sex, disease, and density on survival of bighorn sheep. *Ecology* 78(4): 1019–1032.

Kaya, M. A., and Aksoylar, M.Y. 1992. Bozdağ (Konya)'da yasayan Anadolu yaban koyunu, *Ovis orientalis anatolica* Valeciennes 1856'nin davranislari. *Doga–Turkish Journal of Zoology* 16: 229–241.

Nahlik, A. 2000. Dynamics of a mouflon population and its determinant factors. Abstracts of the International Symposium on Mouflon, Sopron, Hungary. Not available.

Toumazos, P., and Hadjisterkotis, E. 1997. Diseases of *Cyprus mouflon* as determined by gross and histopathalogical methods. Pages 150–161 in E. Hadjisteskotis (ed.), The Mediterranean mouflon: management, genetics, and conservation. *Proceedings of the Second International Symposium on Mediterranean Mouflon.* N.p., Cyprus.

Turan, N. 1984. *Türkiye'nin Av ve Yaban Hayvanları, Memeliler.* Ongun Kardesler Matbaacilik Sanayii, Ankara.

42

Sindh Ibex (*Capra aegagrus blythi*) in Kirthar National Park, Pakistan

Sensitivity of a Habitat and Population Model

KUNIKO YAMADA
MAHBOOB ANSARI
RHIDIAN HARRINGTON
DAVID MORGAN
MARK A. BURGMAN

Ibex are found in most high elevation and relatively extensive arid-zone mountain ranges in Pakistan, occurring at altitudes up to 3350 m. The Sindh ibex, *Capra aegagrus blythi*, is restricted to parks, reserves, and remote areas near Karachi in southern Pakistan. It is a large game animal and the most common native ungulate in Kirthar National Park (KNP) (Edge and Olson-Edge 1990) and has a high public profile not only among the locals but also among overseas trophy hunters. The curving horn of adult males may grow to over 102 cm in length (Roberts 1997).

The ibex population in KNP is largely or completely isolated from other populations of the subspecies in Pakistan. The animals are closely associated with the mountain ranges, and they venture to lowland terrain only rarely, usually when they are moving between ranges. Ibex form loosely connected groups that focus around springs and other freestanding water sources within the mountain ranges. Their main requirements seem to be precipitous crags, which are safe from direct disturbance because the terrain is inaccessible to domestic goats and shepherds. Ibex are able to climb or descend vertical rock cliffs, especially when they are threatened (Roberts 1997). Ibex have experienced pressures of hunting, human encroachment on habitat, and overgrazing of habitat by domestic livestock (Edge and Olson-Edge 1990), and the total population of ibex in KNP decreased to as few as 200 before legal protection was introduced in 1967 (Roberts 1997). Schaller (in Edge and Olson-Edge 1990) estimated the density of animals within patches of habitat in 1972 and 1973 to be between 3.3 and 4.1/km^2. Protection has resulted in significant increases in the population since 1971 (Mirza and Asghar 1980).

KNP (Figure 42.1) is home to a large and growing human population, and there is speculation that oil and gas reserves may also exist within the park boundaries, providing further stimulus for local development. It is thought that more than 70,000 people may live permanently within the boundaries of KNP, and legal hunting had been practiced for over 50 years within the adjacent Kirthar Game Reserve. In addition to legal hunting, poaching incidents occur occasionally within KNP. It may be relatively difficult to enforce game laws in more remote areas in the northern parts of KNP due to access difficulties and lack of human resources. Park managers are faced with a range of potential pressures and alternative management options. The purpose of the work reported here is to develop a habitat and population model for Sindh ibex to understand their habitat requirement, to explore the importance of assumptions, to guide further research, and to help better decision-making on the conservation management of this species.

Habitat Model

A helicopter-based aerial census of ibex populations was conducted in November 2000 (Morgan and Harrington 2001). Flight routes of the helicopter census and the mountains in KNP are visible on a hillshade map of the regional terrain, in which a hypothetical illumination of a surface is shown in gray scale (Figure 42.1). KNP was stratified into regions that were likely to support ibex populations. Because the boundaries between mountain ranges and lowlands were clearly defined in the digital terrain models, simple expert rules (M. Ansari and D. Morgan, unpublished data) were used to delineate ibex habitat by highlighting areas with slopes greater than 10 degrees. In the western half of KNP, patches were delineated as areas with elevation greater than 350 m. In the eastern half, thresholds of 300 m were used, and 250-m-high areas were used to outline habitat for the southernmost ranges within the game reserve.

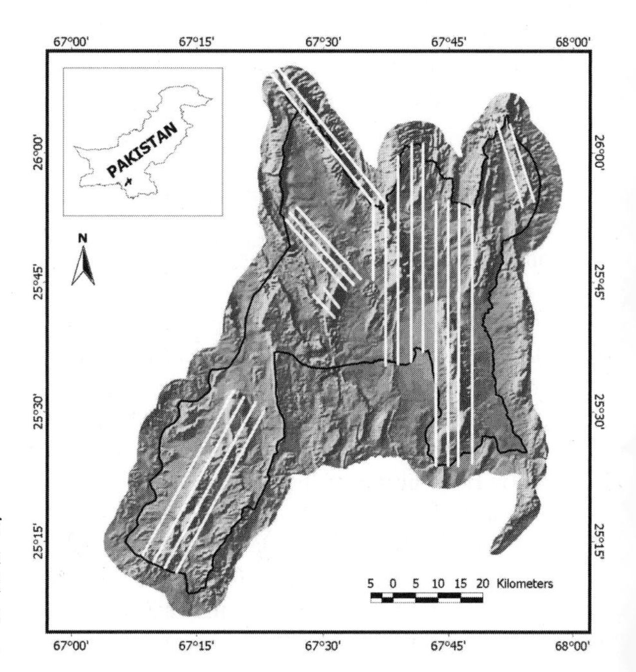

Figure 42.1 Location of Khirthar National Park (*boundary in black lines*) and helicopter census pathways (*in white lines*).

The habitat map resulting from the classification rules was then used to estimate the area of each patch of habitat. Information on the distribution of human populations and associated agricultural activities, aerial population censuses of the ibex, and our expert advice were used to assess the carrying capacity of each habitat patch (animals per km^2) and to set initial conditions for the model. Parameter values in the model were chosen so that the stable age distribution expected from the matrix model was similar to that observed by Morgan and Harrington (2001). Therefore, initial population distributions were set with the assumption that the stage structure of each population was given by the distribution expected under equilibrium conditions. Interpreted habitat polygons for ibex populations in KNP based on expert rules for elevation and slope are shown in Figure 42.2.

Population Model

The dynamics for individual populations were modeled by interpreting published accounts of population dynamics (Edge and Olson-Edge 1987, 1990; Edge et al. 1988; Roberts 1997) and from field observations conducted in 2000. The model uses stages reflecting the important biological stages of the species' life history and an annual time step reflecting the monsoon-driven dynamics of the landscape (see Stage Matrix dialog under the Model menu of the Metapopulation Model subprogram; in the stage names, letters Y, M, and F represent "yearling," "male," and "female," respectively, and numbers represent ages). The first row of the stage matrix represents the average number of juveniles produced per female of each stage per year that are alive to be counted at the next census. Females start giving birth during their third year and may produce as many as two offspring each in good (wet) years.

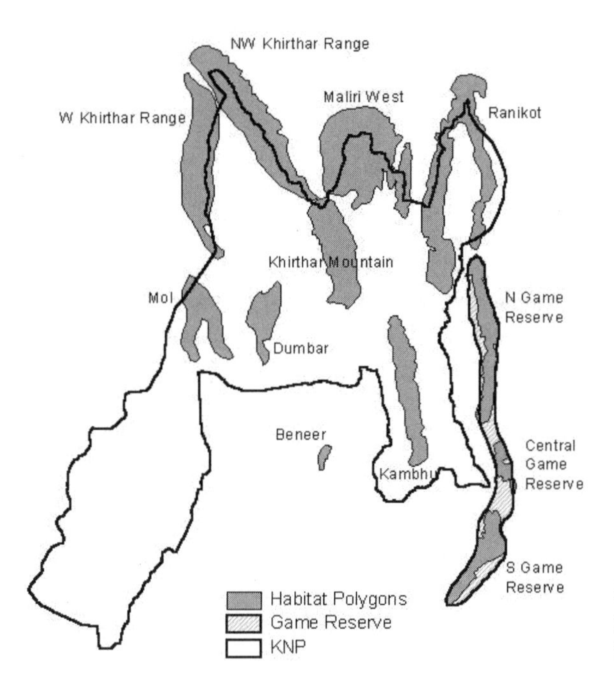

Figure 42.2 Interpreted polygons of Sindh ibex habitat in Kirthar National Park.

It is likely that juveniles of both sexes would experience almost the same kinds of environmental stress as one another when under the care of their mother, but the survivorships of older groups would be more independent. Therefore, juveniles were not differentiated by sex in the model to make the survivorships of male and female juveniles perfectly correlated. The sum of the survival rates in the first column represents the overall survival of juveniles. The model assumes that the survival of females to the yearling stage (0.4) is better than that of males (0.3) (Edge and Olson-Edge 1987, 1990). The transition rates assume that a census is taken just before the breeding season, in which juveniles are not quite 1 year old. Consequently, the fecundities include maternity rates and the number of young born per female, discounted by mortality between birth and the time of the first census (see Burgman et al. 1993). The model included demographic stochasticity (see Chapter 1 in this volume).

Female ibex may live to be around 15 years of age, and males do not live as long as females, with a maximum age around 10 years (Schaller and Laurie 1973). For this reason, ibex populations have a female bias in older age classes, resulting in a sex ratio of 40:60 (male:female) in some larger herds. Several publications—including Schaller and Laurie (1973), Mirza and Asghar (1980), Edge and Olson-Edge (1987, 1990), and Roberts (1997)—provided estimates of the age and sex structure of Sindh ibex populations. Most estimates report fewer yearling males than yearling females. In this model, the stage matrix elements were the median values for this species found in the literature, and the stable age distribution of the population was estimated using these values. This population structure, the longevities of males and females, and the proportions of relatively old animals generated by the model were compared to recent field observations.

Females and males were considered separately in the model due to the following reasons. Ibex populations have a female-biased sex ratio, resulting from higher fecundity and survivorship rates for females. There are some suggestions that males and females may have different propensities to disperse. In addition, males were included because older males are the targets of legal harvesting activities, and it was important to evaluate the consequences of harvesting activities for the probability of population decline.

Next, the structured population model was used to represent the dynamics of populations within spatially explicit patches given by the habitat model. The dispersal-distance function was used to specify the dispersal rates between two populations, and distances were measured as the nearest distance between patch edges. Males may be more likely to disperse than are females of the same age. Migration during the rutting season has been observed, with some moving from 8 to 20 km between mountain ranges (G. S. Jamali, pers. comm.). Our experience is that herbivores in arid environments explore when environmental conditions are favorable but tend to stay in their accustomed range during poorer conditions to remain close to the places where they can find food and water. Thus, environmental variation in rainfall is likely to affect dispersal. The presence of humans and the size and distance to adjacent patches are also likely to play parts in determining dispersal rates between patches.

The dispersal rates were computed and adjusted to include our expert opinion that the total dispersal rate from closely connected patches did not exceed 3% in any year, and was less for distant patches (see Dispersal Matrix dialog). The probability of dispersal was reduced as a linear function of population size relative to the carrying capacity. It was assumed that dispersal rates should reflect the fact that populations are more inclined to explore habitat when the population size is close to the carrying capacity and with the tendency of herbivores described here. Males were given a slightly higher relative dispersal rate than females (0.95 versus 0.9), and males aged 3 to 4 were given a slightly higher relative rate than other males (1 versus 0.95).

Carrying capacity was determined by area and habitat suitability, the latter estimated from our expert evaluation of the ruggedness of the terrain, and the amount of competition from agriculture and domestic animals. To evaluate some of these aspects, relevant parameters in the model such as the carrying capacity and initial population sizes were adjusted systematically, and the effects of these adjustments on the results were examined. Because the animals live in loose social groups and are limited by water and food, a scramble model for competition was used (Burgman et al. 1993).

The growth rate in the absence of competition was set at 1.3. In good years, every female may have twins and a substantial proportion of them may survive, resulting in a relatively high potential growth rate (Edge and Olson-Edge 1987). The values of vital rates and variation in vital rates were adjusted proportionally within reasonable biological bounds to give a maximum growth rate of 1.3. RAMAS itself adjusted these values while running the models, and the transition probabilities and fecundities result in a finite rate of increase (λ) of 1.17. Density dependence was assumed to affect only fecundities, and these were reduced as a function of population size using the Ricker function (Akçakaya 2001).

There were no direct measurements or observations of standard deviations in fecundities or survivorships, and there was no evidence for the strength of correlations between variables. Other studies of ungulates have reported modest levels of variation in the survivorships of ungulates, with larger variation in fecundities. The population dynamics of most arid-zone species are tightly tied to rainfall. Three times in the last 13 years, most ibex in KNP were observed to have maternity rates close to the maximum of two offspring per female after years of relatively high rainfall. Reproduction is negligible in the presence of severe drought, and there have been four such events in the last 13 years. These observations suggest that ibex reproduction is likely to closely track random variation in rainfall. Environmental variation was sampled from lognormal distribution in all cases, which was selected because of the strong association between rainfall and vital rates, and because rainfall is strongly right-skewed (Figure 42.3). The rainfall patterns at Hyderabad are typical of those in KNP.

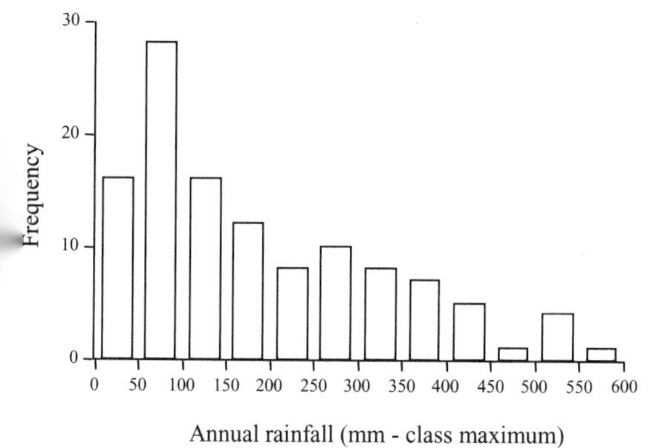

Figure 42.3 Distribution of total annual rainfall at Hyderabad (Pakistan Bureau of Meteorology, Rainfall Records, 2000).

The rainfall in the park has a coefficient of variation of about 70%. More importantly, in 16 of the last 116 years during which rain was recorded at Hyderabad, annual precipitation was less than 50 mm, at which point recruitment of juveniles is likely to be very low. These represent years in which the monsoon fails. A coefficient of variation of 10% was used for survivorships and 20% for fecundities. This provides an outcome such that once in about 20 years, the population approaches full reproductive potential of two successfully recruited offspring per reproductive female. Likewise, once in about 20 years, the population will produce few offspring. The lower level of variation in survivorships reflects the fact that adult survival values are considered to be much less sensitive to lack of water, based on anecdotal field observations of responses to drought. To reflect the occurrence of monsoon failure, a catastrophe was added, with a probability of 0.15, and with the consequence that fecundity is reduced to 0. Modeling rainfall variation and monsoon failure together gives an annual probability of about 16% for rainfall less than 50 mm (resulting in reproductive failure); this figure is close to that in the observed rainfall data.

The model assumed that variations in survivorships and fecundities within populations were perfectly correlated. A year with conditions that leads to high adult survival is also likely to be a year in which reproduction and juvenile survival are high, reflecting rainfall dependence among all stages.

Correlations in environmental conditions from place to place will be reflected in the strength of correlations between survival and fecundity rates in different patches of habitat. Correlations between patches about 1 km apart and about 15 km apart were estimated to be about 0.95 and 0.8, respectively, and the expression $c = \exp(-d^{0.02})$ was used to specify correlations between environmental fluctuations between patches. The function was based on the strength of correlations in average annual rainfall recorded at Karachi Manora, Karachi Airport, and Hyderabad (Table 42.1) since it is approximately 1 km apart between Karachi Manora and Karachi Airport, and about 15 km between Karachi and Hyderabad.

Simulations

A 20-year time horizon was selected to provide a reasonable management planning time horizon. A "base" model that represents the best assumptions for environmental conditions and model parameters was developed first. The objective of sensitivity analysis is to vary uncertain parameters around a range of plausible values to explore the most important parameters in determining the outcome of the model (McCarthy et al. 1995). The uncertainty in the parameter could be ignored if the output is insensitive to the change. If the output is sensitive, bounding estimates of impact could be provided by using the values of the parameter that yield the most extreme estimates of risk (Ferson and Burgman 1995). Sensitivity analysis may also be used to explore alternative model structures (Akçakaya et al. 1999). The model was run with 30 different parameter combinations and assumptions, to explore the sensitivity of the model's results to a range of plausible alternative parameter values and model structures (Table 42.2). Each of the simulations was replicated 500 times for a period of 20 years.

The International Union for the Conservation of Nature (IUCN) (2001) classification system is used in Pakistan to guide government and local stakeholders to set priori-

Table 42.1 Rainfall correlations between three stations around Kirthar National Park

	Karachi Manora	Karachi Airport
Karachi Airport	0.90	1
Hyderabad	0.67	0.74

ties for conservation. When populations were counted in the simulations, juveniles and yearlings were ignored because IUCN protocols recommend that risks be evaluated in terms of the number of reproductively mature adults. Four output statistics were recorded:

1. Number of ibex at the end of 20 years of simulations
2. Number of animals equivalent to an interval extinction risk of 0.5
3. Probability that the population will fall below 4,955 (the population size that gave a 50% chance of interval extinction risk for the base model)
4. Probability that the population will fall below 1,000

Attention was paid to the threshold of 1,000 because the IUCN classifies a species as vulnerable when there are fewer than 1,000 mature individuals. Sensitivity of the four output variables was measured in terms of the proportional change in output given a proportional change in a parameter. To compare the importance of each change, parameters were varied by 20% around the value chosen for the base case where possible. The sensitivity of the model to variations was then evaluated.

Results

The population size of Sindh ibex in KNP was calculated to be between 10,860 and 14,136 (12,500 ± 1,640) (Table 42.3), based on the helicopter census and on interpolation using our expert knowledge of its habitat. For each patch, an assessment was made of the condition of the patch and the extent of human activities that might compete with ibex populations for water, food, or space. Several of the populations are at carrying capacity, most notably those in the center and south of the region that are relatively well policed by wildlife officers. The harvest in game reserves is small and imposes minimal effect on the population (M. Ansari, unpublished data). With these reasons, some patches are estimated to have populations below their potential carrying capacities.

The results of the simulations for the 31 scenarios are provided in Table 42.4. Only simulations with low vital rates, high correlation of environmental conditions, and the combination of six pessimistic factors (i.e., high risks) provided appreciable probabilities that the population will fall below 1,000 at least once in the next 20 years. In addition to the scenario involving a large set of pessimistic assumptions, the simulation with low vital rates resulted in the lowest numbers and the highest risks of population decline. If current management prescriptions and protection measures remain in place, and if the human population does not encroach substantially on remaining habitat, it seems unlikely that the subspecies will be classified as vulnerable within the next 20 years.

Sensitivities, expressed as the proportional change in output resulting from a proportional change in parameters, are shown in Tables 42.5 and 42.6. Magnitude of vital

Table 42.2 Thirty simulations run in RAMAS Metapop

Simulation Label	Changes Made on the Base Model
Low vital rates (VR)	Stage matrix and max. growth rate reduced by 10% (*0.9)[a]
High vital rates	Stage matrix and max. growth rate increased by 10% (*1.1)[a]
Low standard deviation (SD)	SD for stage matrix all reduced by 20% (*0.8)
High standard deviation	SD for stage matrix all increased by +20% (*1.2)
Low carrying capacities (*K*)	*K* reduced by 20% (*0.8)
High carrying capacities	*K* increased by 20% (*1.2)
Low initial abundance (IA)	IA reduced by 20% (*0.8)
High initial abundance	IA increased by 20% (*1.2)
Stochasticity lognormal and uncorrelated	Stochasticity sampled from the lognormal, and fecundities (*F*), survival rates (*S*), and *K* within population all independent as opposed to the base model
Stochasticity normal and correlated	Stochasticity sampled from the normal as opposed to the base model, and perfect correlation between *F*, *S*, and *K* within population
Low coefficient of variation (CV)	CV for dispersal rates decreased by 20% (*0.8)
High coefficient of variation	CV for dispersal rates increased by 20% (*1.2)
Low dispersal (*D*)	*D* rates decreased by 20% (*0.8)
High dispersal	*D* rates increased by 20% (*1.2)
Low correlation	Correlation among population fluctuations all set to 0.5
High correlation	Correlation among population fluctuations all set to 1.0
Low harvest (*H*)	Additional mortality rates in NW and W Kirthar Ranges, Maliri W and Ranikot reduced by 20%
High harvest	Additional mortality rates in NW and W Kirthar Ranges, Maliri W and Ranikot increased by 20%
Harvest_5/6/7_Males	Harvest only on males of year 5–6 and 7+
Harvest 7_Males	Harvest only on males of year 7+
Predation with illegal harvest	Additional harvest of 0.43% was added to all patches on only juvenile to year 2 and year 7+ of both sexes with legal harvest of five males on year 7+ only in three game reserves (GRs)
Predation with no illegal harvest	Harvest set to 0.43% on all patches on only juvenile to year 2 and year 7+ of both sexes with legal harvest in three GRs
No illegal harvest	Illegal harvest ignored with legal harvest in three GRs
Low illegal harvest	Illegal harvest halved with legal harvest in three GRS
High illegal harvest	Illegal harvest doubled with legal harvest in three GRs
Higher harvests	Harvest set to 5% with legal harvest in three GRs tripled
High competition	*K* in flat areas (NW and W Kirthar Ranges, Kirthar Mount., Maliri W and Ranikot) reduced by 25%
High competition and higher harvests	*K* in flat areas (NW and W Kirthar Ranges, Kirthar Mount., Maliri W and Ranikot) reduced by 25%, and harvest was fixed to 5% with legal harvest in three GRs
Six low-risk factors	High VR, high *K*, high IA, low CV, high *D*, and low *H*
Six high-risk factors	Low VR, low *K*, low IA, high CV, low *D*, and high *H*

[a]Vital rates changed by 10% so that growth rate (λ) remains greater than 1.

Table 42.3 Estimated numbers of Sindh ibex in Kirthar National Park

Patches	Area (km^2)	Density (km^{-2})	Population (*D**Area)	Population (by census)	K^b (km^{-2})	*K* (*D**Area)
Mol	72.2	0.0	0	0	2	144
Beneer	5.8	10.0	58	N/A	20	115
Dumbar	47.5	16.9	803	800	20	950
W Kirthar	140	0.6	84	0	2	280
NW Kirthar	175	0.6	105	110	2	350
Kirthar Mount.	102	81.3	8,293	8,310	80	8,160
Maliri West	217	0.6	130	0	2	434
Ranikot	202	0.6	121	0	2	404
Kambhu	85.5	19.2	1,642	1,640	20	1,710
N Game Reserve	74.6	20.0	1,492	N/A	20	1,492
C Game Reserve	17.1	20.0	342	N/A	20	342
S Game Reserve	53.3	20.0	1,066	N/A	20	1,066
Total	1,192	190	14,136	10,860	210	15,448

Note: *D* = density; *K* = carrying capacity. N/A = no simulation run: there were no aerial surveys over the Game Reserve; these estimates are based on the patch area and assume that habitat quality is intermediate.

rates, variation in vital rates, carrying capacity, and initial abundance may be important parameters, depending on the conditions associated with particular management options.

Discussion

The initial size of each population was estimated from the aerial census and our expert knowledge, resulting in a total of approximately 12,500 ibex (±1,640). The aerial census of Ranikot, Maliri West, Northwest Kirthar, and West Kirthar Ranges revealed very few animals, despite apparently suitable habitat. The most likely cause of low populations in these areas is insufficient food and water, exacerbated by competition from domestic animals. The two largest populations, Kirthar Mountain and Kambhu, are centrally located in the park, away from the densest human populations, and are well policed by park rangers. Our experts concluded that though the animals might be most likely limited by avoidance of potential predators and by the availability of food plants (M. Ansari, unpublished data), it may be that patches with current low populations have lower inherent habitat quality or that illegal take has reduced numbers considerably.

Considering that the species prefers to live in remote areas and is easily disturbed by human activities, there is some basis for concern that Sind ibex may become threatened within the next one or two decades. However, this prognosis depends on the maintenance of the status quo within KNP. Any additional impacts that result from increasing human populations, or changes in the kinds of human activities, may precipitate a change in ibex populations outside the expectations for the base case. It is easy to generate scenarios that lead to substantial reductions or even the elimination of the ibex from KNP, simply by assuming a trend in habitat loss that is unrestricted across the park, reflecting unimpeded expansion of agricultural activities. The range of the species will almost certainly contract to the two central large populations if agriculture and other human activities expand in the other patches.

Table 42.4 Results of 31 parameter combinations and scenarios

Simulations	*N* at year 20[a]	*N* when *p* of ier < 0.5[b]	*p* that *N* < 4,955	*p* that *N* < 1,000
Base	7,462	4,955	0.500	< 0.002
Low vital rates (VR) (–10%)	5,270	3,461	0.859	0.012
High VR (+10%)	9,504	6,687	0.114	< 0.002
Low SD (–20%)	7,597	5,260	0.400	< 0.002
High SD (+20%)	7,580	4,550	0.589	< 0.002
Low *K* (–20%)	6,022	4,063	0.762	< 0.002
High *K* (+20%)	9,106	5,657	0.323	< 0.002
Low initial abundance (IA) (–20%)	7,682	4,571	0.614	< 0.002
High IA (+20%)	7,548	5,196	0.442	< 0.002
Stochasticity lognormal uncorrelated	7,628	5,093	0.443	< 0.002
Stochasticity normal correlated	7,748	4,912	0.474	< 0.002
Stochasticity low dispersal CV (–20%)	7,450	5,068	0.473	< 0.002
Stochasticity high dispersal CV (+20%)	7,460	4,952	0.501	< 0.002
Low dispersal (*D*) (–20%)	7,662	4,959	0.498	< 0.002
High *D* (+20%)	7,542	4,950	0.502	< 0.002
Low correlation (0.5)	7,584	5,104	0.454	< 0.002
High correlation (1.0)	7,425	4,866	0.446	0.004
Low harvest (*H*) (–20%)	7,664	4,909	0.512	< 0.002
High *H* (+20%)	7,611	4,893	0.523	< 0.002
Harvest on 5/6/7+-year-old males	8,043	5,019	0.482	< 0.002
Harvest on 7+-year-old males	7,640	5,156	0.455	< 0.002
Predation with illegal *H*	7,497	4,900	0.516	< 0.002
Predation with no illegal *H*	7,814	5,038	0.470	< 0.002
No illegal *H*	7,950	5,183	0.447	< 0.002
Low illegal *H* (halved)	7,585	5,018	0.469	< 0.002
High illegal *H* (doubled)	7,393	4,980	0.495	< 0.002
Higher *H* (5% and 15%)	7,622	4,814	0.543	< 0.002
High competition	6,414	4,366	0.688	< 0.002
High competition and higher *H*	6,292	4,253	0.718	< 0.002
Six low-risk factors	11,704	8,164	0.044	< 0.002
Six high-risk factors	4,243	2,806	0.990	0.014

Note: *N* = number; *P* = probability; ier = interval extinction risk.

Nevertheless, it would be unlikely that the species would be classified as vulnerable within the next 20 years (Table 42.4), if the assumptions in the base case are true, if current management prescriptions remain in place, and if the human population does not further encroach on habitat. There is also some appreciable risk that the species may fall into the vulnerable category because of substantial population decline if vital rates are lower than expected or if illegal harvest rates are higher than believed.

The factors that most readily predispose the population to heightened risk include changes in vital rates, reductions in the carrying capacity and initial abundance, and heightened competition with people and farm animals (Tables 42.5 and 42.6). If management prescriptions or alternative options were to be evaluated, it would be best to explore their consequences primarily in terms of these parameters. For instance, if ibex have lower vital rates than expected because of extraneous factors such as disease or persistent drought, there is an appreciable chance that the taxon could be reclassified as

Table 42.5 Sensitivity of results to changes in parameters, based on the number of ibex at year 20

Parameter	–Ve Change	Sensitivity	+Ve Change	Sensitivity
Vital rates (VR)	10%–	2.94	10%+	2.74
SD	20%–	0.09	20%+	0.08
K	20%–	0.96	20%+	1.10
Initial abundance (IA)	20%–	0.15	20%+	0.06
Stochasticity lognormal	Uncorrelated	0.02		
Stochasticity normal			Correlated	0.04
Stochasticity dispersal CV	20%–	0.01	20%+	0.00
Dispersal (*D*)	20%–	0.13	20%+	0.05
Correlation	0.5	0.02	1.0	0.00
Harvest (*H*) rate	20%–	0.14	20%+	0.10
Harvest (*H*) age structure	5/6/7+-year-old-males	0.08	7+-year-old-males	0.02
Predation	No illegal *H*	0.05	With illegal *H*	0.00
No illegal harvest (*H*)	Illegal *H* ignored	0.07		
Illegal *H*	50%–	0.03	100%+	0.02
Higher *H* (5% & 15)	Legal and illegal *H* increased	0.02		
High competition (*K* 75%)	With normal *H*	0.14	With high *H*	0.16

Note: Ve = variation; sensitivity = the proportional change in expected number of ibex.

a vulnerable species. Substantial population decline and changes in the conservation status of the species are much more likely when risk factors are combined (Table 42.4). Figure 42.4 shows the family of risk curves resulting from the suite of scenarios and parameter options outlined in Table 42.4. The curves are quasi-extinction risk curves, showing the chances of falling below the threshold population size at least once in the next 20 years.

Table 42.6 Sensitivity of results to changes in parameters, based on the population size that gives a 50% chance of interval extinction risk

Parameter	–Ve Change	Sensitivity	+Ve Change	Sensitivity
Vital rates (VR)	10%–	3.02	10%+	3.50
SD	20%–	0.31	20%+	0.41
K	20%–	0.90	20%+	0.71
Initial abundance (IA)	20%–	0.39	20%+	0.24
Stochasticity lognormal	Uncorrelated	0.03		
Stochasticity normal			Correlated	0.01
Stochasticity dispersal CV	20%–	0.11	20%+	0.00
Dispersal (*D*)	20%–	0.00	20%+	0.01
Correlation	0.5	0.03	1.0	0.02
Harvest (*H*) rate	20%–	0.05	20%+	0.06
Harvest (*H*) age structure	5/6/7+-year-old-males	0.01	7+-year-old-males	0.04
Predation	No illegal *H*	0.02	With illegal *H*	0.01
No illegal *H*	Illegal *H* ignored	0.05		
Illegal *H*	50%–	0.03	100%+	0.01
Higher *H* (5% and 15%)	Legal and illegal *H* increased	0.03		
High competition (*K* 75%)	With normal *H*	0.12	With high *H*	0.14

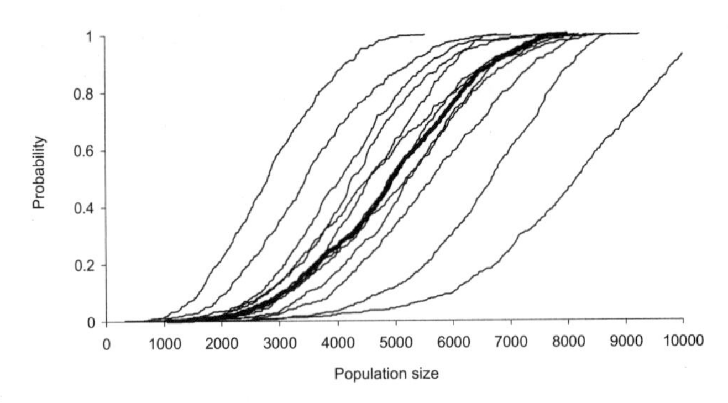

Figure 42.4 The family of risk curves resulting from the 31 simulations (see Table 42.4).

The heavy line in the center of Figure 42.4 is the risk curve for the base case. The cluster of curves close to it represents the outcomes for the parameters to which the model is relatively insensitive. The outermost lines represent the range of results that are possible given a set of six pessimistic and optimistic assumptions. The figure emphasizes the fact that expectations may change quite dramatically if a set of assumptions has been consistently misinterpreted or if conditions within KNP change in such a way as to affect several different model parameters simultaneously.

In some national parks elsewhere, managers are considering the reintroduction of a full range of the original species, including natural predators. Natural predators would generally take either relatively young or old animals. The effect would be to reduce effective fecundity and increase the harvest rates of ibex older than 7 years. The natural predators of ibex include leopards, coyotes, wolves, foxes, and hyenas, but most of these predators have been hunted to extinction in the KNP region. The predation rate of large predators for large herbivores in other arid ecosystems has been estimated at around 0.5% (R. Harrington, unpublished data). These rates imposed on young and old age classes are unlikely to significantly affect the chances of decline of the species in KNP.

Harvest of a modest number (of about 15 per year) of large males has been in place for some time. It is very unlikely that such activities would contribute in any noticeable way to changes in the risks of population decline. Males are essentially in excess and dispersal characteristics are such that at the scale of KNP and its associated populations, the loss of 15 large males is not a significant ecological cost.

Conclusion

Patches of habitat on the northeastern and northwestern borders of KNP straddle the boundary of the park. As a result, contiguous areas of habitat are exposed to different management and protection regimes. The population sizes in these areas are considerably smaller than may be expected on the basis of patch area and terrain alone. In the model, carrying capacities of these patches were thus assumed to be considerably smaller than might be expected as a result of direct competition with domestic stock and agriculture for space and resources. Expected population sizes and chances of persistence could be considerably enhanced if the carrying capacities of these patches were to be effectively increased through the imposition of new management strategies, if these are possible. To this end, it may be worthwhile exploring the illegal harvest rates in these regions and also assessing the relative contributions of these activities compared to direct competition with domestic stock in constraining the sizes of ibex populations.

Population model results and sensitivity analyses concluded that the current ibex population in KNP is likely to remain stable over the next 20 years as long as management of the ibex and the distribution and intensity of human activities do not change appreciably. There is some appreciable risk that the species may fall into the vulnerable category if there are significant changes in vital rates, reductions in the carrying capacity and initial abundance, or heightened competition with people and farm animals. The strength of these interpretations rests on the validity of the underlying assumptions. The model should be used routinely and updated to reflect improvements in knowledge and intuition about the management of ibex in KNP.

Acknowledgments We thank the staff of the Sindh Wildlife Department for their collaboration. We are greatly indebted to Premier Shell Pakistan for providing the resources to the Sindh Wildlife Department that made this work possible and for their logistic support in the field. We furthermore wish to acknowledge Prema Lucas, Stephen Wealands, Ben Miller, Mick McCarthy, Jane Elith, and Andre Zerger of the University of Melbourne for their generous support.

References

Akçakaya, H. R. 2001. *RAMAS GIS: linking landscape data with population viability analysis (ver. 4.0)*. Applied Biomathematics, Setauket, N.Y.

Akçakaya, H. R., Burgman, M. A., and Ginzburg, L. R. 1999. *Applied population ecology: principles and computer exercises using RAMAS® EcoLab*. 2nd ed. Sinauer Associates, Sunderland, Mass.

Burgman, M. A., Ferson, S., and Akçakaya, H. R. 1993. *Risk assessment for conservation biology*. Chapman and Hall, London.

Edge, W. D., and Olson-Edge, S. L. 1987. *Ecology of wild goats and urial in Kirthar National Park: a final report*. Montana Cooperative Wildlife Research Unit, University of Montana, Missoula.

Edge, W. D., and Olson-Edge, S. L. 1990. Population characteristics and group composition of *Capra aegagrus* in Kirthar National Park, Pakistan. *Journal of Mammalogy* 71: 156–160.

Edge, W. D., Olson-Edge, S. L., and Ghani, N. 1988. Response of wild goats to human disturbance near a waterpoint in Kirthar National Park, Pakistan. *Bombay Natural History Society Journal* 85: 315–318.

Ferson, S., and Burgman, M. A. 1995. Correlations, dependency bounds and extinction risks. *Biological Conservation* 73: 101–105.

International Union for the Conservation of Nature (IUCN). 2001. *Red List Categories, version 3.1*. IUCN, Gland, Switzerland.

McCarthy, M. A., Burgman, M. A., and Ferson, S. 1995. Sensitivity analysis for models of population viability. *Biological Conservation* 73: 93–100.

Mirza, Z. B., and Asghar, M. 1980. Census of Sind ibex (*Capra hircus blythi*) and gud (*Ovis orientalis blanfordi*) and some estimate of population of chinkara (*Gazella gazella*) in Kirthar National Park and Sumbak Game Reserve, Sind. *Pakistan Journal of Zoology* 12: 268–271.

Morgan, D. G., and Harrington, R. 2001. *Fauna survey*. A report for Khirthar National Park Baseline Environmental Study, Chapter 5: Fauna, Part A. University of Melbourne and Melbourne Enterprises International, Melbourne.

Roberts, T. J. 1997. *The mammals of Pakistan*. Oxford University Press, Karachi, Pakistan.

Schaller, G. B., and Laurie, A. 1973. Courtship behaviour of the wild goat. *Sonderdruck aus der Zeitung für Saugetierkunde* 39: 115–127.

43

Steller Sea Lions (*Eumetopias jubatus*) in the Pacific Rim

Biological Uncertainty and Extinction Risk

LEAH R. GERBER

In this chapter I develop a simple model for Steller sea lions (*Eumetopias jubatus*) and use this model to examine the sensitivity of population status to uncertainty in population parameters and potential response to management actions. I describe key biological attributes of this marine mammal that may be considered in using RAMAS for modeling extinction risk for marine mammals. Steller sea lions (SSLs) offer an interesting example of using population viability analysis (PVA) as a tool for making conservation decisions. Among marine mammals for which reasonable population estimates are available, Steller sea lions are the most abundant species to be listed as "endangered" pursuant to both the Endangered Species Act (ESA) and the World Conservation Union (IUCN) (Gerber and VanBlaricom 2001). Hypotheses for the decline of Steller sea lions include predation, disease, and variability in abundance and distribution of prey; environmental change; nutritional stress; direct kills; intentional and incidental kills by fisheries; entanglement in marine debris; pollution; and other disturbances at a variety of spatial scales. However, the relative importance of these various processes is far from certain. I use RAMAS to examine the degree to which hypothesized causes might affect different life history stages and compare trajectories for alternate scenarios. I then examine the effect of uncertainty in environmental stochasticity, density dependence, and catastrophes on estimates of extinction risk.

Steller Sea Lion Distribution, Population Structure, Status and Management Criteria

Steller sea lions range from southern California around the Pacific Rim to northern Japan, with most of the world population occurring between the central Gulf of Alaska and the western Aleutian Islands (Loughlin et al. 1987). Two separate populations of Steller sea lions are currently recognized within U.S. waters: an eastern population, which includes animals east of Cape Suckling, Alaska (144° west longitude), and a western U.S. population, which includes animals at and west of Cape Suckling (Figure 43.1). Hereinafter I focus exclusively on the western population of Steller sea lions. This population currently includes 37,801 adults (Sease et al. 2001) but has declined in abundance by as much as 75% since 1975 (Figure 43.2) (Loughlin et al. 1992).

Steller sea lions were listed as a threatened species under the provisions of the ESA in December of 1990 (55 FR 49204). The Steller Sea Lion Recovery Plan of the National Marine Fisheries Service (NMFS) (1992) suggested delisting criteria that were perceived as arbitrary and were not accepted by NMFS. To date, no explicit criteria have been used in establishing the species' status. The Recovery Plan for the species reports that quantitative measures such as PVA or trend analysis would provide a robust estimation of the likelihood of extinction. York et al. (1996) developed three spatially distinct metapopulation models to investigate the population's persistence, assuming a range of population structures and characteristics. These predictions, the IUCN's classification of Stellers as endangered, and other information about population trends from 1990 to 1993 influenced NMFS to reevaluate the status of the species (National Marine Fisheries Service 1995). In October 1995, NMFS proposed that the western population be listed as endangered, while the eastern population remains threatened (69 FR 192). This status determination was approved and finalized in May 1997.

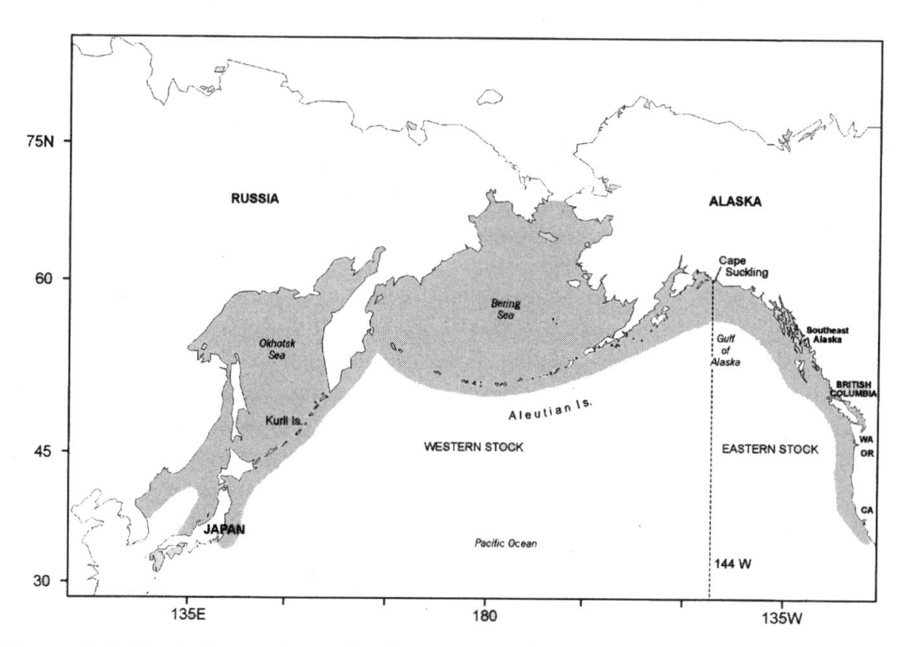

Figure 43.1 The Steller sea lion is distributed around the North Pacific Ocean rim from northern Hokkaiddo, Japan, through the Kuril Islands and Okhotsk Sea, Aleutian Islands, and the central Bering Sea, southern coast of Alaska, and south to the Channel Islands, California. The population is divided into western and eastern stocks at 144° W longitude (National Marine Fisheries Service 1992).

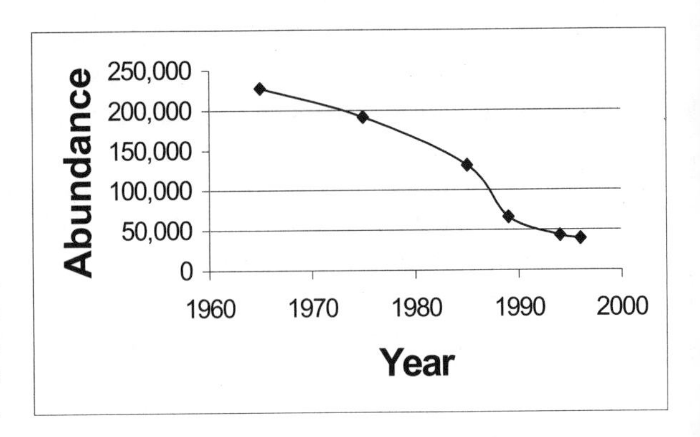

Figure 43.2 Trends in total abundance of the western stock of Steller sea lions (data from National Marine Fisheries Service 1995).

Changes in age structure suggest reduced survival of juvenile and breeding female SSLs (York 1994), and changes in prey availability have been implicated as a potential cause of the SSL decline. SSLs feed on a variety of fishes and invertebrates, particularly demersal and off-bottom schooling fishes (Pitcher 1981). Although historical data document population trends for harvested species such as pollock and herring in large geographic areas, little is known about changes in the local availability of prey in some important feeding areas. Population trends of other important prey species, such as capelin and sand lance, are also poorly known. In a given feeding area, the abundance, nutritional value, and species composition of forage vary considerably over time, not only seasonally but also over longer time scales. Further, the food and energy requirements of SSLs are not well known (National Marine Fisheries Service 1992, 1995). Merrick (1996) found that major dietary changes have occurred in areas with declining sea lion populations. These changes appear to result from declines in the abundance of preferred prey, which is small schooling fish such as capelin and juvenile pollock. Merrick (1996) speculates that causes of this decline may include an interaction between oceanographic changes of the late 1970s and the effects of intense exploitation of marine mammal and fish species beginning in the 1950s. Alternatively, the decline may be related to large-scale atmospheric and oceanic changes in the North Pacific in the late 1970s (Trenberth 1990), which appear to have affected production in North Pacific ecosystems (Francis and Hare 1994) and could have altered abundance and spatial distributions of key prey populations for sea lions (Shima 1996). Removals by the 1960s of most of the large whale biomass in the North Pacific may have also adversely affected the composition of prey in the ecosystem for the SSL by leading to the large increase in the pollock population and a corresponding decrease in the preferred forage fish prey (Merrick 1996) or by causing a shift in the diet of killer whales from baleen whales to Steller sea lions (James Estes, U.S. Geological Service, pers. comm.).

A comparison of measurements from samples of SSL taken during the 1970s and 1980s indicated a reduction in body growth rate, in late-term pregnancy rates, and in juvenile survival that were consistent with the food limitation hypotheses (Bowen et al. 2001). While these results suggest that food availability or quality may be responsible for the declines in SSL, they are based on vital rates that applied more than 15 years ago (see York 1994), when the oceanographic regime, the fishery activities, and the rate of decline of the SSL population were likely to have been substantially different. Thus, vital rate estimates during this period may not be representative of those currently being experienced by the population. In light of uncertainty about current vital rates, my approach here is to test alternative hypotheses about the factors responsible for the current trends in numbers (Table 43.1).

Examining Sensitivity of Viability Estimates to Biological Uncertainty

Assessing Demographic Sensitivity

I used an age-structured population model, parameterized with values from York (1994) and Pascual and Adkinson (1994). York (1994) reported vital rates for the sea lions when population growth rates were stable based on Calkins and Pitcher's (1982) early estimates of life table parameters, and he further found that a 20% reduction in juvenile survival yields a trajectory consistent with the observed decline rate and observed age structure changes. However, the demographic parameter estimates on which this result is based are derived from data from a period before the 1990s; this period was a time when the population decline was considerably steeper than at present, suggesting that the causes of the decline were different than they are now. Thus, rather than assuming that the observed decline is entirely due to reduced juvenile survival, my approach is to use RAMAS to examine the effect of different levels of mortality (simulated as harvest) for particular age classes (pups, juveniles, subadults, adults) on the median time to extinction (MTE) and the expected minimum abundance (EMA). While MTE is a standard risk metric used in PVA, EMA, which represents the average minimum abundance over all simulations, is particularly relevant for Steller sea lions. In particular, EMA can be used as an index of propensity to decline, especially when the population variability and risks of decline are low (McCarthy and Thompson 2001). I therefore use EMA, in addition to MTE, to compare results between alternate models.

Using RAMAS, I performed elasticity analysis to examine effects of proportional changes in demographic transitions on λ (de Kroon 1986). Because elasticity values sum to unity, they can be interpreted as the relative contributions of matrix transitions to λ. The left eigenvector v associated with the dominant eigenvalue λ is the reproductive value, and the right eigenvector w represents the stable age distribution (SAD), where $v'*w = \lambda*v'$.

Table 43.1 Characterization of the principle hypotheses surrounding the Steller sea lion decline, and potential effects on demographic rates

Cause	Hypothesized Effect
Fisheries competition	Increased mortality, reduced reproductive output
Environmental change	Increased mortality, reduced reproductive output
Predation	Increased pup and juvenile mortality
Anthropogenic effects	Increased mortality
Disease	Increased mortality, reduced reproductive output
Contaminants	Increased mortality, reduced reproductive output

Assessing Stochasticity

I used the RAMAS parameter fields for demographic and environmental stochasticity in the calculation of population growth. For all simulations, I incorporated the effects of environmental variability on vital rates by assuming lognormal distributions with standard deviations of 0.01 for all vital rates, and I incorporated demographic stochasticity. Variability in abundance due to sampling error was incorporated based on the reported coefficient of variation (CV) of the abundance estimate of the starting population size (Sease 1993). In the RAMAS parameter field for stochasticity, I therefore assume a CV of 0.06 (sampling error for *N*) to simulate the effect of measurement or sampling error.

For additional scenarios, I examined the effect of catastrophic stochasticity on MTE. RAMAS was thought to be appropriate for modeling Steller sea lions because catastrophic stochasticity may be incorporated by defining the probability of occurrence and maximum impact on population and biomass of different types of catastrophes. Simulations were conducted using the same assumptions as York et al. (1996) and Gerber and VanBlaricom (2002) so that results are comparable to results from these models. Thus, the probability of a catastrophe occurring was 25%, and these catastrophes resulted in a 15% population reduction of all stages. The dramatic impact of El Niño on some otariid (fur seal and sea lion) populations has highlighted the potential for short-term catastrophic impacts on otariids. The assumed probability and intensity of catastrophes are based on estimates made from historical data on catastrophic events for otariids (Gerber and Hilborn 2000).

Assessing Density Dependence

For scenarios in which density dependence was included, I assumed that all vital rates were affected. Because the population is observed to be declining, and the stage matrix yields a $\lambda < 1$, I considered the effect of ceiling-type density dependence. Carrying capacity (*K*) was specified based on historical abundance levels. The estimate of carrying capacity (140 and 115, respectively) was based on counts of adult and juvenile Steller sea lions in the Aleutian Islands and Gulf of Alaska in the late 1950s and early 1960s (Kenyon and Rice 1961, Mathison and Lopp 1963). Because coefficients of variation are not reported for these references, this estimate is highly tentative. In addition, there is evidence that *K* might be changing from year to year due to indirect effects of fishing activities (Bowen et al. 2001). To incorporate this uncertainty in the effect of environmental uncertainty on *K*, I consider the effect of variation by assuming a standard deviation of *K* as 20,000 and sampling *K* from a distribution each year.

Results

Assessing Demographic Sensitivity

The elasticities obtained for Steller sea lions indicate that the population is far more sensitive to changes in survival rates than to those in reproduction (Figure 43.3). These values represent the extent to which individuals of a particular age contribute to ances-

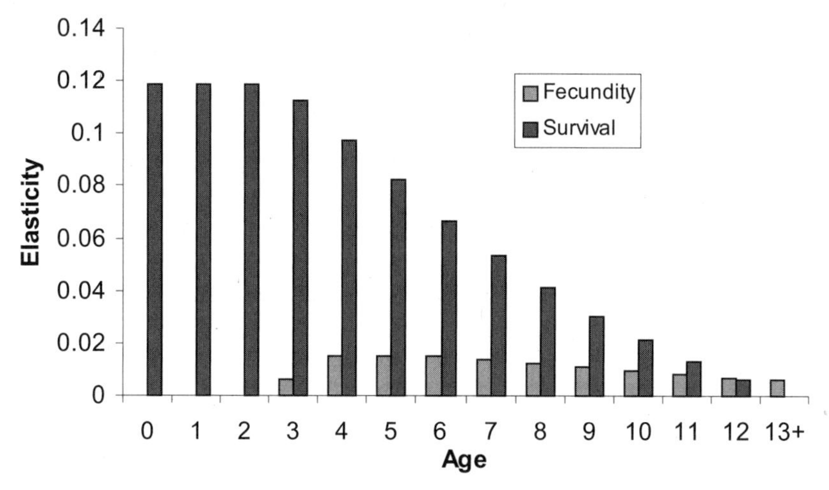

Figure 43.3 Elasticity values for Steller sea lions obtained from RAMAS output.

try of future generations (Caswell 2001). The elasticity of survival rates for pups and young juveniles stands out as especially high when compared to all other demographic entries. Reproductive value for Steller sea lions was highest for animals between the ages of 4 and 5. This value represents the number of offspring an individual in a given age class will produce, including all its descendants, relative to the reproductive value of an individual in the first age class.

With the baseline rate of population growth of $\lambda = 0.984$ estimated from the life table for Steller sea lions, I used RAMAS to examine the effect of different levels of mortality (simulated harvest) for particular age classes (pups, juveniles, subadults, adults) on the MTE and the EMA. Results agree with those from sensitivity analyses and previous work, suggesting that the results are highly sensitive to juvenile survival. In particular, both MTE and EMA were lowest for scenarios that reduced juvenile mortality (Table 43.2). This supports the hypothesis that protecting the juvenile life stage will have relatively strong impacts on reversing the population decline compared to protecting other life stages.

Assessing Stochasticity

Inclusion of catastrophic stochasticity significantly reduced MTE (Table 43.3). This suggests that additional research on existing data for Steller sea lions should be conducted in order to identify the appropriate distribution for environmental and catastrophic variability.

Assessing Density Dependence

MTE varied very little depending on whether density dependence was included in the model (Table 43.3). This result is not surprising, given that the population is declining and is currently at approximately 20% of historical abundance. It is unlikely that density dependence is a relevant factor in the Steller sea lion population dynamics, due to

Table 43.2 Median time to extinction (MTE) and expected minimum abundance (EMA) for 20% increase in mortality (simulated as harvest) for each age class (pups, juveniles, subadults, adults) and extinction threshold of 500.

Age Class	EMA	MTE
No change	6.8	264
Adults	1.5	188
Subadults	1.0	167
Juveniles	0.7	154
Pups	0.9	164

its current size and trend. Certainly, the population may be declining in response to a decline in its carrying capacity (brought about, e.g., by a decline in food resources), but this scenario is not considered in this study. Therefore, it remains unclear whether density dependence is a relevant factor in the population decline of the species (or has been in the past).

Discussion

Developing a PVA requires decisions about what type of model structure represents the dynamics of the population, as well as the choice of a meaningful model output to represent a population's persistence. My results should be no surprise to conservation biologists who are familiar with the paucity of data available for imperiled species. When PVAs are based on limited data, they should be viewed as tools to compare relative risk among populations. In cases in which data are highly uncertain, model output is likely to be less precise, thereby limiting the potential use of such models in conservation decisions. Nonetheless, PVA helps to elucidate those data that are critical to better assess a population's risk of extinction.

More information could be extracted from the counts of SSL by developing spatially explicit models using age-specific data from individual rookeries or haul-outs. Such models could help us understand how demography has changed in different areas over the course of the decline. This information could be used, for example, to evaluate hypotheses concerning which components of the population have recently been affected.

Table 43.3 Median time to extinction for different assumptions about density dependence and stochasticity for the extinction threshold of 500 (lognormal environmental variation (EV))

	No Catastrophe	Catastrophe
No density dependence	263	77
Ceiling	264	77

Note: Catastrophe has 25% probability of occurring with a 15% impact on the population.

RAMAS offers a relatively simple and straightforward tool for evaluating the effect of biological uncertainty on the efficacy of alternate conservation strategies. For long-lived marine species such as Steller sea lions, management actions that focus on improving juvenile survival are likely to have a significantly larger effect on population growth than will actions that focus on other vital rates. This conclusion depends on the cost of the modeled changes being the same. For example, juveniles may be the single most difficult age class on which to focus research and conservation because animals do not consistently haul out on land during breeding seasons. Nonetheless, an important next step will be to empirically test these model predictions. However, measuring the conservation success of a particular management measure for long-lived and wide-ranging animals will not be immediately realized. As such, monitoring should be conducted at biologically meaningful timescales in order to examine the effects of putative management options.

References

Bowen, W. D., Harwood, J., Goodman, D., and Swartzman, G. L. 2001. *Review of the November 2000 Biological Opinion and Incidental Take Statement with Respect to the Western Stock of the Steller Sea Lion, with Comments on the Draft August 2001 Biological Opinion.* Final Report Prepared for North Pacific Fishery Management Council, Seattle, Wa.

Calkins, D., and Pitcher, K. W. 1982. *Population assessment, ecology and trophic relationships of Steller sea lions in the Gulf of Alaska.* Department of Fish and Game, Anchorage.

Caswell, H. 2001. *Matrix population models: construction, analysis, and interpretation.* Sinauer Associates, Sunderland, Mass.

de Kroon, H., van Groenendael, J., and Ehrlen, J. 2000. Elasticities: a review of methods and model limitations. *Ecology* 81: 607–618.

Francis, R. C., and Hare, S. R. 1994. Decadal-scale regime shifts in the large marine ecosystems of the Northeast Pacific: a case for historical science. *Fisheries. Oceanography* 3: 279–291.

Gerber, L. R., and Hilborn, R. 2001. Estimating the frequency of catastrophic events and recovery from low densities: examples from populations of Otariids. *Mammal Review* 31(2): 131–150.

Gerber, L. R., and VanBlaricom, G. R. 2001. Implications of three viability models for the conservation status of the western population of Steller sea lions (*Eumetopias jubatus*). *Biological Conservation* 102(3): 261–269.

Kenyon, K. W., and Rice, D. W. 1961. Abundance and distribution of the Steller sea lion. *Journal of Mammalogy* 42: 223–234.

Loughlin, T. R., Perez, M. A., and Merrick, R. L. 1987. *Eumetopias jubatus. Mammalian Species* 283: 1–7.

Loughlin, T. R., Perlov, A. S., and Vladimirov, V. A. 1992. Range-wide survey and estimation of total number of sea lions in 1989. *Marine Mammal Science* 8: 220–239.

Mathisen, O. A., and Lopp, J. 1963. *Photographic census of the Steller sea lion herds in Alaska, 1956–58.* Special Scientific Report Fisheries 424. U.S. Fish and Wildlife Service, Washington, D.C.

McCarthy, M. A., and Thompson, C. 2001. Expected minimum population size as a measure of threat. *Animal Conservation* 4: 351–355.

Merrick, R. L. 1996. The relationship of the foraging ecology of Steller sea lions (*Eumetopias jubatus*) to their population decline in Alaska. Ph.D. diss., University of Washington, Seattle.

National Marine Fisheries Service (NMFS). 1992. *Final recovery plan for the Steller sea lion* (Eumetopias jubatus). Prepared by the Steller Sea Lion Recovery Team for NMFS, Silver Spring, Md.

National Marine Fisheries Service (NMFS). 1995. *Status review of Steller sea lions* (Eumetopias jubatus). National Marine Mammal Laboratory, Alaska Fisheries Science Center, National Marine Fisheries Service, Seattle.

Pascual, M. A., and Adkinson, M. D. 1994. The decline of the Steller sea lion in the northeast Pacific: demography, harvest, or environment? *Ecological Applications* 4(2): 393–403.

Pitcher, K. W. 1981. Prey of the Steller sea lion, *Eumetopias jubatus,* in the Gulf of Alaska. *Fisheries Bulletin of the U.S.* 79: 467–472.

Sease, J. L. 1993. A*erial and ship-based surveys of Steller sea lions* (Eumetopias jubatus) *in Southeast Alaska, the Gulf of Alaska, July 1992*. NOAA Technical Memorandum NMFS-AFSC. U.S. Department of Commerce, Seattle.

Sease, J. L., Taylor, W. P., Loughlin, T. R., and PItcher, K. W. 2001. *Aerial and land-based surveys of Steller sea lions* (Eumetopias jubatus) *in Alaska, June and July 1999 and 2000.* NOAA Technical Memorandum NMFS-AFSC-122. U.S. Department of Commerce, Seattle, Wa.

Shima, M. 1996. A study of the interaction between walleye pollock and Steller sea lions in the Gulf of Alaska. Ph.D. diss., University of Washington, Seattle.

Trenberth, K. E. 1990. Recent observed interdecadal climate changes in the northern hemisphere. *Bulletin of the American Meteorology Society* 71: 988–993.

York, A. E. 1994. The population dynamics of Northern sea lions 1975–1985. *Marine Mammal Science* 10(1): 38–51.

York, A. E., Merrick, R. L., and Loughlin, T. R 1996. An analysis of the Steller sea lion metapopulation in Alaska. Pages 259–292 in D. R. McCullough (ed.), *Metapopulations and wildlife conservation.* Island Press, Covelo, Calif.

44

Florida Panther (*Puma concolor coryi*)

Using Models to Guide Recovery Efforts

KAREN V. ROOT

South Florida is a rapidly urbanizing region with large amounts of intensive types of agriculture such as citrus and sugarcane. This region, where the temperate and tropic regions meet, is very diverse in flora and fauna. At least 68 federally listed species occur in South Florida, including the endangered Florida panther, *Puma concolor coryi* (U.S. Fish and Wildlife Service 1999). These vulnerable species are under increasing pressure as the human population continues to increase and economic expansion is accompanied by extensive land-use alterations. Cox et al. (1994) estimated that more than 3.2 million ha (or 8 million acres) of forest and wetland habitats have been cleared in Florida to accommodate the expanding human population over the last 50 years. The human population in Florida is expected to continue its growth, reaching 17.8 million people by 2010, with about half of them living in the South Florida ecosystem (Floyd 1997).

The endangered Florida panther is the only representative of the species *Puma concolor* surviving in the eastern United States, and it survives only in a small area in South Florida. The Florida panther is an excellent example of a unique rare Florida species. The high-profile, endangered cat is a habitat generalist that ranges widely, seeking its preferred prey of deer and hogs. Panther numbers, though, are greatly reduced. As of 2001, the verifiable, by radiotelemetry and other field data, number of panthers in South Florida was 78 adults and juveniles (McBride 2001).

This species once ranged from eastern Texas and the lower Mississippi River Valley east through the southeastern states (Young and Goldman 1946). The current distribution, as indicated by radiotelemetry data, is reduced to approximately 810,000 ha in South Florida south of the Caloosahatchee River (U.S. Fish and Wildlife Service 1999). Dispersal northward, either across the Caloosahatchee River or through the urbanized areas east of Lake Okeechobee, has been very rare, often fatal, and by a few males (Maehr et al. 1992, 2002b; Maehr 1997).

This geographic isolation combined with habitat loss and fragmentation, population declines, and inbreeding have led to a decrease in the genetic variability of this species (Roelke et al. 1993). Because of its low abundance and documented traits such as heart defects, cryptorchidism, and reduced sperm viability, there was great concern that this species was suffering from inbreeding depression (Seal and Lacy 1992, Roelke et al. 1993, Barone et al. 1994). In 1995, eight female Texas cougars (*Puma concolor stanleyana*) were introduced into South Florida as part of a genetic restoration program. While this genetic restoration program may have alleviated many of the genetic issues, fragmentation and loss of habitat continue to pose the greatest threats to panther recovery.

To evaluate the viability of the current Florida panther population and to complement ongoing habitat suitability analysis, I developed a spatially explicit metapopulation model for the Florida panther in South Florida. Specifically, I wanted to evaluate the factors affecting the long-term viability of the existing South Florida panther population and explore some potential strategies for recovery of this listed species.

Methods

Since 1981, Florida panthers have been radiocollared and monitored on public and private lands throughout South Florida (Maehr 1997, Maehr et al. 2002a). A total of 108 panthers have been radiocollared and over 100 kittens marked at the den since telemetry research began (Shindle et al. 2001). Florida panther radiotelemetry data collected February 22, 1981, to March 30, 2002, are shown in Figure 44.1. These data were used to estimate survival rates and fecundity for previous viability analyses (Seal and Lacy 1989, 1992; Maehr et al. 2002a). Using these data, combined with detailed geographic information system (GIS) habitat data, I constructed female-only, stochastic, structured, spatially explicit metapopulation models for the Florida panther in RAMAS GIS (Akçakaya 2002).

RAMAS GIS provides a framework for building detailed metapopulation models with complex spatially explicit structure. It is well suited for addressing conservation and management questions for species at risk because of its close integration of demographics with habitat dynamics. Therefore, using this model, I could examine the long-term viability of the panther and explore potential recovery options, such as increasing natural dispersal and translocation.

Metapopulation Models

Linking habitat data to metapopulation models requires three key steps: identify the species-habitat relationship; locate discrete habitat patches; and estimate population- and metapopulation-level parameters. Florida panthers are habitat generalists but prefer areas with cover, such as forests, for feeding, breeding, and shelter, and they avoid urban areas (Belden et al. 1988, Kautz et al. 1993, Cox et al. 1994, Maehr and Cox 1995, Dees et al. 2001). A base map was created by lumping forest cover types from two sets of maps: the Florida water management districts land use and land cover maps for South Florida based on 1995 aerial photography data and the Florida Wildlife

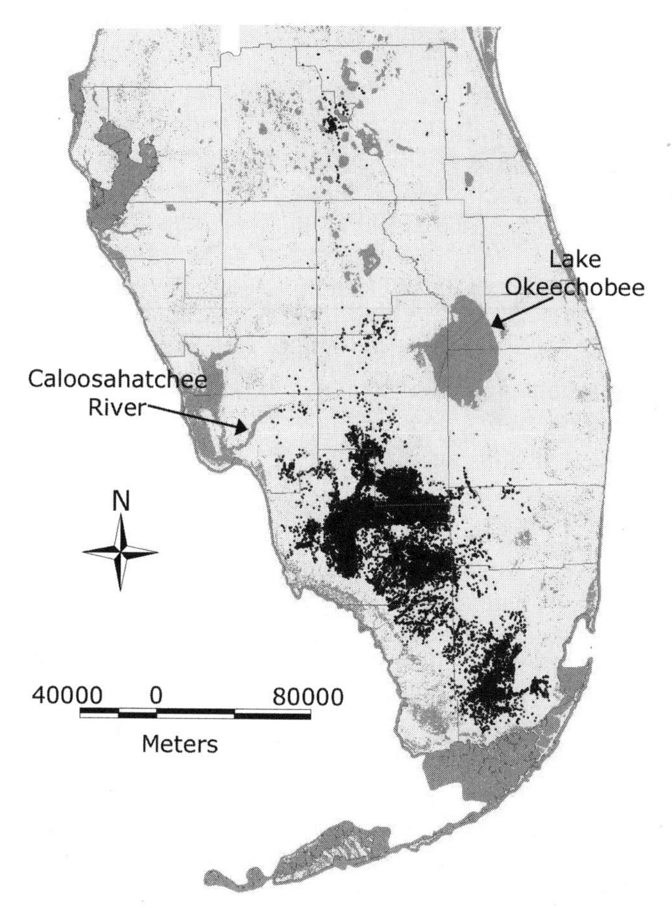

Figure 44.1 Florida panther radiotelemetry data (*black dots*) collected February 22, 1981, to March 30, 2002, by personnel from the Florida Fish and Wildlife Conservation Commission, Big Cypress National Preserve and Everglades National Park (U.S. Fish and Wildlife Service 2002).

Commission's land use and land cover map developed by Kautz et al. (1993) from 1985–1989 Landsat satellite TM imagery updated to 1996 using change detection analysis (U.S. Fish and Wildlife Service 2002). Forest cover patches smaller than 2 ha were eliminated, and a nonurban land cover buffer of 200 m (based on the spatial accuracy of the telemetry data; Belden et al. 1988) was added for each forest patch. Suitable habitat for the Florida panther was assumed to occur in patches greater than 2 ha in size, within 100 m of forest and more than 300 m from urban areas, based on the relationships described in Maehr and Cox (1995). The result was a map of potential panther habitat, as shown in Figure 44.1. The smallest density estimate for panthers is one panther per 110 km^2 of optimal habitat (Maehr et al. 1991). Extracting all forest patches greater than 110 km^2 in size and grouping contiguous patches together created a map of large suitable habitat patches (Figure 44.2) (U.S. Fish and Wildlife Service 2002).

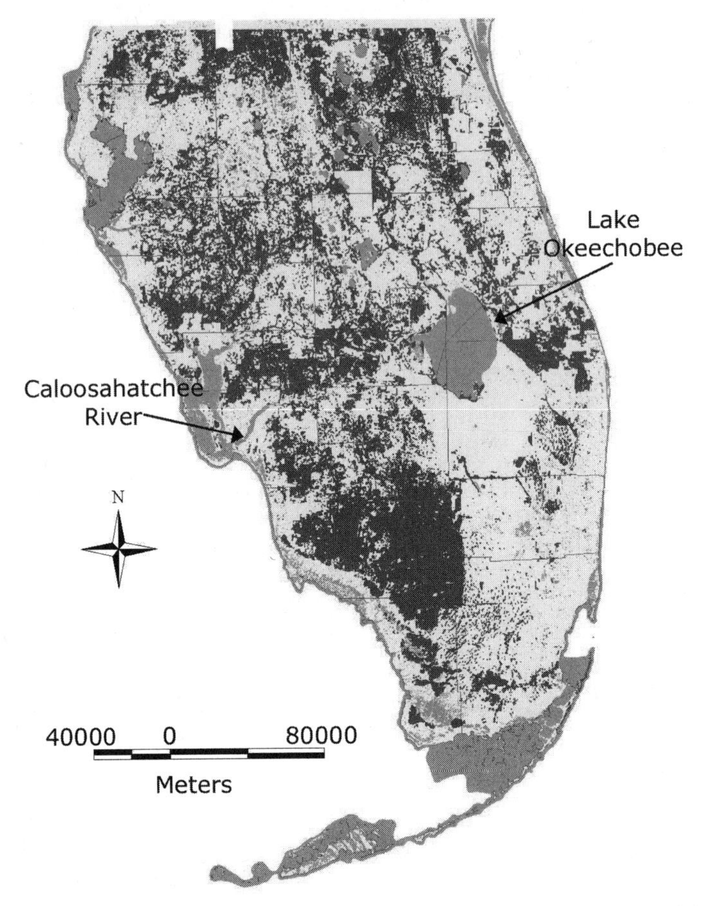

Figure 44.2 Potential panther habitat (*black areas*) in Florida based on Florida Fish and Wildlife Conservation Commission 1985–1989 Landsat data (updated to 1995–1996), combined with Florida's water management districts' 1995 data as forest patches greater than 2 ha in size plus a nonurban buffer of 200 m (U.S. Fish and Wildlife Service 2002).

Using the map of large suitable habitat patches, I delineated discrete populations in RAMAS GIS based on distribution of suitable habitat. This resulted in a metapopulation structure that grouped cells that were within normal dispersal distance into the same population (Figure 44.3). There were 10 potential populations in the model, but currently only the two populations south of the Caloosahatchee River (nos. 9 and 10 in Figure 44.3) are occupied by panthers. These two currently occupied populations are spatially distinct and have only infrequent dispersal among them.

Carrying capacity of each population was based on home range size (one panther per 110 km^2) (Maehr et al. 1991) and habitat area. The dispersal values among populations were based on the distance among populations and the dispersal patterns documented from telemetry studies (Shindle et al. 2001, Maehr et al. 2002b). For models that focused only on the existing panther populations (South Florida), only populations south of the Caloosahatchee River were considered.

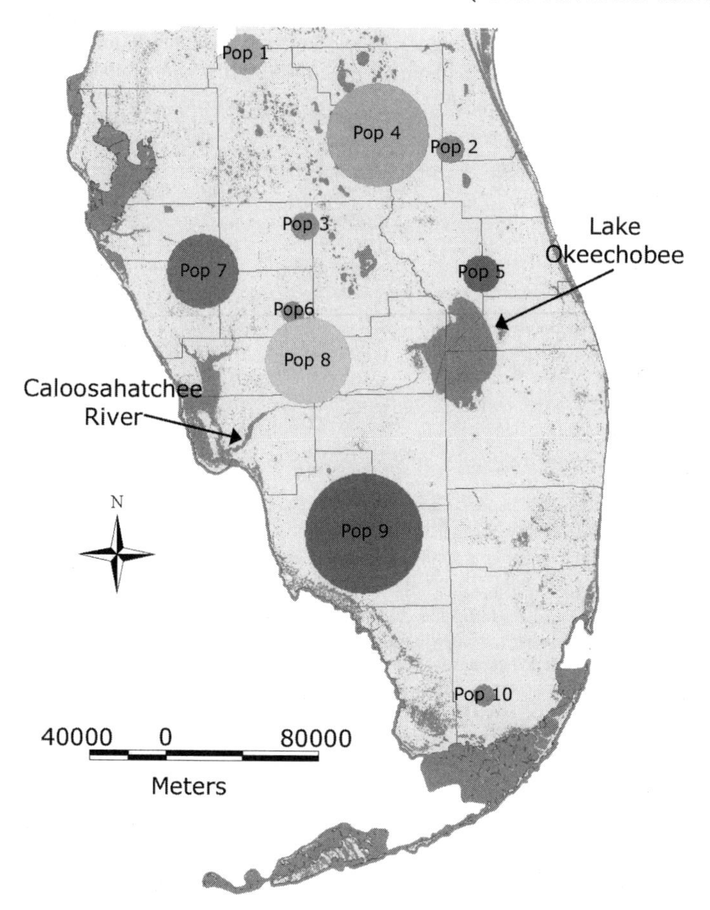

Figure 44.3 Metapopulation structure developed for the RAMAS GIS panther metapopulation model, based on the potential panther habitat (forest patches greater than 2 ha with a 200-m nonurban buffer). Only populations 9 and 10 on this map are currently occupied by Florida panthers.

For comparison, I constructed three general single-sex models, shown in Table 44.1. My Seal 1989 model was based on the analysis by Seal and Lacy (1989), except that annual juvenile mortality was updated based on more recent field data to 38% (D. Land, pers. comm.) instead of 50%. In this three-stage model, annual survival for young adults and adults is 70% and 75%, respectively. On average, females begin reproduction in their third year or later and have a 50% chance of breeding in a given year. Fecundity in this model, based on the product of the probability of breeding, average litter size, and average first-year survival, was 0.375.

I based the second model, labeled Seal 1992, on the analysis by Seal and Lacy (1992) except that juvenile mortality was also updated to 38% instead of 20%. In this two-stage model, annual survival for adults was 80%. On average, females began reproduction in their second year or later and had a 50% chance of breeding in a given year; fecundity in this model was 0.310.

Table 44.1 Vital rates for each of the three metapopulation models constructed for the Florida panther in RAMAS GIS. The annual vital rates are for females only.

Parameter	Seal 1989	Seal 1992	Maehr 2002
Sex ratio	0.50	0.50	0.50
Survival 0–1	0.62	0.62	0.62
Survival 1–2	0.70	0.80	0.80
Survival 2–3	0.70	0.80	0.83
Survival 3+	0.75	0.80	0.83
Age at first reproduction[a]	3	2	2
Females with litter (%)	0.50	0.50	0.50
Litter size	1.50	1.00	1.07
Fecundity[b]	0.465	0.310	0.332
Initial abundance[c]	41	41	41
Carrying capacity[d]	53	53	53
Deterministic growth rate[e]	0.985	1.039	1.077

Note: Seal 1989 based on Seal and Lacy (1989); Seal 1992 based on Seal and Lacy (1992); Maehr 2002 based on Maehr et al. (2002a). See text for discussion of differences.

[a]Age at first reproduction for females only.
[b]Fecundity in the model is (daughters only): % with litter*litter size*first-year survival.
[c]Initial abundance assumes a stable age distribution.
[d]Carrying capacity is on reproductively active stages only (i.e., adults) for the occupied Florida panther populations.
[e]Deterministic growth rate (or λ) is the finite rate of increase, which is the result of matrix analysis (eigenanalysis) ignoring density dependence, dispersal, catastrophes, stochasticity, and the effects of initial age/stage distribution, in RAMAS GIS.

My third model, labeled Maehr 2002, was based on the 1999 consensus model in Maehr et al. (2002 a) except that juvenile mortality was 38% instead of 20%. This was a two-stage model with young adult and adult survivals of 80% and 83%, respectively, and a fecundity of 0.332. Breeding was possible at 2 years of age or older, as in the Seal 1992 model.

Simulation Scenarios

The baseline version of each model had no catastrophes or epidemics, no change in habitat quality or amount, and a ceiling type of density dependence (maximum density 1/110 km^2). All models began with a stable age distribution. Variants of these models had different density dependence or none, various levels of habitat loss, intermittent catastrophes or epidemics, or scheduled translocations or reintroductions. I assumed that the existing South Florida panther population consisted of 41 females in two populations south of the Caloosahatchee River (populations 9 and 10 in Figure 44.3) and that the populations north of the river were "empty" or unpopulated at the start of the simulations, except when a hypothetical fully populated metapopulation was modeled. Each simulation was run with 10,000 replications for 100 years.

For models that included habitat loss, I used a 1% reduction in habitat for each of the first 25 years of the 100-year simulation. This rate of habitat loss corresponds to the estimated rate of loss from 1986 to 1996 for five southwest counties based on land use changes (R. Kautz, pers. comm.). The habitat loss only occurred in the currently occupied populations.

Historically, dispersal northward out of the currently occupied populations has rarely occurred. Recently, though, three collared male panthers have crossed the Caloosahatchee River; two of the panthers died, and third's fate is unknown due to collar failure (Shindle et al. 2001). All three crossed in the same general region. This "natural" corridor was included in some versions of the models as if a female might cross naturally or is assisted by active management. Alternatively, in some versions I modeled reintroduction as the addition of individuals from a hypothetical external population (such as the Texas cougars used for genetic restoration) to populations north of the Caloosahatchee River.

As in any model of metapopulation dynamics, the model of the Florida panther makes a number of assumptions. These assumptions were necessary largely because of data limitations but also to keep the model simple enough to be reasonably functional. Following is a list of the major assumptions of the model:

1. Either 2 (existing only) or 10 (existing plus potential) populations functioned as discrete populations loosely connected through dispersal, forming a metapopulation.
2. The vital rates of the past (as measured through telemetry data) reflect the values in the future. This assumes that monitoring the population has had no effect on the survival or fecundity rates.
3. The density within a population was assumed uniform throughout the entire area, and only suitable habitat (based on the GIS analysis) for the panther was included in estimates of population area, density, and carrying capacity.
4. The model assumes (except in the scenarios where carrying capacity was changed) that the habitat remains in exactly the same shape and condition that it was at the time of the habitat suitability analysis. In other words, there was no change in the amount or quality or configuration of the habitat during the 100 years of the simulation unless explicitly specified in the scenario.
5. Habitat within a population was assumed to be contiguous and readily accessible.
6. Dispersal was considered as permanent movement of a proportion of individuals from one population to another in a single year. This was dependent on the distance among the populations, although travel across the Caloosahatchee River was very infrequent.
7. For the purposes of reproduction, mates were assumed to be readily available and non-limiting.
8. The density ceiling only applied to adults to simulate territoriality.

Results

Under assumptions of no change in habitat and no catastrophes, the models suggest that the Florida panther was likely to persist (i.e., probability of extinction >0) over the course of the 100-year simulation. For the Seal 1989 model, the probability of extinction was 78.5% in 100 years, with a mean final abundance of 3.5 females (Figure 44.4). Also, the probability of a large decline in abundance (50%) was 94.1%. Under this model any perturbation such as habitat loss or catastrophes greatly increased the probability of extinction and resulted in mean final abundances near 0.

The probability of extinction for the existing panther populations in South Florida (Figure 44.4b) was quite low under both the Maehr 2002 and Seal 1992 models: ap-

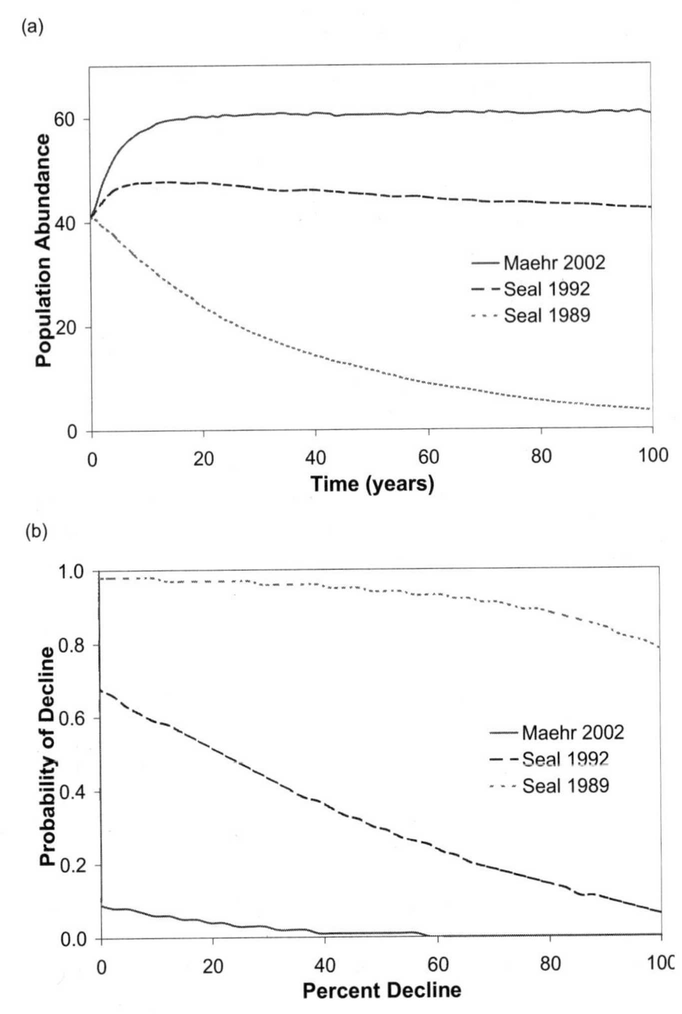

Figure 44.4 Three Florida panther metapopulation models for only the currently occupied South Florida populations: **(a)** mean final abundance over time and **(b)** probability of a decline, as a percentage of the initial abundance. The models assume no habitat changes and no catastrophes.

proximately 2% under Maehr 2002 and 5% under Seal 1992. However, the probability that the population size will decline was much greater. For example, there was a 9% probability and a 20% probability that the number of panthers would decline by half for the Maehr 2002 and Seal 1992 models, respectively. The mean final abundance of females, shown in Figure 44.4a, was 42.3 females and 51.2 females for the Seal 1992 and Maehr 2002 models, respectively. When the model included all of the potential populations north of the Caloosahatchee River (assuming these potential populations are unpopulated at the beginning of the simulation) and allowed infrequent distance-dependent dispersal among all of the populations, the probability of extinction was reduced (by 1%–2%); the probability of a 50% decline was reduced (by 5%–9%); and the mean final abundance was much larger (111%–220%), as shown in Figure 44.5.

(a)

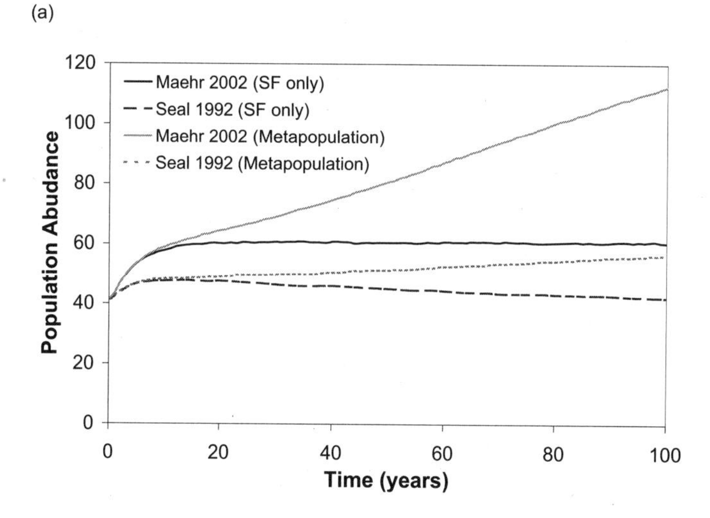

(b)

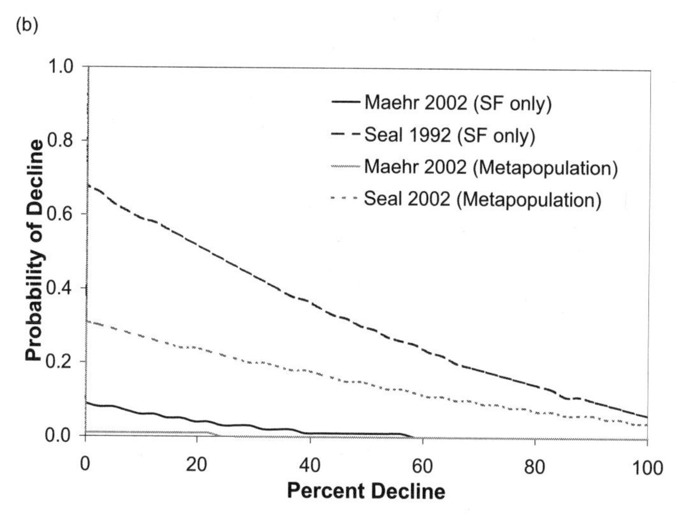

Figure 44.5 Two Florida panther metapopulation models for only the currently occupied South Florida populations (SF only) or all potential populations (Metapopulation): **(a)** mean final abundance over time and **(b)** probability of a decline, as a percentage of the initial abundance. The models assume no habitat changes and no catastrophes.

If 25% of the habitat is lost over the first 25 years of the simulation (i.e., 1% lost per year), the probability of extinction is increased approximately 1% (Figure 44.6a). The mean final abundance with habitat loss, though, is reduced by 26% to 37.9 and 31.2 females for the Maehr 2002 and Seal 1992 models, respectively. Similarly, the probability of extinction is only slightly increased when all potential populations are included and if habitat loss is restricted to the two southern populations (Figure 44.6b). Even with the additional populations, though, the mean final abundance with habitat loss is reduced by 19% to 24%, compared to the same models without the habitat loss.

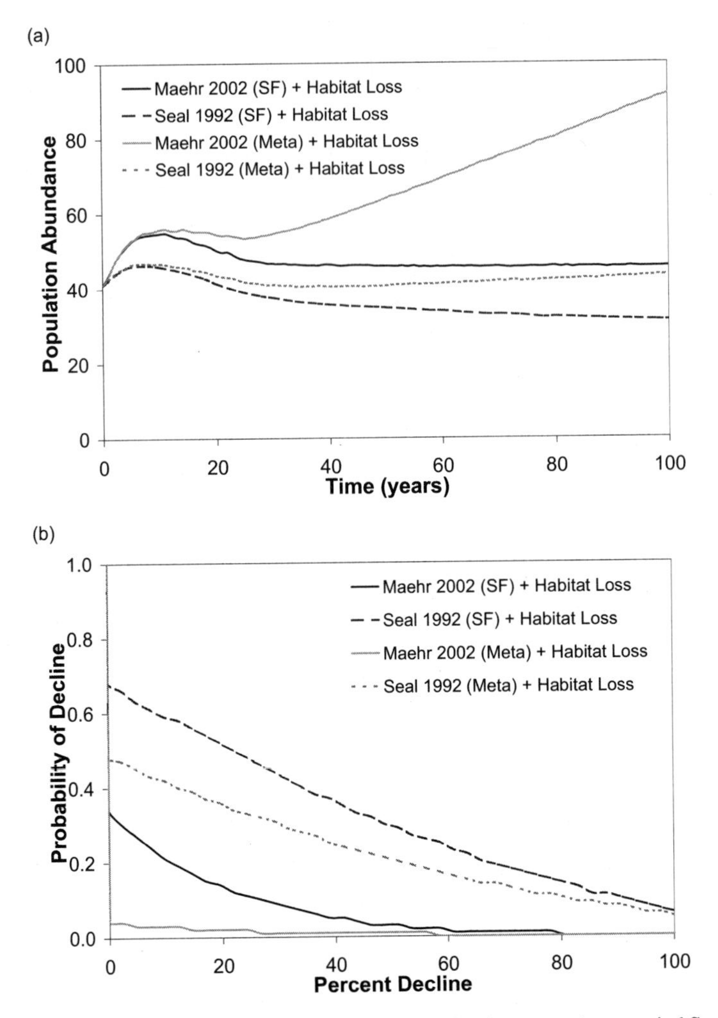

Figure 44.6 Two Florida panther metapopulation models for the currently occupied South Florida populations (SF) or all potential populations (Meta), with and without habitat loss: **(a)** mean final abundance over time and **(b)** probability of a decline, as a percentage of the initial abundance. Habitat loss, when included, was modeled as a 1% loss of habitat, in the currently occupied populations only, for the first 25 years of the 100-year simulation.

If one female adult, each year, crossed the Caloosahatchee northward at a natural corridor, the mean final number of females increased substantially from the initial two southern populations (Figure 44.7a), and the probability of extinction decreased (Figure 44.7b) with the Maehr 2002 set of parameters. With the corridor, the number of females increases as the northern populations are filled, increasing by 66 additional panthers. It is interesting to note that, with the Seal 1992 set of parameters, the increase in the mean final abundance is not present at the end of the 100 years and the probability of extinction actually increases slightly. In contrast, with either the Seal 1992 or the Maehr 2002 parameters, if the additional female adult added to the population north of

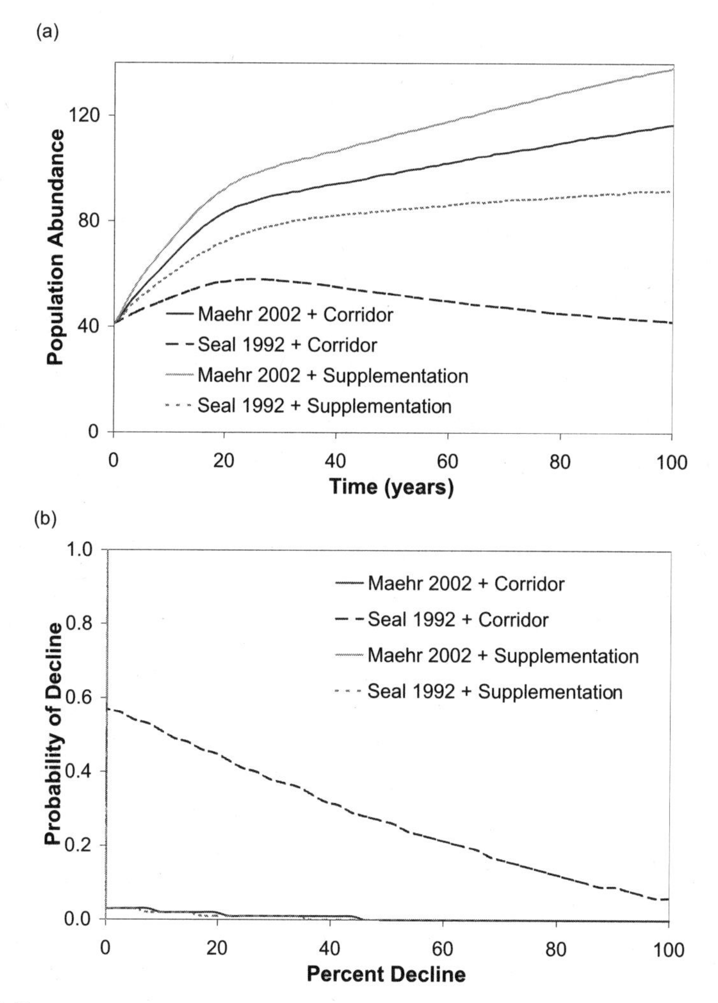

Figure 44.7 Two Florida panther metapopulation models with one female adult annually added to the population just north of the Caloosahatchee River (i.e., Pop. 8), either from the existing occupied southern populations (Corridor) or from a external source (Supplementation): **(a)** mean final abundance over time and **(b)** probability of a decline, as a percentage of the initial abundance.

the Caloosahatchee River was introduced from a population external to existing populations, such as from a captive Florida panther population, the reintroduction reduced the probability of extinction to 0 (Figure 44.7b). Reintroduction increased the number of females at the end of the 100 years by 78 to 87 (Figure 44.7a).

Discussion

Small populations, in general, are susceptible to a number of problems such as inbreeding depression, genetic drift, Allee effects, population bottlenecks, and catastrophic effects (Soulé 1980). Loss of genetic variability may reduce a species' ability to adapt to a changing environment. Gilpin and Soulé (1986) coined the term "extinction vortex" to describe the tendency of small populations to decline toward extinction. Therefore, it is important in a population viability analysis to focus not just on the probability of extinction but also on possible genetic effects, such as loss of heterozygosity, and the probability of large declines in abundance.

The results of my panther metapopulation models suggest that the long-term survival of the South Florida panther population requires maintenance of the current habitat configuration and condition indefinitely. Under most scenarios, establishing additional populations decreases the overall risk of extinction for the species if sufficient habitat is available and if there is adequate dispersal. Additional habitat loss or catastrophes would significantly increase the risk of extinction for this species and certainly lead to a decrease in abundance. In particular, the metapopulation models suggest that reducing habitat by 25% slowly (1% per year) will substantially reduce the number of panthers that persist for the next 100 years.

These results closely correspond to previous models constructed in Vortex based on early monitoring data—that is, before the introduction of the Texas cougars (Seal and Lacy 1989, 1992). In 1989 Seal and Lacy developed a simple model based on a small number of individuals that was quite pessimistic. That model predicted that there was a 100% chance of the existing panther population (estimated at 30–50 adults) becoming extinct in the next 100 years. It also predicted that the population would have greatly reduced genetic viability. The Seal 1989 model also predicted nearly 100% probability of extinction even with a larger initial abundance and static habitat conditions.

Panther population modeling was later improved, based on a larger sample of monitored panthers (Seal and Lacy 1992), but the newer model also suggested that the panther had reduced genetic viability and a significant chance of extinction in 100 years. The most optimistic scenario assumed an initial population of 50 adults, no change in carrying capacity, first reproduction at 2 years of age, and 20% juvenile mortality. Under these parameters, populations were nearly always driven to extinction due to the interacting effects of demographic variability and inbreeding when inbreeding effects on juvenile mortality were similar to those seen in other mammals (Ralls et al. 1988). The Captive Breeding Specialist Group recommended an introduction of genetic material from another population, which was later implemented using Texas cougars (Seal and Workshop Participants 1994). In the Seal 1992 model in RAMAS GIS, similarly, there was only a small chance (less than 5%) of extinction under static habitat conditions, and perturbations such as habitat loss significantly increased the probability of a decline from initial abundance.

Maehr et al. (2002a) developed a panther model using a consensus approach for parameter estimation. The model, run in Vortex, resulted in a 99% or greater probability of the panther population persisting for the next 100 years, although the final median population sized varied, depending on the conditions. For example, a 25% loss in habitat (over 25 years) did not increase the probability of extinction, but the final population size was 46.7 panthers compared to 65.6 panthers without habitat loss. The results for the Maehr 2002 model were very similar, with a greater than 99% probability of persistence over 100 years but reduced population sizes with habitat loss.

These models suggest that the probability of a large decline is likely for the Florida panther population unless the population increases substantially in size or its growth rate is increased. The Maehr 2002 model, which had an annual growth rate of approximately 7.7%, has the lowest probability of extinction and the largest mean final abundance. The model was also quite sensitive to assumptions about density dependence. If the carrying capacity was increased, the probability of extinction also decreased. Therefore, restoration of habitat or habitat improvement to make it more suitable for panthers would increase the chances of long-term panther viability. Two major issues complicate habitat improvement for the Florida panther, though: a large portion of the habitat that panthers currently used is privately owned (Maehr 1990), and the currently unoccupied, but suitable, habitat north of the Caloosahatchee River is characterized by a high degree of fragmentation based on a GIS analysis (U.S. Fish and Wildlife Service 2002).

Based on these modeling results and those of the past, a number of important management strategies are recommended. Establishing additional populations, all other things being equal, reduces the overall risk of a decline. Expansion to the north of the Caloosahatchee River could improve the probability of long-term viability and sustainability; a corridor would increase the rate of expansion. The key point, though, is that there must be sufficient "excess" individuals in the existing populations for dispersal, or the probability of large declines increases substantially. Under the Seal 1992 model, a simple corridor, with one adult female moving annually (naturally or through active management), actually increases the risk of extinction and lowers the mean final abundance. The current panther populations in South Florida may not be large enough or growing fast enough to compensate for the loss of panthers regularly moving north over the Caloosahatchee River, which highlights the need to protect and enhance the existing populations of Florida panthers, while further evaluating potential expansion habitat in unoccupied areas.

Habitat loss greatly increases the risk of a decline or extinction even under the most optimistic assumptions. These models clearly indicate that unless we are able to safeguard the current condition, amount, and configuration of the currently occupied panther habitat, the long-term viability of the Florida panther is not secure. While Florida panthers may continue to persist, as habitat is lost, without management interventions the populations will become more vulnerable to the problems of small populations—for example, inbreeding and Allee effects, as the number of individuals dwindles. It cannot be overemphasized that these models assume that there will be no loss of habitat (unless specifically mentioned), no degradation in quality, no difficulties in finding mates, no additional human-induced mortality, and no intermittent catastrophic events. In addition, if recovery goals are to expand the population of panthers, more habitat will be needed to allow for population expansion and subsequent dispersal.

References

Akçakaya, H. R. 2002. *RAMAS GIS: linking spatial data with population viability analysis (version 4.0)*. Applied Biomathematics, Setauket, N.Y.

Barone, M. A., Roelke, M. E., Howard, J., Brown, J. L., Anderson, A. E., and Wildt, D. E. 1994. Reproductive characteristics of male Florida panthers: comparative studies from Florida, Texas, Colorado, Latin America, and North American zoos. *Journal of Mammalogy* 75: 150–162.

Belden, R. C., Frankenberger, W. B., McBride, R. T., and Schwikert, S. T. 1988. Panther habitat use in southern Florida. *Journal of Wildlife Management* 52: 660–663.

Cox, J., Kautz, R., MacLaughlin, M., and Gilbert, T. 1994. *Closing the gaps in Florida's wildlife habitat conservation system*. Florida Game and Fresh Water Fish Commission, Tallahassee.

Dees, C. S., Clark, J. D., and Van Manen, F. T. 2001. Florida panther habitat use in response to prescribed fire. *Journal of Wildlife Management* 65: 141–147.

Floyd, S. S. (Ed.). 1997. *Florida county rankings 1997*. Bureau of Economic and Business Research, Warrington College of Business Administration, University of Florida, Gainesville.

Gilpin, M. E., and Soulé, M. E. 1986. Minimum viable populations: processes of species extinction. Pages 19–34 in M. E. Soulé (ed.), *Conservation biology: the science of scarcity and diversity*. Sinauer Associates, Sunderland, Mass.

Kautz, R. S., Gilbert, D. T., and Mauldin, G. M. 1993. Vegetative cover in Florida based on 1985–1989 Landsat Thematic Mapper imagery. *Florida Scientist* 56: 135–154.

Maehr, D. S. 1990. The Florida panther and private lands. *Conservation Biology* 4: 167–170.

Maehr, D. S. 1997. *The Florida panther: life and death of a vanishing carnivore*. Island Press, Washington D.C.

Maehr, D. S., and Cox, J. A. 1995. Landscape features and panthers in Florida. *Conservation Biology* 9: 1008–1019.

Maehr, D. S., Land, E. D., and Roof, J. C. 1991. Social ecology of Florida panthers. *National Geographic Research and Exploration* 7: 414–431.

Maehr, D. S., Root, J. C., Land, E. D., McCown, J. W., and McBride, R. T. 1992. Home range characteristics of a panther in south central Florida. *Florida Field Naturalist* 20: 97–103.

Maehr, D. S., Lacy, R. C., Land, E. D., Bass Jr., O. L., and Hoctor, T. S. 2002a. Evolution of population viability assessments for the Florida panther: a multi-perspective approach. Pages 284–311 in S. R. Beissinger and D. R. McCullough (eds.), *Population viability analysis*. University of Chicago Press, Chicago.

Maehr, D. S., Land, E. D., Shindle, D. B., Bass, O. L., and Hoctor, T. S. 2002b. Florida panther dispersal and conservation. *Biological Conservation* 106: 187–197.

McBride, R. T. 2001. *Current panther distribution, population trends, and habitat use: report of field work: fall 2000—winter 2001*. Report to Florida Panther SubTeam of MERIT, U.S. Fish and Wildlife Service, South Florida Ecosystem Office, Vero Beach.

Ralls, K., Ballow, J. D., and Templeton, A. 1988. Estimates of lethal equivalents and the cost of inbreeding in mammals. *Conservation Biology* 2: 185–193.

Roelke, M. E., Martenson, J. S., and O'Brien, S. J. 1993. The consequences of demographic reduction and genetic depletion in the endangered Florida panther. *Current Biology* 3: 340–350.

Seal, U. S., and Lacy, R. C. 1989. *Florida panther viability analysis*. Report to the U.S. Fish and Wildlife Service, Captive Breeding Specialist Group, IUCN, Apple Valley, Minn.

Seal, U. S., and Lacy, R. C. 1992. *Genetic management strategies and population viability of the Florida panther* (Felis concolor coryi). Report to the U.S. Fish and Wildlife Service, Captive Breeding Specialist Group, IUCN, Apple Valley, Minn.

Seal, U. S., and Workshop Participants. 1994. *A plan for genetic restoration and management of the Florida panther* (Felis concolor coryi). Report to the Florida Game and Freshwater Fish Commission, by the Conservation Breeding Specialist Group, Species Survival Commission, IUCN, Apple Valley, Minn.

Shindle, D., Land, D., Cunningham, M., and Lotz, M. 2001. *Florida panther genetic restoration and management*. Annual Report. Florida Fish and Wildlife Conservation Commission, Tallahassee.

Soulé, M. E. 1980. Thresholds for survival: maintaining fitness and evolutionary potential. Pages 151–169 in M. E. Soulé and B. A. Wilcox (eds.), *Conservation biology: an evolutionary-ecological perspective*. Sinauer Associates, Sunderland, Mass.

U.S. Fish and Wildlife Service. 1999. *South Florida multi-species recovery plan*. Atlanta, Ga.

U.S. Fish and Wildlife Service. 2002. *Landscape conservation strategy for the Florida panther in South Florida*. Draft Report. Atlanta, Ga.

Young, S. P., and Goldman, E. A. 1946. *The puma: mysterious American cat*. American Wildlife Institute, Washington, D.C.

Appendix

Using RAMAS GIS

This book comes with a CD-ROM that includes a demonstration version of RAMAS GIS 4.0, along with the input files for all the models included in this book. The demonstration version is intended only for viewing models developed with the regular version and not for developing, creating, or modifying models. It differs from the regular version in three ways:

1. It does not include a manual, but a complete set of help files is provided; these files contain most of the material in the manual.
2. It does not allow either creating a new model or modifying a model by changing its parameters. You can change parameters (e.g., to see the list of options available for a parameter), but you must click the "Cancel" button to exit a dialog box or window.
3. You can run simulations, and review the results, but you cannot save the results.

The rest of this appendix lists hardware and software requirements and then briefly describes how to use the program. For more detailed information, see the help files.

Required Hardware and Software

The program requires a personal computer with a CD drive, running Microsoft Windows 95/98/2000/NT 4.0/XP. The program will not work under Windows 3 or 3.1. The program will run on a 486 processor, although we recommend a processor that is 100 MHz or faster. Memory requirements depend on the operating system (OS):

OS	*Recommended memory*
Windows 95	48 MB or higher
Windows 98/NT	64 MB or higher
Windows 2000	104 MB or higher
Windows XP	168 MB or higher

Starting the Program

The program will *not* be installed on your computer's fixed drive; it must be run from the CD-ROM. Put the CD-ROM in a drive. If the program does not start automatically, double-click on the file RAMASGIS.EXE on the CD-ROM. This will start a shell program that provides access to all programs of RAMAS GIS:

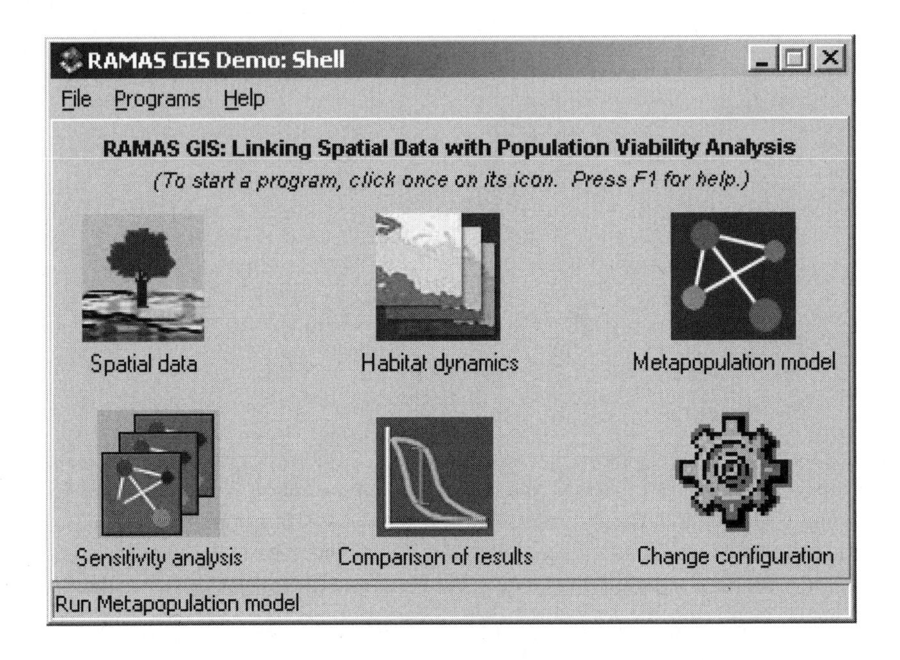

RAMAS GIS consists of five programs, represented by icons in this shell program. The main program is named *Metapopulation Model* (or *RAMAS Metapop*); it is used to build stage-structured, spatially explicit metapopulation models; to run simulations with these models; and to predict the risk of species extinction, time to extinction, expected metapopulation abundance and its variation and spatial distribution. The main features and parameters of this program are summarized in Chapter 1 and in the help files. This program requires both demographic data (e.g., survival rates, fecundities, and density dependence) and spatial data (e.g., location and size of subpopulations). Metapopulation Model files used by this program have the extension .MP. All models described in this book are represented by metapopulation model (.MP) files.

Exploring and Running Metapopulation Models

In the RAMAS GIS Shell program (see the preceding figure), click on the icon for Metapopulation Model to start this program. The main window of RAMAS Metapop consists of the title bar, menu bar, tool bar, model summary, and status bar. The menu bar includes File, View, Model, Simulation, Results, and Help:

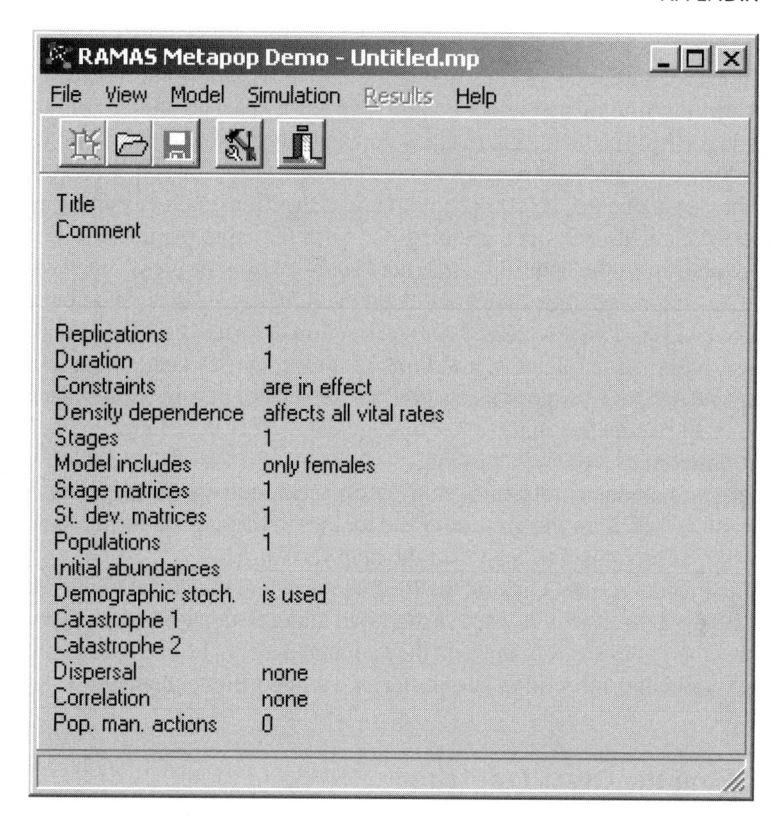

Click on the File menu, and select Open. If the "Open" dialog box does not point to the CD drive in which you put the CD-ROM, then navigate to this drive by using "Look in" on the toolbar:

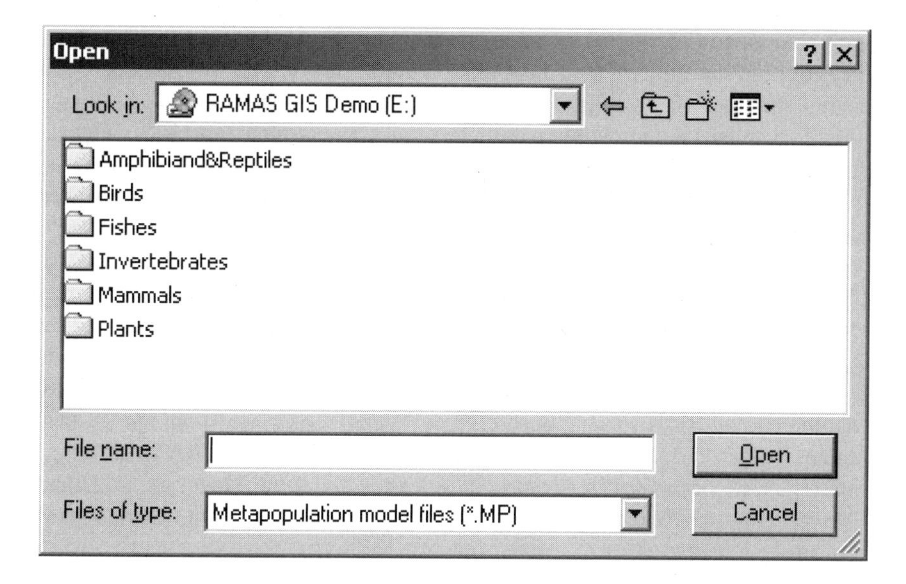

The sample files are arranged in folders according to taxonomic groups, as shown above. To examine the model files, navigate to one of these folders after you select "Open" from the File menu, select a subfolder for a species, and double click on a Metapopulation Model (.MP) file. In many cases, there is a single model in the species' folder, with more in an "Additional Models" folder.

Once the file is opened, select each item under the Model menu, starting with *General Information.* This will open a dialog box with the input parameters. The parameters are explained in the help file (click the "Help" button, or press F1).

In the General Information dialog, click on the *Additional Information* button if it is present. This will start your web browser and open an associated HTM file that includes background information, pictures, and links about the species and the model.

Return to RAMAS Metapop (using Alt-Tab, or selecting from the taskbar), and examine the input parameters in the other dialog boxes under the Model menu. Click the "Cancel" button to exit each dialog box.

You can run a model by selecting "Run" under the Simulation menu. While the simulation is running, click on the buttons in the toolbar to display the metapopulation trajectory or the metapopulation map. See the help file for details.

To see the model results, click on the Results menu and select one of the results. See the help file for a discussion of each result type, along with its interpretation.

Note that the sample files or the data they contain may not be used in any publication or research without prior written permission of Applied Biomathematics.

Using the Other Programs

In some models described in this book, the spatial structure of the metapopulation model is based on habitat data. This link between habitat data and the metapopulation model is made possible by the *Spatial Data* program. This program uses spatial data on habitat requirements of a species, such as GIS-generated maps of vegetation cover, microclimate, land use, and so on. It combines these data into a map of habitat suitability (HS) with a user-defined function. This map is then used to find habitat patches by identifying areas of high suitability where a population might survive. The program determines the spatial structure of the metapopulation (locations of patches and distances among them) and calculates demographic parameters (such as carrying capacities, vital rates, and initial abundances) of populations in each patch, with user-defined functions of the HS in that patch. Both the spatial structure and the demographic parameters are used as inputs for the metapopulation model.

To use this program, click on the icon for *Spatial Data* in the RAMAS GIS Shell program. The main window consists of title bar, menu bar, toolbar, model summary, and status bar. The menu bar includes File, View, Model, Simulation, Results, and Help. Click on the File menu and select Open. The file type (extension) for this program is .PTC. Note that spatial data files are available only for some of the species, including capercaillie (in the Birds folder); bush cricket, *Echinococcus,* and *Lopinga* (in the Invertebrates folder); and snowshoe hare (in the Mammals folder). As in RAMAS Metapop, the input data are organized in several dialog boxes under the Model menu. Click on the "Help" button in each of these dialog boxes for more information about the program (also see Chapter 1).

Another program (*Habitat Dynamics*) allows the metapopulation model to incorporate temporal dynamics in the habitat, with maps input as time series. For an example, see Chapter 15 (*Lopinga*). Please see Chapter 1 and read the help file of this program to learn more about how it is used.

Two additional programs can be used to support population viability analysis, risk assessment, and sensitivity analysis. The *Sensitivity Analysis* program is not functional in the demo version. In the regular release version, it is used to run several simulations of a metapopulation model to analyze the sensitivity of results to parameters.

The *Comparison of Results* program is used to compare different metapopulation models by superimposing their results (such as graphs of metapopulation abundance and occupancy, risk curves, and time-to-extinction distributions). It also allows statistical comparison of different risk curves. It can be used to view the results of a sensitivity analysis, to compare management options, or to assess anthropogenic impact. Start this program by clicking on its icon in the RAMAS GIS shell program. From the main program window, press Ctrl-O (i.e., hold the "Ctrl" key down, then press "O"). You'll see the "Load files" dialog box. If there are files listed in this window, highlight the first file, then click "Remove" until the list is empty. Click "Add." In the "Open" dialog box, navigate into a species folder that has at least two metapopulation model (.MP) files. Click on the first filename, and click "Open." Click "Add" again, and this time select the other filename; click "Open" (you can repeat this to add up to a total of five files). Finally, click "OK" (in the "Load files" dialog box).

Select the Results menu (or press Alt-R). Select each type of result, especially risk-related results, and examine the graphs. While displaying a result, you can press F1 to learn about it.

Index